May 17 1988

RAPID SOLIDIFICATION TECHNOLOGY

Source Book

A collection of outstanding articles from the technical literature

Other Source Books in This Series:

Production to Near Net Shape
Tool and Die Failures
Quality Control
Foundry Technology
Titanium and Titanium Alloys
The Ferritic Stainless Steels
Applications of the Laser in Metalworking
Innovative Welding Processes
Electron Beam and Laser Welding
Brazing and Brazing Technology
Gear Design, Technology and Performance
Copper and Copper Alloys
Materials for Elevated-Temperature Applications
Maraging Steels
Powder Metallurgy
Wear Control Technology
Selection and Fabrication of Aluminum Alloys
Materials Selection (Volumes I and II)

Metals Science Source Book Series:

The Metallurgical Evolution of Stainless Steels
Hydrogen Damage

RAPID SOLIDIFICATION TECHNOLOGY

Source Book

A collection of outstanding articles from the technical literature

---

Compiled by
Consulting Editor

R. L. ASHBROOK, Ph.D.
President, R. L. Ashbrook & Associates, Inc.
Metallurgical Consultants
Formerly Head, Alloys Section and Ceramics Section
NASA Lewis Research Center

American Society for Metals
Metals Park, Ohio 44073

Library of Congress Catalog Card No.: 82-74298
ISBN: 0-87170-162-6
SAN 204-7586

PRINTED IN THE UNITED STATES OF AMERICA

Contributors to This Source Book

K. I. ARAI
Tohoku University

H. BODE
Battelle Geneva

E. M. BREINAN
United Technologies Research Center

JOHN C. BITTENCE
Editor, *Materials Engineering*

A. R. COX
Pratt & Whitney Aircraft Group

POL DUWEZ
California Institute of Technology

R. D. FIELD
University of Illinois

H. L. FRASER
UTRC

F. S. GARDNER
Office of Naval Research

BILL C. GIESSEN
Northeastern University

JOHN J. GILMAN
Allied Chemical Corp.

N. J. GRANT
Massachusetts Institute of Technology

G. HAOUR
Battelle Geneva

P. R. HOLIDAY
United Technologies Research Center

R. W. K. HONEYCOMBE
University of Cambridge

S. C. HSU
University of Illinois

AKIHISA INOUE
Tohoku University

W. L. JOHNSON
California Institute of Technology

SHELDON KAVESH
Allied Chemical Corp.

B. H. KEAR
United Technologies Research Center

S. KOU
University of Illinois

A. KUROIWA
TDK Electronics Co. Ltd.

ALAN LAWLEY
Drexel University

C. G. LEVI
University of Illinois

R. E. MARINGER
Battelle Columbus Laboratories

TSUYOSHI MASUMOTO
Tohoku University

ROBERT MEHRABIAN
National Bureau of Standards

D. G. MORRIS
Institut CERAC

K. OHMORI
Tohoku University

T. OJIMA
TDK Electronics Co. Ltd.

R. J. PATTERSON II
Pratt & Whitney Aircraft Group

R. M. PELLOUX
Massachusetts Institute of Technology

DONALD E. POLK
Northeastern University

RANJAN RAY
Marko Materials Inc.

T. H. SANDERS, JR.
Georgia Institute of Technology

I. R. SARE
CSIRO – Division of Materials Science

D. B. SNOW
United Technologies Research Center

E. A. STARKE, JR.
Georgia Institute of Technology

C. SURYANARAYANA
Bnaras Hindu University

T. TAKEUCHI
Tohoku University

C. C. TSUEI
IBM Thomas J. Watson Research Center

N. TSUYA
Tohoku University

DAVID TURNBULL
Harvard University

E. C. VAN RUETH
Defense Advanced Research Projects Agency

NOTE: Affiliations given were applicable at date of contribution.

PREFACE

The very idea that rapid solidification could turn metals into glass attracted tremendous interest when first reported by Pol Duwez some twenty years ago. Other results of rapid solidification, the formation of new phases or extended ranges of solid solubility, although less startling, also commanded attention.

Rapid solidification has usually been defined as cooling from the melt to the solid at some arbitrary rate (e.g., > 10^6 °C/s). More recently, however, rapid solidification has been defined as cooling from the liquid to the solid at rates fast enough to achieve compositions, phases or microstructures not normally obtained at solidification rates of conventional ingots or castings.

Rapid solidification is most often associated with the work of Duwez in which high cooling rates were achieved in metal droplets accelerated to high speed and spread out on a conductive substrate. This process is sometimes called splat cooling. Since this initial work, interest in rapid solidification has increased and so have research and development in this field. Rapid solidification technology (RST), at first a laboratory curiosity capable of producing only tiny splats or foil specimens, has now been scaled up to cast continuous ribbons or strip of amorphous alloys and to atomize crystalline powders of steel and superalloys on a tonnage basis.

Much of the interest in glassy or amorphous metals arose because of their magnetic or electrical properties, and these properties of glasses continue to be of great importance. However, some amorphous ribbons or foils are converted into powder for subsequent consolidation into microcrystalline bulk products. These materials, which have great hardness and resistance to thermal shock, are used as extrusion dies or as inserts in die-casting dies.

Amorphous brazing alloy foils and amorphous transformer sheet are being produced in large quantities. These materials are being cast directly from the melt onto cooled rolls. Interest is now growing in the use of this direct strip casting (DSC) process as a means of producing crystalline alloy strip at great savings in energy and other processing costs.

Rapid solidification rate processing by radiation and convective cooling to produce very fine grained crystalline metal powders constitutes a major portion of RST. It has been developed to a high degree by Cox and his associates at Pratt & Whitney Aircraft (P&WA), with the support of the Defense Advanced Research Projects Agency (DARPA). In the P&WA process, high-velocity jets of helium are used to further atomize and cool molten alloy that is initially atomized centrifugally when poured on a rotating disk.

The interest generated by this work has caused a re-examination of atomized powders in general. Some powders atomized by inert gas or by the soluble-gas process or the rotating-electrode process were at first classified merely as powder metallurgy (P/M) products. Now, however, they are recognized as having structures and properties akin to those of the powders produced by more sophisticated techniques. As a result, some of these powders are now considered to be rapidly solidified materials.

Rapidly solidified crystalline or microcrystalline powders are finding many applications. Rapidly solidified superalloy powders are being used for gas turbine disks and are now even being made into turbine blades. High speed steel powders are replacing material made from cast ingots. Forgings or billets made from rapidly solidified aluminum powders have potential applications in gas turbine engines where high specific stiffness or specific strength is required. These aluminum alloys may displace titanium in some applications.

Purpose of This Source Book

This Source Book is not aimed at the experts in the field of rapid solidification. Rather, it is aimed at those who wish to become knowledgeable in the field, or at those who are working in one section of this growing field and wish to broaden their view. Because of the varied backgrounds of the readers, articles have been selected from sources ranging from engineering magazines to scientific journals. Emphasis has been given to review papers that provide a broad view of the subject and contain many references.

Organization of This Book

It seemed appropriate to start with a paper by Pol Duwez, because his name more than anyone else's is associated with rapid solidification technology. His paper and one by David Turnbull make up the first section, which is concerned with the *metastable structures* that may result from rapid solidification. Metastability increases with the severity of the quench from the melt. As solidification rate increases, the phases found may be supersaturated solid solutions, then new crystalline phases and, finally, amorphous phases — i.e., glasses.

The second section deals with the *processes* by which rapidly solidified materials are produced. As mentioned above, these processes include the atomization and convective cooling of metal from droplets to powder by high-velocity gas streams, and they also include conductive cooling of ribbons or foils on cold substrates. A special variant of the cold-substrate cooling is the self-substrate cooling found in laser or electron beam surface melting.

Surface melting of a thin layer of a thicker specimen, whether by a laser or an electron beam, results in rapid conductive cooling. Heat is rapidly dissipated to the unmelted portion of the specimen. In some processes, bulk material can be produced by feeding material into the molten pool and making repeated passes over the substrate.

A paper is included on yet another type of conductive cooling — that which results from dynamic compaction. Dynamic compaction of a cold pressed powder compact, whether by explosives or a high-velocity projectile, can result in local melting at the interfaces between adjacent powder particles. Although much work remains to be done, this process holds promise for compacting metastable powders and retaining their metastable structure.

Slow high-temperature processes that require a long time for completion, like hot isostatic pressing (HIP), make metastable structures transform to more nearly equilibrium structures. Only low-bulk-temperature processes or processes of short duration will be able to retain the metastable structures.

An understanding of *heat flow* during rapid solidification is essential to an understanding of the structures obtained. The third section consists of two papers by Robert Mehrabian, who has done so much to advance the knowledge of the fundamentals of RST. He makes clear the important distinction between the rate of cooling of the liquid (or cooling during noncrystalline solidification) and during crystalline solidification. Because of the heat of fusion, the rate of cooling, at equivalent rates of heat extraction, is considerably lower during crystalline solidification.

The remainder of the book is devoted to the compositions, structure, properties and applications of materials produced by rapid solidification processes. It is divided into two sections, one dealing with *amorphous or glassy materials* and one with *microcrystalline and crystalline materials*. Not all alloys can be made to solidify as a glass, regardless of how fast they are cooled from the melt. Some materials that solidify as a glass are processed to convert them into microcrystalline materials to enhance their strength or hardness. Glassy materials are usually of interest because of their magnetic or electrical properties. Microcrystalline or crystalline materials are usually of interest as structural materials.

Other Sources

For the benefit of the reader who wishes to dig deeper, a list of other sources is given at the end of this preface. The series of publications from the National Materials Advisory Board (NMAB) gives an excellent review of several topics that fall under the broad heading of RST. The NMAB reports cover amorphous, metastable and microcrystalline materials; aluminum alloys; and superalloys. They summarize the science and technology of these three important areas of RST and also discuss existing and potential applications. A fourth report in this series, which has just been issued, deals with the status of and facilities for RST.

The most concise introduction to RST is the "Review of Rapid Solidification Technology," published in 1981 by the Metals and Ceramics Information Center as MCIC-81-45. These MCIC and NMAB documents are subject to export controls and cannot be made available to foreign nationals without a validated export license. A bibliography, *Rapidly Quenched Metals, a Bibliography 1973-1979,* has been compiled by C. Suryanarayana. It has a very useful format, being cross-indexed by author, alloy system and subject. Suryanarayana is also the co-author of "Review: A Decade of Quenching from the Melt," published in 1971.

Detailed papers on RST can be found in many different scientific journals. Concentrations of papers on various aspects of RST are contained in the published proceedings of the dozen or so RST conferences that have been held since 1970.

"Amorphous and Metastable Microcrystalline Rapidly Solidified Alloys: Status and Potential," National Materials Advisory Board, NMAB-358 (May 1980).

"Rapidly Solidified (RS) Aluminum Alloys — Status and Prospects," National Materials Advisory Board, NMAB-368 (1981).

"Superalloys from Powder: Production and Properties," National Materials Advisory Board, NMAB-369 (1981).

"Review of Rapid Solidification Technology," Metals and Ceramics Information Center, MCIC-81-45 (1981).

Rapidly Quenched Metals, a Bibliography 1973-1979, compiled by C. Suryanarayana, IFI/Plenum Press, New York, New York.

"Review: A Decade of Quenching from the Melt," T. R. Anantharaman and C. Suryanarayana, Journal of Material Science, 6 (1971), 1111-1135.

International Conference on Metastable Metallic Alloys, Brela, Yugoslavia, 1970, Fizika, 2, Supplement 2 (1970).

Rapidly Quenched Metals, edited by N. J. Grant and B. C. Giessen, Massachusetts Institute of Technology, Cambridge, Massachusetts (1976).

Rapidly Quenched Metals Section II, edited by N. J. Grant and B. C. Giessen, Elsevier Sequoia S.A., Lausanne, Switzerland (1976).

Rapidly Quenched Metals III, Vols. 1 and 2, edited by B. Cantor, Chameleon Press Ltd., London (1978).

Amorphous Magnetism, edited by H. O. Hooper and H. M. DeGraaf, Plenum Press, New York, New York (1973).

Amorphous Magnetism II, edited by R. A. Levy and R. Hasegawa, Plenum Press, New York, New York (1977).

Metallic Glasses, ASM Seminar, September 18-19, 1976, edited by J. J. Gilman and H. J. Leamy, American Society for Metals, Metals Park, Ohio (1978).

Rapid Solidification Processing — Principles and Technologies, edited by Kear, Mehrabian and Cohen, Claitors Publishing Division, Baton Rouge, Louisiana (1977).

Rapid Solidification Processing — Principles and Technologies II, edited by R. Mehrabian et al., Claitors Publishing Division, Baton Rouge, Louisiana (1980).

Fourth International Conference on Liquid and Amorphous Metals (LAM 4), edited by F. Cryot-Lackman and P. Desre, Les Editions de Physique, Orsay, France (August 1980).

* * * * * *

The American Society for Metals extends most grateful acknowledgment to the many authors whose work is presented in this book, and to their publishers.

R. L. ASHBROOK
R. L. Ashbrook & Associates, Inc.
Cleveland, Ohio
10 January 1983

CONTENTS

SECTION I: METASTABLE STRUCTURES

SECTION II: PROCESSES

SECTION III: HEAT FLOW

SECTION IV: AMORPHOUS OR GLASSY MATERIALS

SECTION V: MICROCRYSTALLINE AND CRYSTALLINE MATERIALS

RAPID SOLIDIFICATION TECHNOLOGY

Source Book

A collection of outstanding articles from the technical literature

SECTION I
Metastable Structures

Structure and Properties of Alloys Rapidly Quenched from the Liquid State

POL DUWEZ

We are assembled here today to honor the name of Edward deMille Campbell, a great metallurgist and a great teacher. It was not my privilege to know Professor Campbell, but as a young student of metallurgical engineering in Belgium I knew of the work of three American metallurgists, Campbell, Howe and Sauveur. What I did not know at that time is that the largest portion of Professor Campbell's pioneering work in research and teaching was achieved after he had lost his eyesight. In honoring him, the American Society for Metals pays tribute not only to an outstanding metallurgist but also to a man of indomitable courage. I feel highly privileged and honored to have been chosen by the American Society for Metals to deliver this forty-second Campbell Memorial Lecture.

Metallurgists are very familiar with the term "quenching," which is generally defined as a process of rapid cooling from an elevated temperature by contact with liquids, gases or solids. The main purpose of quenching is to cool an alloy at a high enough rate so that phases stable at high temperatures are either partially retained, transformed into non-equilibrium phases or both. Subsequent heat treatment (tempering, aging, etc.) is then used to control the relative amounts as well as the microstructure of the desired phases and achieve the optimum physical properties of the final product. In this definition of quenching, nothing is said about the initial state of the material to be quenched, although it is taken for granted that it is in the solid state. This need not be the case and high cooling rates can indeed be applied to alloys in the liquid state as well. The purpose of this lecture is to discuss what has been and eventually could be achieved by cooling liquid alloys at very high rates.

In quenching solid alloys, the purpose is to cross the phase boundaries fast enough to totally, or at least partially, prevent the equilibrium reactions from taking place. Quenching from the liquid state has the same purpose, but the critical phase boundaries in this case are the liquidus and the solidus in the phase diagram. Since the atomic mobility in a liquid is far greater than in a solid, the rates of cooling required to influence the crystallization of an alloy are obviously much greater than those necessary to prevent a phase change in the solid state. As a result, the conventional techniques of quenching a solid do not lead to any significant results when applied to a liquid alloy. Recently, very simple techniques based on cooling by conduction of the liquid onto a solid substrate have been developed and crystallization of liquid alloys can definitely be modified and in some cases completely suppressed.

A brief review of the short history of the liquid quenching developments and why it was initiated in 1959 may be of interest. Whether the intention of a researcher is to gain new knowledge (which is the justification for fundamental research) or develop new material having some specified properties (which may be called applied research) there should always be a definite goal to initiate a new research program. The motivation for achieving extreme rates of cooling in liquid alloys was to try to prevent the separation into two phases in binary alloy systems in which, according to the generally accepted Hume-Rothery rules, the two metals should form a complete series of solid solutions, and yet a eutectic system is found under equilibrium conditions. The test case chosen for evaluating the efficiency of the quenching techniques was the copper-silver system, and in September 1959, a complete series of solid solutions in this system was obtained at Caltech. At that time, the rate of quenching was not known, and no time was devoted to try to measure it, because more exciting results immediately followed.

Source: ASM Transactions, Vol. 60, 1967, 607-633, © 1967 American Society for Metals

These unexpected results were the synthesis of a new crystalline nonequilibrium phase in the silver-germanium system which under equilibrium conditions is of the simple eutectic type. Shortly after that, the ultimate goal in quenching from the liquid state was reached when a noncrystalline (amorphous) structure was obtained in gold-silicon alloys. Total suppression of the crystallization process during solidification had been achieved.

RAPID QUENCHING TECHNIQUES

In quenching from the liquid state, the most important condition to satisfy is a very rapid cooling rate at the time the liquid to solid transition occurs and this means a very high rate of heat transfer at that critical time. There are only three basic mechanisms of heat transfer: radiation, convection and conduction. In radiation heat transfer, the only factors affecting the rate of cooling are the initial temperature and the size of the globule, since other physical conditions such as the emissivity of the alloys, etc., are fixed parameters. The initial temperature is obviously limited by the evaporation of the alloy constituents. In addition, increasing this temperature by several hundred degrees will not appreciably affect the rate of cooling at the liquid to solid transition temperature. It can be shown that the size of a radiating liquid globule should be about one micron or less in order to achieve rates of cooling of the order of 1000 C/sec at about 1000 C. It will be shown later that these rates are too low to have any effect on the solidification process.

Rapid removal of heat by convection is the classical approach to the problem of retaining metastable phases in solid alloys. In addition to increasing the initial temperature of the sample and decreasing its size (as in radiation cooling), it is also possible to drastically increase the rate of heat removal by increasing the velocity of the cooling fluid in contact with the sample. It can be shown that a cooling gas moving at high velocity around the sample is better than any liquid coolant (including liquid helium). By using gases having a high heat conductivity such as hydrogen or helium, velocities of several hundred feet per second and specimens about 50 microns thick and a few millimeters in diameter, rates of cooling of the order of 10,000 to 20,000 C/sec have been attained (1). Fine gold wire, 0.5 mm in diameter, has been quenched at rates of 50,000 C/sec (2). The convection type of quenching has been applied to liquid globules. Olsen and Hultgren have used this technique to study the effect of quenching rate on the segregation in nickel-copper solid solutions (3). A liquid drop was ejected (at an unspecified rate) into an iced sodium hydroxide solution and the calculated rate of cooling was nearly 100,000 C/sec. This technique was tried in this laboratory with Ag – Cu eutectic alloys and the globule was propelled into a liquid bath at several hundred feet per second. The globule disintegrated into a fine powder and an x-ray analysis showed that the two equilibrium phases were present. To obtain higher rates of quenching, the heat removal mechanism must therefore be based on conduction.

The advantages of conduction cooling were recognized in 1952 by Falkenhagen and Hofmann (4). They devised a very ingenious technique of rapidly filling a refrigerated copper mold (maintained at liquid nitrogen temperature) with molten aluminum-base alloys. The mold was wedge shaped, about 1 mm at the bottom and 8 mm at the top. The bottom of the mold was closed by a very thin sheet of aluminum. The mold was evacuated and then placed into contact with the liquid alloy bath. As soon as the aluminum sheet melted, the liquid alloy was sucked into the mold and the rapid solidification took place by conduction to the copper walls. In designing their apparatus, Falkenhagen and Hofmann did not take full advantage of the principle of conduction cooling, which, in addition to using a substrate made of a good heat conductor, also requires that the following conditions be satisfied: 1) A very good thermal contact must exist between the liquid and the substrate, 2) the liquid layer must be relatively thin, and 3) the time between initial contact of liquid with substrate and the end of solidification must be as short as possible. The first condition is obvious and need not be discussed. Condition number 2 is based on the assumption that nonequilibrium phases are likely to occur during solidification of an alloy if the solid-liquid interface can be forced to move at sufficiently high velocity. Under simplified assumptions, it can be shown that if an infinite liquid layer is suddenly placed into contact with a heat conducting substrate (one dimensional problem) the motion of the solid-liquid interface is governed by the diffusion equation and its velocity will decrease as it propagates into the remaining liquid. Assuming that there is a certain critical interface velocity below which equilibrium phases will nucleate and grow, the thickness of the liquid layer on the substrate should not exceed a certain critical value. Condition number 3 takes into account the fact that as soon as the globule makes contact with the substrate, nucleation will start, and any time spent between the first contact and the complete spreading of the liquid is likely to induce the growth of equilibrium phases. These basic principles have been introduced in three types of rapid quenching methods which will be referred to as the "Gun," the "Piston and Anvil" and the "Plasma Spray" techniques.

THE GUN TECHNIQUE

The basic concept of this technique is to accelerate a liquid globule to high speed and spread it on a heat conductive substrate at a proper angle under such conditions that centrifugal force provides a good thermal contact between the liquid layer and the

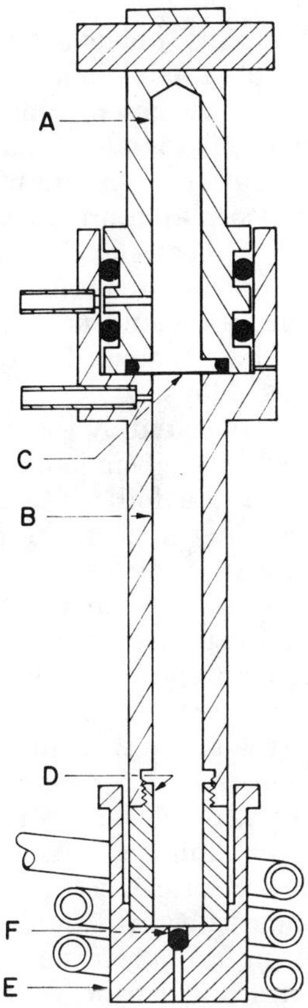

FIG. 1. Gun quenching device.

substrate (5). A simplified sketch of the main components of this apparatus is shown in Fig. 1. The propelling device consists of a shock tube, with a high-pressure chamber A, and a low-pressure chamber B, separated by a mylar diaphragm C. The high-pressure chamber is filled with helium gas and argon is used in the low-pressure chamber. The shock tube is made of stainless steel, but its bottom part D, exposed to high temperature, consists of a tantalum cylinder. The crucible E, containing the specimen to be quenched F, is heated by induction. This crucible is generally made of graphite, but a small insert of alumina can be introduced in order to avoid contact between the molten alloy and the graphite.

This design of a quenching gun has been used very extensively to produce a large number of metastable phases. It has two important limitations. First, the molten globule must be contained in a crucible, and this requirement excludes the study of reactive alloys such as titanium, zirconium, etc. Second, the relatively long time (of the order of 30 sec) to heat the charge above its melting point makes it difficult to prevent evaporation of volatile components in the alloy. These limitations have been recently removed in an improved model of the quenching gun due to Willens and Buehler (6). The principle of propelling the liquid globule by means of a shock tube has been retained, but the method of melting is entirely different. The material to be melted (about 100 mg) is supported on a water-cooled silver hearth. The radio frequency field (450 kcps) is concentrated on the charge by means of a silver conical susceptor of special design. The lower silver hearth, which is also loosely coupled to the rf coil, develops a levitation effect on the specimen, so that there is no close contact between the melt and the silver hearth. The energy concentration is very high, and specimens such as tantalum carbide (3877 C) can be melted in approximately 1 sec. With such short times, specimens containing volatile components such as chromium or manganese

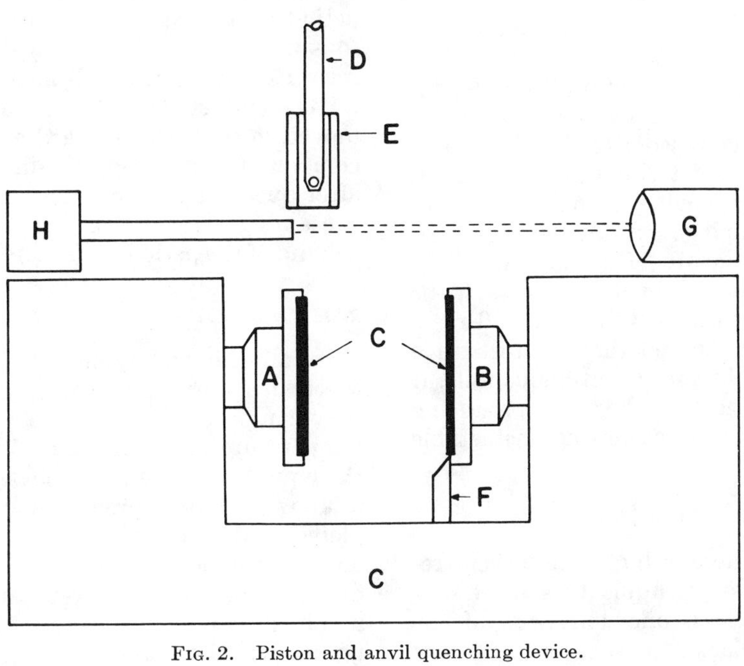

FIG. 2. Piston and anvil quenching device.

can be handled without appreciable change in composition.

THE PISTON AND ANVIL TECHNIQUE

The rapid spreading of a liquid globule against the surface of a metal having a high thermal conductivity, which is the essential feature of the gun technique, can also be achieved by squeezing the globule between two heat-conducting surfaces brought in contact very rapidly (7). A quenching apparatus using this principle is shown schematically in Fig. 2. The alloy is contained in a tube (generally made of fused silica) having a small opening at its bottom end shaped more or less like an eye dropper. The alloy can be heated by any conventional method; induction melting through a susceptor E is shown in Fig. 2. Levitation melting can be used and has the advantage of eliminating the crucible problems in the study of reactive alloys. The liquid globule is ejected at a negligible speed by applying a small gas pressure to the fused silica tube. Rapid cooling is achieved by squeezing the globule between a fixed anvil A and a fast-moving piston B actuated by high-pressure gas (200 psi). Both piston and anvil are lined with replaceable copper discs C. The piston B under pressure is retained by a solenoid-actuated finger F. When the globule is released, it intercepts a beam of light between a source G and a photocell H, and an adjustable delay circuit triggers the solenoid releasing the piston. Several variations from this basic design are possible. The rapid motion of the piston, for example, can be achieved by an electromagnetic drive, rather than gas pressure (8). The instrument can also be located inside a high-vacuum chamber, and both piston and anvil could be cooled to cryogenic temperatures if necessary.

THE PLASMA JET TECHNIQUE

A technique involving the use of a plasma jet for rapidly cooling liquid droplets has been described by Moss, et al. (9). The powder of the alloy to be quenched is introduced into an inert gas plasma-jet, and the molten droplets moving at high speed (not specified in Ref. 9) impinge on a cold substrate. Nonequilibrium phases in Mo – Ru alloys were obtained by this technique (9), which appears to be quite interesting for quenching refractory and reactive alloys. In spite of the fact that the technique does not lend itself to obtaining continuous foils, it might be a very practical method for producing relatively large amounts of interesting metastable phases in powder form.

SHAPE AND SIZE OF QUENCHED MATERIALS

Since, as explained before, a high rate of heat removal by conduction from a liquid to a solid substrate requires that at least one dimension of the liquid layer (perpendicular to the substrate) be very small, materials rapidly quenched from the liquid state will never be obtainable in any shape than thin foils or powder. At the present time, the width and length of quenched foils is also limited to approximately 1 or 2 in., but this limitation is not a consequence of the physical requirements necessary to achieve a high rate of heat transfer to the substrate. It is conceivable that, with suitable engineering development, a technique could be developed leading to the production of continuous, but necessarily thin, ribbons of quenched alloys.

The foil of alloy deposited by the gun technique is about 1 cm wide and 2 or 3 cm long. The thickness is very uneven and the foil contains many holes. Depending on the alloy and on the pressure in the shock tube, the thickness may vary from practically zero (on the edges and around the holes) to a few microns in the thicker parts. Very often portions of a foil are thin enough (probably less than 2000 A) for direct observation by transmission electronmicroscopy (10, 11). Because of their fine grain size, the gun foils are ideally suited for x-ray diffraction investigation either with a powder camera or a diffractometer. Their nonuniform thickness precludes studies of mechanical properties. The foils do not adhere to the copper substrate and can be easily removed. By pouring a liquid resin such as "Quickmount" on top of a foil, it can be transferred onto a nonconducting support. The desired specimen shape can be obtained by contour milling and measurements such as electrical resistance, thermoelectric power, Hall coefficient, and superconducting transition temperature can be carried out with more or less conventional methods.

The foils obtained with the piston and anvil device are quite uniform in thickness. Depending on the type of alloy, the size of the liquid drop and the actuating pressure in the driving cylinder, circular foils up to about 25 mm in diameter and 20 to 75 microns thick can be obtained. In most cases the foil does not stick to the copper liners and can be removed in one piece. Hard alloys and intermetallic compounds are generally divided in small pieces after quenching. The same is true for brittle alloys containing elements of the third, fourth and fifth column of the periodic table.

RATE OF COOLING

In all the techniques of liquid quenching, the thermal history of the alloy during solidification is very complex. It is probable that the most critical temperature range in the cooling process is between the equilibrium melting point of the alloy and some unknown temperature at which solidification is completed. Measuring both the rate of cooling and the degree of undercooling of the liquid alloy is a very difficult problem, mostly because of the small mass involved and the very short time allowed for measurements. The order of magnitude of the cooling rate in the gun technique was measured by Predecki, et al. (12). They embedded two dissimilar metals close to one another in an insulating base. When the liquid globule is spread on the substrate, an electrical contact is made between the two dissimilar metals nickel and silver and a thermoelectric potential is generated and is recorded on an oscilloscope. The main results were that the average cooling rates were 1.5 to 3.0 $\times$ 10^7 C/sec for a gold alloy containing 14 at. % antimony. A theoretical analysis of the cooling rate has been published recently by Ruhl (13). Many assumptions must, of course, be made to carry out such computations, one of the most important being the value of the heat transfer coefficient at the interface. In an ideal case, the temperature of the liquid would drop instantaneously to that of the substrate as soon as contact is established and the heat transfer coefficient would be infinite. The other extreme case corresponds to a completely interface-controlled cooling, in which case the temperature gradients in the alloy and in the substrate are neglected. Calculations were also carried out for intermediate cases. Several factors such as initial temperature of the liquid globule, temperature of the substrate and thickness of the foil were included in the analysis. A summary of the results of Ruhl's calculations is presented in Table 1. For an infinite heat transfer coefficient between the foil and substrate, the average cooling rate varies as the square of the foil thickness. The effect of the interface coefficient is also quite important. Assuming an intermediate heat transfer coefficient (100 in Table 1) and a foil thickness of 1 to 10 microns (experimentally observed thickness range), the calculated average rate of cooling of 3 $\times$ 10^7 to 8 $\times$ 10^8 agrees quite well with the measurements of Predecki, et al. (12).

TABLE 1. Calculating Cooling Rates (C/sec) in the Gun Technique for Different Thicknesses of Foils and Different Interface Heat Transfer Coefficients—Liquid Iron on Copper Substrate (13)

Foil thickness, microns	Heat transfer coefficient, cal/cm²-sec-°C		
	Infinite	100	1
1	8 $\times$ 10^9	6 $\times$ 10^8	7 $\times$ 10^6
10	8 $\times$ 10^7	3 $\times$ 10^7	7 $\times$ 10^5
100	8 $\times$ 10^5	7 $\times$ 10^5	6 $\times$ 10^4

The rate of cooling in the piston and anvil technique is difficult to calculate, because of the complexity of the process. In addition, it is difficult to achieve identical quenching conditions from specimen to specimen. The main reason for the lack of reproducibility is the inherent difficulty of having the liquid globule located exactly in the midpoint between the anvil and the moving piston at the critical time at which the fast-moving piston is about one globule diameter away from the anvil, and this critical time is of the order of 1 msec. If, for some reason, the globule happens to touch the anvil before that time, local cooling occurs and nucleation of the solid phases in the globule cannot be prevented. If the liquid globule falls toward the moving piston and is hit too soon, the same effect is achieved and the quenching rate is probably decreased by several orders of magnitude. Until a more sophisticated instrument is devised involving a better control of the trajectory of the liquid globule, the piston and anvil technique will remain more or less a "hit and miss" method in which from 50 to 75% of the foils must be rejected after their structure has been analyzed.

Experimental determination of an average rate of cooling has been reported recently (8). The principle of the method was to measure the thermoelectric potential generated between two electrodes attached to the piston and the anvil between the time they made contact with the liquid globule and the end of solidification. The two electrodes were made of dissimilar metals (iron and aluminum) and were electrically insulated from the piston and anvil. Since the thermoelectric power generated between the liquid globule and either one of the electrodes was unknown, an approximate calibration was obtained by matching the thermal arrest on the oscillograph record with the melting point of the alloy. This approximation is not too bad, if it is assumed that the alloy does not drastically undercool under these conditions. The average cooling rates obtained by these experiments are summarized in Table 2. It should be pointed out that these results apply to the particular apparatus used by the authors, for which the speed of the piston at the time of impact (measured by high-speed photography) was 12.6 ft/sec. In any case, if the average rates of cooling in Table 2 are compared with those in Table 1, it is apparent that the piston and anvil technique is second to the gun technique in quenching rate. This fact has been indirectly established in a comparative study of the two techniques in their ability to achieve a complete series of solid solutions in copper-silver alloys (14). In this system, compositions from 40 to 60 at. % are most difficult to obtain into truly one-phase structures, but this can be achieved in a large percentage of the gun foils whereas a mixed structure is always obtained in the piston and anvil foils.

TABLE 2. Measured Cooling Rates in the Piston and Anvil Technique (8)

Liquid	Cooling rate, C/sec
Al	6.9 $\times$ 10^5
Pb	1.2 $\times$ 10^6
50-50 solder	7.8 $\times$ 10^5

TYPES OF ALLOYS RESULTING FROM LIQUID QUENCHING

The structure of alloys rapidly quenched from the liquid state is always unusual in some respect, even if the phases present are crystallographically the same as in the equilibrium alloys. For example, binary alloys in which a complete series of solid solutions exists can be obtained with the highest degree of homogeneity. This particular feature could be of interest in preparing homogeneous samples for calibrating an electronmicroprobe for quantitative analysis. Quenching from the liquid state also results in a high concentration of defects; the technique has been used to study vacancy concentrations in quenched aluminum (15). However, the most interesting application of the technique is, by far, the synthesis of new alloy phases which cannot be obtained either under equilibrium conditions nor by quenching from the solid state. These new phases can be classified into three types: solid solutions extending beyond the equilibrium concentrations, crystalline phases and noncrystalline (amorphous) phases.

SUPERSATURATED SOLID SOLUTIONS

Supersaturated solid solutions can easily be obtained by quenching solid alloys as exemplified by the classical metallurgical processes performed everyday on aluminum alloys. It is quite logical to expect that by rapid cooling from the liquid state, liquidus-solidus transition might be crossed fast enough to prevent nucleation and growth of a second solid phase, and the solute atoms might be trapped into the parent lattice in excess of the equilibrium concentration. This effect was found by Falkenhagen and Hofmann in aluminum alloys quenched into a refrigerated copper mold (4). The solubility of the elements of the first transition series in aluminum were increased as follows: Ti, 0.09 to 0.19 at. %; V, 0.18 to 0.55 at. %; Cr, 0.45 to 2.85 at. %; Mn, 0.7 to 4.7 at. %; Fe, 0.026 to 0.082 at. %. With higher rates of quenching available now, solubility limits can be drastically increased in many alloy systems. Complete series of solid solutions can be obtained in otherwise eutectic systems, providing the Hume-Rothery conditions are satisfied. Briefly stated these conditions are: same crystal structure, small difference in valence and electronegativity, and not more than approximately 13% difference in atomic radii. In several binary alloys for which the conditions are satisfied, a miscibility gap is found in the solid state. One of the most flagrant exceptions to the rule of solid solubility is the case of silver-copper alloys. The early results obtained by liquid quenching (16, 17) were not quite convincing for alloys in the range of 40 to 60 at. % because of the difficulty in obtaining truly single-phase alloys. In a more thorough investigation (18), a large number of foils were quenched and only a few were selected

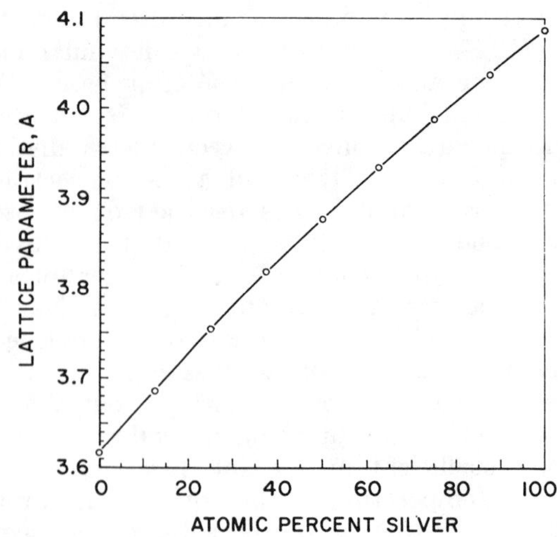

FIG. 3. Lattice parameters of nonequilibrium copper-silver solid solutions.

for lattice parameter measurements, on the basis of the sharpness of the x-ray diffraction peaks. As shown in Fig. 3, the lattice parameter of the supersaturated solid solution varies almost linearly with concentration. Because of the small size of the figure, the data points corresponding to alloy compositions within the equilibrium range are not shown, but the solid lines near 0 to 100% concentrations have been drawn on the basis of published data of equilibrium solid solutions and the positive deviation from Vegard's law was predicted by extrapolating the equilibrium data. The miscibility gap was also suppressed by rapid quenching in three other binary systems, namely Ag – Pt (20) Cu – Rh and Ni – Rh (21). It is probable that this could also be achieved in Ti – W alloys.

Another typical example of complete solid solubility resulting from fast solidification has been found in pseudo-binary alloys containing germanium and gallium-antimonide (22). Here again, the similarity in electronic configuration, crystal structure and unit cell size of the two components would suggest complete solubility. Under ordinary rates of cooling, however, negligible solubilities are observed. With the exception of one point, the measurements reported in Ref. 22 seem to fall on a straight line, indicating that in these pseudo-binary alloys Vegard's law is very nearly satisfied.

All these results agree quite well with the predictions based on the Hume-Rothery rules. However, it would not be justified to generalize them to all binary systems satisfying these conditions. At least one exception has been found so far, namely, Cd – Zn alloys, for which liquid quenching always results in a two-phase alloy. These unsuccessful results were attributed at first to the fact that perhaps room temperature was high enough (relative to the melting points of the components) to promote phase separation after quenching. Many experiments were then performed in a dry box with a copper substrate cooled to liquid-nitrogen temperature. The deposited foils were kept at low temperature and transferred to the x-ray diffractometer. The diffraction patterns of both cadmium and zinc were always present and rather crude measurements of lattice parameters (limited in accuracy because only low-angle reflections were observed) corresponded to the equilibrium concentration limits. These negative results have not yet been satisfactorily explained.

The case of binary alloys in which complete solid solubility is *a priori* excluded because of differences in crystal structure or atomic size will be considered now. The first alloy system chosen to investigate this case was silver-germanium. Under equilibrium conditions, the solubility of Ge in Ag is 9.6 at. % and the phase diagram is of the eutectic type with no measurable solubility of silver in germanium. Rapid quenching from the liquid state extended the Ag terminal solid solution up to approximately 13 at. % Ge and the electron-to-atom ratio corresponding to this concentration is 1.39 ± 3 (23, 24). This value is in excellent agreement with the theoretical predictions of Hume-Rothery and Raynor for the limits of stability of solid solutions in fcc noble metals with higher valence elements such as germanium (25). Here again a generalization of these results is dangerous because it has been found that in the three alloy systems Ag–Si, Au–Si and Au–Ge, whose phase diagrams are similar to that of Ag–Ge, no measurable extension of the terminal solid solution could be obtained by liquid quenching. It is possible that the failure in extending solubility limits in alloys other than Ag–Ge may be attributed to the present limitation in the rate of cooling obtainable with the gun technique. If this rate could be increased by one or two orders of magnitude, metastable solid solutions might be extended up to the theoretical limit in the four homologous alloys mentioned above, and if so, there will be no need for sophisticated explanations.

The extension of solubility limits was systematically investigated in nickel-base alloys with the elements of the IV B column of the periodic chart (Si, Ge and Sn) (26). A similar study was made of cobalt-base alloys with the IV B and III B elements (27). The results obtained for these alloys are summarized in Table 3. The increase in solubility brought about by rapid quenching is quite appreciable in some cases (Ga in Co, from about 10 to 21 at. %) and negligible in other cases (Si in Ni). The extent of solid solutions in nickel and cobalt-base alloys under equilibrium conditions has not yet received any theoretical treatment and as a consequence very little can be said about nonequilibrium alloys. One interesting correlation, however, can be obtained between the extent of the metastable solid solution and the location of the eutectic or peritectic in the phase diagrams. The eutectic or peritectic concentrations are given in Table 3, and it is apparent that, in no case, the maximum concentration of the nonequilibrium solid solution exceeded that of the eutectic, although these two concentrations were relatively close in some cases. This observation indicates that for concentrations beyond the eutectic (or peritectic) the first intermediate equilibrium phase nucleates very fast in the undercooled alloys and cannot be suppressed.

TABLE 3. Extension of Terminal Solid Solubility Limits in Nickel- and Cobalt-Base Alloys (26, 27)

Solute	Equilibrium maximum solubility, at. %	Extended solubility limit, at. %	Concentration of eutectic or peritectic, at. %
	Ni alloys		
Si	17.6	18.0	21.4
Ge	12.0	20.5	22.0
Sn	10.4	18.0	19.3
	Co alloys		
Si	≈14.0	15.0	22.0
Ge	≈15.0	19.0	25.0
Sn	≈ 3.0	5.0	20.5
Al	≈15.0	18.5	19.5
Ga	≈10.0	21.0	27.0

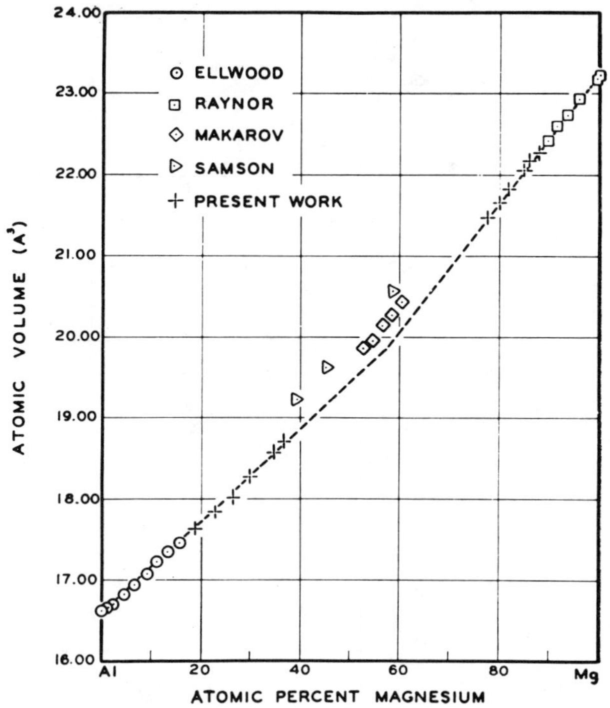

FIG. 4. Variation of atomic volume with concentration in Al–Mg alloys.

A study of the aluminum-magnesium system gives further support to these conclusions (28). By liquid quenching, the solubility of magnesium in aluminum is increased from 18.9 to 37 at. %, and that of aluminum in magnesium from 11.6 to 23 at. %. Here again the maximum concentrations in nonequilibrium solid solutions fall just short of the eutectic compositions. The lattice parameters of the supersaturated solid solutions follow the same linear relationships reported by previous investi-

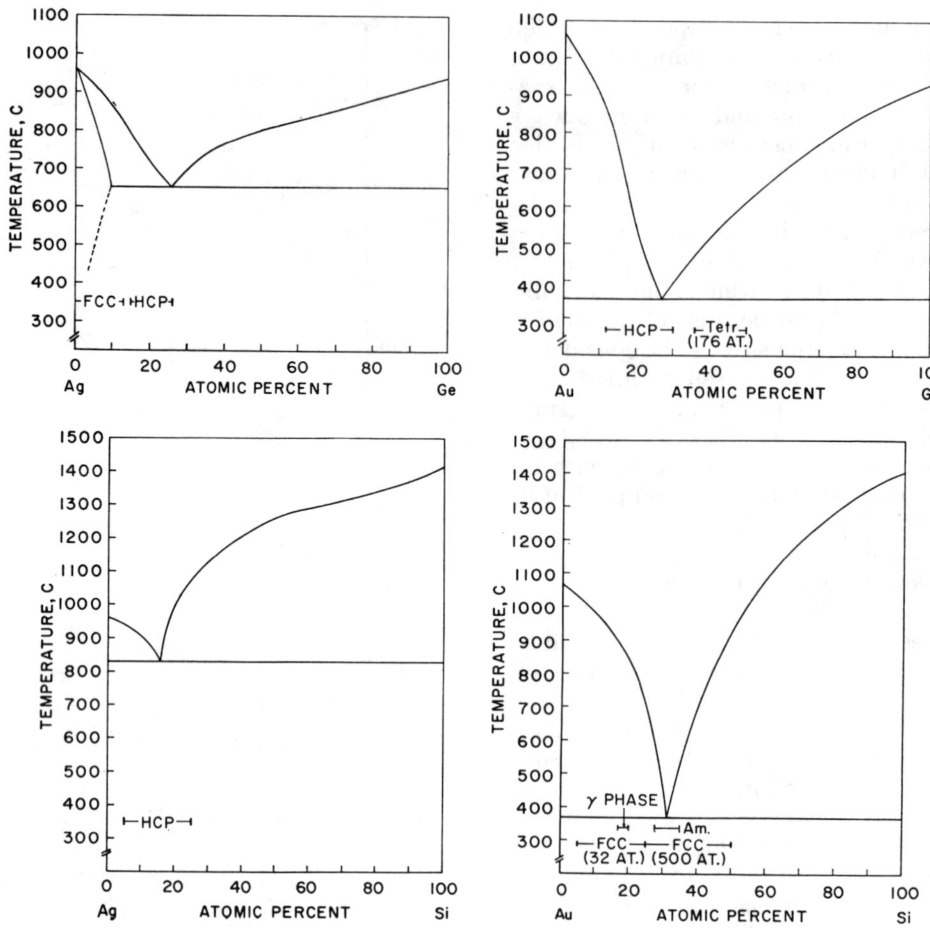

FIG. 5. Nonequilibrium phases in binary alloys.

gators for equilibrium alloys (29, 30). These results are shown in terms of atomic volumes in Fig. 4. The two straight lines meet at about 58 at. % Mg and the atomic volumes of the intermediate equilibrium phases, beta and gamma (31, 32), are above the extrapolated lines, indicating that these phases are, as expected, less close packed than the fcc and hcp solid solutions. It is interesting to speculate on the significance of the intersection of the two extrapolated straight lines at about 58 at. % Mg. The axial ratio c/a of the magnesium-rich phase increases with aluminum content, and if a graph is made of this ratio vs concentration, an acceptable extrapolation to the ideal value of 1.633 can be obtained at 58 at. % Mg in Al. If this extrapolation is accepted, it would then be logical to expect that if 42 at. % Al could be retained in hcp Mg solid solution, it would have the same atomic volume as a fcc Al solid solution containing 58 at. % Mg and the two structures could eventually coexist at the same concentration.

NONEQUILIBRIUM CRYSTALLINE PHASES

A limited number of binary alloys are known to possess stable and metastable phase diagrams and both of these can be obtained with conventional rates of cooling. A typical example of such a behavior is found in cadmium-antimony alloys. The two diagrams are very similar and consist of two eutectics separated by an intermediate phase. The stable phase corresponds to CdSb and its melting point is 456 C; the metastable phase is located around Cd_3Sb_2 and has a melting point of 420 C (33). The familiar iron-carbon phase diagram is another example of metastability, but in this case, the stable alloys (Fe plus graphite) are most difficult to obtain under practical rates of cooling, and the metastable phase Fe_3C (cementite) is generally present. These two examples show that it is possible: 1) to create an intermediate phase in systems in which there is none (Fe – C case); 2) to replace an intermediate phase by another one of different composition and crystal structure (Cd – Sb case). It was logical to expect that similar results could be obtained by rapid quenching from the melt.

The first nonequilibrium crystalline phase in an alloy system in which there is none under equilibrium was found in silver-germanium (23). This phase exists in quenched alloys containing from about 15 to 26 at. % Ge and is hexagonal close packed with an axial ratio decreasing with increasing germanium concentration. Since the range covered by the hexagonal phase contains the electron ratio

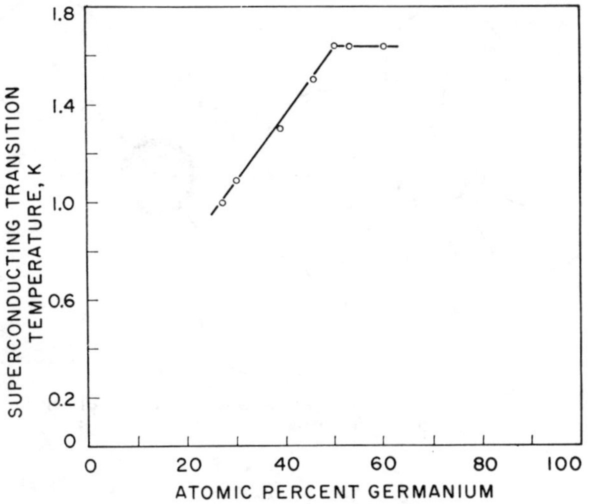

FIG. 6. Superconducting transition temperature in metastable Au–Ge phase.

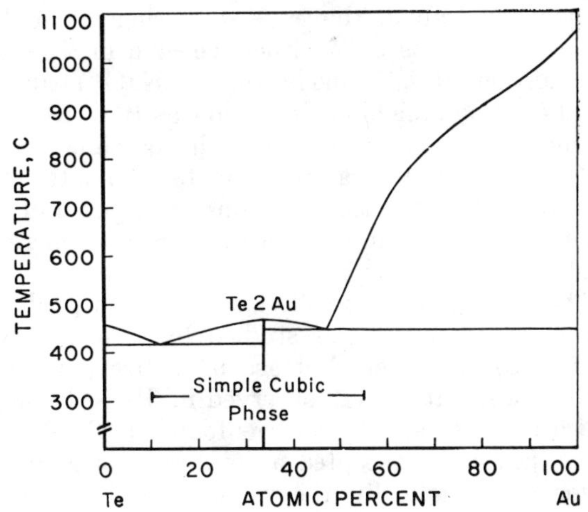

FIG. 7. Extent of the simple cubic nonequilibrium phase in Te–Au alloys (38).

$^3/_2$, it can be considered as a nonequilibrium Hume-Rothery type electron compound. As shown in Fig. 5A to 5D, the phase diagrams of Ag – Ge, Ag – Si, Au – Ge and Au – Si are very similar and it was expected that a hcp phase would exist in the four systems. This is true for Ag – Si and Au – Ge, but not for Au – Si. The range of existence of these phases indicated on the figures always includes the atomic concentration of 16.6% corresponding to an electron-to-atom ratio of $^3/_2$. In Au – Ge alloys, another nonequilibrium phase exists between about 35 and 50 at. % Ge. The complex diffraction pattern of this phase was studied by Anantharaman, et al. (34) who concluded that the crystal structure might be tetragonal with 176 atoms per unit cell. In spite of its rather complicated atomic arrangement, this Au – Ge phase has been found to be superconducting (35), with a transition temperature varying from about 1 to 1.63 K, as shown in Fig. 6. It is believed that this constitutes the first example of a superconducting metastable phase obtained by rapid cooling from the melt in an alloy system in which not only neither of the components are superconducting, but no intermediate phase (superconducting or not) is present under equilibrium conditions.

Quenched Au – Si alloys do not follow the pattern of the three other analogous systems just discussed. An amorphous phase can be obtained in alloys near eutectic composition (31 at. % Si) and this phase will be discussed in a later section. In one investigation of the system, a fcc phase with 32 atoms per unit cell was found between 10 and 25 at. % Si, followed by another fcc phase with about 500 atoms per unit cell extending up to 50 at. % Si (36). In another investigation, another cubic phase was found to have the gamma brass-type of crystal structure and extended from 17 to 20 at. % Si (37). These conflicting results are rather disappointing, but it must be kept in mind that the phases resulting from rapid quenching greatly depend on the average rate of cooling which is difficult to control and cannot be measured accurately enough to be meaningful. Several different phases could, therefore, be obtained at a given composition, depending on the rate at which the liquid solid transformations takes place. To illustrate the relationship between the type of alloy phase and the rate of cooling, Anantharaman, et al. (34) suggested a kind of constitution diagram in which rate of cooling rather than temperature be plotted on the ordinate. It is possible that this approach to a complex problem will eventually explain some of the seemingly contradictory results obtained in different laboratories.

As previously stated, nonequilibrium crystalline phases can also be obtained in alloy systems in which one or more equilibrium phases are present under equilibrium conditions. The case of tellurium-base alloys with gold and silver was selected as a typical example of what can happen in such systems. The equilibrium diagrams of Te – Au and Te – Ag alloys consist of two eutectics separated by an intermediate phase, Te_2Au or Te_2Ag_3. After rapid cooling from the melt, a new phase occurs in both systems with the most unusual simple cubic structure with one atom per unit cell (38). The only element crystallizing in this structure is polonium (19), although an allotropic form of selenium has also been proposed (39). No report has been found on simple cubic alloy phases under equilibrium conditions.

The simple cubic phase exists between 10 and 45 at. % Au in Te and between 10 and 34 at. % Ag in Te. In the case of Au, this new phase replaces a stable compound Te_2Au (Fig. 7). For silver, however, the simple cubic phase does not reach the composition of the first equilibrium compound, Te_2Ag_3. No other reflections than those corresponding to a simple cubic structure were detected in the x-ray diffraction patterns. In a stoichiometric AB compound having a NaCl-type structure, however, a simple cubic pattern could be obtained if the two elements had about the same scattering factor, but this is not the case for either Te – Au or Te – Ag. In addition, it is unlikely that a NaCl structure could exist off stoichiometry as low as 10 or 15 at. % of one of the two elements. The possibility of a NaCl-type structure can therefore be eliminated, and a simple cubic structure with one atom per unit cell must be accepted, in spite of its unusual occurrence in alloy systems.

The change in unit cell size with both gold and silver concentration was studied by Luo and Klement (38). A large decrease in lattice parameter with concentration was observed in Te – Au alloys, and a much smaller effect was found in Te – Ag alloys. Since the accepted atomic radii of gold and silver are practically the same, these results are rather surprising but the large difference in lattice spacings between the two types of alloys is real and beyond experimental errors. Recent measurements made on ternary alloys in which copper is partially replacing gold or silver indicate that in these metastable phases it is possible that the atomic size of the tellurium atom is not a constant and drastically depends on the concentrations of the noble metals. Such extreme deviations from Vegard's law have not been reported before in equilibrium alloys.

A review of crystalline phases obtained by liquid quenching would not be complete without mentioning what has been obtained so far in an alloy system most familiar to metallurgists, namely iron-carbon. Interstitial atoms such as carbon can diffuse very rapidly and hence extreme rates of quenching are required to prevent nucleation and growth of equilibrium phases. The first successful experiment in Fe – C alloys was reported recently by Ruhl and Cohen (40). Iron-carbon alloys containing from 3.3 to 4.2 wt % C were quenched from the liquid state by the gun technique. A new metastable hcp phase designated "epsilon" was found in binary alloys as well as in ternary alloys containing some silicon, chromium or nickel. The lattice parameters of this phase are $a = 2.622$ A and $c = 4.318$, with a $c/a = 1.646$. The epsilon phase is not isomorphous with the hexagonal ϵ-carbide since its unit cell is much smaller than that of the carbide. The authors of Ref. 40 suggest that the new phase is similar to the hcp form of iron which exists only at high pressure (41). If so, this would tend to indicate that it might be possible to obtain high-pressure phases by quenching from the melt. The next step would be to subject liquid alloys to both a high rate of pressure release, probably of the order of 10 kilobars per millisecond and a high rate of cooling and the number of new metastable phases might become very impressive. No suggestions on how to achieve this goal can be given at this time.

AMORPHOUS ALLOYS

The Amorphous State. In this paper dealing with inorganic solid and excluding polymers, the term amorphous is used as synonymous with noncrystalline. Hence, a definition of the crystalline state will automatically define the amorphous state. Quoting from Ref. 42, "Atoms in a crystal are arrayed in a pattern that repeats itself in three dimensions throughout the interior of the crystal." This repetition of the pattern is obtained by the translation of three vectors forming a parallelepiped that defines the space lattice and is called the unit cell. In principle, this cell may contain any number of atoms which may be located at various distances from each other. For the purpose of illustration, a two dimensional equivalent of a space lattice (two dimensional network) will be discussed. Starting with a certain number of atoms arranged in a given configuration, for example six of one kind and two of another, as shown in Fig. 8, a crystalline structure is obtained by first defining a unit cell of the lattice and repeating the structure by translation of the unit cell in two

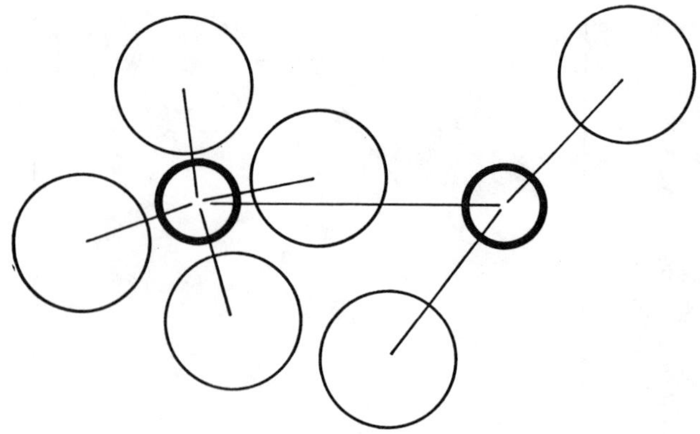

FIG. 8. Group of 8 atoms used as a motif for two dimensional models of crystalline and amorphous structures.

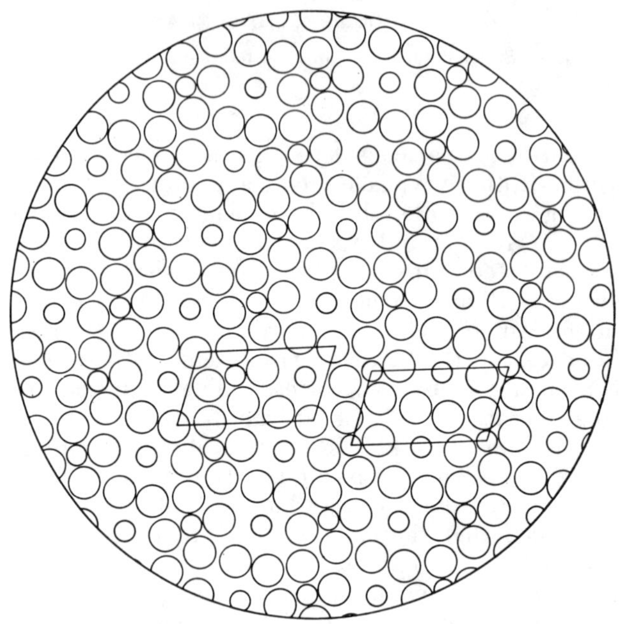

FIG. 9. Two dimensional representation of a crystalline structure without symmetry elements.

directions (Fig. 9). Once the unit cell has been defined, its origin is arbitrary, and the two cells shown in Fig. 8 are equivalent. This two dimensional network has no symmetry and corresponds to the three dimensional space group $P\ \bar{1}$. Crystals without symmetry are not unusual, and a few alloy phases such as δ Al_{11} Mn_4 and Al_4 Re belong to that class (39).

The definition of a crystal as given above appears to be very precise. However, it is vague in one respect, because it does not state how many times the lattice cell should be translated before the structure so obtained can be called a crystal. Hence the definition of a microcrystal should not be based on the size of the crystal but on the number of unit cells contained within its volume. The minimal crystal size may vary within large limits since unit cell dimensions can be as small as 2.28 A, the lattice parameter of beryllium, and as large as 28.239 A, the lattice parameter of cubic Al_3Mg_2 with 1168 atoms per unit cell (32). The distinction between the amorphous and the microcrystalline state ultimately depends on the sensitivity of the experimental methods used for measuring the actual number of perfectly aligned unit cells in a given solid. It is well known that as the average crystal size in a polycrystalline solid decreases, the x-ray diffraction peaks become progressively wider and ultimately the pattern is much like that of an amorphous substance.

The broadening effect decreases with decreasing wave length and, hence, electron diffraction is a more powerful tool than x-rays for studying microcrystalline and amorphous alloys. If only very broad rings are present in an electron diffraction pattern obtained with 100 kv, it is highly probable that the structure is amorphous. Complete absence of diffraction contrast in transmission electron-microscopy at high magnification is another direct proof of lack of crystallinity.

The atomic arrangement in an amorphous solid or in a liquid is not necessarily completely random and in most cases the grouping of atoms around any given atom has some characteristic geometrical features. A simple two dimensional model of the amorphous state can be obtained by considering the group of eight atoms used previously to generate a crystal (Fig. 8). If instead of translating this group of atoms in two directions, it is displaced more or less at random, a picture shown in Fig. 10 is obtained. In constructing the model it was assumed that the relative position and the bond distances between the eight atoms constituting the original grouping would remain approximately the same while the groups were displaced at random with respect to each other. The resulting structure is definitely not crystalline but yet the group of eight atoms which was used as a unit cell to build the crystal of Fig. 9 can still be found in the amorphous structure.

These concepts taken from the studies of silicate glasses can be extended to liquid metals and alloys.

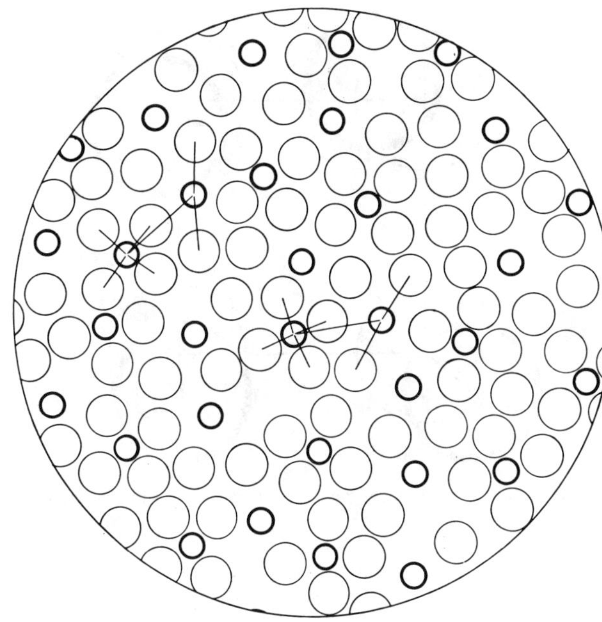

FIG. 10. Illustration of an amorphous structure based on a group of 8 atoms shown in Fig. 8.

There is some evidence to support the idea that in many cases the short-range atomic distribution in liquid metals just above the melting point is very much like that in the crystalline solid (43). This appears to be true for most of the fcc close-packed structures such as aluminum, gold and lead, for which the number of nearest neighbors in the liquid is very close to twelve, and for the bcc structures such as sodium and potassium for which this number is close to eight. Both germanium and tin, which have four nearest neighbors in the crystalline state, do not follow this rule and are relatively more close packed in the liquid state with about eight neighbors for germanium and eleven for tin. Recent studies of the structure of liquid mercury and lead by Kaplow, et al. (44) lead to the conclusion that "Comparison of radial distribution for the solid and the corresponding liquid near the melting temperature show that the radial density functions for the liquid and the corresponding crystalline solid are strikingly similar." Taking this statement as a guide, a few very simple two dimensional models of the liquid (or amorphous) state will be presented. In all cases, it is assumed that the local order present in the solid state has been somewhat retained in the liquid state.

The first case to be discussed is a regular two dimensional square network of atoms, equivalent to a simple cubic structure with one atom per unit cell in three dimensions. The atomic positions were then disturbed by displacing each atom from its original position along two vectors x and y shown in Fig. 11. These displacements were limited to a maximum value corresponding to making contact between two atoms. Within these limits, the values of x and y, as well as their signs (plus or minus) were taken from

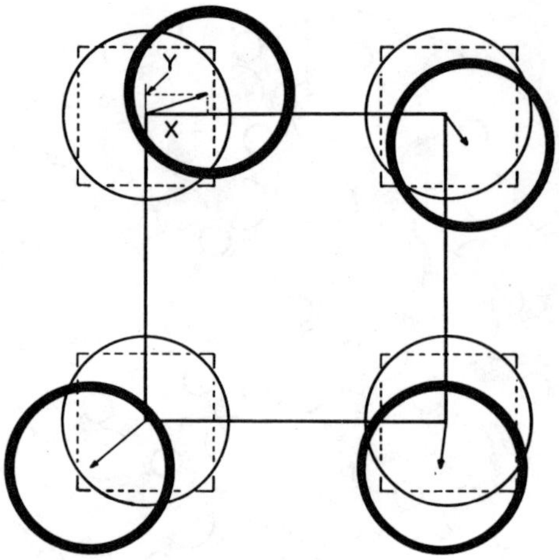

FIG. 11. Random displacement of atoms in a square network.

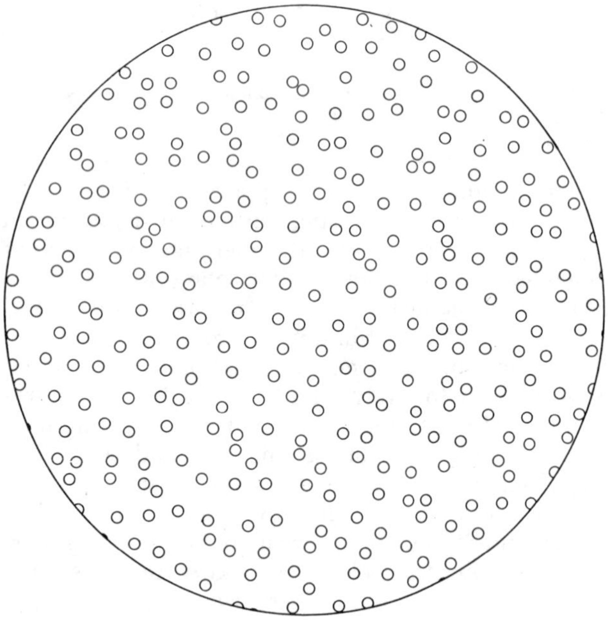

FIG. 12. Random configuration obtained from a square two dimensional network.

a table of random numbers (sets of four numbers). The resulting configuration is shown in Fig. 12. The atomic distribution in this case looks indeed quite random. In spite of the fact that the process of transforming the square network into a random arrangement allowed contact between two circles, this did not occur simply because the probability of such an event is about one in 56,000 and only approximately 1000 circles are shown in Fig. 12.

The same random process was applied to a hexagonal close-packed arrangement and the results are shown in Fig. 13. In constructing these models it was assumed that the number of atoms per unit area remained the same, which corresponds to neglecting the change in density between solid and liquid state, this change being generally of the order of a few per-

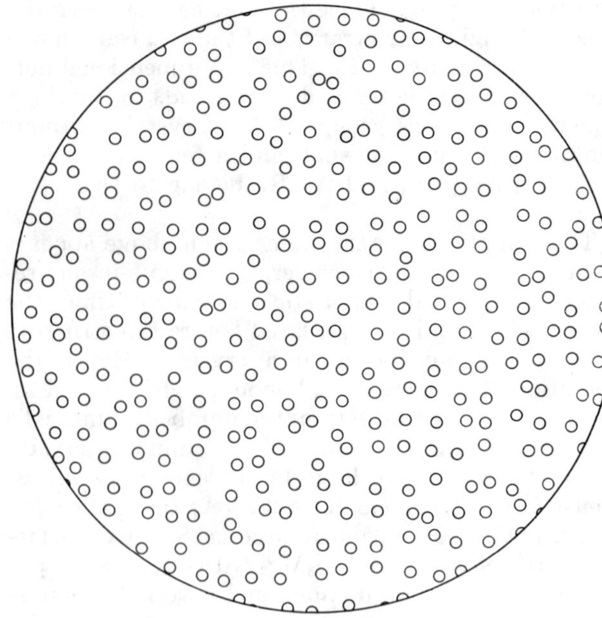

FIG. 13. Random configuration obtained from a hexagonal two dimensional network.

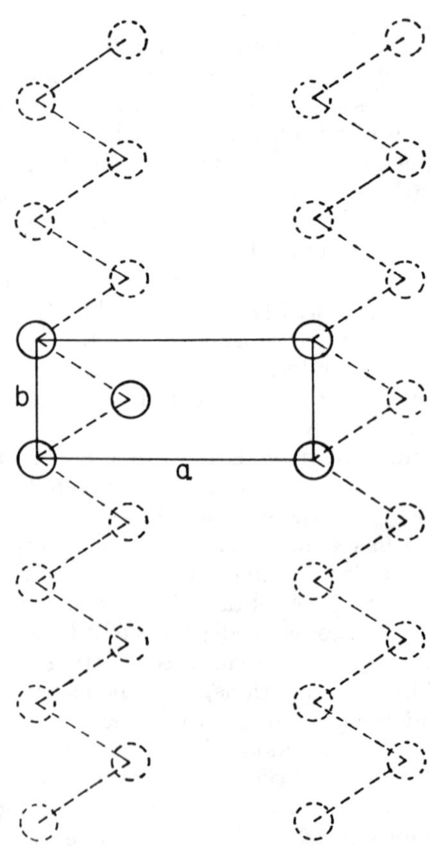

FIG. 14. Unit cell of a two dimensional rectangular network with two atoms per unit cell leading to a chain-like structure.

centage points. The two random structures look very much the same and it is not possible to detect any obvious relationship between these structures and the corresponding crystalline ones, although it

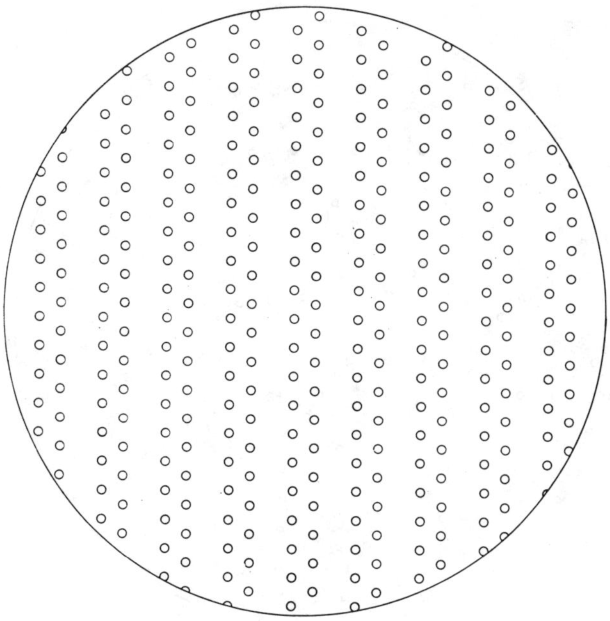

FIG. 15. Two dimensional network built on unit cell shown in Fig. 14.

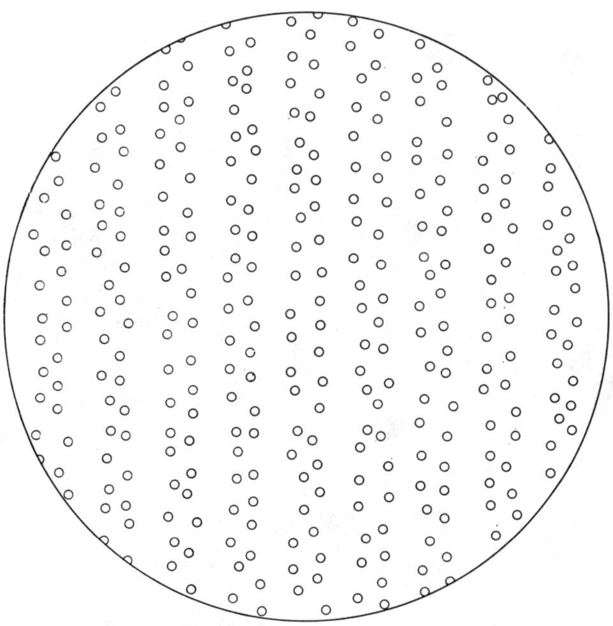

FIG. 16. Random displacement of atoms in rectangular network of Fig. 15.

is probable that a mathematical analysis of the average number of nearest neighbors in the two cases would lead to two different answers.

The next illustration deals with a rectangular network with two atoms per unit cell located as shown in Fig. 14. This simple cell can serve as a model for a tetragonal, hexagonal or orthorhombic space lattice in three dimensions. It was chosen because, as shown in Fig. 15, it is a picture of a chain-like atomic arrangement with two nearest neighbors typical of tellurium, which will be discussed later in this

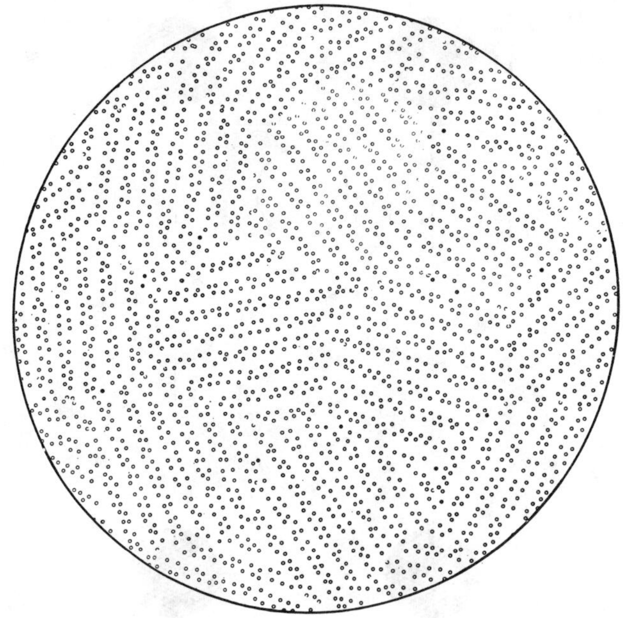

FIG. 17. Random arrangement of polygons containing the oriented amorphous structure shown in Fig. 16.

lecture. When the rectangular crystalline structure is randomly disturbed following the same rules as for the square and hexagonal networks, the atomic arrangement shown in Fig. 16 is obtained. It is readily seen that the chain-like structure has been conserved, but the distances between two adjacent chains vary throughout the structure. The fact that the chains are still present is not surprising since, according to the rules set forth for atomic displacements, no atom could have moved into the empty space already present in the crystalline structure. Obviously, this model does not prove anything concerning the structure of the liquid state, but it demonstrates that if it is assumed that the type of local atomic bonding (in this case the presence of two nearest neighbors in the crystalline structure) is preserved in the liquid state, some characteristic features of the crystal structure will be observed above the melting temperature.

A structure such as that shown in Fig. 16 is definitely oriented in a certain direction which is that of the chains of atoms. There is no reason to believe that this direction will not change either progressively or abruptly throughout the solid. If this is so, an amorphous solid may be divided into regions or domains having different orientations. In order to illustrate this possibility, a certain number of pictures similar to that of Fig. 16 were cut into polygonal shapes and reassembled side by side in a random manner. The resulting structure is shown in Fig. 17. Except for the fact that the domains in this case are noncrystalline, such a structure is very similar to the familiar grain structure in crystalline metals and alloys. Perhaps some of the properties of amorphous alloys could also depend on grain size, and be affected by such phenomena as grain growth,

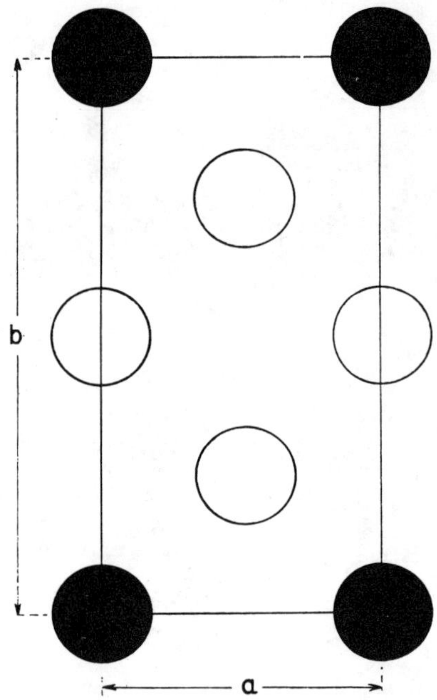

FIG. 18. Rectangular network based on a square network but with two kinds of atoms per unit cell, representing an AB_3 compound.

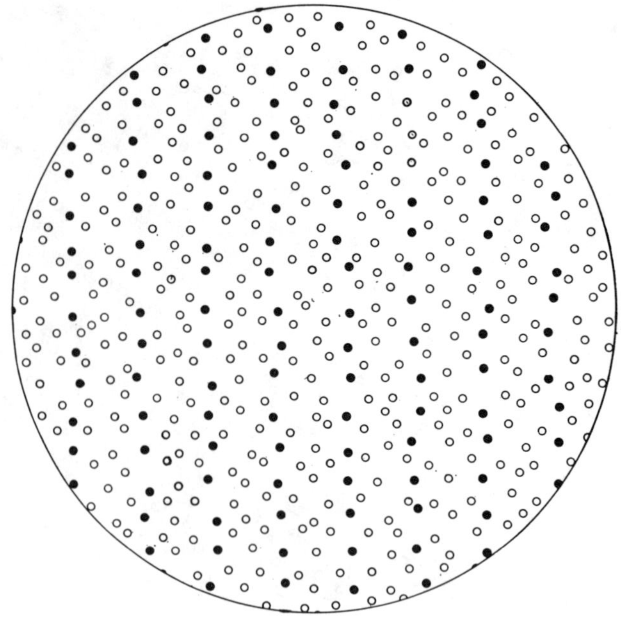

FIG. 20. Random displacement of atoms in rectangular network of Fig. 18 with two kinds of atoms (AB_3 compound).

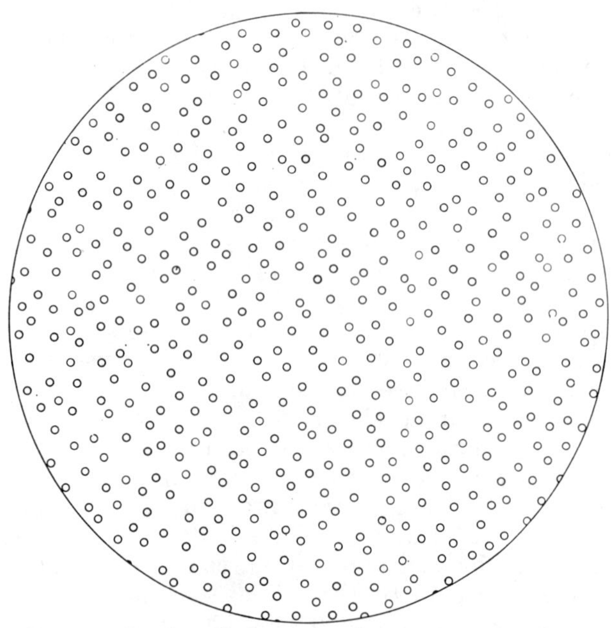

FIG. 19. Random displacement of atoms in rectangular network of Fig. 18 with one kind of atom only.

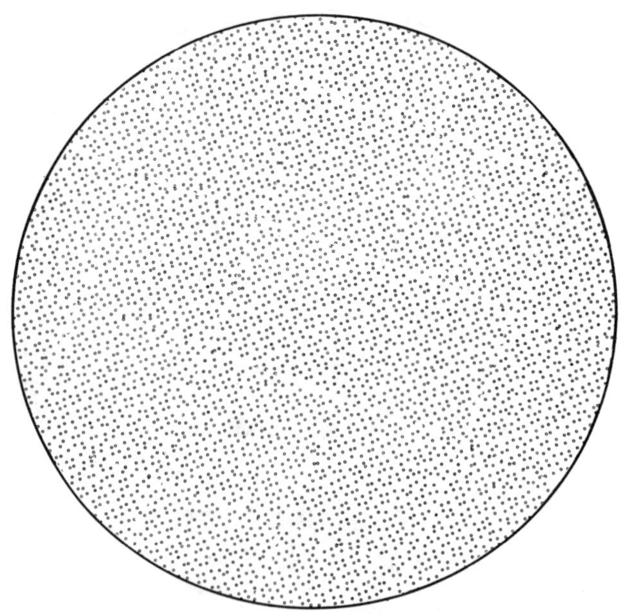

FIG. 21. Random arrangement of polygons containing the amorphous structure shown in Fig. 19.

grain boundary precipitation, grain boundary relaxation, etc.

A chain-like atomic distribution is not the only configuration leading to the possible existence of orientation in amorphous structures. Similar results can be obtained starting with a close-packed structure, providing there are more than one kind of atoms. Consider a rectangular network made of two square-centered unit cells (Fig. 18). If all positions were occupied by the same kind of atoms, this configuration would be identical with the square lattice shown in Fig. 11. When the random process previously described is applied to this structure, the random structure shown in Fig. 19 is obtained. This configuration is essentially identical with that of Fig. 12, but when the two kinds of atoms are introduced in the model, a definite alignment appears and the structure is oriented (Fig. 20). Polygonal areas were cut and assembled side by side as previously

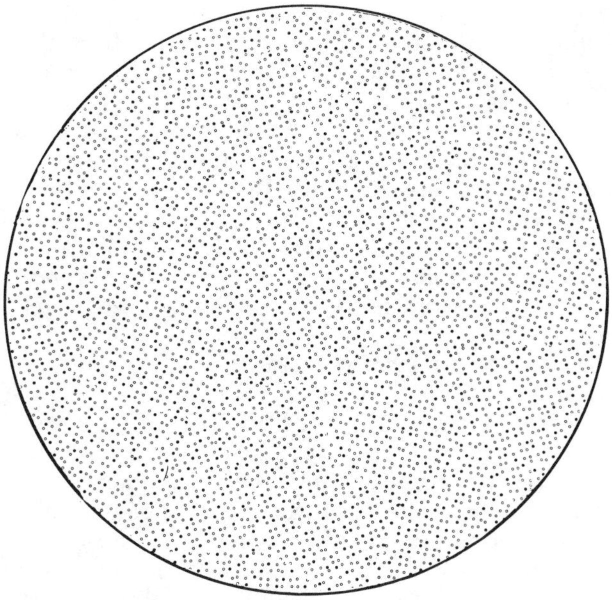

FIG. 22. Random arrangement of polygons containing the amorphous structure shown in Fig. 20.

described. With only one kind of atom, the composite picture looks like a perfectly random arrangement (Fig. 21), but preferred directions and domain boundaries are immediately apparent when two kinds of atoms are introduced in the structure (Fig. 22).

The simple pictorial representation of an amorphous structure just described is not intended to explain the nature of the amorphous state. However, it does show that it is possible to conceive the existence of a preferred direction in an otherwise random atomic arrangement. This concept then leads to that of boundaries between regions having different orientations. These two possible structural characteristics of the amorphous state will be referred to later in discussing some properties of both tellurium-base alloys and ferromagnetic iron-base alloys.

The amorphous alloys obtained so far by quenching from the liquid state can be classified in two categories, according to their physical properties. The first category includes alloys having a low electrical conductivity. Many of these are semiconductors. The second category includes what might be called "metallic" amorphous alloys on the basis of their electrical, and in some cases magnetic properties. Most of the alloys reported so far in the first category are tellurium-base alloys with the elements of the fourth and fifth column of the periodic table. In the second category are alloys rich in gold, palladium, platinum, and iron. In this case, the alloying elements always have a high valence (silicon, carbon or phosphorus).

Tellurium-Base Alloys. The choice of tellurium as a constituent of alloys in which amorphous structures could be obtained was guided by its similitude with selenium, which is amorphous when quenched from the melt at relatively slow rates of cooling. Pure tellurium, however, cannot be quenched into an amorphous structure, even with the present techniques. The addition of germanium, gallium or indium retard the crystallization rates and amorphous structures

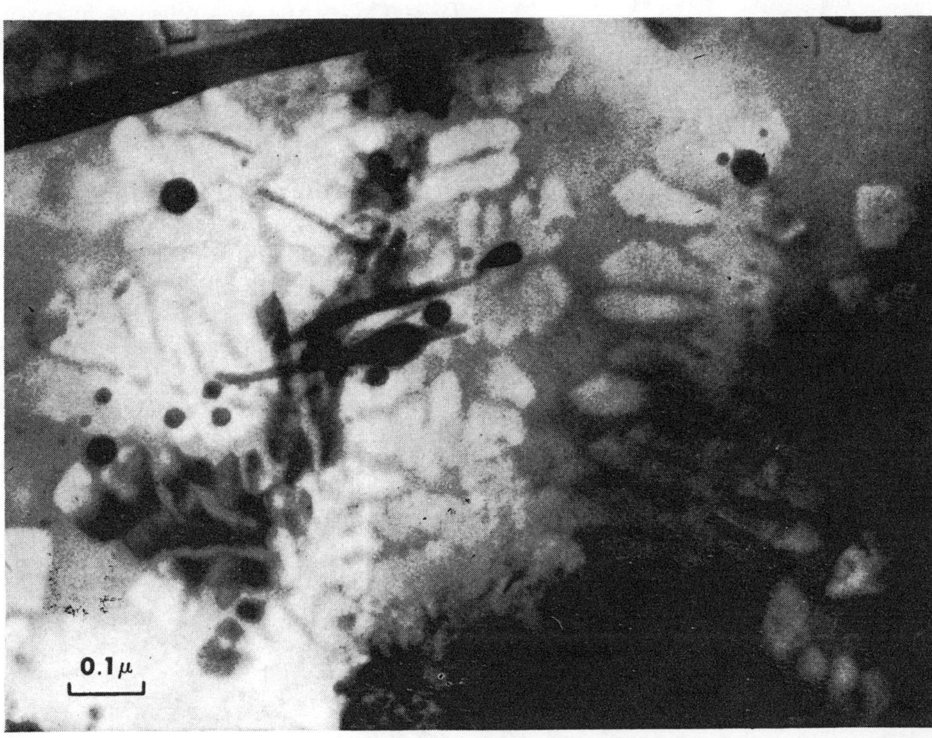

FIG. 23. Random arrangement of polygons containing the amorphous structure shown in Fig. 20.

are obtained in alloys containing from 10 to 30 at. % Ga or In and from 10 to 25 at. % Ge (45).

The amorphous nature of an alloy containing 15 at. % Ge was established by electron diffraction and electronmicroscopy (10). An electron diffraction pattern taken with 100 kv electrons has three or four diffuse rings typical of the liquid state. Electronmicroscopy of the same solid does not show any contrast. In addition the crystallization of this alloy was observed while the specimen was being heated in the electronmicroscope. Well-defined dendrites developed during heating, as shown in Fig. 23. Further heating of the foil resulted in dendritic growth at the expense of the amorphous matrix. These results provide a conclusive proof for the amorphous nature of the quenched alloy.

The results of x-ray diffraction analysis of amorphous tellurium-base alloys are described in detail in Ref. 46. Diffraction patterns were obtained using a Debye-Scherrer camera and MoK_α radiation monochromatized by the (200) reflection of LiF crystal. All patterns consisted of diffuse rings, and a typical microphotometer trace is shown in Fig. 24. The intensity vs $\sin\theta/\lambda$ curves were corrected for polarization, absorption of the specimen and background following generally accepted techniques and the radial distribution functions were determined with the aid of a computer. Within the limits of uncertainty, the results were the same, irrespective of the nature of the alloying element. The number of nearest neighbors and second nearest neighbors, as well as the interatomic distances resulting from this investigation are listed in Table 4. For the purpose of comparison, the results of a study of liquid tellurium (47) and the data relative to solid tellurium are also given. Taking into account the uncertainties in the measurements of radial distribution functions, it can be concluded that the atomic arrangement in the amorphous alloys is very similar to that observed in liquid tellurium, which in turn is very much the same as that in crystalline hexagonal tellurium.

TABLE 4. Structure of Amorphous Tellurium-Base Alloys

Nearest neighbor order	Number of atoms	Interatomic distance
	Amorphous alloys (46)	
1st	1.3–2.20	2.65–2.80
2nd	10.8–12.7	4.00–4.25
	Liquid tellurium (465 C) (47)	
1st	2.3	2.93
2nd	...	4.4
	Crystalline tellurium	
1st	2	2.864
2nd	4	3.468
3rd	6	4.457
4th	2	4.463

The electrical properties of the amorphous tellurium-base alloys with germanium, gallium or indium were difficult to study because of their very low electrical conductivity. However, such measurements were made possible when amorphous structures were discovered in ternary alloys of tellurium with copper and gold (48). As described in a previous chapter, simple cubic phases with one atom per unit cell are found in binary alloys of tellurium with either gold or silver. A similar phase does not exist in binary Te – Cu alloys and, instead, an amorphous structure is obtained in alloys containing about 30 at. % Cu. Since at this concentration alloys of tellurium with silver and gold are simple cubic, a study of the pseudo-ternary systems involving $Te_{70}-Au_{30}$, $Te_{70}-Cu_{30}$ and $Te_{70}-Ag_{30}$ provides a means of investigating the progressive transition from a simple cubic crystalline structure to an amorphous one, the fields of existence of these two types of phases being separated by regions in which alloys containing two phases are obtained.

One of the amorphous ternary alloys, namely $Te_{70}-Cu_{25}-Au_5$, was selected for extensive studies of its structure as well as its electrical properties. The analysis of the diffraction pattern of this alloy and of the radial distribution function leads to the conclusion that, at least as a first approximation, the atomic arrangement is similar to that previously described for the binary tellurium base alloys with Ge (see Table 4).

The electrical properties of the amorphous $Te_{70}-Cu_{25}-Au_5$ phase definitely confirm the semiconducting nature of this alloy. Previous observations had

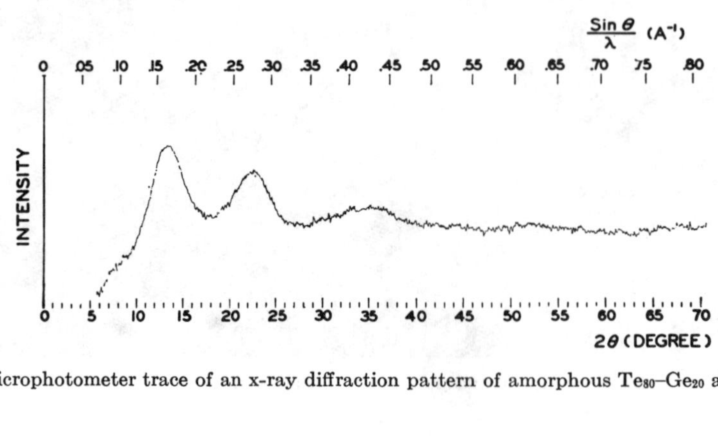

FIG. 24. Microphotometer trace of an x-ray diffraction pattern of amorphous $Te_{80}-Ge_{20}$ alloy (46).

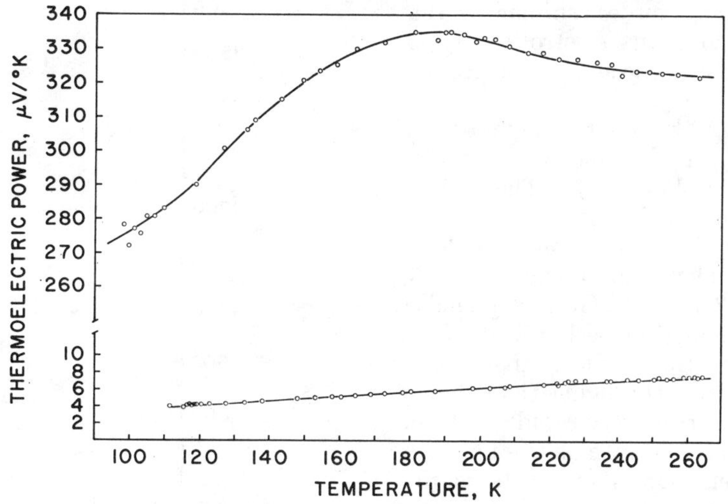

FIG. 25. Thermoelectric power vs temperature of an amorphous (upper curve) and crystalline (lower curve alloy Te_{70}–Cu_{25}–Au_5 (48).

already pointed out that semiconducting properties could exist in both liquids and amorphous solids. Theoretically, Gubanov (49) has shown that a band structure can exist in an amorphous solid as long as there is short-range order, and this is the case of the chain-like structure of amorphous tellurium alloys. The intrinsic energy gap of the amorphous tellurium-base alloy with copper and gold was deduced from measurements of electrical resistivity vs temperature and is approximately 0.2 ev. Measurements of thermoelectric power from 100 K to room temperature also confirmed the semiconducting behavior of these alloys (48). As shown in Fig. 25, the thermoelectric power increases with temperature in the extrinsic range and decreases slowly with temperature in the intrinsic range. The magnitude of the thermoelectric power is also rather large, and is about fifty times that of the same alloy in the equilibrium crystalline state (Fig. 25). Further studies of amorphous semiconductors obtained by liquid quenching may prove very useful in extending the band theory to noncrystalline solids.

A recent investigation of the Mössbauer effect (50) confirms the existence of a chain structure in an amorphous tellurium-base alloy. It was indeed found that the well resolved quadrupole splitting observed for the amorphous alloy (0.74 ± 0.02 cm/sec) is the same as that in pure crystalline tellurium (0.73 ± 0.02 cm/sec). This leads to the conclusion that each tellurium atom still has, in average, two tellurium nearest neighbors, and the gold and copper atoms, which represent 30% of the total, must be located outside of the tellurium chains. These atoms might be located between the chains, which would then be further apart than in crystalline tellurium. Another possibility is for the noble metal atoms to segregate into the "domain boundaries" previously described in connection with a hypothetical amorphous chain-like structure shown in Fig. 17. This is an intriguing problem whose solution

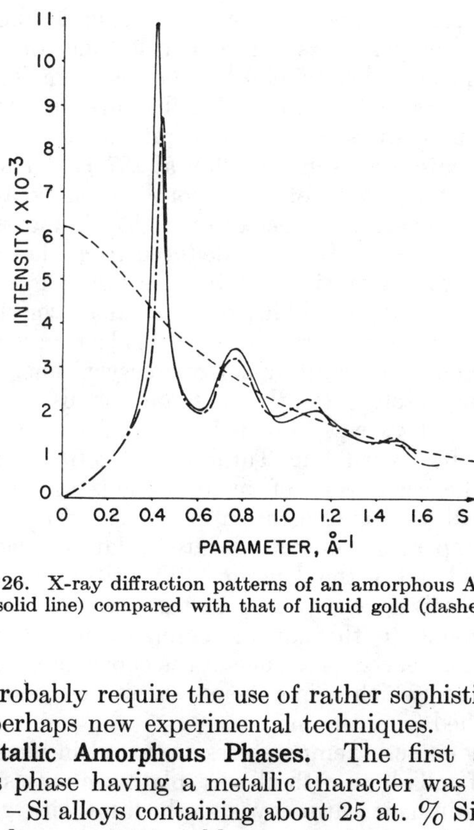

FIG. 26. X-ray diffraction patterns of an amorphous Au_{70}–Si_{30} alloy (solid line) compared with that of liquid gold (dashed line) (52).

will probably require the use of rather sophisticated and perhaps new experimental techniques.

Metallic Amorphous Phases. The first amorphous phase having a metallic character was found in Au – Si alloys containing about 25 at. % Si (51). This phase was not stable at room temperature and transformed into a crystalline metastable form (not yet identified) after about 24 hr. A detailed x-ray diffraction investigation of an amorphous Au – Si alloy containing 30 at. % Si has been reported by Dixmier and Guinier (52). These authors concluded that, as shown in Fig. 26, the difference between the x-ray diffraction pattern of the amorphous alloy and that of liquid gold was not very pronounced. Their analysis of the radial distribution function also led to the conclusion that in both cases the two curves were very similar. They also pointed out that a two-dimensional short-range order, similar to that previously found in amorphous electroless Ni – P alloys (53), could also exist in the amorphous Au – Si alloy.

Additional evidence for the amorphous nature of the Au – Si liquid quenched alloys has been given recently by Chen and Turnbull (54). The rate of crystallization of the binary alloy at room temperature was decreased by replacing some silicon by germanium. A ternary alloy containing 76.9 at. % Au, 9.45 at. % Si and 13.66 at. % Ge was quenched by the gun technique, and the specific heat of the amorphous foils was measured. Up to about 10 C, the specific heat is only slightly higher than that of the equilibrium alloy. It rises very rapidly with increasing temperature and then falls as the temperature is increased further. Onset of crystallization takes place at about 30 C, a few degrees above the maximum specific heat. In addition, the heat capacity above this transition region can be smoothly extrapolated to the values obtained for liquid alloys at higher temperatures. The conclusions of these studies can be best stated by quoting from Ref. 54, "It appears that the most plausible interpretation of our results is that a transition from the glass to the liquid state occurs in this alloy at 297 K. The specific heat behavior of the amorphous alloy at this temperature is as expected for a glass-liquid transition." These statements, deduced from thermodynamic studies of the Au – Si – Ge liquid quenched alloys, provide an additional and most convincing proof that a structure which, in this lecture is called amorphous, is definitely not microcrystalline.

Immediately after the first amorphous phase in Au – Si alloys was obtained by rapid cooling from the melt, Cohen and Turnbull (55, 56) suggested that the occurrence of an amorphous structure in these alloys was due (in part) to the existence of a low-temperature eutectic located relatively close to the gold end of the diagram. This situation results in a relatively sharp drop in temperature from the pure metal to the eutectic composition. This observation served as a guide in the choice of alloy systems in which amorphous structures could be quenched from the melt.

Low eutectic temperatures are found in binary alloys of palladium with silicon; platinum with silicon, germanium or antimony and rhodium with germanium. In all these systems, amorphous structures can be obtained in alloys whose compositions are located near the eutectic concentration (57, 58). An amorphous phase can also be obtained in Pd – Ge alloys (around 20 at. % Ge) for which the equilibrium diagram is not known. Unsuccessful attempts to obtain amorphous phases in nickel and cobalt-base alloys seem to give support to the low eutectic empirical rule, since in all binary alloys of nickel or cobalt with the groups III B and IV B of the periodic table, eutectic temperatures are relatively high compared with the melting points of nickel and cobalt. Among the binary alloys involving palladium, platinum or rhodium as the main constituent, the Pd – Si alloys were found to be the most practical ones to quench into an amorphous phase, the term practical refering to the relatively large percentage of successful quenching with the piston and anvil technique. For this reason, most of the studies of physical properties and transformation kinetics were carried out on a Pd – Si alloy containing 20 at. % Si.

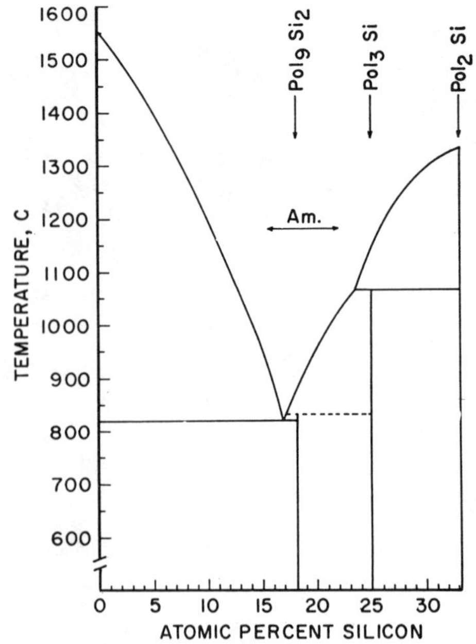

FIG. 27. Revised Pd–Si phase diagram.

The Pd rich end of the published Pd – Si diagram (33) consists of a eutectic (at about 15.5 at. % Si and 800 C) between palladium and an intermediate phase $Pd_3 - Si$ formed by peritectic reaction (59). The next phase, $Pd_2 - Si$, melts congruently at 1250 C. In order to explain some inconsistencies observed in the crystallization of the amorphous $Pd_{80} - Si_{20}$ alloy, the palladium-rich portion of the diagram was reinvestigated. The alloys in the range of from 10 to 30 at. % Si have a strong tendency to undercool, a fact already mentioned in the early paper of Lebeau and Jolibois (60). To obtain meaningful cooling curves, it is necessary to inoculate the melt near the liquidus transition and continue the process near a eutectic or peritectic reaction because the presence of the first crystalline phase will not nucleate the reaction. These studies, still in progress, lead to the conclusion that another intermediate phase exists between Pd and $Pd_3 - Si$. In a recent paper, Nylund (61) assigns the stoichiometric formula $Pd_9 - Si_2$ to this phase, whose crystal structure would be orthorhombic with 44 atoms per unit cell. On the basis of these investigations, a revised version of the Pd – Si diagram is shown in Fig. 27. Additional work is needed for locating exactly the peritectic temperature for the $Pd_9 - Si_2$ phase which is extremely close to the eutectic temperature.

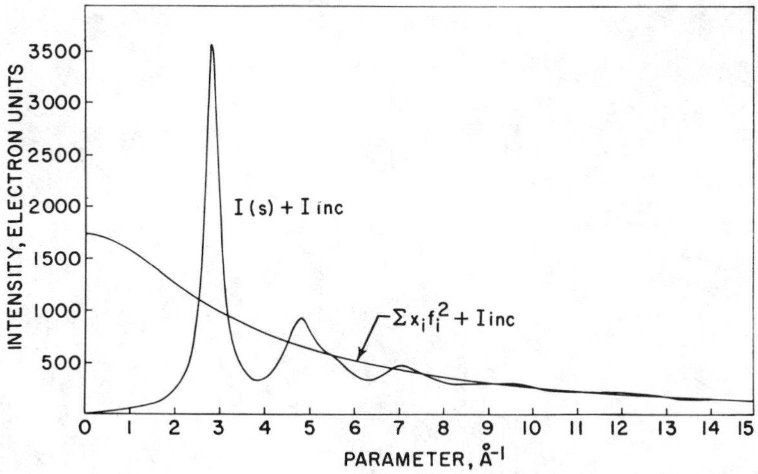

FIG. 28. Corrected diffraction intensity for amorphous Pd_{80}–Si_{20} alloy.

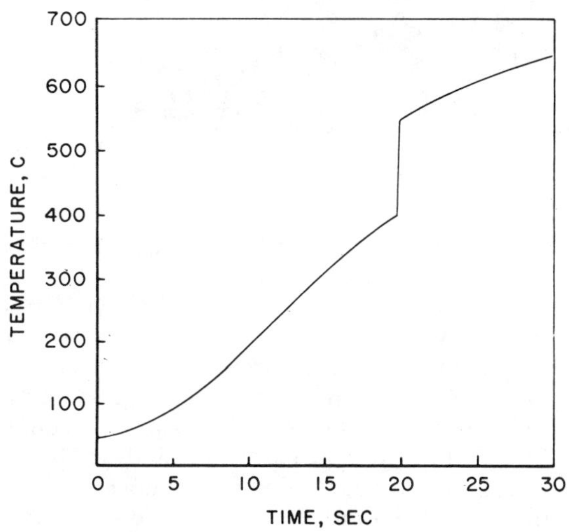

FIG. 29. Temperature vs time curve of amorphous Pd_{80}–Si_{20} alloy.

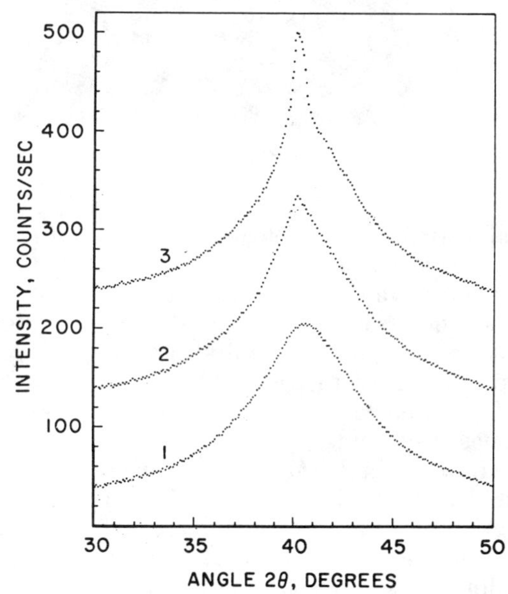

FIG. 30. Changes in the diffraction patterns of a Pd_{80}–Si_{20} amorphous alloy after aging at 250 C for 27 and 87 hr.

The amorphous phase can be obtained in alloys containing from 15 to 23 at. % Si. Thin foils for x-ray diffraction and electronmicroscopy can be prepared by the gun technique, and foils having a uniform thickness of about 50 microns and as large as 25 mm in diameter can be obtained with the piston and anvil technique. These are quite suitable for diffractometer work and for electrical resistivity measurements. The structure of an alloy containing 20 at. % Si was studied by x-ray diffraction, with MoK_α radiation monochromatized by a LiF crystal. A typical intensity vs sin θ/λ curve is shown in Fig. 28. Preliminary results of an analysis of the radial distribution function, indicate that the number of first neighbors is approximately 11.6, at an average distance of 2.80 A. In pure palladium (fcc), there are 12 neighbors at 2.75 A. It is, therefore, apparent that the structure of the amorphous alloy is very closely related to that of palladium, and would probably be very similar to that of liquid palladium, which unfortunately has not been studied. These conclusions are remarkably similar to those reached for amorphous tellurium-base alloys. For palladium, as for tellurium, the alloying element seems to be required to prevent crystallization of the melt, but has little effect on the structure of the amorphous solid, whose atomic arrangement is dictated by the coordination of the host element.

A study of the kinetics of decomposition of an amorphous $Pd_{80}-Si_{20}$ alloy into a crystalline structure has not yet been completed, but some of the preliminary results appear reliable enough to be discussed. When heated at a rate greater than about 20 C/min, an amorphous specimen undergoes a strong exothermic reaction at 400 ± 10 C (Fig. 29). The heat released is of the order of 1000 cal/mole, which is comparable to a latent heat of solidification. The amorphous-to-crystalline transition is obviously not as fast as it appears to be from measurements made with high rates of heating. The progress of

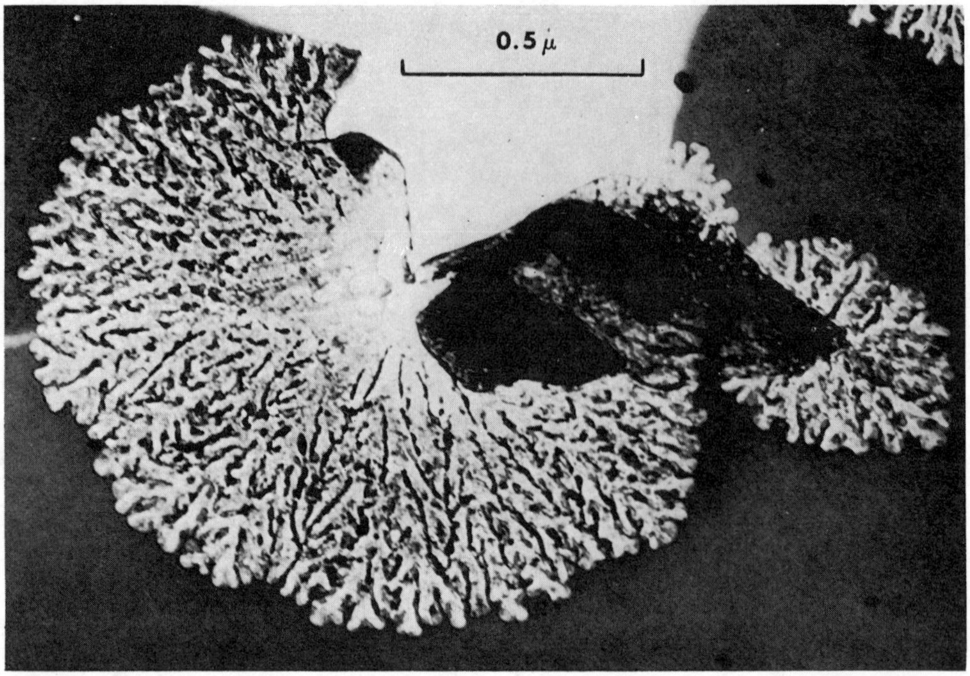

FIG. 31. Electronmicrograph of a Pd_{80}–Si_{20} amorphous alloy in the early stage of crystallization (58).

crystallization can be detected by x-ray diffraction in specimens aged at temperatures below 400 C. In order to observe the very early stages of crystallization, it is necessary to carefully scan the angular 2θ range corresponding to the first diffuse band of the amorphous structure (from 30 to 50° in 2θ). At 200 C, no change in the shape of the amorphous band has been found after more than 15,000 hr at temperature. At 250 C, the beginning of crystallization is detectable after about 24 hr. Typical x-ray diffraction patterns are shown in Fig. 30 for a specimen maintained at 250 C for 27 and 87 hr. A small crystalline peak corresponding to the (111) reflection of palladium is observed at a 2θ angle of 40.15°. This peak is superimposed upon the diffraction band of the amorphous matrix, which is still as broad as before heat treatment. With further aging, the crystalline peak grows in intensity and the amorphous band decreases in intensity without sharpening. The growth of palladium crystals at temperatures near 250 C was confirmed by electronmicroscopy. The kind of crystalline growth shown in Fig. 31 was found in a $Pd_{80}-Si_{20}$ alloy heated by the electron beam in the microscope. An electron diffraction pattern of the crystallized region consisted of continuous rings which were indexed as those of fcc palladium. The conclusion is that at temperatures near 250 C, the first crystalline phase nucleating in the amorphous matrix is a palladium phase and not the equilibrium phases Pd_9-Si_2 and Pd_3-Si. The growth of palladium crystals with a very simple fcc structure, instead of the complicated crystal structures of the equilibrium silicides, may be related to the fact that in the structure of the amorphous alloy, each atom is surrounded by approximately twelve nearest neighbors, and only very small diffusion distances are required to align these atoms into a fcc crystalline nucleus. Much longer aging times at 250 C are still in progress, but after about 15,000 hr the palladium phase is still the predominant one and the equilibrium silicides are barely detectable. At aging temperatures of 300 to 350 C, crystallization rates are obviously increased, but a palladium phase still nucleates first and grows for at least 3000 hr. In spite of the fact that isothermal heat treatments have been in progress for more than 15,000 hr, quantitative results are still only preliminary and it is expected that an accurate description of the crystallization process will not be obtained before about 30,000 hr (approximately 3.5 years).

Measurements of electrical resistivity give interesting information on the nature of an amorphous alloy as well as its transformation into the crystalline state. Preliminary results (57) have shown that the room-temperature resistivity of the amorphous alloy is about 2.3 times that of the crystalline alloy of the same composition. The resistivity decreases with temperature, but the temperature coefficient is very small (about 0.00017/C) and at 2 K about 96% of the room-temperature value is retained. Evidently, the electrical resistivity of the amorphous alloy is mainly governed by scattering resulting from a random structure, and the phonon scattering contribution is of secondary importance. More recently, Crewdson (58) has made additional measurements on a $Pd_{80}-Si_{20}$ amorphous alloy covering a range of temperatures from liquid nitrogen to the melting temperature and beyond. His results are

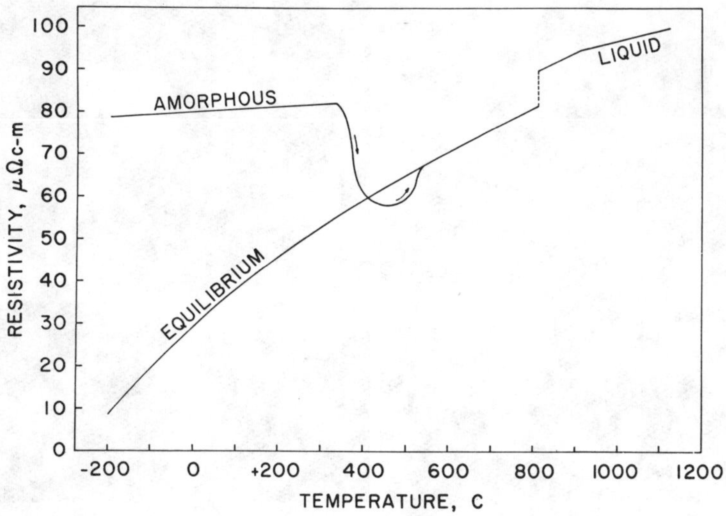

Fig. 32. Electrical resistivity of amorphous, crystalline and liquid Pd_{80}–Si_{20} alloys; heating rate 1.5 C/min (58).

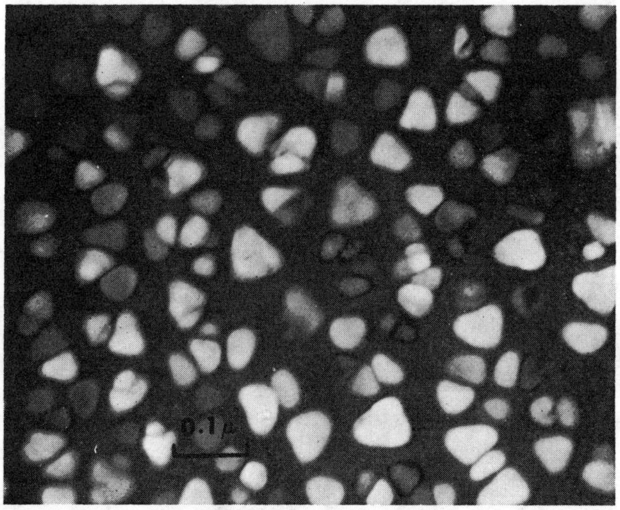

Fig. 33. Amorphous and crystalline phases in a quenched Pd_{88}–Si_{12} alloy (58).

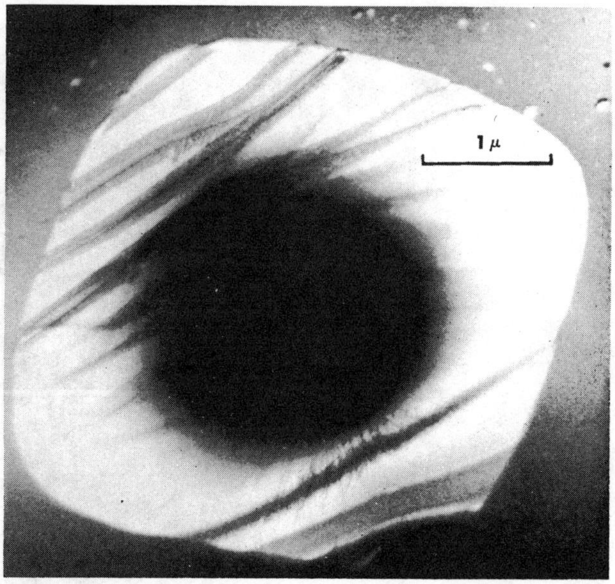

Fig. 34. Isolated crystal of a metastable bcc phase in a quenched Pd_{82}–Ge_{18} alloy (58).

shown in the graph of Fig. 32. The equilibrium alloy shows a normal behavior up to the eutectic temperature, where, as expected, a sudden resistivity increase occurs. The slope of the curve in the two-phase regions between the eutectic (800 C) and the liquidus (920 C) is smaller than in the solid state, and is still smaller above the liquidus. It is, of course, impossible to carry on the electrical resistivity measurements of the amorphous phase up to the melting temperature. However, it is significant that the portions of the curve relative to the amorphous and the liquid states shown in Fig. 32 can be joined together into a single, rather smooth curve. This would suggest that the structure of the amorphous alloy is somewhat related to that of the liquid state.

The devitrification of amorphous alloys can be observed in transmission electronmicroscopy. However, quantitative measurements of growth rate with time and temperature are difficult to obtain because of the uncertainty in determining the actual temperature of the region within the field of view. In spite of these limitations, it may be of interest to show a few examples of typical amorphous-to-crystalline transitions in some of the alloys of palladium or platinum with silicon and germanium.

When an alloy of composition near, but outside, the amorphous range is quenched, crystalline and amorphous phases are generally present. This is illustrated for a $Pd_{88}-Si_{12}$ alloy in Fig. 33. Small crystals approximately 0.1 micron wide are dispersed in an amorphous matrix. Since the foil is also approximately 0.1 micron thick (on the basis of its transparency to electrons), most of the crystals must be aligned in a single layer. Electron diffraction shows that these crystals are fcc palladium, which, at this concentration, is an equilibrium phase. In

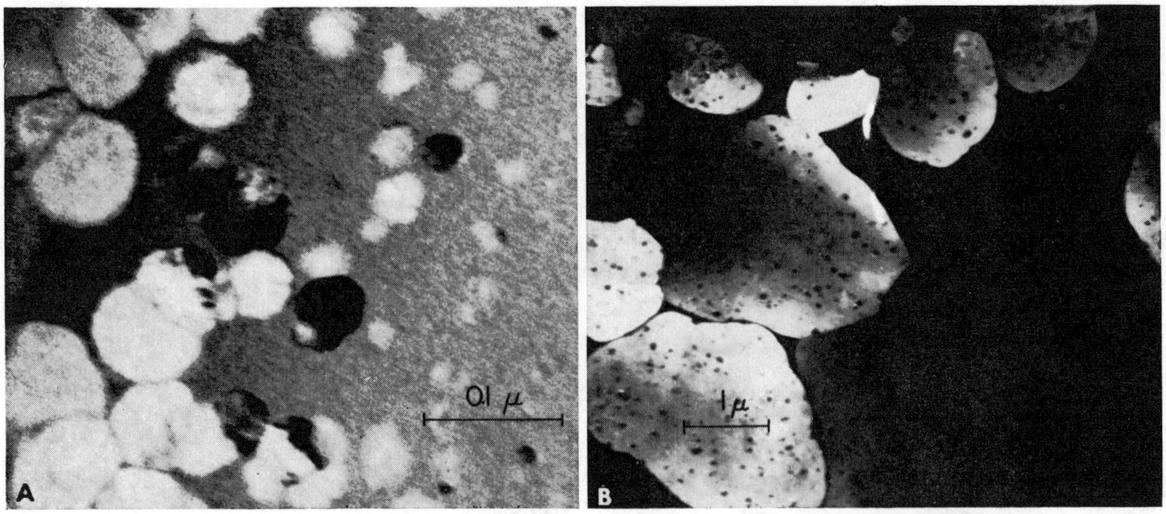

FIG. 35. Electronmicrographs of: (A) transition from amorphous to crystalline structure in a quenched Pt_{82}–Ge_{18} alloy (58); (B) intergranular melting in bismuth (62).

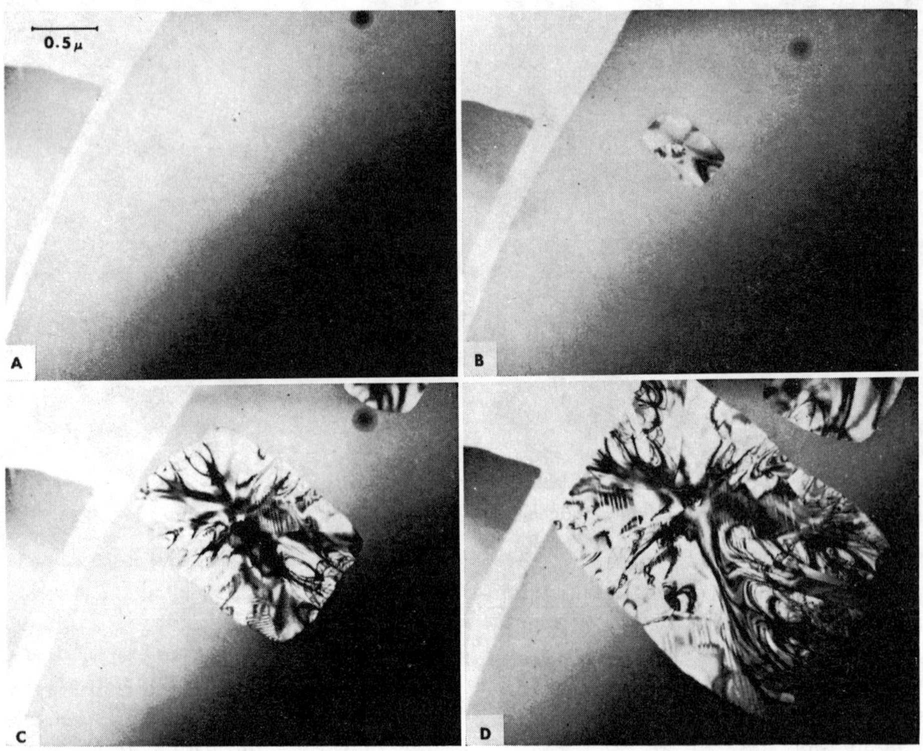

FIG. 36. Progressive growth of a crystalline phase in amorphous Pt_{77}–Si_{23} alloy.

addition, they are oriented with the (111) direction perpendicular to the foil thickness. Occasionally, a rather large isolated crystal has been found in an amorphous matrix. This is illustrated in Fig. 34 showing a bcc crystal typical of a metastable phase $Pd_{82}-Ge_{18}$ in a quenched alloy.

A transition from a crystalline to an amorphous region is sometimes found in a quenched foil in which, for some experimental reason, the entire foil was not cooled at the same average rate. Such a case found in a $Pt_{82}-Ge_{18}$ alloy is illustrated in Fig. 35A. The transition from crystalline to amorphous is gradual, and the amorphous matrix can be seen between grains and also around other grains which are completely isolated. This type of microstructure is compared with a recently published electronmicrograph (Fig. 35B) showing the melting of grain boundaries in bismuth observed in transmission electronmicroscopy (62). Except for the large difference in magnification (a factor of about 10), there is a striking similitude between the two microstructures.

The progressive crystallization of an amorphous phase can also be followed by heating the specimen in the electronmicroscope and the case of a $Pt_{77}-Si_{23}$

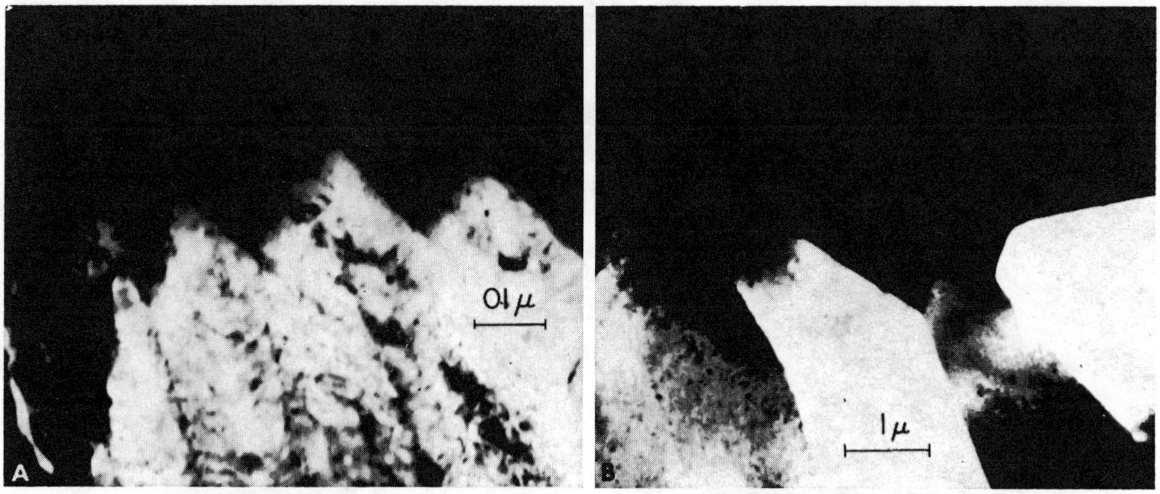

FIG. 37. Electronmicrographs of a crystallization front in: (A) amorphous Pd_{80}–Si_{20} alloy (58); (B) liquid bismuth (62).

alloy is illustrated in Fig. 36. A very small nucleus can be seen in the amorphous matrix in Fig. 36A. Further heating leads to the progressive growth of this nucleus as shown in Fig. 36B–36D. The very sharp and almost straight boundaries found in this case are rather unusual, since, in most cases, a saw tooth-type of boundary has been observed. A typical microstructure of this type, found in a $Pd_{80} - Si_{20}$ alloy, is shown in Fig. 37A. Figure 37B illustrates a transition from solid to liquid bismuth taken from Ref. 62, and here again a strong analogy can be seen between the two microstructures.

These rather cursory results on the morphology of crystallization of amorphous alloys have been shown here for the purpose of illustrating the complexity of the problem and the magnitude of the experimental difficulties. Definite conclusions cannot be drawn at this time and the problem should be a challenging one for electronmicroscopists.

Ferromagnetic Amorphous Alloys. The existence of ferromagnetism in amorphous structures has been theoretically predicted (63). The model used by Gubanov does not require a periodic arrangement of the atoms in a lattice and takes into consideration only the exchange integral of neighboring ferromagnetic atoms and the radial distribution function of the structure. Ferromagnetism in amorphous alloys has been confirmed experimentally. By evaporating cobalt and gold on a substrate maintained at liquid nitrogen temperature, Mader and Nowick (64) succeeded in depositing amorphous alloy films with thicknesses ranging from 200 to 600 A. The films containing from 25 to 60 at. % Au were ferromagnetic. These thin films are relatively unstable: crystallization is noticeable at room temperature and proceeds very rapidly around 100 C.

Two different approaches were taken in trying to synthesize amorphous ferromagnetic alloys by liquid state quenching. The first was to introduce a limited number of ferromagnetic atoms into an amorphous structure; the second was to start with an alloy rich in either Fe, Co, or Ni and find alloying elements which would generate an amorphous structure. Both approaches have led to successful results.

Palladium-silicon alloys were selected as the basic composition to which iron, cobalt or nickel could be added. The reason for this choice was simply that palladium forms complete series of solid solutions with either one of the three ferromagnetic metals and it was anticipated that these elements could be substituted to palladium up to a certain critical concentration without preventing the occurrence of an amorphous phase after quenching. The atomic concentration of silicon was kept constant at 20% and amorphous alloys were obtained in alloys containing up to 5 at. % Fe, 12 at. % Co and 15 at. % Ni (65). The magnetic properties of these alloys have not yet been fully determined. Some data obtained on an alloy containing 12 at. % Co are: coercive force 466 Oe, and remanence about 0.1 Gauss. These ferromagnetic alloys differ from those obtained by high-vacuum vapor deposition in two respects: first, they can be obtained in rather massive foils about 50 microns thick (0.002 in.) instead of less than 0.05 microns; second, they are definitely more stable than evaporated films, since preliminary kinetics studies have shown that the amorphous Pd – Si plus Fe, Co or Ni alloys would be stable for 10 to 100 years at 175 C.

Since the maximum concentration of a ferromagnetic atom in the Pd – Si amorphous structure is limited, these alloys are obviously rather weak ferromagnets. A strongly ferromagnetic amorphous alloy should contain a rather large concentration of one of the four ferromagnetic elements in the periodic table, namely iron, cobalt, nickel or gadolinium. Remembering that most of the amorphous alloys obtained so far by liquid quenching are near a eutectic composition and the eutectic temperature is always low with respect to the melting temperature of the main component, the choice of alloying elements was

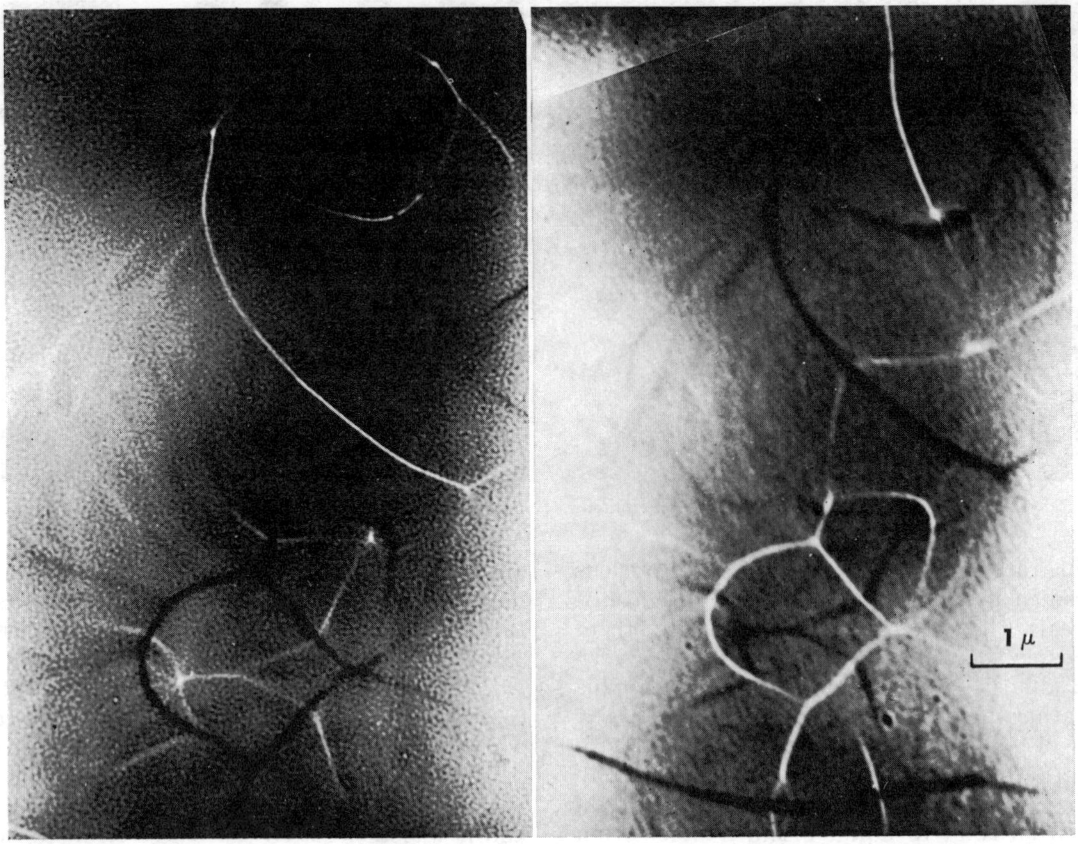

FIG. 38. Lorentz electronmicrographs of ferromagnetic domain boundaries in an amorphous Fe–P–C alloy (67).

limited to those satisfying these conditions. No suitable alloying elements were found for gadolinium, but phosphorus and sulfur appeared to satisfy the low eutectic condition for iron, cobalt or nickel. Binary Fe – P alloys of near eutectic composition (17.5 at. % P) were not entirely amorphous after quenching. The addition of carbon (which probably further depresses the eutectic temperature) resulted in an amorphous structure. Although the amorphous phase can be obtained within a certain range of compositions, most of the alloys studied (and analyzed after quenching) contained approximately 6.9 wt % P and 1.7 wt % C (66).

The diffraction patterns of the quenched foils, taken with either a diffractometer or a powder camera and using monochromatic MoK_α radiation, show three very broad peaks at 2θ angles of 19.8, 33.7 and 49.5°. The results of transmission electronmicroscopy and electron diffraction are similar to those previously reported for Pd – Si alloys, namely complete lack of contrast in microscopy and existence of three very broad diffraction rings. The hysteresis cycle of quenched foils is typical of a soft ferromagnetic alloy with a saturation magnetization of about 6.8 kG and a coercive force of approximately 3 Oe. As in the case of Pd – Si, the Fe – P – C amorphous phase transforms very rapidly into the equilibrium crystalline phases (Fe – C solid solution, Fe_3P and Fe_3C) when heated at rates exceeding about 100 C/min. Under these conditions, a sudden release of heat can be detected by thermal analysis at 420 C and the temperature increase due to the amorphous-to-crystalline transition corresponds to approximately 900 cal/mole. A study of the kinetics of transformation has not been completed, but it is established that the amorphous phase is stable for more than 24 hr at 350 C. As a result of these observations, it was possible to measure the Curie temperature, which was found to be 320 C. It is believed that this is the first amorphous alloy for which the Curie temperature is such that it can be observed before the alloy transforms rapidly into a crystalline structure.

As a result of these studies, it is quite clear that ferromagnetism does exist in the absence of crystallinity, even in massive specimens such as those obtained by rapid quenching from the liquid state. Many puzzling questions remain to be answered, and one of these is concerned with the nature of domain boundaries in an amorphous solid. The technique of Lorentz electronmicroscopy was used to establish the existence of such boundaries in amorphous Au – Co thin films (64). The same technique applied to the Fe – P – C quenched alloys also revealed domain boundaries, as shown in Fig. 38 (67). Magnetic domains are the result of spin orientation, and in describing domains in a crystalline solid, these orientations are generally related to crystallographic direc-

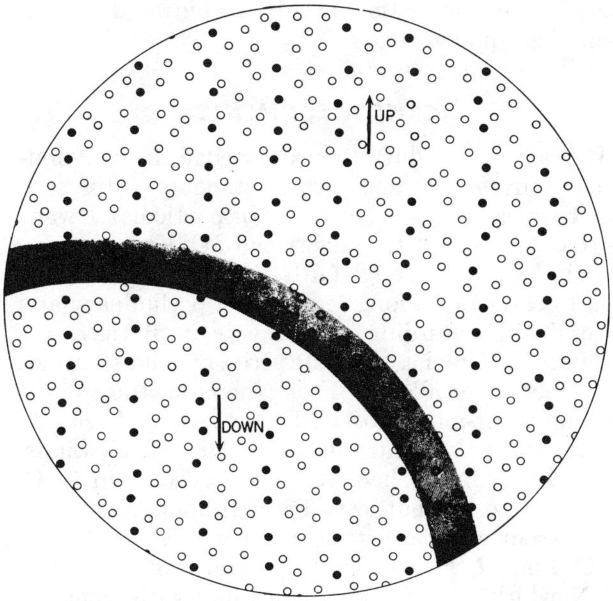

FIG. 39. Illustration of a ferromagnetic domain boundary separating two regions of opposite spins, in an amorphous alloy.

tions. If in an amorphous solid, the atomic positions were really random, it would be difficult to conceive of the existence of domains. However, as explained in the section of this lecture devoted to the amorphous state, complete randomness is probably never found in a noncrystalline solid, and it is possible to conceive of a preferred direction within a limited volume of the solid. The hypothetical crystalline structure shown in Fig. 18 in which there are three atoms of one kind and one of another could represent Fe_3-P, which in turn is a simplified model of the crystalline phase in the ternary alloy $Fe-P-C$. After the atoms were randomly displaced from the original positions, the resulting (amorphous) structure exhibited a preferred direction (Fig. 20). Assuming that this structure represents very crudely that of the amorphous $Fe-P-C$ alloy, it is possible to conceive that all the spins of the iron atoms can be oriented parallel to the preferred direction. If the picture shown in Fig. 20 is now divided in two sections by an arbitrary line, it is conceivable that the spins on the two sides of this line might remain parallel but take opposite directions. This line could than be considered as a domain boundary, and, as shown schematically in Fig. 39, a domain wall would then separate the two regions of opposite spins. This oversimplified picture of domain boundaries in an amorphous ferromagnetic alloy may not correspond at all to the physical reality. In spite of their shortcomings, simple models can inspire new experiments and in so doing may serve a useful purpose, even if they happen to be found completely wrong. The existence of ferromagnetism in amorphous solids is an undeniable fact and it is hoped that further studies of these unusual alloys will shed some light on the alloying behavior of metals as well as on the nature of ferromagnetism.

CONCLUDING REMARKS

In this lecture, I have summarized the results obtained in alloys quenched from the liquid state at rates of the order of 10^5 C/sec and higher. The nonequilibrium phases resulting from such drastic quenching are either supersaturated solid solutions, new crystalline phases or amorphous phases. Although the technique of ultra-rapid quenching from the melt is only seven years old, the number of new phases (excluding a large number of supersaturated solid solutions) has passed the fifty mark. The study of the physical properties of these phases is obviously lagging behind their structure determinations, but a limited number of results have been published and show that some of these phases have unusual electrical, semiconducting and thermoelectric properties. A few of these phases are superconducting and one of the newest findings is the existence of ferromagnetism in amorphous iron-base alloys. I must apologize to those of you who are particularly interested in the mechanical properties of metals and alloys for not having even mentioned the subject in this lecture. The reason for this voluntary omission is that there is a definite gap in our knowledge of the mechanical properties of nonequilibrium alloys, and yet it seems that the strength properties of crystalline phases with crystal sizes in the range of 50 to 1000 A should be of great interest. The elastic and anelastic properties of amorphous alloys appear to be a fascinating subject for those interested in the atomic mechanisms involved in internal friction. The small size of the specimens obtained by rapid quenching should not constitute a handicap in the study of their mechanical properties, since techniques have been developed for studying mechanical properties of metallic as well as nonmetallic whiskers which are, at least in one of their dimensions, much smaller than the presently available quenched foils.

Traditionally, the Campbell Memorial lecturer does not expect questions from the audience, so I decided to bring up one question which I assume many of you would ask if you had a chance. Will any of the new alloys obtained by rapid quenching from the melt find useful applications? My answer to that question is a very definite yes, but I would not venture to say when this application will actually occur. As you all know, there is always a long time interval between what could be called a metallurgical laboratory curiosity and a large-scale development. Many examples of such a delay time could be given but I like to refer to the Campbell Memorial Lecture given by my immediate predecessor Professor J. J. Gilman who told you about "Monocrystals in Mechanical Technology." As Professor Gilman mentioned in his lecture, Professor Campbell was probably aware at the end of his career in 1925, that German and British metallurgists were studying individual metal crystals, and I am sure that at that time, they were motivated more by the beauty and the mysteries surrounding a new subject than by the preoccupation of developing new and immediately useful materials. As Professor Gilman told you, fifty years later, alloy monocrystals are a reality and may be the next generation of turbine blades in jet engines. I would not dare to predict such a spectacular future for alloys quenched from the liquid state, and I am sure they will never be used in jet engines. The requirements imposed for the rapid removal of heat from a liquid alloy by conduction will always limit the thickness of the material to about 0.010 in. at the most. I venture to predict that if any use is found for these materials, it will be in the field of electronics, or in very small electromechanical devices. In spite of the fact that these materials may be produced in grams rather than tons, they still belong to the field of metallurgy, and I am sure that many of you in the audience are right now concerned with some aspect of what may be called micrometallurgy. So I have faith in the possible usefulness of what a prominent Russian physicist visiting Caltech a few years ago called "Duwez's stupid alloys." If the time lag is 30 to 50 years, I do not expect to have the pleasure of witnessing such applications, but I hope that some of my students will then remember the early days when they helped me to try to fool the atoms by freezing them into unconventional positions by rapid cooling from the liquid state.

ACKNOWLEDGMENTS

It is a pleasant duty to acknowledge the contributions of my former and present students to the subject I covered in this lecture. In particular I want to thank Drs. R. H. Willens and W. Klement, Jr., who, in 1959, had enough faith in the liquid quenching idea to carry out a long sequence of preliminary and generally unrewarding experiments until they successfully obtained a complete series of solid solutions in copper-silver alloys. From there on, the subject became a leit-motiv for Ph.D. theses and I wish to acknowledge the contributions of Drs. R. K. Linde, H. L. Luo, C. C. Chao, R. C. Crewdson and C. C. Tsuei. Last but not least, special thanks are due to the present graduate students Messrs. R. D. Ayers, C. C. Lin, M. E. Weiner, P. L. Maitrepierre and L. R. Newkirk who are still struggling with some aspects of the problem.

The research carried out at the California Institute of Technology on rapid quenching from the liquid state would have been impossible without the support of the U. S. Atomic Energy Commission. I would like to take this opportunity to express my sincere appreciation to Drs. D. K. Stevens and W. F. Sheely of the U. S. Atomic Energy Commission for their continued interest in the program and their helpful suggestions on many aspects of the problem.

REFERENCES

1. Pol Duwez, Effect of Rate of Cooling on the Alpha-Beta Transformations in Titanium and Titanium-Molybdenum Alloys, Trans AIME, 191 (1951) 765.
2. J. W. Kauffman and J. S. Koehler, The Quenching of Lattice Vacancies in Pure Gold, Phys Rev, 88 (1952) 149.
3. W. T. Olsen, Jr., and Ralph Hultgren, Effect of Rate of Freezing on Degree of Segregation in Alloys, Trans AIME, 188 (1950) 1323.
4. G. Falkenhagen and W. Hofmann, Die Auswirkung Extrem Hoher Abkuhlungsgeschwin-digkeit auf die Erstarrung and Gefuge Binarer Legierungen, Z Metallkunde, 43 (1952) 69.
5. Pol Duwez and R. H. Willens, Rapid Quenching of Liquid Alloys, Trans AIME, 227 (1963) 362.
6. R. H. Willens and E. Buehler, Rapid Quenching of Reactive and Refractory Alloys from the Liquid State, Trans AIME, 236 (1966) 171.
7. Paul Pietrokowsky, Novel Mechanical Device for Producing Rapidly Cooled Metals and Alloys of Uniform Thickness, Rev Sci Instr, 34 (1963) 445.
8. D. R. Harbur, J. W. Anderson and W. J. Maraman, Rapid Quenching Drop Smasher, Los Alamos Scientific Laboratory of the University of California, Report No. LA-3584, July 1966.
9. Marvin Moss, D. L. Smith and R. A. Lefever, Metastable Phases and Superconductors Produced by Plasma Jet Spraying, Appl Phys Letters 5 (1964) 120.
10. R. H. Willens, Dendritic Crystallization of an Amorphous Alloy, J Appl Phys, 33 (1962) 3269.
11. R. H. Willens, A New Method for Preparing Samples for Electron Microscopy, Proc Fifth International Meeting for Electron Microscopy, Academic Press, New York, Vol 1 (1962), EE-6.
12. Paul Predecki, A. W. Mullendore and N. J. Grant, A Study of the Splat Cooling Technique, Trans AIME, 233 (1965) 1581.
13. R. C. Ruhl, Cooling Rates in Splat Cooling, Matls Sci & Engng, 1 (1967) 313.
14. R. K. Linde, Kinetics of Transformation of Metastable Silver-Copper Solid Solutions, Ph.D. Thesis, California Institute of Technology, Pasadena, Calif. (1964).
15. G. Thomas and R. H. Willens, Defects in Aluminum Quenched from the Liquid State, Acta Met, 12 (1964) 191.
16. W. Klement, Jr., Metastable Structures in Alloys Rapidly Cooled from the Melt, Ph.D. Thesis, California institute of Technology, Pasadena, Calif. (1962).
17. Pol Duwez, R. H. Willens and W. Klement, Jr., Continuous Series of Solid Solutions in Silver-Copper Alloys, J Appl. Phys, 31 (1960) 1136.
18. R. K. Linde, Lattice Parameters of Metastable Silver-Copper Solid Solutions, J Appl Phys, 37 (1966) 934.
19. W. B. Pearson, Handbook of Lattice Spacings and Structures of Metals and Alloys, Pergamon Press, New York (1958).
20. W. Klement, Jr., and H. L. Luo, Metastable Solid Solutions in Silver-Platinum Alloys, Trans AIME, 227 (1963) 1253.
21. H. L. Luo and Pol Duwez, Solid Solutions of Rhodium with Copper and Nickel, J Less Common Met, 6 (1964) 248.
22. Pol Duwez, R. H. Willens and W. Klement, Jr., Metastable Solid Solutions in the Gallium Antimonide-Germanium Pseudo-Binary System, J Appl Phys, 31 (1960) 1500.
23. Pol Duwez, R. H. Willens and W. Klement, Jr., Metastable Electron Compound in Ag-Ge Alloys, J Appl Phys, 31 (1960) 1137.
24. W. Klement, Jr., Lattice Parameters of the Metastable Close-Packed Structures in Silver-Germanium Alloys, J Inst Met, 90 (1961) 27.
25. W. Hume-Rothery and G. V. Raynor, The Structure of Metals and Alloys, Institute of Metals, London (1956).
26. W. Klement, Jr., Nickel-Rich Solid Solutions in Binary Alloys with Tin, Germanium and Silicon, Can J Phys, 40 (1962) 1397.

27. H. L. Luo and Pol Duwez, Face-Centered Cubic Cobalt-Rich Solid Solutions in Binary Alloys with Aluminum, Gallium, Silicon, Germanium and Tin, Can J Phys, 41 (1963) 758.
28. H. L. Luo, C. C. Chao and Pol Duwez, Metastable Solid Solutions in Aluminum-Magnesium Alloys, Trans AIME 230 (1963) 1488.
29. E. C. Ellwood, Factors Affecting Equilibrium in Certain Aluminum Alloys, J Inst Metals, 80 (1952) 107.
30. G. V. Raynor, The Lattice Springs of the Primary Solid Solutions in Magnesium of the Metals of Group III B and Tin and Lead, Proc Roy Soc, A 180 (1942) 107.
31. E. S. Makarov, Structure of Aluminum-Magnesium Alloys, Doklady Akad Nauk USSR, 74 (1950) 935.
32. S. Samson, The Crystal Structure of the Phase β Mg_2 Al_3, Acta Cryst, 19 (1965) 401.
33. M. Hansen and K. Anderko, Constitution of Binary Alloys, McGraw-Hill Book Co., New York (1958).
34. T. R. Anantharaman, H. L. Luo and W. Klement, Jr., Non-Equilibrium Structures in Gold-Germanium Alloys, Trans AIME, 233 (1965) 2014.
35. H. L. Luo, M. F. Merriam and D. C. Hamilton, Superconducting Metastable Compounds, Science, 145 (1964) 581.
36. T. R. Anantharaman, H. L. Luo and W. Klement, Jr., Formation of New Intermetallic Phases in Binary Eutectic Systems by Drastic Under-Cooling of the Melt, Nature, 210 (1966) 1040.
37. Paul Predecki, B. C. Giessen and N. J. Grant, New Metastable Alloy Phases of Gold, Silver and Aluminum, Trans AIME, 233 (1965) 1438.
38. H. L. Luo and W. Klement, Jr., Metastable Simple Cubic Structures in Gold-Tellurium and Silver-Tellurium Alloys, J Chem Phys, 36 (1962) 1870.
39. W. B. Pearson, Handbook of Lattice Spacings and Structures of Metals and Alloys, Vol 2, Pergamon Press, New York (1967).
40. R. C. Ruhl and M. Cohen, A New Metastable hcp Phase in the Iron-Carbon System, Acta Met, 15 (1967) 159.
41. R. L. Clendenen and H. G. Drickamer, The Effect of Pressure on the Volume and Lattice Parameters of Ruthenium and Iron, J Phys Chem Solids, 25 (1964) 865.
42. C. S. Barrett and T. B. Massalski, Structure of Metals, McGraw-Hill, New York (1966).
43. G. H. Vineyard, The Theory and Structure of Liquids; in, Liquid Metals and Solidification, American Society for Metals, Metals Park, Ohio (1958).
44. Roy Kaplow, S. L. Strong and B. L. Averbach, Radial Density Functions for Liquid Mercury and Lead, Phys Rev, 138 (1965) A 1336.
45. H. L. Luo and Pol Duwez, Metastable Amorphous Phases in Tellurium Base Alloys, Appl Phys Letter, 2 (1963) 21.
46. H. L. Luo, Metastable Amorphous Phases in Tellurium-Base Alloys, Ph.D. Thesis, California Institute of Technology, Pasadena, Calif. (1964).
47. R. Buschert, I. G. Geib and K. Lark-Horowitz, Structure of Liquid Tellurium, Phys Rev, 98 (1955) 1157.
48. C. C. Tsuei, Electrical Resistance and Thermoelectric Power of Metastable Phases in Tellurium-Base Alloys, Thesis, California Institute of Technology, Pasadena, Calif. (1966).
49. A. I. Gubanov, Quasi-Classical Theory of Amorphous Ferromagnetics, Soviet Physics, Solid State, 2 (1960) 468.
50. C. C. Tsuei and E. Kankeleit, Mossbauer Effect of Te^{125} in Simple-Cubic and Amorphous Tellurium-Base Alloys, to be published in Oct. 10 issue of Phys Rev, 1967.
51. W. Klement, Jr., R. H. Willens and Pol Duwez, Non-Crystalline Structure in Solidified Gold-Silicon Alloys, Nature, 187 (1960) 869.
52. J. Dixmier and A. Guinier, Obtention de Phases Nouvelles par Trempe Ultra-Rapide d'Alliages Liquides, Rev Metallurgie, 64 (1967)53.
53. J. Dixmier, K. Doi and A. Guinier, Structure de l'Alliage Nickel-Phosphore Amorphe, Physics of Non-Crystalline Solids, J. A. Prins, John Wiley and Sons, Inc., New York (1965) 67.
54. H. S. Chen and D. Turnbull, Thermal Evidence of a Glass Transition in Gold-Silicon-Germanium Alloy, Appl Phys Letters, 10 (1967) 284.
55. M. H. Cohen and D. Turnbull, Composition Requirements for Glass Formation in Metallic and Ionic Systems, Nature, 189 (1961) 131.
56. D. Turnbull, The Liquid State and the Liquid-Solid Transition, Trans AIME, 221 (1961) 422.
57. Pol Duwez, R. H. Willens and R. C. Crewdson, Amorphous Phase in Palladium-Silicon Alloys, J Appl Phys, 36 (1965) 2267.
58. R. C. Crewdson, Structure and Properties of an Amorphous Palladium-Silicon Alloy, Ph.D. Thesis, California Institute of Technology, Pasadena, Calif. (1966).
59. B. Aronsson and A. Nylund, The Crystal Structure of Pd_3Si, Acta Chemica Scandinavia, 14 (1960) 1011.
60. Paul Lebeau and P. Jolibois, Sur les Composes du Silicium et du Palladium, Comptes Rendus Acad Sci, Paris, 146 (1908) 1028.
61. Anna Nylund, Some Notes on the Palladium-Silicon System, Acta Chemica Scandinavia, 20 (1966) 2381.
62. M. E. Glicksman and C. L. Vold, Observation of Solidification and Melting in Metals Using the Electronmicroscope, Acta Met, 15 (1967) 1409.
63. A. I. Gubanov, Quasi-Classical Theory of Amorphous Ferromagnetics, Soviet Phys, Solid State, 2 (1960) 468.
64. S. Mader and A. S. Nowick, Metastable Co-Au Alloys; Example of an Amorphous Ferromagnet, Appl Phys Letters, 7 (1965) 57.
65. C. C. Tsuei and Pol Duwez, Metastable Amorphous Ferromagnetic Phases in Palladium-Base Alloys, J Appl Phys, 37 (1966) 435.
66. Pol Duwez and S. C. H. Lin, Amorphous Ferromagnetic Phase in Iron-Carbon-Phosphorus Alloys, J Appl Phys, 38 (1967) 4096.
67. A. J. Ardell, private communication, California Institute of Technology, June 1967.

Metastable Structures in Metallurgy

DAVID TURNBULL

Metastable structures or, more accurately, configurationally frozen metastable structures are no novelty in metallurgy. Indeed, much of the traditional practice of metallurgy has centered on the formation, characterization, understanding and control of structures which are either compositionally, topologically and/or morphologically metastable. However, in the past two to three decades we have seen a great upsurge in the production and study of new metastable structures in metallurgy, as well as in other condensed phase sciences. This upsurge reflects developments in the techniques of melt quenching, condensation and irradiation of materials, as well as in the kinetic understanding of structure evolution; and it has brought us nearer to making the concept of ''ultramolecular engineering'' viable. Among the new materials produced are glassy metals, highly supersaturated crystalline alloys and new alloys with exceptionally high interfacial densities. An overview of these new developments will be offered, following a discussion of the principles of metastable structure synthesis.

ALONG with other members of the present generation of Campbell Lecturers I have had no association with Professor Campbell. I did note, with interest, the allusion to his contributions in Saveur's article on ''Metallurgical Reminiscences'' which was reprinted in the *Journal of Metals*[1] about a year ago. From this I

The Edward DeMille Campbell Memorial Lecture was established in 1926 as an annual lecture in memory of and in recognition of the outstanding scientific contributions to the metallurgical profession by a distinguished educator who was blind for all but two years of his professional life. It recognizes demonstrated ability in metallurgical science and engineering.

DAVID TURNBULL, Gordon McKay Professor of Applied Physics at Harvard University, has contributed to the fields of nucleation and growth of crystals, diffusion in solids and liquids, solid state reactions, nature of the glassy state and thermionic emission.

A native of Kewanee, Illinois, Dr. Turnbull received his B.S. degree from Monmouth College in 1936 and his Ph.D. degree in physical chemistry from the University of Illinois in 1939. In 1958 he received an Honorary Sc.D. from Monmouth College and was the recipient of an Honorary M.A. from Harvard in 1962.

From 1939 to 1946 Dr. Turnbull was on the faculty of Case Institute of Technology. He was a research scientist at the General Electric Research Laboratory from 1946 to 1962, where he served as manager of the Chemical Metallurgy section from 1950 to 1958. Adjunct professor of metallurgy at Rensselaer Polytechnic Institute from 1954 to 1962, he spent the academic year 1957–58 on leave at Cavendish Laboratory, Cambridge University. He joined the faculty of Harvard in 1962.

Dr. Turnbull is a member of the National Academy of Sciences and a Fellow of the American Academy of Arts and Sciences, the American Association for the Advancement of Science, The Metallurgical Society of AIME, and of the American Physical Society. He is co-editor, with Henry Ehrenreich and Frederick Seitz, of the *Solid State Physics* series. He was Institute of Metals Division (AIME) Lecturer in 1961 and Robert S. Williams (MIT) Lecturer in 1969. He was awarded the von Hippel Prize of the Materials Research Society and the Acta Metallurgica Gold Medal in 1979.

 Reprinted with permission from Metallurgical Transactions A, Vol. 12A, May 1981, 695-708, learned that Campbell was among the distinguished metallurgists who played key roles in the early characterization of one of the types of metastability, the morphological, about which I will speak.

Indeed, metastable structures which have frozen configurationally are no novelty in metallurgy. The most familiar are structures, like those studied by Campbell and others, which may be said to be morphologically metastable because of their considerable density of such extended imperfections as intergrain and interphase boundaries and dislocations. Other examples are compositionally metastable structures—*e.g.*, supersaturated solutions of carbon in steel—and structures which are "topologically" metastable such as the fcc phase of iron dispersed, at low temperatures, in a copper matrix. It is evident that much of the traditional practice of metallurgy has centered on the formation, characterization, understanding and control of metastable structures. Indeed, it is, perhaps, this focus on metastability which most uniquely sets metallurgy apart from other condensed phase sciences. Thus, it might seem that lecturing an audience of metallurgists on metastable structures is only belaboring the obvious.

However, important developments in the last two to three decades have shown that the ranges of metastability, as delineated by structural variety and energy departures from stable equilibrium, can be extended far beyond what we had thought possible. These developments have stemmed from advances in the techniques of quenching, condensation and irradiation of materials and in the understanding of the kinetics of structural evolution. They have stimulated a tremendous upsurge in activity directed at the synthesis, study and exploitation of novel metastable structures. Much of the early stimulus to this activity, especially in metallurgy, came from the demonstration by Pol Duwez and his students that some metal alloys can be melt quenched to amorphous solid or "glassy" form. The earlier results on the formation of glassy and other metastable alloys by rapid melt quenching and on the interesting properties of these new alloys were reviewed by Professor Duwez in his 1967 Campbell Lecture.[2] Since then activity in the field—exploiting condensation and irradiation as well as melt quenching techniques—has continued to burgeon and has led to the synthesis of other new and interesting metastable metallic, as well as semiconducting, structures. In view of these developments I thought it appropriate in this lecture to attempt an overview of the principles and methods of metastable structure synthesis and of some of the more recent results.

What distinguishes the more recent from the older studies of metastable structures is, I think, the synthesis of materials which are much further out of equilibrium—as measured by their energy in excess, ΔU_{sm}, of that of the stable structure—than those known hitherto. Table I shows a broad, and somewhat arbitrary, classification of metastable structures and the excess energies—in units of $R\mathbf{T}_m$, where $\mathbf{T}_m$ is the average of the equilibrium melting temperatures, T_m, of the elementary constituents—approached by some recently synthesized members of each class. The maximum excess energies are roughly double those of the more familiar metastable materials and the amorphous metal class, though existent, was hardly recognized prior to the 1960 paper of Duwez and colleagues. Morphological metastability increases with the volume density, $A_{if}/V = \mathbf{A}_{if}$, of interfacial area, A_{if}, and with the interfacial tension, σ. The maximum $\mathbf{A}_{if}$ which is physically significant, $\sim 10^7$ cm^{-1}, had been approached in structures phase separated by spinodal decomposition or by copious coherent nucleation. However, a relatively small σ, or small gradient coefficient, is required for the operation of either of these mechanisms. The newer developments are notable for the synthesis of structures with exceptionally high densities of incoherent high energy interfaces such that the excess energy/volume, $\sim \mathbf{A}_{if}\sigma$, is substantially larger than in the more familiar interphase or intergranular dispersions.

The general procedure in metastable synthesis is to energize and then quench a material. The energization might consist of evaporation, dissolution, irradiation, plastic deformation or simply heating the material into a state, solid or liquid, in which its volume is not much expanded. Quenching is normally characterized by a cooling rate. If a body is energized, *e.g.*, by im-

Table I.

Nature of meta (or "in")-stabilities	Materials Examples	$\frac{\text{Interfacial Area}}{\text{Volume}}$ (cm^{-1})	Typical Excess Energy ($R\mathbf{T}_m$)
Compositional	Supersaturated Solutions		$\lesssim 1$
Structural	Intermetallic Cpds. Amorphous Solids		$\lesssim 0.5$
Morphological	Microcrystalline Compositionally Modulated Films Interphase Dispersions	$10^4 - 10^7$	$\lesssim 0.1$

Table II. Some Methods of Metastable Structure Synthesis In General: Energize and Quench

Method		"Quench" Rate (K-s^{-1})
Melt or Solid Quench	Conventional "Splat" Laser Pulse	10^3 10^5 to 10^8 to 10^{13}
Condensation	Vapor Sputter Electro Chemical	10^{12}
Irradiation	Particle Bombardment Ion Implantation Ion Beam Mixing	10^{12}
Mechanical Cold Working		

planting an ion or atom the temperature in the implanted region may not be well defined but in this case we might view quenching as the restoration of local thermal equilibrium following the implantation. The principal methods of metastable structure synthesis and the maximum quench rates approached in each are noted in Table II.

During the quench the system may go through one or more metastable or unstable intermediate states in which the atomic mobility is quite high. At the end of the quench the system may be "frozen," we will say "configurationally frozen," into one of these intermediate states. Upon annealing this state often evolves into other interesting and useful metastable states.

THERMODYNAMICS OF METASTABILITY AND CONFIGURATIONAL FREEZING

A truly metastable body, *e.g.*, an undercooled liquid, is in internal equilibrium, though its free energy is above that of the same mass in its stable state. Within it, local atomic rearrangements occur very rapidly so that it passes freely through the many microstates which constitute its macrostate. A more stable phase is formed by the migration of an interface between it and the metastable phase. While the atomic mobility may be high enough to permit rapid movement of this interface the metastable phase can, and often does, persist for long, and even indefinite, periods because of its high resistance to the *ab initio* formation of the interface *i.e.*, to nucleation. However, even if the probability of homogeneous nucleation within it is negligible, a truly metastable body, whether liquid or solid, is rarely of any technical use because of the difficulty of guarding against heterogeneous nucleation of a more stable phase.

We consider that a body is unstable if its free energy can be lowered by localized homophase atomic rearrangements. Clearly in its high mobility range such a body would be of even less use than a metastable one.

Actually nearly all solids used practically are in states which are, in some essential respect, configurationally frozen. In configurational freezing the times required for thermodynamically favored local rearrangments become long, owing to a need for thermal activation, compared with the periods during which the body is observed or used. Unstable, metastable and even stable phases may be quenched into configurationally frozen states. An obvious example of a configurationally frozen unstable phase is a homogeneous alloy with a composition in the spinodal range which has been quenched to a temperature where the frequency of the interatomic exchanges which effect the composition fluctuations is negligible. However, metastable and stable phases may be frozen into configurations characteristic of equilibria at temperatures above the freezing. One example is the freezing of the compositional short range order in alloys with compositions well within the solubility limits.[3] A more striking example is the transition from the undercooled liquid to the glass state ("glass transition") in which the liquid goes out of configurational equilibrium[4] as it homogeneously solidifies, *i.e.*, attains continuously a certain high shear resistance, without crystallization. All of these frozen states are unstable in the sense that their free energy could be lowered by certain local atomic rearrangements ("structural relaxation"). However, it is common practice to apply to a solid the label of the state it was in when it went out of configurational equilibrium.

In configurational freezing the energy associated with structural relaxation stops changing with falling temperature; an effect which is reflected by a rapid drop in heat capacity and other thermal coefficients during the process. The temperatures at which these changes occur decrease somewhat with decrease in the rate of cooling.

An instructive example of the kinetic and thermodynamic manifestations of configurational freezing is that of the transition from an undercooled melt to a glass. Generally the heat capacity of a body is higher in its undercooled melt than in its crystalline state and the difference, ΔC_p, increases with falling temperature until the glass transition, centering at T_g, in which ΔC_p decreases almost to zero. Thus the entropy change, ΔS_m, on melting must decrease with undercooling until the glass transition is reached, as shown schematically in Fig. 1. Also shown schematically in the figure are the isobaric courses with temperature, computed from the thermal data, of the chemical potentials μ of the various states (liquid, crystal and

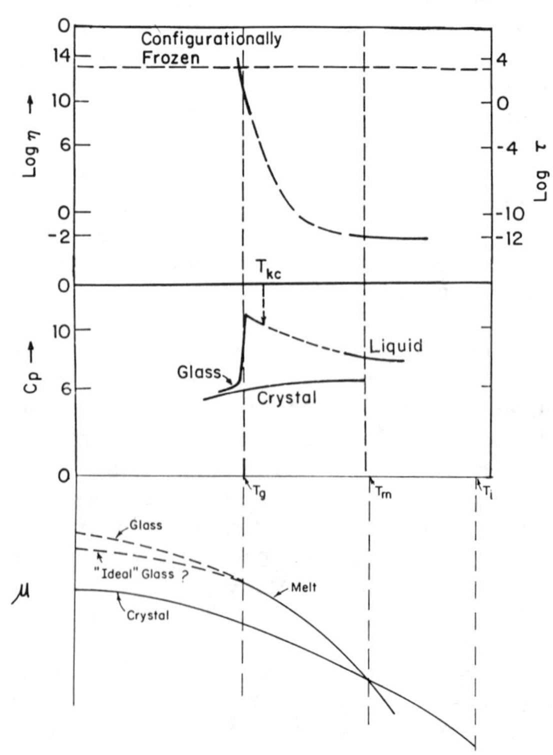

Fig. 1—Rheological (top) and thermal (middle) manifestations of the configurational freezing (melt → glass transition) of a liquid. The lowest diagram shows schematically the temperature dependence of the chemical potential, μ, of a substance in its crystal, melt and glass forms. It is presumed that the crystal becomes unstable (from a metastable state) at a temperature, T_i, well above its thermodynamic melting point, T_m.

glass) of the body. We note that, at any given temperature, the glass state exhibits a narrow continuum of chemical potentials which reflect varying degrees of structural relaxation reached in various quenches. The origin of this dispersion can be understood from the temperature dependence of the atomic transport resistance in glass forming liquids. Figure 1 shows the temperature variation of the shear viscosity $\eta(T)$ typical of metallic glass forming melts. Also shown is the corresponding time constant, $\tau_\eta(T)$, for flow which we will here assume corresponds also to that, τ_υ, for structural relaxation.

As the liquid is undercooled it will, if crystallization is bypassed, reach a temperature, which is lower the lesser is the cooling rate, where it goes out of configurational equilibrium; *i.e.*, it undergoes the glass transition, and also, becomes mechanically solid. In the fixed or frozen configuration the contribution, Q_R, to the apparent activation energy for flow associated with structural relaxation is absent and, provided the measurement times are short compared with τ_υ, the shear viscosity will follow an isoconfigurational course with temperature, characterized by a much lower activation energy, Q_{IC}, than that, $Q_N = Q_R + Q_{IC}$, for flow of the fully relaxed body.[5,6] The temperature of intersection of the isoconfigurational viscosity curve, $\eta_{IC}(T)$, with the viscosity curve, $\eta(T)$, for the fully relaxed body is known as the fictive temperature, T_f. This temperature and the energy associated with departure from configurational equilibrium will be greater the greater was the cooling rate through the configurational freezing range. This energy dispersion, associated with different cooling rates, is reflected by the chemical potential dispersion of the glass state shown in Fig. 1.

The configurational freezing of other types of systems roughly parallels that of undercooled melts excepting that it is generally more diffuse in temperature and the accompanying changes in heat capacity and other thermal coefficients are much smaller.

It is interesting to compare the temperatures of "freezing" of various types of metallurgical configurations. Consider, for example, compositional short range order in substitutional fcc alloys. We may suppose that this order becomes frozen when the atomic jump frequency falls below one per day corresponding to a diffusivity $D_l \lesssim 10^{-21}$ cm²s⁻¹. According to the melting point scaling relation[7] the isobaric diffusivity, $D_l(T_r)$, of the substitutional impurity is a function only of the reduced temperature defined as $T_r = T/T_l$ where T is the actual and T_l the liquidus temperature in degrees Kelvin. The data are roughly fitted by

$$D_l \sim 1 \times 10^{-8/T_r} \text{ cm}^2\text{s}^{-1} \qquad [1]$$

which gives a reduced configurational freezing temperature, $T_{rcf} = 0.4$. Configurations of interstitial impurities will, of course, freeze at considerably lower T_r because of the lower relative activation energies required for their diffusion.

Correspondence relations seem not to have been established for grain boundary and free surface diffusivities, D_B and D_S, respectively. However, the existing information suggests that typical activation energies for these diffusivities would be, respectively, about one half and one fourth of that for bulk diffusion.[8] Then to an order of magnitude the diffusivities may be expressed as:

$$D_B \sim 0.1 \times 10^{-4/T_r} \text{ cm}^2\text{s}^{-1} \qquad [2]$$

and

$$D_S \sim 0.01 \times 10^{-2/T_r} \text{ cm}^2\text{s}^{-1} \qquad [3]$$

from which we estimate:

$T_{rcf} \sim 0.2$ for grain boundaries

and

$T_{rcf} \sim 0.1$ for free surfaces.

The dependences of the diffusive jump times on T_r calculated from these relations are shown in Fig. 2. Also shown in the figure is the T_r dependence of the time constant for viscous flow in the molten glass forming alloy $Au_{78}Ge_{14}Si_8$ (reduced glass temperature $T_{rg} \sim 0.45$) in its glass transition range.[9] The time constant at configurational freezing was taken to be 10^5 s. (See dashed horizontal line).

It is possible for a morphology to be frozen even when the atomic mobilities are not frozen. For example, the dispersion of oxide particles in an internally oxidized alloy can remain virtually fixed to the melting point of the alloy provided the solubility (C_0 moles/volume) of the oxide in the alloy is sufficiently small. The dispersion coarsens by Ostwald ripening whereby, according to the theory of Lifschitz and Slyozov[10] (also see Wagner),[11] the mean particle radius, $\langle r \rangle$, should, after a steady scaled size distribution is reached, increase at the rate:

$$\langle \dot{r} \rangle = \frac{8\sigma \mathbf{V}^2 D C_0}{9RT\langle r^2 \rangle} \qquad [4]$$

where

σ = interfacial tension
$\mathbf{V}$ = molar volume of precipitate
D = diffusivity of least mobile precipitate component in the matrix.

For the noble metals at temperatures near T_m the first factor in this expression is of the order of 10^{-6} cm⁴/mol. In this case:

$$\langle \dot{r} \rangle \sim 10^{-6} \frac{DC_0}{\langle r^2 \rangle} .$$

We may suppose that the dispersion is frozen if 10^{-13} cm/s $> \langle \dot{r} \rangle$ or:

$$10^{-7}\langle r^2 \rangle \gtrsim DC_0 .$$

The temperature dependence of DC_0 will be determined by the factor

$$DC_0 \propto e^{-(Q + \Delta \mathbf{H}_S)/RT}$$

where Q is the activation energy for diffusion and $\Delta \mathbf{H}_s$ is the heat of solution of the precipitate. If $D \cong 10^{-8}$ cm²/s and $\langle r \rangle \cong 1$ μm the dispersion would be frozen if $C_0 < 10^{-7}$ mol/cm³.

One apparent example of a particle dispersion that

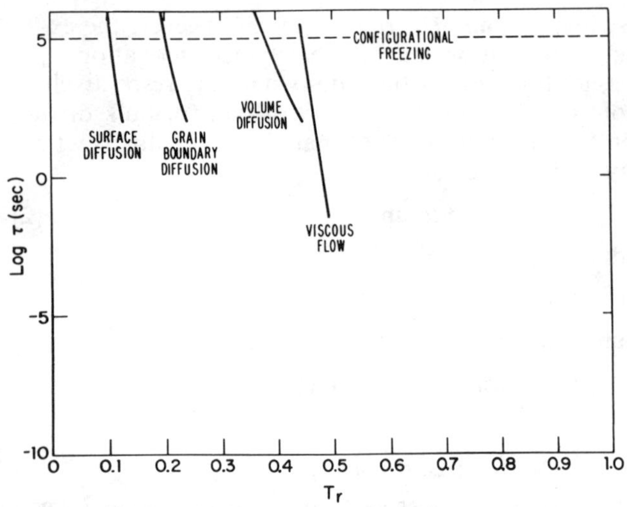

Fig. 2—Dependence of time constants for various atom transport processes on reduced temperature ($T_r = \tau / T_l$) in the configurational freezing range.

remains virtually fixed to the matrix melting temperature is that of 1 vol pct BeO, $\langle r \rangle \sim 0.02\ \mu$m, in copper formed by Bahk and Ashby,[12] by internal oxidation and annealing. This dispersion was so stable that a substantial part of the stored energy of cold working could be removed only by melting the matrix. Indeed, the melting temperature of the cold worked matrix was depressed 4 °C below T_m of the recrystallized matrix.

There are well known examples in metallurgy of configurations which apparently cannot be frozen by quenching, *i.e.*, there are atom movements which require no thermal activation. Familiar examples are interfaces which apparently advance by dislocation glide as in martensitic type transformations.[13] Another possible example which will be considered below are the amorphous phase-crystal interfaces in pure metals.

COMMENTARY ON STEP "RULE" OF STRUCTURAL EVOLUTION

In structure synthesis the quench often opens up to the system several thermodynamically permissible courses of structural evolution including, of course, that to the most stable state. Thus we would like to understand what is kinetically preferred within the range of thermodynamic possibilities. Clearly, metastable structure synthesis can occur only when the kinetically preferred course differs from the one most favored thermodynamically.

Ostwald noted long ago that systems seemed generally to transform, when the option was available, to a metastable rather than to their most stable phase and he suggested that this behavior might reflect the operation of a step rule. For example, in many metallurgical systems coherent rather than more stable incoherent phases precipitate initially from supersaturated alloys. Another interesting example of consistency with a step rule was Cech's demonstration[14] that undercooled molten droplets of certain Fe-Ni alloys crystallized to the bcc before transforming to the more stable fcc structure. However, we hardly expect that a superheated metastable phase would form in preference to a stable one. Therefore, the Ostwald generalization probably has more validity as a step energy, or step entropy, than as a step free energy "rule".

Actually, as Stranski[15] and Volmer[16] emphasized, there is no fundamental justification for a step principle. There must be a particular kinetic explanation for each instance where it applies. In the examples cited it can be argued plausibly that, if the initial system is far out of equilibrium, nucleation of the metastable phase is preferred because the tension of the interface it forms with the initial phase should be lower than that formed by the more stable phase. In melt crystallization the relative magnitudes of the interfacial tensions would reflect that they often scale as the energy differences between phases. In the precipitation example the structural component of the interfacial tension is eliminated by coherency.

A guiding principle for synthesis which may have some qualitative generality is that the paths most favored kinetically in structural evolution are those requiring the lesser correlations of atomic positions and motions. These would also be the paths of lesser, but not necessarily minimum, entropy change. Thus, short range order, whether compositional or topological, can develop more rapidly than long range order. In spinodal decomposition, for example, the correlation range is proportional to the wave-length of the composition fluctuations.[17,18] Kinetic analysis showed that the maximum rate of amplification of these fluctuations should occur initially at wavelengths which are relatively very short though still somewhat longer than the minimum permitted thermodynamically.

Also in accord with the idea that paths of lesser correlation are kinetically favored is the experience that materials generally condense, at temperatures below their configurational freezing points, as amorphous solids, where the positional correlations are short range, rather than in their more stable crystalline forms. It is significant that amorphous solid structures are also generated in computer simulated condensation, of hard spheres[19,20] or tetrahedrally bonded elements,[21] in which only short range interatomic forces operate and no post condensation atomic rearrangements are permitted. In these instances certain topologically non-crystalline local arrangements which are compatible with or even favored by the shortrange interatomic forces are likely to form and their inclusion leads to the generation of an amorphous structure.[19]

METHODS OF METASTABLE STRUCTURE SYNTHESIS

Bulk Phase Quenching

Consider the formation of metastable structures by thermally energizing and then quenching a bulk phase. Energization brings the body into some stable high temperature state. In the first stage of the quench this state becomes metastable, or even unstable, but the atomic mobility remains high. As the quench proceeds the state may persist through configurational freezing or it may transform to a different state, metastable or stable, and then freeze. The general conditions for bypassing transformation to more stable phases in the high mobility regime of the quench have been set forth elsewhere.[22,23] These are determined by a set of process and a set of materials variables. The process variables are the quench rate, $\dot{T}$, the volume, v, of the body and the number density, ρ_N of heterogeneous nucleation sites. Avoidance of transformation is more likely the larger is $|\dot{T}|$ and the smaller are v and ρ_N. Important materials variables include the thermal conductivity, $\varkappa$, an appropriate atomic transport frequency—*e.g.*, k_D for diffusive transport or k_i for interfacial rearrangement—and some measure, α, of the interfacial tension between the more stable and the initial phase. Large α, leading to infrequent nucleation, and/or low k_D or k_i are the conditions most favorable for quenching through the high mobility regime without transformation.

If α is so large that the frequency of homophase nucleation of the more stable phase remains negligible at all temperatures it becomes possible to quench large masses of the high temperature phase through the high mobility regime at relatively low rates.[22] This method can be feasible even when the atomic mobilities are so high that the transformation times after seeding are very short. Its successful application requires that the concentration, ρ_N, of heterogeneous nucleation sites be reduced to very low levels or eliminated.

In metallurgical practice quench rates up to 10^4 to 10^5 K/s are sometimes used. It was the experiments of Duwez and colleagues which demonstrated the wide range of new opportunities for structure synthesis at quench rates of 10^6 K/s and higher. The general technique[2,23] is to quickly spread molten metal into a thin overlay on the surface of a massive heat sink with high thermal conductivity. It has been reported that quench rates as high as 10^8 to 10^9 K/s can be obtained by variants of this method.

Even higher quench rates are possible when thin molten overlays are formed *in situ* on metal surfaces by short high energy laser pulses. Bloembergen[24] has shown that quench rates as high as 10^{13} to 10^{14} K/s might result from such pulses in the picosecond range. The structures that can be produced by this method, unless it is combined with a procedure like ion implantation, are limited to those which can derive from the substrate material. Also their form is limited to that of a thin overlay strongly bonded to a substrate. Of course, such overlays can effect useful alterations of the surface properties of the substrates.

At quench rates of 10^6 K/s transformation with nuclei concentrations as large as $10^{10}/\text{cm}^3$ would be suppressed at all interface speeds up to 1 cm/s. Use of these and higher quench rates led to the discovery of metallic glasses and exposure of glass formation by alloys exhibiting reduced glass temperatures[25] ranging from $^2/_3$ to as low as 0.45.

In simple, or "classical", nucleation theory the steady frequency/volume, I, of homogeneous nucleation is related to the scaled interfacial tension, α, by an equation of the form:[26,22]

$$I \cong Ak_i(T)\exp[-16\pi\alpha^3\beta/3(\Delta T_r)^2T_r] \qquad [5]$$

where α is defined by:

$$\alpha = (\mathbf{N}\mathbf{V}^2)^{1/3}\frac{\sigma}{|\Delta H|} \qquad [6]$$

and

$\mathbf{N}$ = Avogadro's number
$\mathbf{V}$ = molar volume of the crystal
σ = interfacial tension between the high temperature and forming phase, here assumed to be isotropic
ΔH = molar heat of transformation
$\beta = \frac{|\Delta S|}{\mathbf{N}k}$, where ΔS is the entropy of transformation, here assumed constant
$\Delta T_r = \frac{\mathbf{T} - T}{\mathbf{T}}$, where $\mathbf{T}$ is the equilibrium temperature, in degrees Kelvin of transformation
$T_r = \frac{T}{\mathbf{T}}$
A = constant specified by the kinetic analysis.

From this relation we calculate that I should never reach measurable levels when $\alpha\beta^{\frac{1}{3}} \gtrsim 0.9$. If $k_i(T)$ scales with the frequency of configurational rearrangement the magnitude of $\alpha\beta^{\frac{1}{3}}$ at the nucleation cutoff decreases from 0.9 as the temperature, T_{rcf}, at configurational freezing increases. Theoretical estimates of the structural contribution to the tension of incoherent interfaces[27] generally lead to $\alpha\beta^{\frac{1}{3}}$ values well above 1 and the few available measurements qualitatively support these estimates. Therefore, we expect that homogeneous nucleation of incoherent phases within perfect crystals should rarely, if ever, reach detectable levels.[28] This generalization is supported by many observations and not contradicted by any experience I am aware of. Exploitation of this nucleation resistance to retain metastable structures from high temperatures is, of course, limited by copious heterogeneous nucleation on extended imperfections, *e.g.*, dislocations and grain boundaries.

In the linear kinetic regime the velocity, u, of interface (*e.g.*, crystal-melt or solid-solid) motion may be expressed as follows:[29,30]

$$u \cong fk_i\,\lambda\beta\Delta T_{ri} \qquad [7]$$

where

$$\Delta T_{ri} = \frac{\mathbf{T} - T_i}{\mathbf{T}}$$

f = fraction of interface sites at which atom rearrangement may occur
λ = interface displacement/rearrangement.

The heat of crystallization is transported away at the rate (an analogous relation would apply for impurity transport):

$$\dot{Q} \cong -\varkappa(\text{grad }T)_i \cong u\frac{\Delta H}{\mathbf{V}} \qquad [8]$$

where $(\text{grad } T)_i$ is the thermal gradient at the interface. By combining these equations we get:[30]

$$\mathbf{T} - T_i \cong B(T_{ri}) \frac{\varkappa}{fk_i} (\text{grad } T)_i \quad [9]$$

where

$$B(T_{ri}) = \frac{\mathbf{V}T_{ri}}{R\beta^2}.$$

The experiments of Walker[31] and of Colligan and Bayles[32] showed that dendrite tips in pure Ni or Co melts may move at velocities ~50 m/s at *ambient* undercooling $\Delta T_r \sim 0.1$. These enormous speeds suggested to the author and his colleagues[29,30] that crystal growth in pure metal melts might be limited only by the collision rate of atoms from the melt with the interface. Then fk_i would be identified with z_1, the collision frequency/interface site and the upper limiting growth speed [from Eq. (7)] would become

$$u \lesssim z_1 \lambda \beta \Delta T_{ri} \quad [10]$$

or

$$u \lesssim u_s \beta \Delta T_{ri} \quad [11]$$

where u_s is the sound speed in the liquid. If these relations are valid the results of Walker *et al*[31,32] indicate (with u_s ~2000 m/s, $z_1 \sim 10^{13}$ s^{-1}) that $\Delta T_{ri} \gtrsim 0.025$ at ambient undercooling $\Delta T_r = 0.1$.

Taking $z_1 \sim 10^{13}$ s^{-1} and $B(T_{ri})\varkappa \sim 10^7$ cm s^{-1} as typical *I* estimate:

$$T_m - T_i \gtrsim 10^{-6}(\text{grad } T)_i$$

for pure liquid metals. With gradients of the usual magnitude the displacement of T_i from T_m should be almost negligible and we say that interface motion is "diffusion controlled". In contrast, the thermal gradients in the experiments of Walker *et al*[31,32] were extremely high (~10^6 K/cm), owing to the large curvatures of the dendrite tips, so that $T_m - T_i$ must have been several 10s of degrees, as the growth speeds, indeed, indicated.

We note in passing[30] that Eq. [11] also should give the upper limiting speed of the interface in crystal melting where $-\Delta T_{ri}$ would be the scaled interfacial superheating. In heterogeneous melting the crystal at the interface must be superheated (see Fig. 1), to some extent, into its regime of metastability. The minimum superheating should be related to the measured u, from Eq. [11], by:

$$|\Delta T_{ri}| \gtrsim \frac{u}{u_s\beta} \quad [12]$$

This superheating can be substantial in the melting of crystals during short high energy laser pulses.

Interface movement in rapid quenching may generate structures which depart widely from morphological and compositional equilibrium. These departures arise[30] because of the large displacements of the interface temperatures from equilibrium which attend, as noted already see Eq. [9], the high thermal gradients during rapid quenching. Such displacements can provide the free energy for development of crystal shape instabilities as well as for the ultra-equilibrium incorporation or "trapping" of impurity by the moving interface.

The thermodynamic limits on impurity trapping in melt crystallization have been set forth by Baker and Cahn.[33,34] However, the kinetics of the process are still not clear. Suppose that the interface is undercooled sufficiently for all the melt impurity to be trapped in the growing crystal. Complete entrapment might then occur if the speed of the crystal-melt interface greatly exceeds the diffusive speed, u_{iD} of the impurity in the melt.[35,36,30] If the impurity concentration is sufficiently small for the laws of the dilute solution to apply u_{iD} should not exceed D_i/λ, *i.e.*,

$$u_{iD} \lesssim \frac{D_i}{\lambda}, \quad [13]$$

where D_i is the diffusivity of the impurity in the immediate vicinity of the interface, and for trapping:[30]

$$u_s\beta\Delta T_{ri} > u_{iD} \lesssim D_i/\lambda. \quad [14]$$

It is not likely that D_i/λ will exceed 10 m/s in a metal melt. Thus, it appears that trapping should be nearly complete when the undercooling, ΔT_{ri}, is sufficient to drive the interface at speeds in the range 1 to 10 m/s, which are well below those reached in the experiments of Walker *et al.* [31,32]

The ultimate limit to the supersaturation of solid solutions achievable in melt quenching would be set by the miscibility of the constituents in their stable liquid states. Liquid-liquid phase separation often becomes thermodynamically possible in the metastable liquid regime. In fluid metal melts the time constants for the early stages of this process usually lie in the nanosecond range. Under these conditions the quenching of a metal melt to a homogeneous solid solution, glassy or crystalline, would require quench rates of order 10^{10} K/s. Much lower quench rates might suffice for forming homogeneous glassy solutions if the miscibility temperature was not much above the glass temperature.

Condensation

Condensation techniques include vapor, electro-, chemical and sputter deposition. Essentially, material is condensed on a surface maintained at a temperature well below that of configurational freezing. The times required for thermal equilibration of the condensing atoms are estimated to fall in the picosecond range so it might be said that the quench rate is of order 10^{13} K/s. Ideally, the condensing atom will lodge at the first binding site, *i.e.*, position of local minimum of potential energy, it encounters. Thus the favored configurations would be those with short range positional correlations which assemble to topologically and compositionally disordered structures. The tendency toward disorder will be countered by atomic rearrangements occurring on the surface after deposition. The rates of these rearrangements presumably will be limited by surface diffusion. They would be negligible only at temperatures below that of surface configurational freezing which may be far lower than the T_{rcf} of the interior of the deposit. Thus the probability of formation of disordered structures should be higher the lower is the surface temperature and the greater is the deposition rate.

The potential for novelty in metastable structure synthesis should be much greater in deposition, where the quench rates are, in principle, the highest, than in melt quenching. This expectation is generally born out by experience. For example, in contrast with melt quenching, there should be no limit on the composition ranges of solid solutions formable by condensation since attractive interactions between unlike, as well as like, atoms are always longer range than repulsive ones. Thus, it should be possible to codeposit, over the entire composition range, elements with only limited liquid state miscibility. An interesting example of such a system is the Ag-Ni one where in the liquid state the solubilities are only of the order of 2 to 3 at. pct at the melting point of Ni and complete miscibility is not approached at the highest temperatures of experimental observation. The solid solubilities are of the order of only 0.2 at. pct at the melting temperature of Ag. However, Dahlgren and coworkers[37] succeeded in sputter co-depositing at room temperature homogeneous fcc solutions with up to 13 at. pct Ni in Ag and 8 pct Ag in Ni. Because of the small solid state solubilities the morphology of a precipitate dispersion of one of these elements in the other, once formed, should remain frozen to quite high temperatures.

Also it is well established that the component variety and composition ranges of alloys which form amorphous solids are much wider in condensation than in melt quenching processes. For example, Cu-Sn,[38] CuAg,[39] and Au-Co[39] are examples of alloys which have been vapor or sputter co-deposited, but not yet melt-quenched, in amorphous solid forms. We expect that amorphous solids formed by condensation should be less relaxed structurally, or generally more disordered, than those with the same composition made by melt-quenching. This expectation is supported by the limited experience available which indicates that *e.g.*, atom transport processes are more rapid in condensed than in melt-quenched solids prior to structural relaxation.[40] However, in the few instances investigated[40,41] the structure and atomic transport behavior become the same, within experimental uncertainty, for the two kinds of isocompositional preparations following structural relaxation anneals.

Some of the other unique and interesting deposited materials are layered or composiitonally modulated films with modulation wavelengths as small as two atom spacings, corresponding to interfacial densities as high as 2×10^7 cm^{-1}. The deposition and characterization of crystalline metal films, *e.g.*, Cu/Ni, of this type was pioneered by Hilliard and associates.[42] Certain of their films exhibited quite unexpected and seemingly anomalous mechanical[43] and magnetic behavior.[44] For example Young's moduli of Cu/Pd films with the shortest wavelengths were three to four times the average of that of the pure bulk metals.

Recently Rosenblum *et al.* [45] sputter deposited compositionally modulated films, Pd based/Fe based, with wavelengths of order 20Å which were amorphous. The very small interdiffusivities, *e.g.*, 10^{-22} cm^2-s^{-1} in these structures were measured by monitoring the decay with annealing of the X-ray satellite peak associated with the modulation.

Irradiation

Among the principal irradiation methods of structure synthesis are structural damage by irradiation and the more recently developed ion implantation and ion beam mixing techniques. In these methods a solid is bombarded by beams of highly energetic particles. Much of the energy of each particle is imparted to a highly localized region of the solid. New configurations are produced in the energized region and these or their metastable derivatives may be frozen in during the extremely rapid ($\sim 10^{-13}$ s) deenergization. Ion implantation occurs when the bombarding ion becomes lodged in the solid following deenergization.

Amorphous overlays have been produced on semiconductor[46] and some alloy crystals[47] by radiation damage. In this process local configurations similar to those which predominate in the bulk amorphous solid presumably are produced in the local energy pulse. As bombardment continues the number density of these disordered regions increases until the regions eventually overlap and occupy the entire overlay. The structure[48] of amorphous silicon overlays produced by radiation damage appears to be the same, within the experimental resolution, as that[49] of amorphous silicon deposited from the vapor.

A wide variety of metastable solid solutions, in which the supersaturation levels are extreme, have now been prepared by ion implantation of thin overlays on metals and semiconductors.[50] Often the original structure is heavily damaged by the implantation but most of this damage can be removed by short laser pulses in which the overlay melts and then is crystallized by expitaxial regrowth of the substrate.[51] This regrowth occurs so rapidly that the implanted impurities are often "trapped" in the regrown crystal at their average implant concentrations.[52] By this means semiconductor films can be prepared in which the concentrations of p or n type impurities are far above their equilibrium levels while the crystal perfection remains satisfactory.

In ion beam mixing[53] (*in situ* dispersion technique) contacting phases of different composition are homogenized by the high energy pulses to produce a solution which is supersaturated at a sub T_{rcf} ambient temperature. One technique is to deposit, by sequential *e*-gun deposition, alternating layers (layer thicknesses ~20 to 100Å) of two elements and then bombard the resulting film with heavy ions, *e.g.*, Xe^+. In this way Mayer and associates[54] formed homogeneous Ag-Cu solid solutions at all compositions and homogeneous Ag-Ni solutions to 4.5 at. pct Ni in Ag and 16 at. pct Ag in Ni. These results, as well as the formation of supersaturated solutions by ion implantation, indicate that the local configurations produced by particle irradiation are much more disordered and energetic than those occurring in liquids at stable or metastable equilibrium.

Cold Working (*In Situ* Dispersion Technique)

Cold working increases structural metastability or instability principally by increasing the number density of extended imperfections (dislocations and interfaces) in the system. It may also produce point defects but these are annihilated quickly at $T_r \gtrsim 0.2$. The formation of dislocated structures by cold working has, of course, been studied intensively for many decades.

Recently structures with extremely high densities (10^6 cm^{-1}) of interphasial area have been produced by the cold deformation of certain two-phase alloys.[55,56] The essential requirements are that the two phases be ductile and have mechanical properties which are not greatly disparate. Then during deformation the two phases will be reduced in area to the same degree and by very large amounts without cracking. After deformation the internal phase is distributed as a set of parallel and closely spaced ribbons or filaments. What minimum interfilamentary spacings are attainable with this technique has not been determined but Cu-Nb composites have been fabricated[57] in which the mean spacing of ribbons of the internal Nb phase was ~250Å. If the two structures, are dissimilar, as are Cu and Nb, an extraordinarily high dislocation density can be trapped in the external phase. Indeed, Cu-Nb composite specimens have been prepared[58] with dislocation densities ~10^{13} cm/cm^3 in the Cu matrix and these have exhibited tensile strength to modulus ratios,[57] 2 to 3 pct, approaching the theoretical limit.

Structures with high area densities of interphasial boundaries also may be formed, of course, by solid state transformation, *e.g.*, by spinodal decomposition. However, the achievement of high interphasial areas in this way requires that the structural component of the interphasial tension be zero or very small. Thus the method can be applied successfully only to those two phase systems in which the structural mismatch is relatively small. Also in two phase structures, *e.g.*, eutectics,[59] formed by rapid melt quenching the interphasial area density generally scales inversely with the interphasial tension. In contrast, the cold deformation method can effect very high interphasial dispersion even when the structural mismatch and interphasial tension are very high. Also it is interphase boundaries of this type which are likely to be most effective in impeding dislocation passage and in flux pinning in superconducting alloys.

Comparison of Methods

The relative merits of the various methods of metastable structure synthesis may be summarized as follows. For forming structures which are highly disordered, compositionally and topologically, the deposition and irradiation methods appear to have far greater flexibility and range than does rapid melt quenching. For example, the degree of miscibility achieved in some alloy systems, *e.g.*, Ag-Ni, by deposition and irradiation could not have been reached by melt quenching. However melt quenching has been much superior to the other methods for producing large amounts of metastable materials cheaply. Also melt quenching is the preferred technique for forming ordered metastable phases, *e.g.*, intermetallic compounds.

Thick specimens of certain amorphous alloys have been formed by electrodeposition.[60,61] However, the procedures for finding the proper conditions for electro-codeposition of elements are generally laborious and difficult. Consequently, the method has been applied successfully to only a few systems (*e.g.*, Ni-P, Co-P). Dahlgren and his colleagues[62,63] have formed massive bodies of amorphous and other metastable alloy structures by their high speed sputter deposition technique. This technique appears to have wide applicability to structure synthesis but it is difficult and expensive compared with melt quenching.

Thus it seems that the deposition and irradiation methods, though having far reaching potentialities for synthesis of novel structures, are limited, for the most part, to producing these structures in the form of thin films or overlays. However, such methods should prove highly useful in both semiconductor and metal thin film technology and for improving the surface properties (*e.g.*, corrosion and wear resistance) of massive bodies.

METALLIC GLASSES

Glassy metal alloys are among the most unique and interesting of the recently discovered metastable materials. Their high potential for technological application has been discussed extensively.[64] The conditions leading to and most favorable for metal glass formation have been surveyed in a number of papers by the author and his colleagues[64,65] and others.[25,66,67] Here I will review these conditions with particular attention to the questions of the universality of glass formation by metals, alloying effects, kinetic stability of metal glasses and the possibility of bulk metal glass formation by melt quenching.

Impurity Effects on Metal Glass Formation

In the free volume model of atomic transport even pure monatomic, including metallic, liquids should, provided crystallization does not intervene, undergo transition to a glass at sufficiently low temperature.[68] This transition occurs because the atomic configurations become "locked" by the operation of short range repulsive forces, when the system is sufficiently collapsed. An amorphous structure results, it is thought, because these short range interactions favor tetrahedral configurations[19] which cannot by themselves form a crystal without systematic distortions. It is the tetrahedral configurations which predominate in the Bernal dense random packed hard sphere structure.[69] These random packed structures appear to be internally stable and capable of crystallizing only by reconstruction, *i.e.*, nucleation and growth. In, at least, seveal pure metal melts the frequency of homogeneous crystal nucleation is too small to measure until the reduced temperature has fallen below $T_r \sim 0.75$ to 0.70. Since the T_{rg} of pure metals seem to be much smaller than 0.7, this resistance to nucleation indicates that the scaled crystal-melt tension, α, must be 0.6 or larger. However, there have been no reports of successful quenching of pure metal melts to glasses.

Computer simulation and physical model studies have indicated that metal or rare gas atoms should condense to amorphous structures of the polytetrahedral type if all atoms lodge at the first binding site encountered and no atomic rearrangements occur in the surface or body of the deposit.[19,20] Actually, pure metal vapors generally condense in fine grained crystalline form on chilled substrates.[70] It has been reported that vapors of some transition metals form amorphous solid deposits which exhibit quite low kinetic crystallization temperatures.[71] However, certain impurity admixtures facilitate, and may be essential for the deposition of these and other metals in amorphous form. The glass temperature, T_g, of transition and noble metal based glasses are generally well above room temperature even when the impurity amounts are quite small. Therefore, it seems unlikely that the tendency of pure metal vapors to form crystalline deposits at low temperatures reflect low bulk T_g's. Rather, the behavior may indicate that the surface configurational freezing temperatures of pure metals is exceptionally low, perhaps well below my $T_r \sim 0.1$ estimate.

Alternatively, if crystal growth in pure amorphous metals is not thermally activated, but only collision limited as supposed in the kinetic analysis outlined in an earlier section, there would be no interfacial configurational freezing and the metal would always deposit in crystalline form provided seed or heterogeneous crystallization nuclei were available.

In contrast, when crystal growth requires impurity redistribution the growth rate will be limited by the transport rate of the impurity in the amorphous or crystalline phase and k_i will correspond to k_D, the diffusive jump frequency of the impurity, rather than to z_1. This transport frequency falls with decreasing temperature and generally becomes negligible at temperatures not far below the glass temperature. Growth of compositionally ordered intermetallic compounds, phase separated crystalline structures or crystals of higher purity than the amorphous phase all might be limited by impurity transport rates. Thus, some admixture of an impurity, to play one of these limiting roles, may be essential for glass formation in both condensation and melt quenching.

The deposition results generally support the view that, with some impurity admixture, most metals can form an amorphous solid phase. However, as we noted, the composition requirements for ready glass formation in melt quenching are rather more stringent than in deposition. The most favorable impurity admixtures for melt formation of glasses appear to be those which most increase T_{rg}, as well as restrain crystal growth. Since T_g is not sharply composition dependent in the glass forming range it is alloys with the lowest liquidus temperatures, T_l, relative to those of the pure constituents, which are most likely to have the highest $T_{rg}(= T_g/T_l)$ and form glasses at the lesser cooling rates.[72] Indeed, it is found that the compositions of melt quenched glasses are generally those with low T_l, often at or near low lying eutectic points. The elementary combinations generally fall in one or the other of the following categories:[64,66] noble or transition metal-metalloid, late-early transition metal, and simple metal. Their heats of mixing in the liquid state are usually negative. However, the successful deposition of alloys with positive heats of mixing, *e.g.*, Au-Co and Cu-Ag, in amorphous solid forms[39,64] shows that a negative heat of mixing, or compound forming tendency, is not a requirement for glass formation. It should be possible to extend the range of glass forming compositions still further with the higher quench rates available in laser pulsing. Actually, Baras *et al.* [73] have shown that with laser pulsing in the nanosecond range a Pd-Cu alloy with as little as 5.1 at. pct Si can be quenched to a glass. Also von Allmen[74] has reported formation of Au-Si glasses with as little as 10 at. pct Si by nanosecond laser pulsing of compositionally modulated Au/Si films. With quench rates in the 10^6 K/s range > 15 at. pct Si is usually required for glass formation by these alloys.

Interpretation of Alloying Effects

The microscopic theory for the effect of alloying on glass forming tendency—really the theory for T_l, T_g and α—is not well developed. In the following certain of the alloying effects will be considered at the phenomenological level.

It has been observed that the stable crystalline forms of a considerable number of glass forming alloys have Frank-Kasper structures.[64,75] These structures can be described[76] as assemblies of periodically packed polytetrahedral units in which the tetrahedral configurations have been systematicaly distorted to effect space filling. The polytetrahedral unit is necessarily large and may contain fifty or more atoms. Thus the topology of a Frank-Kasper phase is qualitatively, at least, similar to that of the polytetrahedral models (*e.g.*, the Bernal DRP) for metal glass structure. It is possible that disordering a Frank-Kasper structure compositionally with some small readjustments of the interatomic spacings of the redistributed atoms would produce a structure closely resembling that of a metallic glass.

In any amorphous structure, liquid or solid, there must be some dispersion in the geometry—*i.e.*, number, spacings and angular distributions of near neighbors about an atom—of its constituent local configurations. Thus it is reasonable to suppose that the elemental combinations, AB, most conducive to glass formation should be those which permit such dispersion, at constant density, with minimum energy expenditure.[77,64] It seems that these combinations would be the ones in which the AB interatomic potential is least sensitive around its minimum to the AB separation and the local configuration. Actually there is substantial dispersion in the AB near neighbor separations in the Frank-Kasper and in the metal-metalloid intermetallic phases at compositions near the glass forming ones.

The atomic volumes, $\mathbf{V}$, of relaxed metallic glasses have proven to be remarkably close (~0.5 pct higher) to those of the crystallized alloys at the same composition.[78] Further, the author has noted that the $\mathbf{V}$ of many of the metal (A)-metalloid (B) alloys (especially where B is Si, P or Ge) in both their glassy and crystalline intermetallic states are within ±2 pct of the atomic volume, $\mathbf{V}_A^0$, of the pure metal host element in its crystalline state. The limited data on composition dependence are, within experimental error, consistent with a linear dependence of $\mathbf{V}$ on the atom fraction x_B, of metalloid, thus:

$$\mathbf{V} = \mathbf{V}_A^0 + (\mathbf{V}_B - \mathbf{V}_A^0)x_B$$

where $\mathbf{V}_B$ is constant.

The values of $\mathbf{V}_B$ for the metalloids Si, P and Ge, computed from the volumetric data were generally equivalent, within ±10 pct to $\mathbf{V}_A^0$, which ranged from 6.6 cm³/gm · atom for Ni through 10.22 cm³/gm · atom for Au. This behavior exhibits a quite unusual scaling, or adjustment, of the impurity atomic volume to that of the host. Recently the author has found,[79] by analysis of the lattice parameters tabulated by Pearson,[76] that this scaling tendency and equivalence relation ($\mathbf{V}_B \sim \mathbf{V}_A^0$) extends also to the crystalline solutions (fcc or bcc) of B in the host. Figure 3 shows the atomic volume of each metalloid in its various states (intermetallic compound, crystalline or glassy solution) plotted against $\mathbf{V}_A^0$. Also indicated are the estimated volumes $\mathbf{V}_B^0$ of the pure *metallic* states of the metalloids. We note that there is little variation of $\mathbf{V}_B$ with the structure of a given host. This rough constancy means that the AB near neighbor spacings must vary substantially with structure in order to accommodate to the different AB coordination numbers, especially between the crystalline solutions and intermetallic phases. Also, we see that $\mathbf{V}_B$ for each metalloid roughly follows the 45 deg line, as $\mathbf{V}_A^0$

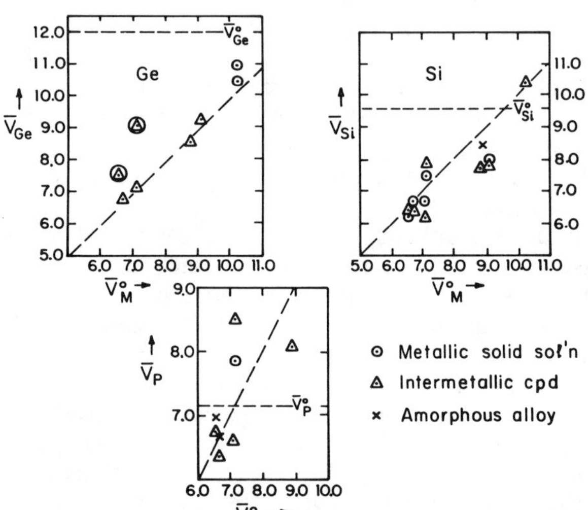

Fig. 3—Dependence of partial gram-atomic volume of metalloid elements in various alloys on the gram-atomic volume, $\mathbf{V}_M^0$, of the host element (noble or late transition) in its pure crystalline state at room temperature.

changes from 6.6 to 10.3 cm³/mol, and for Si and Ge is, in the alloys, generally well below the metallic volume, $\mathbf{V}_B^0$, of the pure metalloid. This scaling and equivalence behavior of $\mathbf{V}_B$, whatever its microscopic explanation, indicates that the metalloids do accommodate to a variety of local metallic environments, at a given density, with little energy variation. Such accommodation may favor glass formation by reducing the energy of local configurational dispersion in amorphous structures.

Crystallization Behavior and Bulk Glass Formation

The present experience on crystal nucleation in glass forming alloy melts is mostly qualitative and insufficient to establish the measure or even occurrence of homogeneous nucleation. There is evidence that in some alloys, *e.g.*, Fe_4B, many nuclei form in the quench but their growth is aborted by the glass transition.[80] This nucleation might be homophase but, alternatively, it could have occurred homogenously in liquid which phase separated in the quench or heterogeneously, *e.g.*, on finely dispersed oxide particles formed by internal oxidation. In either event the resulting structure would be one of many microcrystals dispersed in an amorphous matrix. When the alloy is reheated to temperatures near T_{rg} the growth of these crystals may be rapid enough to completely obscure the glass transition behavior. At the end of the first stage of crystallization the structure would be microcrystalline and further annealing would lead to normal grain growth or, possibly, secondary recrystallization.

Other alloy glasses clearly exhibit both the thermal and rheological manifestations of the glass → melt transition. Upon reheating at moderate rates the temperature, T_{rkc}, at which they begin to crystallize is often well above T_{rg}. In these alloys there could have been little nucleation during the quench. On the basis of our simple kinetic considerations we would expect $T_{rkc} - T_{rg}$ to be larger the larger is T_{rg}. Some results are in accord with this expectation but others are not. However, no definitive interpretation of the crystallization kinetics of metallic glasses can be made until the extent of the contribution of heterogeneities to the crystal nulceation processes is known.

The T_{rg}'s of some representative metal glasses formed by melt quenching are listed in Table III. They range from 0.70 down to 0.45. Presumably alloy glasses with still lower T_{rg} might be exposed with higher quench rates. If the scaled crystal-melt tension, α, were as high in glass forming alloys as in pure metals the frequency of homogeneous crystal nucleation would be negligible when $T_{rg} \gtrsim 2/3$. Thus there is the interesting possibility that alloy melts with such high T_{rg}'s could, under certain conditions, be slow cooled in massive form to glasses.

It is instructive to survey this possibility in the light of the centuries old processes of making the common nonmetallic glasses, in which large liquid masses are slowly undercooled through the mobile metastable regime to the glass transition temperature which, for

Table III.

Substance	T_m (K)		T_g (K)	T_{rg}	Ref.
		Non-Metals			
SiO_2	1780	(quartz)	1480	0.835	81
	1990	(cristobalite)	1480	0.745	
B_2O_3	~725		552	0.76	82, 83
o-terphenyl	328.5		240	0.73	84
GeO_2	1348	(rutile)	870	0.65	85
	1384	(quartz)		0.63	
BeF_2	813		590	0.73	83,86
Se	500		303	0.605	40
		Metals			
$Au_{77}Ge_{14}Si_9$	630		297	0.475	67
$Pd_{77.5}Cu_6Si_{16.5}$	1010		636	0.63	67
$Fe_{80}B_{20}$	1425		~700	~0.49	80
$Cu_{60}Zr_{40}$	1152		735	0.635	67, 87
$Ca_{65}Al_{35}$	820		563	0.69	87

these materials, is generally 0.6 or higher on the reduced scale. When glass if formed under these conditions the homogeneous nucleation frequency can hardly have exceeded 1 to 10/s × cm³. Indeed, there is evidence that in many of the common glass forming melts—*e.g.*, SiO_2, As_2S_3, B_2O_3, Se, o-terphenyl—I never attains measurable levels at any temperature. When $T_{rg} = 2/3$ such behavior would imply that $\alpha\beta^{1/3} \geq 0.5$. Glass formation by these materials is facilitated by relatively slow crystal growth rates which are typically of the order of nm/s and interface limited and thermally activated.

It would, perhaps, not be surprising if massive glass formation in slow-cooling occurred only in materials with T_{rg} greater than some lower limit, such as 2/3. However, there is a widespread impression that T_{rg} is constant at 2/3 for all such glass formers. That this impression is incorrect is evident from the T_{rg}'s, here defined as the temperature at which the viscosity is 10^{13} poise, for some typical glass formers listed in Table III. We see that the T_{rg} values actually range[81–86] from 0.60 to 0.84.

If the crystal growth rates are low a considerable density of heterogeneous nucleation centers can be tolerated in the normal processing to produce massive glass bodies. However, at moderate undercooling the crystal growth rates in molten glass forming alloys are in the range centimeters, or even m/s, rather than nm/s, as in the nonmetallic bulk glass formers. This means that for bulk glass formation the allowable concentration, ρ_N, of heterogeneous nuclei will be 10 to 20 orders of magnitude smaller for the metal than for the nonmetallic glass formers. Thus for bulk metal glass formation in slow cooling virtually all heterogeneous nuclei must be removed from the melt. This removal might be effected by fluxing the alloy with a molten silicate. It is known[88,31,32] that with such fluxing several hundred gram masses of Fe, Ni, Co or Ag may be undercooled over 200 °C without crystallization.

Atomic Transport

It appears that metallic glasses are more susceptible than other glasses to the occurrence of local atomic rearrangements and crystallization at temperatures well below T_{rg}.[89] This behavior indicates that the atomic transport frequencies in metal glasses are much higher than the flow frequency k_η in the fully relaxed glasses. Actually, given the ultra-rapid quench rates used for their formation, we expect that metallic glasses, as formed, will be much further out of configurational equilibrium, *i.e.*, their fictive temperatures will be relatively higher, than are ordinary glasses. Thus the atomic transport frequencies should be very high initially and then drop precipitously in relaxation or stabilization annealing. Indeed, it has been shown[90] that the shear viscosity ($\propto 1/k_\eta$) of $Pd_{82}Si_{18}$ glass can, without crystallization, increase by at least 4 orders of magnitude during relaxation annealing. In fully relaxed alloy melts at T_r around T_{rg} the diffusive frequencies of the major constituent determined from diffusivity measurements[91] or estimated from the kinetics of crystallization or liquid-liquid phase separation[92] are roughly equal to and scale as the flow frequency.

However, a surprising feature of atomic mobility behavior in metallic glasses is that the diffusive and flow frequencies, while scaling together (*i.e.*, $\dot{k}_D/k_D \sim \dot{k}_\eta/k_\eta$ at given T) fall much more slowly in relaxation annealing than would be expected if the structural relaxation frequency, k_v, scaled with them.[89] For directly measurable diffusivity, even with the highest resolution, every atom must jump, on average, hundreds of times. Yet the measured diffusivities of major constituents of metallic glasses, while decreasing slowly, often remain for long periods at levels which are orders

Table IV. Some Problems and Opportunities in Metastable Structure Synthesis

Alloy Glass Formation
- Crystal Nucleation: hetero-or homo-phase?
- Bulk Formation?
- Minimum impurity requirement:
  - melt quenching?
  - condensation?

Impurity Trapping in Crystals
- Maximum:
  - melt quenching?
  - other techniques?

Potentialities of Fluxing and Containerless (e.g. drop tube) Techniques

Superheating in Laser Pulsing
- Extent and Duration?

Compositionally Modulated Films
- Why enhanced Young's Modulus?
- Magnetization
- Transport-atomic and electrical

Dispersion by Plastic Deformation
- Limiting interfacial densities?
- Limiting dislocation content?
- Dislocation core properties.

of magnitude higher than in the fully relaxed melts. Quite parallel behavior is exhibited by the diffusion controlled crystallization rates.

Thus it appears that in unrelaxed metallic glasses the atomic arrangements leading to structural relaxation are of order 100 times less frequent than those effecting flow or diffusive transport. In the free volume model of atomic transport the departure from configurational equilibrium would be measured by the free volume, Δv_f, in excess of that in the fully relaxed glass. Thus the isothermal decay of the specific volume, **v**, with time might provide a measure of the structural relaxation rate and k_v. Presumably a high proportion of the local density fluctuations which lead, also according to the free volume model, to diffusive transport or flow can take place without any overall shrinkage of the body.

CONCLUSION

The field of metastable structure synthesis is still in a phase of rapid development. Some current problems and research opportunities which seem to me to be especially challenging are listed in Table IV.

REFERENCES

1. A. Saveur: *J. Metals,* June 1979, pp. 20–27.
2. P. Duwez: *Trans. ASM,* 1967, vol. 60, p. 607.
3. *See e.g.* R. H. Kernrohan and M. S. Wechsler: *J. Phys. Chem. Solids,* 1961, vol. 18, p. 178.
4. W. Kauzman: *Chem. Rev.,* 1948, vol. 43, p. 219.
5. F. Spaepen and D. Turnbull: *Metallic Glasses,* pp. 114–27, ASM, Metals Park, OH, 1978.
6. D. R. Uhlmann and R. W. Hopper: *Metallic Glasses,* pp. 128–60, ASM, Metals Park, OH, 1978.
7. C. E. Birchenall: *Atom Movements,* pp. 112–28, ASM, Cleveland, OH, 1951.
8. N. A. Gjostein: *Diffusion,* pp. 241–74, ASM, Metals Park, OH, 1973.
9. H. S. Chen and D. Turnbull: *J. Chem. Phys.*, 1968, vol. 48, p. 2560.
10. I. M. Lifschitz and V. V. Slyozov: *J. Phys. Chem. Solids,* 1961, vol. 19, p. 35.
11. C. Wagner: *Z. Elektrochem.*, 1961, vol. 65, p. 681.
12. S. Bahk and M. F. Ashby: *Scr. Met.*, 1975, vol. 9, p. 129.
13. L. Kaufman and M. Cohen: *Progress in Metal Physics,* B. Chalmers and R. King, eds., vol. 7, pp. 165–246, Pergamon Press, London, U.K., 1958.
14. R. E. Cech: *Trans AIME,* 1956, vol. 206, p. 585.
15. I. N. Stranski and D. Totomanow: *Z. Physik Chem.*, 1933, vol. 163 A, p. 399.
16. M. Volmer: *Kinetik der Phasenbildung*, pp. 200–05, Steinkopff, Dresden, 1939.
17. J. W. Cahn: *Trans. TMS-AIME*, 1968, vol. 242, p. 166.
18. J. E. Hilliard: *Phase Transformations,* pp. 497–560, ASM, Metals Park, OH, 1970.
19. C. H. Bennett: *J. Appl. Phys,* 1972, vol. 43, p. 2727.
20. D. J. Adams and A. J. Matheson: *J. Chem. Phys.*, 1972, vol. 56, p. 1989.
21. N. J. Shevchik and W. Paul: *J. Non-Cryst Solids*, 1972, vol. 8–10, p. 381.
22. D. Turnbull: *Contemp. Phys,* 1969, vol. 10, p. 473.
23. H. Jones: *Rapidly Quenched Metals,* N. J. Grant and B. C. Giessen, eds., pp. 1–28, M.I.T. Press, Cambridge, MA, 1976.
24. N. Bloembergen: *Laser-Solid Interactions and Laser Processing,* S. D. Ferris, H. J. Leamy and J. M. Poate, eds., A.I.P. Conference Proc., vol. 50, pp. 1–10, 1979.
25. *See e.g.* H. S. Chen and K. A. Jackson: *Metallic Glasses,* pp. 74–96, ASM, Metals Park, OH, 1978.
26. D. Turnbull: *Physics of Non-Cryst. Solids,* J. W. Prins, ed., pp. 41–56, North Holland, Amsterdam, 1964.
27. H. Brooks: *Metal Interfaces,* pp. 20–64, ASM, Cleveland, OH, 1952.
28. D. Turnbull: *Impurities and Imperfections,* pp. 121–44, ASM, Cleveland, OH, 1955.
29. D. Turnbull and B. G. Bagley: *Treatise on Solid State Chemistry,* N. B. Hannay, ed., vol. 5, pp. 513–54, Plenum Press, NY, 1975.
30. F. Spaepen and D. Turnbull: to be published in *Laser and Electron Beam Processing of Semiconductor Surfaces*, J. M. Poate and J. W. Mayer, eds., Academic Press, NY.
31. J. L. Walker, *see* B. Chalmers: *Principles of Solidification*, pp. 114–15, 122–24, Wiley, NY, 1964.
32. G. A. Colligan and B. S. Bayles: *Acta Met.*, 1962, vol. 10, p. 895.
33. J. C. Baker and J. W. Cahn: *Solidification,* pp. 23–58, ASM, Metals Park, OH, 1971.
34. J. W. Cahn, S. R. Coriell and W. J. Boettinger: *Laser and Electron Beam Processing of Materials,* C. W. White and P. S. Peercy, eds., pp. 89–103, Academic Press, NY, 1980.
35. D. Turnbull: *J. de Physique*, Colloque C-4, suppl. 5, 1980, vol. 41, p. C4-109.
36. K. A. Jackson, G. H. Gilmer and H. J. Leamy: *Laser and Electron Beam Processing of Materials,* C. W. White and P. S. Peercy, eds., pp. 104–10, Academic Press, NY, 1980.
37. S. D. Dahlgren, J. W. Patten and M. T. Thomas: *Thin Solid Films,* 1978, vol. 53, p. 41.
38. W. Buckel and R. Hilsch: *Z. Phys.*, 1954, vol. 138, p. 109.
39. S. Mader and A. S. Nowick: *Appl. Phys. Letters*, vol. 7, p. 57, 1965; *J. Vac Sci. Tech.*, 1965, vol. 2, p. 35; *see also*: A. S. Nowick and S. Mader: *I.B.M. J. Res. and Dev.*, 1965, vol. 9, p. 358.
40. R. B. Stephens: *J. Appl. Phys.*, 1978, vol. 49, p. 5855.
41. M. P. Rosenblum and D. Turnbull: *J. Non-Cryst. Solids*, 1980, vol. 37, p. 45.
42. H. E. Cook and J. E. Hilliard: *J. Appl. Phys.*, 1969, vol. 40, p. 2191.
43. W. M. C. Yang, T. Tsakalakos and J. E. Hilliard: *J. Appl. Phys.*, 1977, vol. 48, p. 874.
44. B. J. Thaler, J. B. Ketterson and J. E. Hilliard: *Phys. Rev. Letters,* 1978, vol. 41, p. 336.
45. M. P. Rosenblum, F. Spaepen, and D. Turnbull: *Appl. Phys. Letters,* 1980, vol. 37, p. 184.
46. L. N. Large and R. V. Bicknell: *J. Mater. Sci.*, 1967, vol. 2, p. 589.
47. A. G. Cullis, J. M. Poate and J. A. Borders: *Appl. Phys. Letters,* 1976, vol. 28, p. 314; *see also* review: S. M. Myers in *Thin Films-Interdiffusion, and Reactions,* J. M. Poate, K. N. Tu and J. W. Mayer, eds., pp. 533–66, Wiley, NY, 1978.
48. J. E. Smith, Jr., M. H. Brodsky, B. L. Crowder, M. I. Nathan and A. Pinczuk: *Phys. Rev. Letters,* 1971, vol. 26, p. 642; *see also* J. F. Graczyk and P. Chaudhari: *Phys. Status Solidi,* 1973, vol. 58B, pp. 163, 501.
49. S. C. Moss and J. F. Graczyk: *Phys. Rev. Letters,* 1969, vol. 23, p. 1167.
50. *See* J. W. Mayer, L. Eriksson and J. A. Davies: *Ion Implantation in Semiconductors,* Academic Press, NY, 1970.
51. J. M. Poate: *J. de Physique*, 1980, vol. 41, p. C4–1.
52. S. R. Wilson, C. W. White, F. W. Young, Jr., B. R. Appleton and J. Nardyan: *J. de Physique*, 1980, vol. 41, p. C4–91.
53. B. Y. Tsaur, S. S. Lau, Z. L. Liau and J. W. Mayer: *Thin Solid Films,* 1979, vol. 63, p. 31.
54. B. Y. Tsaur and J. W. Mayer: *Appl. Phys. Letter*, 1980, vol. 37, p. 389.
55. H. P. Wahl and G. Wasserman: *Z. Metallkund.*, 1970, vol. 61, p. 326; C. C. Tsuei: *Science,* 1973, vol. 180, p. 57.
56. J. P. Harbison and J. Bevk: *J. Appl. Phys.*, 1977, vol. 48, p. 5180.
57. J. Bevk, J. P. Harbison and J. L. Bell: *J. Appl. Phys.*, 1978, vol. 49, p. 6031.
58. K. Karasek and J. Bevk: *Scr. Met.*, 1980, vol. 14, p. 431.
59. L. E. Greenwald, E. M. Breinan and B. H. Kear: *Laser-Solid Interactions and Laser Processing,* S. D. Ferris, H. J. Leamy and J. M. Poate, eds., A.I.P. Conf. Proceeding, vol. 50, pp. 189–204, A.I.P., NY, 1979.
60. A. Brenner, D. E. Couch and E. K. Williams: *J. Res. Nat.*

Bur. Stds., 1950, vol. 44, p. 109; *also see*: A. Brenner: *Electrodeposition of Alloys, Principles and Practice*, Academic Press, NY, 1963.
61. G. S. Cargill, III: *J. Appl. Phys.*, 1970, vol. 41, p. 2248.
62. S. D. Dahlgren: *J. Vac. Sci. Tech.*, 1974, vol. 11, p. 832.
63. M. D. Merz and S. D. Dahlgren: *J. Appl. Phys.*, 1975, vol. 46, p. 3235.
64. P. Chaudhari and D. Turnbull: *Science*, 1978, vol. 199, p. 11.
65. F. Spaepen and D. Turnbull: *Rapidly Quenched Metals*, N. J. Grant and B. C. Giessen, eds., pp. 205–29, M.I.T. Press, Cambridge, MA, 1976.
66. D. E. Polk and B. C. Giessen: *Metallic Glasses*, pp. 1–35, ASM, Metals Park, OH, 1978.
67. H. S. Chen: *Reports on Progress in Physics*, 1980, vol. 43, p. 353.
68. M. H. Cohen and D. Turnbull: *J. Chem. Phys.*, 1959, vol. 31, p. 1164.
69. J. D. Bernal: *Nature*, 1960, vol. 185, p. 68.
70. R. Hilsch: *Non-Crystalline Solids*, R. D. Fréchette, ed., pp. 348–73, Wiley, NY, 1960.
71. M. R. Bennett and J. G. Wright, *Phys. Status Solidi*, 1974, vol. 13A, p. 135; P. K. Leung and J. G. Wright: *Phil. Mag.*, 1974, vol. 8, no. 30, pp. 185, 995.
72. M. Marcus and D. Turnbull: *Mat. Sci. Eng.*, 1976, vol. 23, p. 211.
73. E. M. Breinan, B. H. Kear, C. M. Banas and L. E. Greenwald: *Proc. Third Int. Symp. on Super. Alloys*, p. 435, Claitors Publ. Div., Baton Rouge, LA, 1976.
74. M. Von Allmen: *Physical Processes in Laser-Materials Interactions*, M. Bertolotti, ed., NATO-ASI Series, in press.
75. The structural similarity between metallic glasses and Frank-Kasper phases has been pointed out to me by several persons, independently, in private conversations during the past fifteen years. The possible importance of the similarity was especially stressed by W. L. Johnson; *see also*: Ref. 64.
76. *See* review in W. B. Pearson: *The Crystal Chemistry and Physics of Metals and Alloys*, pp. 31–45, Wiley, NY, 1972.
77. D. Turnbull: *J. de Physique*, 1974, vol. 35, p. C4–1.
78. D. Turnbull: *Sri. Met.*, 1977, vol. 11, p. 1131.
79. D. Turnbull: to be published.
80. A. L. Greer: *Kinetics of Crystallization of $Fe_{80}B_{20}$ Glass*, submitted to Acta Met.
81. B. Hetherington, K. H. Jack and J. C. Kennedy: *Phys. Chem. Glasses*, 1964, vol. 5, p. 130.
82. H. Tweer, N. Laberge and P. B. Macedo: *J. Amer. Cer. Soc.*, 1971, vol. 54, p. 121.
83. J. D. Mackenzie: *J. Phys. Chem.*, 1960, vol. 32, p. 1150.
84. R. J. Greet and D. Turnbull: *J. Chem. Phys.*, 1967, vol. 46, p. 1243.
85. E. H. Fontana and W. A. Plummer: *Phys. Chem. Glasses*, 1966, vol. 7, p. 139. J. P. DeNeufville, Ph.D. Thesis, Harvard University, Cambridge, MA, 1969.
86. C. T. Moyhihan and S. Cantor: *J. Chem. Phys.*, 1968, vol. 48, p. 115.
87. B. C. Giessen, J. Hong, L. Kadacoff, D. E. Polk, R. Ramon and R. St. Amand: *Rapidly Quenched Metals III*, B. Cantor, ed., vol. 1, pp. 249–60, The Metals Society, London, England, 1978.
88. Bardenheuer and Bleckman: *Stahl u Eisen*, 1941, vol. 61, p. 49.
89. F. Spaepen: *Defects in Amorphous Metals*, Proc. XXXVth Session of Summer School for Theoretical Physics on *Physics of Defects*, Les Houches, France, in press.
90. A. I. Taub and F. Spaepen: *Scr. Met.*, 1979, vol. 13, p. 195.
91. H. S. Chen, L. C. Kimerling, J. M. Poate and W. L. Brown: *Appl. Phys. Letters*, 1978, vol. 32, p. 461.
92. C-P. Chou and D. Turnbull: *J. Non-Cryst. Solids*, 1975, vol. 17, p. 168.

SECTION II

Processes

FUNDAMENTALS OF PARTICULATE METALLURGY

Alan Lawley
Department of Materials Engineering
Drexel University
Philadelphia, Pa. 19104

ABSTRACT

High performance powder metallurgy reflects significant advances in the science and technology of powder production and subsequent powder consolidation. Powder processing methods involving the atomization of a stream of liquid metal are reviewed with particular reference to relationships between processing parameters and powder properties. These methods include gas and water atomization, several forms of centrifugal atomization and recent innovative approaches at the prototype or laboratory-stage of development. Where available, models and mechanisms of the atomization processes are detailed. To conclude, the role of cooling rate in powder processing, and implications re the characteristics of rapidly solidified powders are examined.

INTRODUCTION

The Sagamore Army Materials Conference held in 1971 focussed on powder metallurgy for high-performance applications. Since that time, a better understanding of the various technologies involved has been achieved. Also, the spectrum of materials amenable to powder processing to full density has increased significantly. In the area of powder production most of the interest has centered on liquid metal atomization (1). More refined empirical relationships between processing parameter and powder properties have been documented and the general level of powder quality, in terms of size, shape, chemistry and purity, has improved (1). Innovative techniques of atomization now exist which may become challengers to the conventional processes of gas or water atomization (2,3). Perhaps the most significant progress has been made in the area of atomization mechanisms. Understanding of the fundamentals of the atomization process provides a basis for optimal design vis a vis powder quality and yield.

Stimulus for further development and understanding of high performance powder metallurgy resides in factors such as:

- materials conservation
- compositional flexibility
- fine grain size
- isotropy
- energy conservation
- homogeneity (structure, composition)

The initial step involves the preparation of a suitable powder. While there are several broad approaches of commercial importance (e.g. chemical, electrolytic, comminution, atomization) it is in the area of liquid metal atomization that most research has been directed over the last decade - with the objective of increasing yields, lowering particle size, tightening chemistry, and reducing impurities such as oxide inclusions. The thrust toward atomization reflects the need to develop viable powder processing routes for materials such as alloy steels, titanium and the superalloys.

In this paper, commercial and more recent prototype-scale methods of liquid metal atomization are reviewed. Consideration is given to processing variables, resulting powder properties and operative mechanisms of atomization. To conclude, the effect of cooling rate on particle structure is examined covering the range from gas atomization ($\sim 10^2$°C/s) to rapid solidification processing ($\gtrsim 10^5$°C/s).

ATOMIZATION METHODS

Gas Atomization

The principle of gas atomization is simple - a continuous stream of liquid metal is broken down into droplets by means of a subsonic or supersonic gas stream or jet. Atomization is due to the kinetic energy transfer from the atomizing medium to the metal. In practice, nitrogen, argon or air are used; mixtures of these gases and helium are also effective. The number and geometry of gas-metal configurations are unlimited; typical arrangements involving multiple jets or an annular ring are illustrated schematically in Figure 1. More detail concerning jet configurations and design of atomizing units is available in the open literature (4-9).

There are several interrelated processing and material variables involved in gas atomization, Table I. In consequence

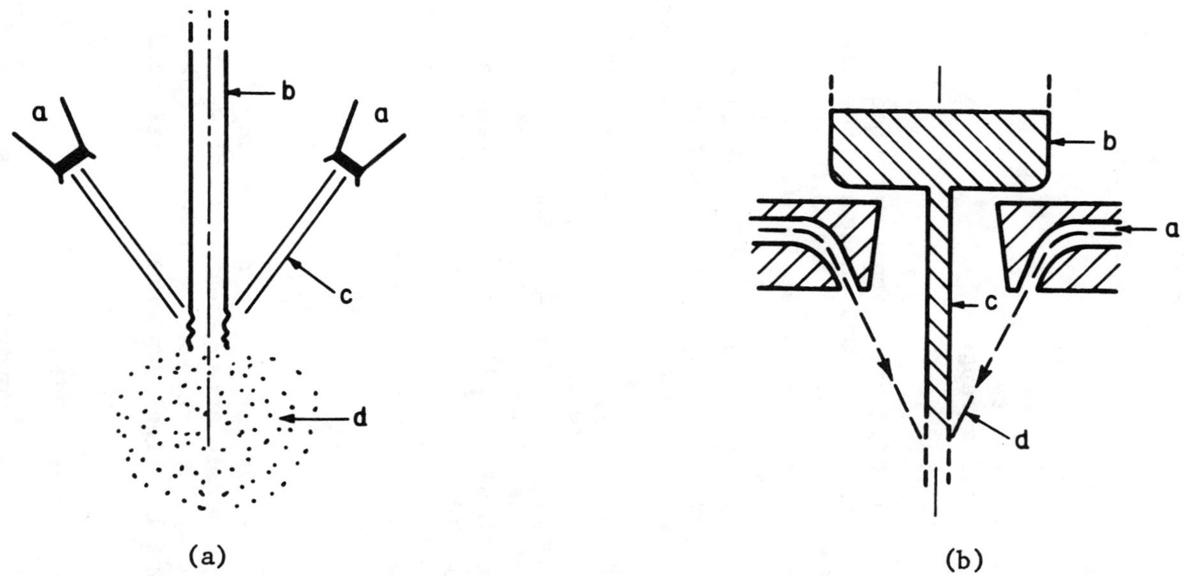

Figure 1. Representative gas-atomization configurations: (a) Two-jet configuration; a = jets, b = liquid metal stream, c = gas stream, d = atomized powder. (b) Annular-ring configuration; a = ring orifice, b = liquid metal reservoir, c = liquid metal stream, d = gas stream.

Table I

Variables in Gas Atomization

Gas jet distance and pressure
Gas velocity and mass flow rate
Metal velocity and mass flow rate
Nozzle geometry
Angle of impingement
Superheat
Metal surface tension
Metal melting range

numerous empirical relationships have been documented or proposed for the prediction of particle size and shape and the distributions of size and shape (1,9). It is always found that as gas pressure increases, as gas mass flow rate increases and/or as the jet to metal stream distance decreases, the average particle diameter decreases (8-12). The power law relating particle size to gas pressure is shown in Figure 2. Typical operating ranges of some of the above variables in gas atomization are summarized in Table II.

Gas-atomized powders are typically spherical and have smooth surfaces. Particle cooling rate is 10^2°C/s. A representative size distribution for superalloy powders is shown in Figure 3, curve a. The mean particle diameter is ∿150μm; yields up to ∿40% are achieved at -325 mesh size. Alloy steels, titanium and superalloy compositions are amenable to inert gas atomization (13-16). Overall, the process has a low energy efficiency and is expensive if gases other than nitrogen have to be used.

Water Atomization

Water atomization is similar to gas atomization in that it is a two-fluid process; geometric arrangements of the water stream relative to the liquid metal stream are similar to those used in gas atomization, Figure 1. Process parameters correspond to those

Table II

Representative Operating Conditions in Gas Atomization

Gas Pressure	14×10^5 - 42×10^5Pa
Gas Velocity	50-150 m/s
Superheat	100-200°C
Angle of Impingement	15-90°

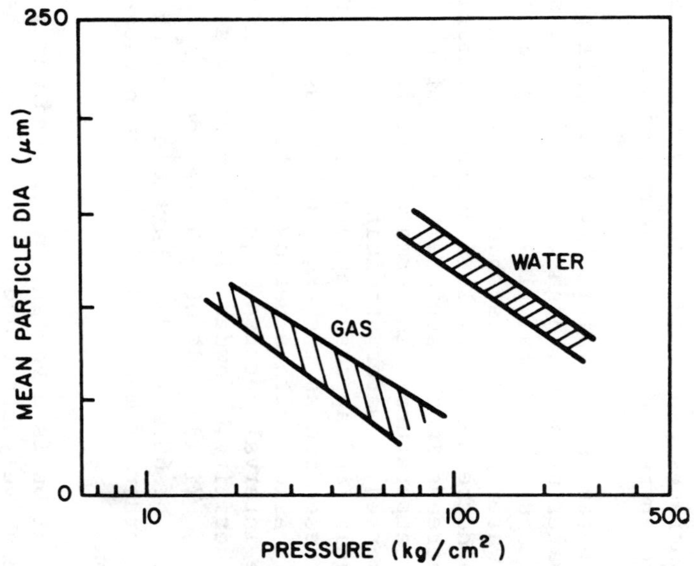

Figure 2. The dependence of particle size on fluid (gas or water) pressure. 1 $kg/cm^2 = 97.9 \times 10^3$ Pa. (11)

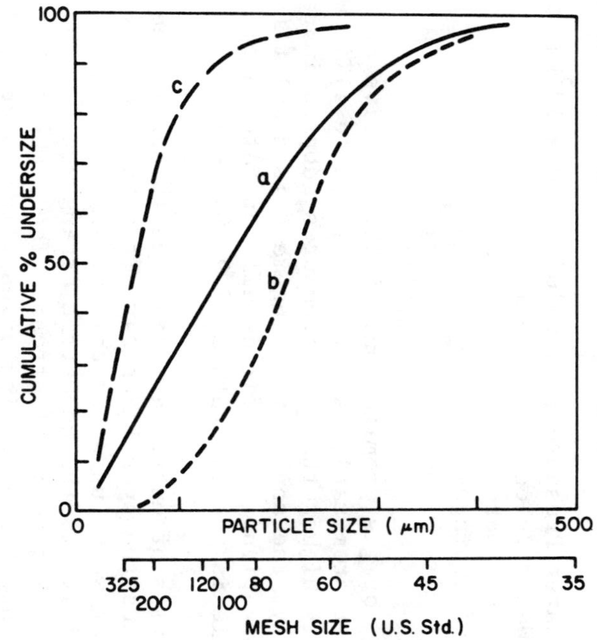

Figure 3. Representative particle size distributions for powders prepared by: a) Inert gas atomization, b) The rotating electrode process, c) Vacuum atomization. (13)

Table III

Representative Operating Conditions in Water Atomization

Water Pressure	35×10^5 - 210×10^5Pa
Water Velocity	40 - 150 m/s
Superheat	100-250°C
Angle of Impingement	≤30°

listed in Table I for gas atomization. Representative operating conditions in water atomization are listed in Table III. Water pressure is usually higher than the pressures adopted in gas atomization. Generally, the angle of impingement is smaller for water than for gas atomization. Specific details on water atomization units are available in the literature (7,9,10,17-21).

Large quantities of energy are required to supply the water at high pressure and the overall energy efficiency is very low, ∿4%. The amount of water used per unit weight of powder produced is extremely sensitive to the geometric arrangement of the water jet relative to the liquid metal stream. The process is significantly cheaper than gas atomization unless air can be used as the gas medium.

Water atomized powders are generally irregular in shape with rough oxidized surfaces. Low and high-alloy steels, including stainless, are amenable to water atomization. Excess oxide on the powder particle surfaces can be subsequently removed or diminished in amount by hydrogen reduction or various forms of acid leaching.

It has been established empirically that small particles are favored by high water pressure, high water velocity, a short jet to metal distance and a low metal flow rate. The dependence of particle size on water pressure is shown in Figure 2. Superheating of the metal also decreases particle diameter, primarily through the associated decrease in the surface tension of the metal at the actual atomizing temperature. In terms of particle shape, some control can be exerted in water atomization. Powder particles become less irregular and approach sphericity as the degree of superheat is increased and as the angle of impingement of the water jet is increased. Alloys with a narrow melting range (i.e., a small temperature interval between liquidus and solidus) result in some degree of sphericity. Powder yields increase as the water pressure increases. Yields ∿60% of -35 mesh powder can be achieved at a water pressure of 60×10^5 Pa. Particle cooling rates (∿10^3°C/s) are an order of magnitude higher than those achieved in gas atomization.

Steam atomization is a variant of gas and water atomization. The process is relatively cheap and can be used in the production of several classes of alloys: carbon and low alloy steels, stainless steels, nickel and cobalt-base superalloys. Subsequent treatment of steam atomized powders is usually necessary to lower surface oxide content. This may take the form of milling or chemical leaching. The virtues of low-pressure steam atomization could result in expanded use of this technique on a commercial scale.

Centrifugal Atomization

In centrifugal atomization, molten metal is ejected in the form of droplets from a rapidly spinning crucible plate or disc. Particles either cool in the environment of the collection chamber or can be gas-quenched as they leave the rotating vehicle. Many configurations have been examined (22-26). The arrangement illustrated schematically in Figure 4 is suitable for titanium alloy powders. Erosion, dissolution, oxidation and creep of the spinner pose severe materials problems; the advent of improved ceramics has been a key factor in the development of this process. Energy requirements are relatively low.

Particle shape can be varied from spherical to flake. Spherical powders produced by centrifugal atomization are relatively coarse with average particle diameters $\gtrsim 200\mu m$.

Other forms of centrifugal atomization include the rotating electrode process (13,27), the Durarc® process (28) and the several techniques termed melt spinning and melt extraction. Since spinning and extraction from the melt produce primarily filaments or ribbons, these techniques will not be considered further. Maringer et al. (29,30) have recently described melt extraction and melt spinning. Cooling rates in excess of 10^6°C/s are reported. Amorphous and/or microcrystalline materials in filament or ribbon form offering unique magnetic properties are now produced on a commercial scale by melt extraction.

The rotating electrode process (13,27) illustrated in Figure 5 is in commercial operation though production rates are low. Material in the form of a rod electrode is rotated rapidly while it is melted at one end by an electric arc. Molten metal spins off the bar and solidifies before hitting the walls of the inert gas-filled chamber. The process has been developed primarily for atomization of titanium alloys and for superalloys. Tungsten contamination from the stationary electrode can be avoided by replacement with a titanium cathode; both cathode and anode rotate in the most recent version.

Rotating electrode powders are usually spherical, of high surface quality but have a large average particle diameter $\gtrsim 200\mu m$. Size range is $\sim$50-400μm, Figure 3 (curve b). Typically, yields run

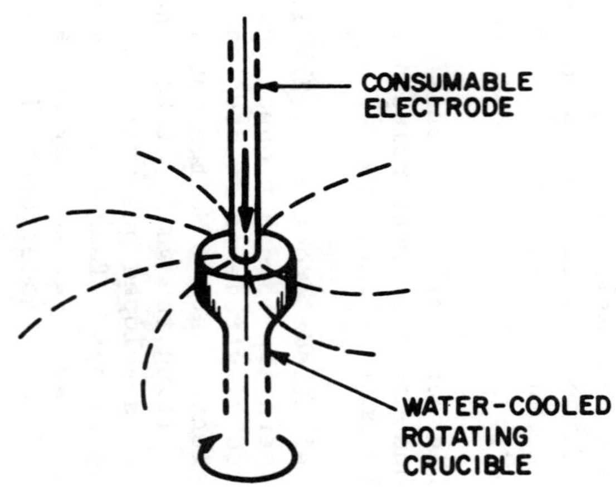

Figure 4. Schematic representation of a centrifugal atomization configuration.

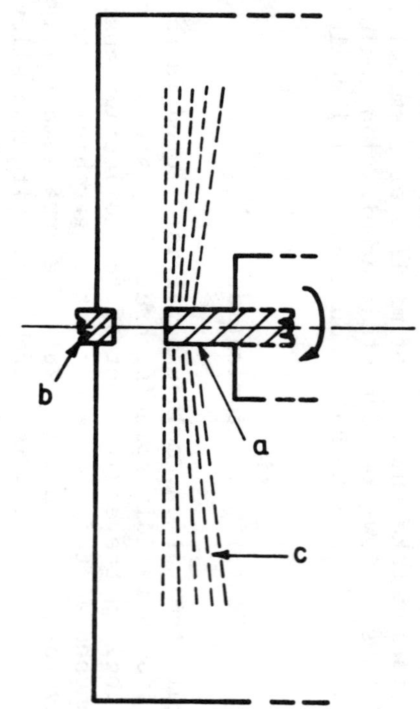

Figure 5. A schematic to illustrate the rotating electrode process. a and b are electrodes, c = powder.

90% for -35 mesh powder. Cooling rate is low and comparable to that achieved in gas atomization, i.e. ∿10^2°C/s.

Other Atomization Techniques

A variety of atomization techniques has been reported in the literature which do not fit the categories of gas, water, or centrifugal atomization. These range from laboratory scale curiosities to prototype or semi-production units. Three of the most advanced/promising methods will be considered.

In roller atomization (31), a stream of molten metal is fed between rapidly rotating rolls (up to 200 rev/s). A schematic is given in Figure 6. The roll gap is ∿50μm. Heat transfer to the rolls is minimized by a suitable insulating coating so that the atomization event actually takes place by break up of the molten sheet of metal below the roll gap. Singer (31) has examined process variables for copper, tin and lead. A median particle size of 220μm has been obtained for tin powders with a yield ∿70%. Median particle size is inversely proportional to roll speed. A wide spectrum of forms can be atomized: flake, acicular, irregular, spheroidal. The method is attractive in terms of energy consumption and should be amenable to scale-up. Cooling rates have not been determined.

Ultrasonic atomization appears attractive for the production of small diameter (<50μm) particles with a narrow size distribution (3) at high quench rates (∿10^5°C/s). Hartman shock wave tubes are used to deliver high velocity pulses of gas (up to Mach 2) at frequencies in the range 60,000-80,000 cps. There is no contact between the ultrasonic 'die' and the liquid metal stream and the configuration is similar to that in gas atomization. Grant (3) has obtained yields in the 80-90% range for -325 mesh powder of stainless steel. Ultrasonic atomization has been used on a commercial scale for low melting point alloys and more recently interest has centered on superalloys. Like roller atomization, there do not appear to be any major deterents to scale-up.

Vacuum or soluble gas atomization (13,32) is a commercial batch process based on the principle that when a molten metal supersaturated with gas under pressure is suddenly exposed to vacuum, the gas expands, comes out of solution, and causes the liquid metal to be atomized. Alloy powders based on Ni, Cu, Co, Fe and Al have been vacuum atomized with hydrogen as the gas. The powders are spherical, clean and are of a high-purity relative to other powder processing methods. A schematic is given in Figure 7 and a typical particle size distirubtion for a superalloy in Figure 3 (curve c).

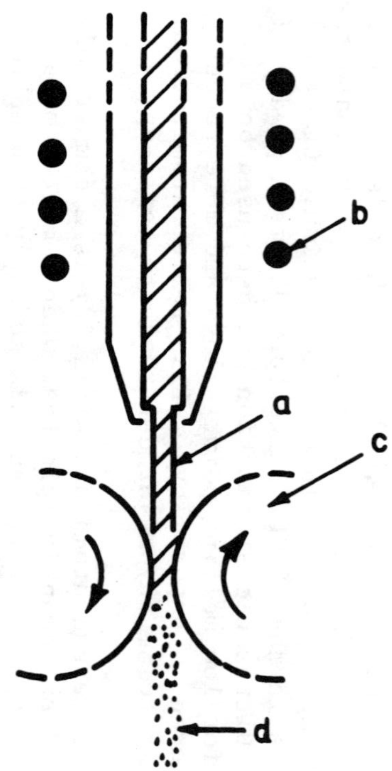

Figure 6. Schematic representation of roller atomization: a = liquid metal stream, b = heat source, c = rolls, d = metal powder.

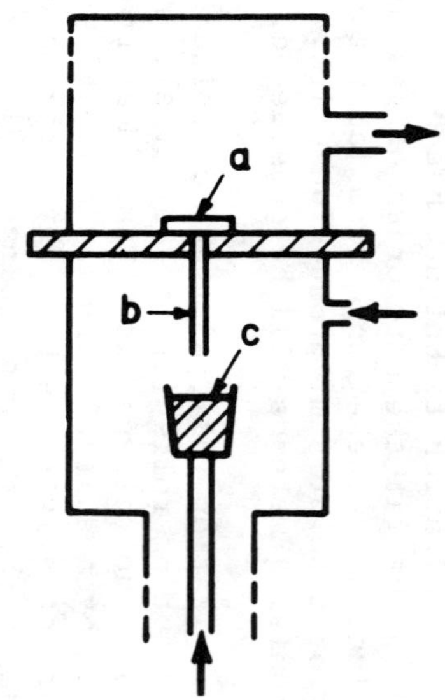

Figure 7. Schematic representation of vacuum atomization: a = trap door, b = transfer tube, c = molten metal.

ATOMIZATION MECHANISMS

Two-fluid atomization does not lend itself to simple modelling or analysis. However, several models have been advanced for explaining the sequence of events leading to liquid droplet formation from the metal stream. These pertain to gas, water, ultrasonic and roller atomization.

Gas Atomization

The basis of the current understanding of the fundamentals of gas atomization must be credited primarily to See et al. (12,33,34). In a series of elegant experiments these workers utilized a variable area meter to map gas flow rate around the liquid metal stream and Pitot tubes to generate gas velocity profiles. The velocity measurements were complemented by Schlieren photography of the gas-metal stream interactions. It is shown convincingly that the overall process of gas atomization is a three-stage process, as illustrated in Figure 8. Primary break-up of the liquid stream occurs in stage I. In stage II the droplets are still molten so that secondary disintegration is possible. Finally in stage III, the particles solidify. It is important to note that the actual atomization event takes place above the point of focus of the gas stream, Figure 8. The reason for this resides in the vacuum created around the metal stream coupled with the stream acceleration which causes the liquid metal to expand into a hollow conical configuration of molten metal. The actual atomization event then occurs at the circumferential periphery of the cone.

What are the details of break up of the hollow cone formed during stage I? Here there is general agreement that a sinuous wave is initiated which increases rapidly in amplitude. Then the wave becomes detached from the bulk liquid, producing a ligament whose dimensions depend on the wavelength at disintegration. Various physical models are available with which to describe these events; stability criteria all involve the Weber number of the liquid metal. The model developed by Dombrowski and Johns (35) for the disintegration of a liquid sheet is illustrated in Figure 9. See, Runkle and King (34) analyzed their experimental data on the nitrogen atomization of lead in terms of this model of ligament formation and break up. From the theory and the known (experimental) values of mass mean particle diameter, a parameter termed the mean relative gas-wave velocity (U) was calculated and compared to the measured values of U parallel to the surface of the hollow cone at the point of atomization. Reasonable agreement existed. The several experimental studies of gas atomization (8,9,12,33,36) involving high-speed photography confirm that break-up of the liquid metal stream involves ligament formation. Subsequently, the ligaments break up into more spherical particles before solidification.

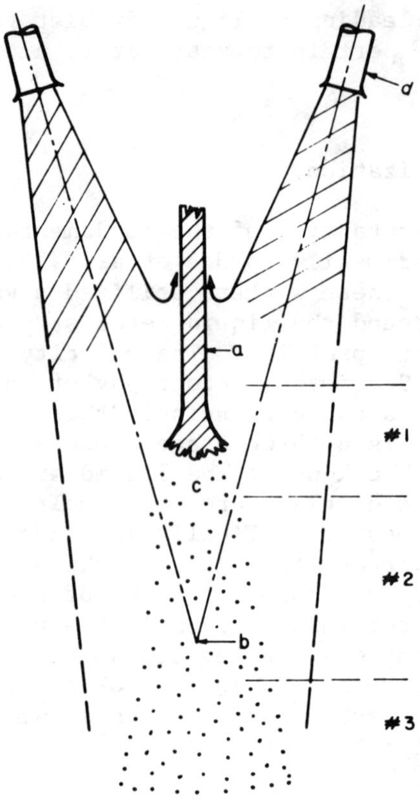

Figure 8. The three stages of liquid metal stream disintegration in gas atomization (12,33).

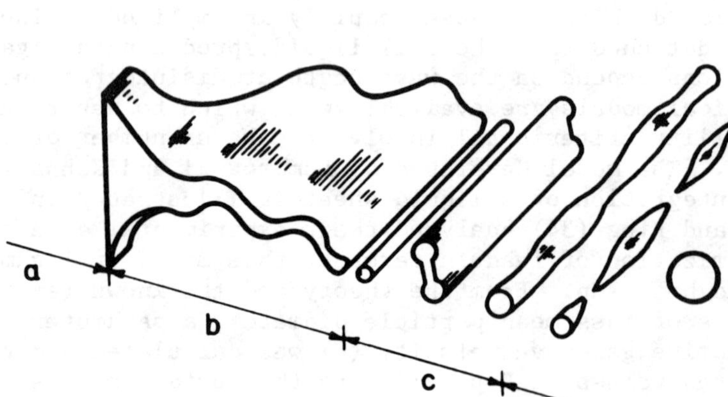

Figure 9. A model to show disintegration of a sheet of liquid: a = stable sheet, b = growth of waves in sheet, c = ligament formation, d = ligament breakdown (35).

Bradley (37) has developed a rigorous mathematical model of gas atomization for the prediction of drop size at subsonic and sonic gas velocities; the model also takes into account the compressibility of the gas and the viscosity of the liquid. It is based on the physical model of ligament formation and break-up proposed by Castleman (38). From Bradley's analysis it is clear that compressibility of the gas stream is important - to the extent that drop size is dependent only sensibly on the Mach number of the gas flow and on the surface tension of the molten metal. Drop size decreases with increasing gas velocity but is linearly dependent on surface tension. The dimensional analysis leads to a single universal curve for the prediction of powder particle size as a function of gas velocity. This curve is given in Figure 10. M is the Mach number and L is a function of ρ, M and N where:

$$\rho = \rho_G/\rho_L; \quad M = U/U_S; \quad N = T/\rho_G U_S V \qquad (1)$$

and ρ_G = gas density; ρ_L = liquid density; T = surface tension of liquid; U = gas velocity; U_S = sonic gas velocity; V = kinematic viscosity of liquid

To predict the diameter of the most frequently occuring particle for a particular gas velocity (in terms of Mach number), the corresponding value of L is obtained from Figure 10. Particle diameter is then given by:

$$d = \frac{2.95\ T}{L\rho_G U_S^2} \qquad (2)$$

A rigorous test of Bradley's analysis awaits more extensive experimental data of the type developed by See et al. (12,33,34).

Water Atomization

There have been only a few studies aimed at elucidating the atomization mechanism in water atomization. As a general observation, there is no evidence to suggest that ligaments occur as an intermediate step during particle formation.

Grandzol (21) has shown that a simple functional relationship exists between median particle diameter d_m and water velocity V_W. Of importance here is the observation that the water breaks up into droplets prior to impacting the liquid metal stream; hence V_W refers to water droplet velocity. The relationship is of the form:

$$d_m = K/V_W \qquad (3)$$

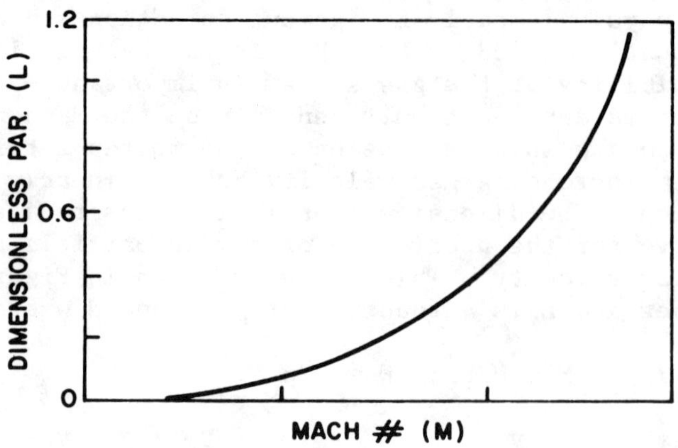

Figure 10. Bradley's universal plot for the determination of particle size (37).

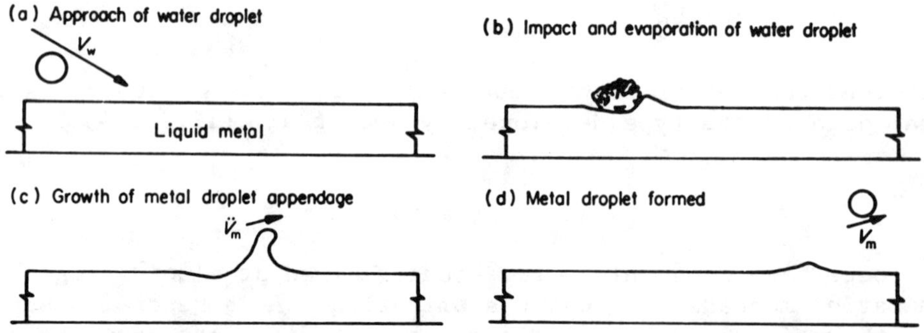

Figure 11. The 'scrape' mechanism of water atomization (21).

Grandzol (21) and Grandzol and Tallmadge (39) also derived a relationship of this form by means of a simple two-liquid atomization model involving: momentum-exchange for determination of droplet size, empirical relations to relate water droplet size to water jet velocity, kinetic energy exchange for the ratio of water droplet velocity to metal particle velocity, and high-speed photography for the determination of the actual metal particle velocity. From their experimental observations, Grandzol and Tallmadge (40) also showed that it is the component of water velocity normal to the metal stream that is the dominant factor that controls metal particle size. Specifically:

$$d_m = \frac{K}{V_W \sin \alpha} \quad (4)$$

where α is the angle between the metal-stream axis and the water jet axis (i.e. half the total impingement angle). This gives credence to a mechanism of droplet formation that involves impact as opposed to shear, as compared to gas atomization. A possible sequence of events consistent with the above model is illustrated schematically in Figure 11. This so-called "scrape" mechanism allows for almost instantaneous formation of the metal droplet. The momentum-interchange model is simple and does not require a one to one ratio between water droplets and resulting metal droplets.

Clearly, more work is needed to evaluate the generality of the Grandzol and Tallmadge model. The importance of water and steam atomization as a low-cost process for powder production on a commercial scale should serve to stimulate such research.

Ultrasonic Atomization

In contrast to gas atomization, ultrasonic atomization appears to be a single step process (3). Because of the high velocity of the gas pulse, the liquid metal stream responds on impact as though it were a solid with low shear resistance. Preliminary calculations indicate a low overall efficiency, due primarily to energy losses by reflection of the gas pulse and through heat generation.

Roller Atomization

Roller atomization is amenable to modelling and analysis with reference to break up of the liquid metal sheet as it moves downward past the roll gap (31). From high-speed photography, it is concluded that the actual atomization mechanism is analogous to that occuring in lubricant films; it is known as cavitation. Thus, the liquid metal actually cavitates a small distance downstream from the nip of the rolls and gives rise to perforations and streamers.

At high roller speeds there is evidence that the liquid metal flows around the points of cavitation and forms thin threads or slivers of liquid. These subsequently break up due to the propagation of liquid-jet instabilities. On-going research (41) is aimed at a quantitative description of the primary break up of the liquid sheet by cavitation and subsequent secondary break up and coalescence of liquid streamers and perforations. This involves establishing the correct boundary conditions for cavitation and solving the Reynolds equation taken from lubrication theory.

RAPIDLY SOLIDIFIED POWDERS

Since the 1971 Sagamore conference on powder metallurgy increasing interest has developed in the area of rapidly solidified powders (RSP) - to the extent that a recent conference was devoted exclusively to the principles and technologies associated with rapid solidification processing (42). The central tenet is that rapid cooling rates offer unique possibilities in terms of the development and control of microstructure in metal powders. To put this in perspective, dendrite arm spacing, which is a measure of the scale of the liquid-to-solid segregation pattern, is listed as a function of cooling rate in Table IV. In turn, the cooling rate is controlled by the physical size of the atomized metal powder particle. Obvious benefits appear to exist in terms of a minimum of segregation within individual powder particles, inhibition of embrittling compounds or films (i.e. a greater tolerance for impurities) and ultrafine grain size.

Two conditions which must be met in order to rapidly solidify powders are small droplet size and high heat transfer at the metal droplet interface. To this end, Cox et al. (26,43) have developed a unique centrifugal atomization facility in which particles are cooled in a blast of helium as they leave the periphery of a rotating disc. In order to achieve cooling rates in excess of 10^5°C/s, particles sizes must be below ∿60μm diameter and this requires rotational speeds in the range 250-600 rev/s. Some

Table IV

Dendrite Arm Spacing as a Function of Cooling Rate

Cooling Rate (°C/s)	Dendrite Arm Spacing (μm)	Process
10^{-2}	5×10^2	conventional ingot solidification
10^2	10^2	gas atomization
10^3	5	water atomization
10^5	1	substrate cooling
10^7	0.1	splat cooling

experimental data for IN100 are shown in Figure 12; cooling rate was determined from measured dendrite arm spacings in the powder particles.

At the present time RSP powders, primarily superalloy compositions, are being characterized with particular reference to structure (dendrite, microcrystalline). Interest also centers on possible changes in microstructure during consolidation to full density. There is a need to compare RSP and standard gas atomized powders consolidated under similar conditions; only in this way can the expected benefits of rapid cooling be truly appraised.

SUMMARY

A broad spectrum of atomization techniques exist. Gas, water, steam and several forms of centrifugal atomization are commercially viable. Roller and ultrasonic atomization appear promising but require further development. Rapidly solidified powders (>10^5°C/s) are now available possessing unique microstructures; it remains to be seen whether or not the properties of consolidated RSP powders are superior to those derived from a conventional gas atomized powder. Significant progress has been made toward an understanding of the mechanism(s) involved in gas and water atomization; this is essential in order to achieve optimal design of each process.

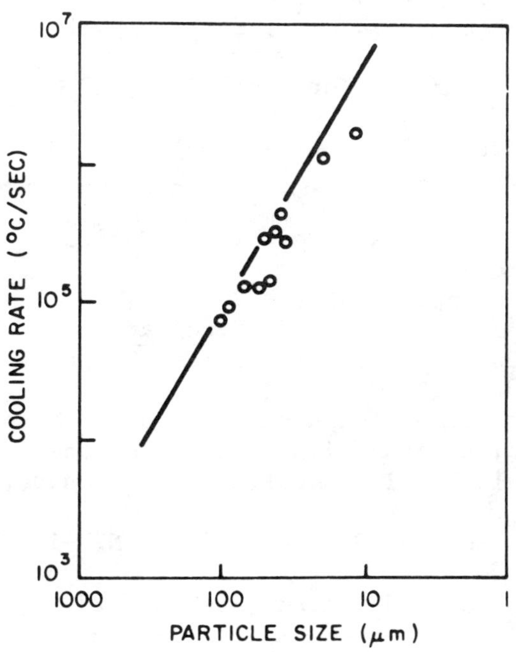

Figure 12. Experimentally determined cooling rates as a function of powder particle size (26).

REFERENCES

1. A. Lawley, The Int. Journal of Powder Met. and Powder Tech., 13, #3, p. 169 (1977).
2. A. Lawley, "Preparation of Metal Powders" in Ann. Rev. Mater. Sci., Vol. 8, p. 49, Ann. Revs. Inc. (1978).
3. N. J. Grant, in Rapid Solidification Processing, Editors: R. Mehrabian, B. H. Kear and M. Cohen, Claitor's Pub. Div., p. 230 (1978).
4. H. Lubanska, J. Met. 22, 45-9 (1970).
5. C. F. Dixon, Can. Met. Quart. 12, 809 (1973).
6. E. Klar and W. M. Shafer, Powder Metallurgy for High Performance Applications, ed. J. J. Burke and V. Weiss, pp. 57-68, New York, Syracuse Univ. Press (1972).
7. P. U. Gummeson, Powder Met. 15, 67-94 (1972).
8. P. Rao, Shape and Other Properties of Gas Atomized Powders, Ph.D. Thesis, Drexel Univ., Philadelphia, Pa., 262pp (1973).
9. J. K. Beddow, The Production of Metal Powders, Heyden and Son, Ltd., Philadelphia, Pa. (1978).
10. S. Small and T. J. Bruce, Int. J. Powder Metall., 4, pp. 7-17 (1973).
11. P. Rao, R. G. Bowrey and J. A. Tallmadge, Chemeca '70, London, Butterworth, pp. 1-16 (1971).
12. G. H. Johnston and J. B. See, Formation of Liquid Metal Droplets, Paper presented at Symp. on Process Eng. of Pyrometall., Imperial College, London (1974).
13. L. P. Clark, Advanced Manufacturing Methods and their Economic Implications: Some Pilot Papers on Powder Metallurgy and Joining, AGARD Report #627, pp. 1-18 (1975).
14. P. B. Wallis, Powder Met. 8, 167-69 (1976).
15. J. S. Benjamin and J. M. Larson, J. Aircraft 14, pp. 613-23 (1977).
16. G. H. Gessinger and M. J. Bomford, Int. Met. Rev., 19, 51-76 (1974).
17. J. F. Watkinson, Powder Met. 1(2), 13-72 (1958).
18. O. Vorsa, Pikroky Praskove Metalurgie, 1, 47 (Brutcher translation #6572) (1963).
19. K. Tamura and T. Takeda, Trans. Nat. Res. Inst. Metals. 5 (5), 252-56 (1963).
20. P. U. Gummeson, See Ref. 6, pp. 27-55 (1972).
21. R. J. Grandzol, Water Atomization of 4620 Steel and Other Metals, Ph.D. Thesis, Drexel Univ., Philadelphia, Pa. 272pp (1973).
22. T. S. Daugherty, Powder Met. 11, 342-57 (1968).
23. H. Stephan, Advanced Fabrication Techniques in Powder Metallurgy and their Economic Implications, 42nd Meeting of the Structures and Materials Panel (AGARD), Ottawa, Canada, p. N77-15156 (1976).
24. P. W. Sutcliffe and P. H. Morton, See Ref. 23, p. N77-15157 (1976).
25. J. Decours, J. Devillard and G. Sainfort, See Ref. 23, p. N77-15154 (1976).

26. A. R. Cox, J. B. Moore and E. C. Van Reuth, Proc. 3rd Int. Conf. on Superalloys, Claitor's Pub. Div., pp. 45-53 (1976).
27. G. Friedman, See Ref. 23, p. N77-15155 (1976).
28. The Evaluation of Ti-6Al-6V-2Sn Pre-Alloyed Powder Processing, Westinghouse Electric Corporation Technical Report AFML-TR-76-65, 135pp (1975).
29. R. E. Maringer and C. E. Mobley, See Ref. 3, p. 208.
30. C. E. Mobley, R. E. Maringer and L. Dillinger, See Ref. 3, p. 222.
31. A. R. E. Singer and A. D. Roche, in Modern Developments in Powder Met., Editors: H. H. Hausner and P. W. Taubenblat, Vol. 9, p. 127, MPIF, Princeton, N. J. (1977).
32. J. M. Wentzell, See Ref. 23, p. N77-15165 (1976).
33. G. H. Johnston and J. B. See, Unpublished work, MIT (1974).
34. J. B. See, J. Runkle and T. B. King, Met. Trans. 4, pp.2669-73 (1973).
35. N. Dombrowski and W. R. Johns, Chem. Eng. Sci. 18, pp. 203-14, (1963).
36. P. Karinthi, G. Raisson, Y. Honnorat and J. Morlet, See Ref. 31, pp. 115-25 (1977).
37. D. Bradley, Appl. Phys. 6, 1724-36, 2267-72 (1973).
38. R. A. Castleman, Jr., Nat. Bur. [illegible] 369-76 (1931).
39. R. J. Grandzol and J. A. Tallmadge, AIChE J. 19 (6) pp. 1149-58 (1973).
40. R. J. Grandzol and J. A. Tallmadge, Int. J. Powder Met. Powder Tech. 11 (1), pp. 103-14 (1975).
41. A. R. E. Singer, Private Communication (1977).
42. Rapid Solidification Processing, Editors: R. Mehrabian, B. H. Kear and M. Cohen, Claitor's Pub. Div. (1978).
43. P. R. Holliday, A. R. Cox and R. J. Patterson, See Ref. 3, p.246.

On the Microstructure of Rapidly Solidified IN-100 Powders

B. H. KEAR, P. R. HOLIDAY, AND A. R. COX

Powder particles generated by the RSR centrifugal atomization process exhibit at least two distinct types of microstructure. When a high melt superheat is maintained during atomization, the powder particles possess the familiar coarse grained dendritic microstructures. On the other hand, when the melt superheat is reduced by increasing the heat flow to the disc of the rotary atomizer, the powder particles are predominantly microcrystalline in character, with typically one small dendrite particle per grain. It is proposed that the unusual microcrystalline material has its origin in dendrite erosion occurring in a "mushy zone" of dynamic solidification on the disc of the atomizer.

DURING a routine examination of the microstructure of powder particles produced by a newly developed rapid solidification rate (RSR) centrifugal atomization process,[1] it was noted that the particles frequently exhibited a seemingly dendrite free microcrystalline structure, in addition to the more characteristic and familiar coarse grained dendritic structure. This intriguing result led quite naturally to much speculation as to the origin of the microcrystalline material. An important clue to the actual mechanism involved was provided by the observation that the fraction of particles having the microcrystalline structure increased with decreasing melt superheat during atomization. Accordingly, it was decided to prepare two batches of powder representing two extremes of melt superheat during atomization and compare their microstructures. This paper documents the principal findings of the study.

EXPERIMENTAL

A complete description of the apparatus used, including calculations of the cooling rates attainable under different operating conditions, has been given elsewhere.[1] A schematic of the apparatus is shown in Fig. 1. As indicated, it consists essentially of a central rotary atomizer, a system of gas nozzles for quenching the atomized spray, and a convenient arrangement for separation and collection of the atomized product.

For the purpose of the present study, two batches of IN-100* powder were processed according to the specifications given in Table I. One batch of powders (HS-43) was prepared under conditions of high superheat (HS), and another batch of powders (LS-26) was prepared under conditions of low superheat (LS). As indicated in Table I, the rate of heat removal by the water-cooled disc was about three times faster in LS-26 than in HS-43. This translates in the case of LS-26 into a significant reduction in the initial melt superheat during atomization, *i.e.* during the short residence time of the melt on the disc of the rotary atomizer. Calculations based on the information presented in Table I indicated that the melt temperature must have fallen below the liquidus temperature in LS-26, corresponding with about 17 pct solid in the atomized spray. On the other hand, similar calculations for the case of HS-43 showed that the temperature must have remained above the liquidus temperature during atomization; the estimated reduction in superheat in this case was only ~100°C. The observed build-up of solid skull on the disc of the rotary atomizer in the two cases was consistent with the calculations.

The two batches of powders were carefully screened, using fine wire mesh sieves, and separated into differently sized fractions. The yield of particles <105 μ diam was 69 pct in HS-43 powder and 34 pct in LS-26 powder, Table II. The various powder fractions were examined by standard metallographic and SEM techniques.

*9.5Cr, 15.0Co, 5.5Al, 4.5Ti, 3.1Mo, 0.9V, 0.07C, 0.020B, 0.06Zr, Balance Ni.

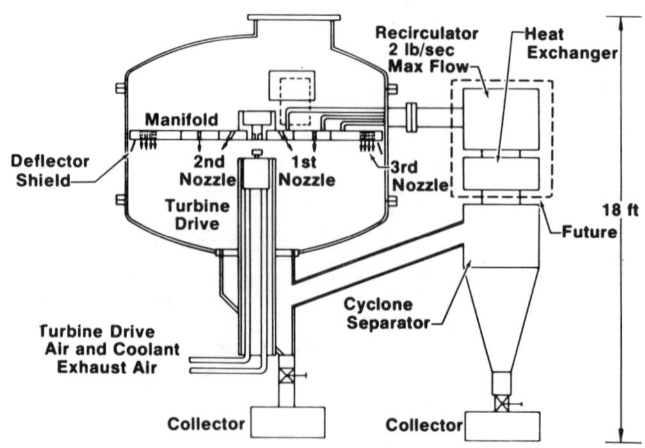

Fig. 1—Schematic of centrifugal atomizer, showing arrangement for high mass flow gas quenching.

Table I. Operating Conditions of Rotary Atomizer

Powder Batch Number	Melt Atomization Rate, kg/s	He Flow Rate, kg/s	Rim Speed, m/s	Initial Pour Temp, °C	Heat Flow to Water, kJ/s
HS-43	0.152	0.98	105.3	1604	12.2
LS-26	0.156	0.78	92.2	1538	31.8

B. H. KEAR is with United Technologies Research Center, East Hartford, CT 06108. A. R. COX is with Government Products Division, United Technologies, West Palm Beach, FL 33402, and P. R. HOLIDAY, formerly with Government Products Division, is now with United Technologies Research Center.

Manuscript submitted August 25, 1977.

Table II. Particle Size Distribution

Mesh Size (U.S. Standard)	Particle Size, μ	Wt. Pct Powder HS-43	Wt. Pct Powder LS-26
+20	>850	0.72	16.92
-20 + 80	-850 + 177	2.07	15.87
-80 + 140	-177 + 105	28.17	33.50
-140 + 230	-105 + 62	43.42	19.75
-230 + 325	-62 + 45	16.98	7.82
-325	<45	8.64	6.14

RESULTS

Initial observations were conducted by scanning electron microscopy. Most of the powder particles, irrespective of particle size, had a distinctly rounded, and frequently near perfect spheroidal morphology. Powder particles from batch HS-43 exhibited a well defined dendritic structure, Fig. 2(*a*), whereas LS-26 powders had a more granular appearance, with what appeared to be intergranular surface porosity, Fig. 2(*b*). Some particle aggregation was noted, which was apparently the result of 'sticking' collisions between solid or semisolid particles, Fig. 3. In addition, collisions involving molten droplets and previously solidified, chilled particles also produced some genuinely splat quenched material. In HS-43, the splats typically took the form of very thin, featureless caps, Fig. 4(*a*), whereas in LS-26 they were more irregular in shape and contained visible particles, Fig. 4(*b*).

The same powders examined by conventional metallographic techniques showed internal structures that correlated well with the surface structures revealed by SEM. Thus, HS-43 powder particles exhibited

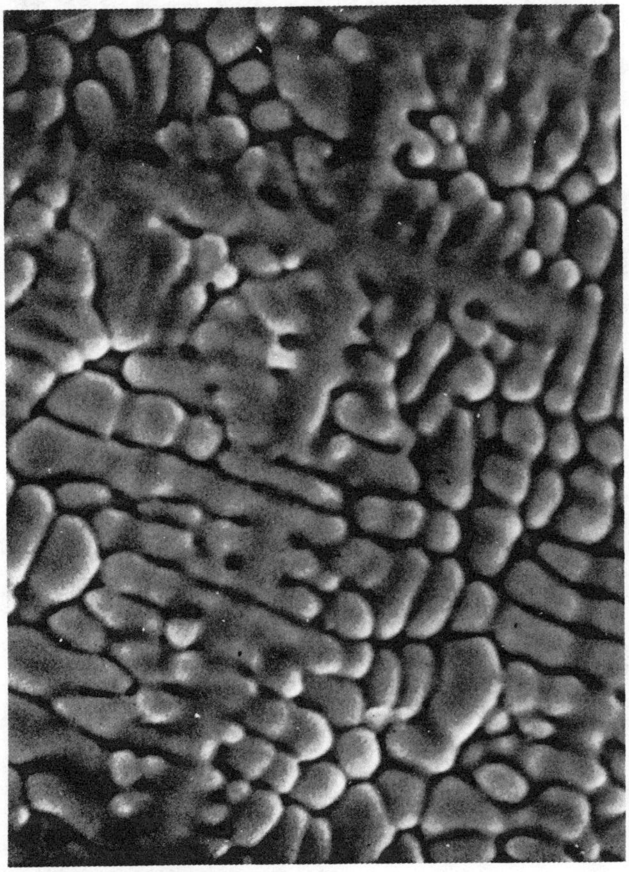

(*a*)

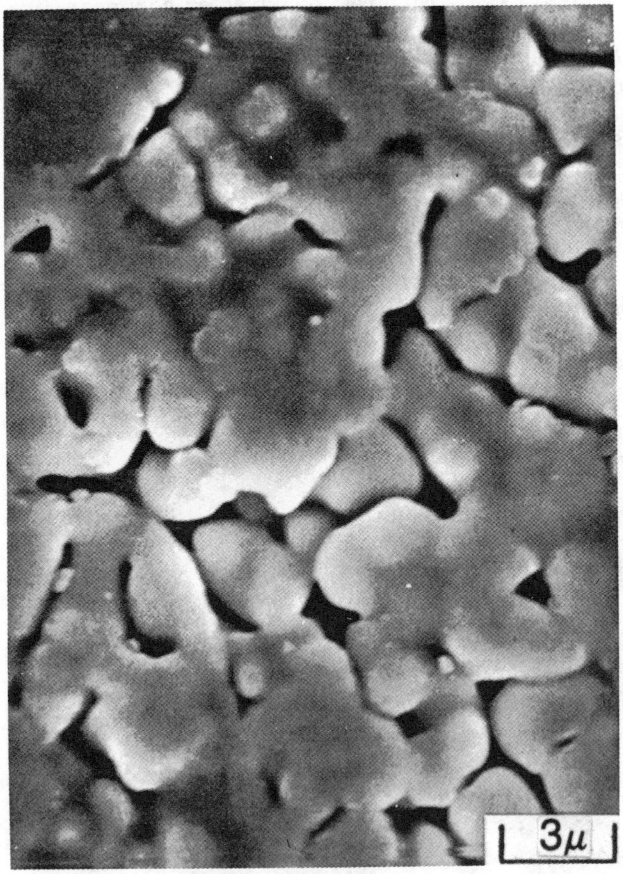

(*b*)

Fig. 2—Representative SEM photomicrographs showing: (*a*) dendritic structure of HS-43 powder, and (*b*) granular structure of LS-26 powder.

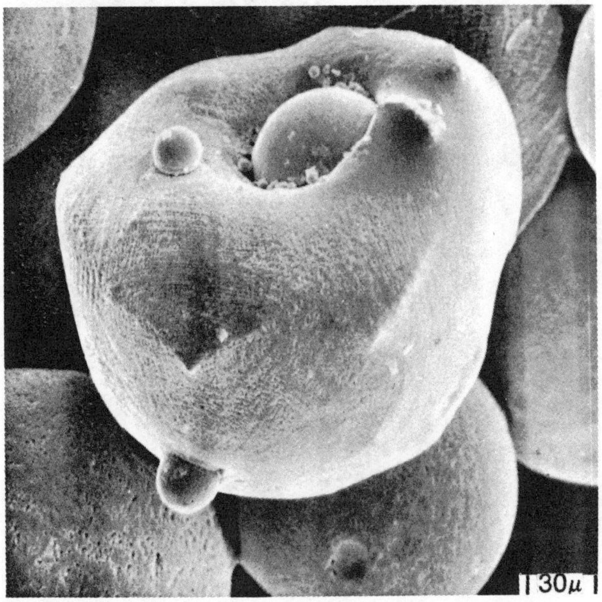

Fig. 3—Evidence for "sticking" collisions between several small solid particles and a much larger semisolid, "mushy" particle.

coarse grained structures, equiaxed or columnar-like, composed of well organized colonies of dendrites, Fig. 5(*a*), whereas LS-26 powder particles showed fine, equiaxed grain structures, with typically one dendrite particle per grain, Fig. 5(*b*). Moreover, there was a discernible trend towards finer dendritic structures, *i.e.* smaller secondary dendrite arm spacings, with decreasing particle size in HS-43 powder, but this was not the case in LS-26 powder. On the contrary, it appeared that in the LS-26 powder there was much less of a change, if any at all, in the appearance of individual dendrites and in the scale of the corresponding grain structure with decreasing particle size. The grain size, however, did vary somewhat from particle to particle in all size ranges of the screened powder. Direct imaging of the two essentially different types of grain structure was obtained by the electron channeling technique, Fig. 6.

Under the optical microscope, HS-43 splats were seen to be fairly uniform in thickness, Fig. 7(*a*), although varying in thickness from one splat to the next. The thinnest sections were generally quite featureless in appearance, whereas thicker sections

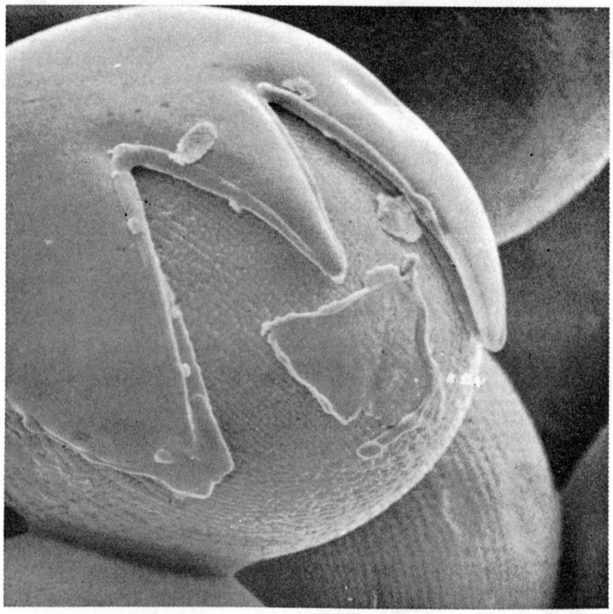

(*a*)

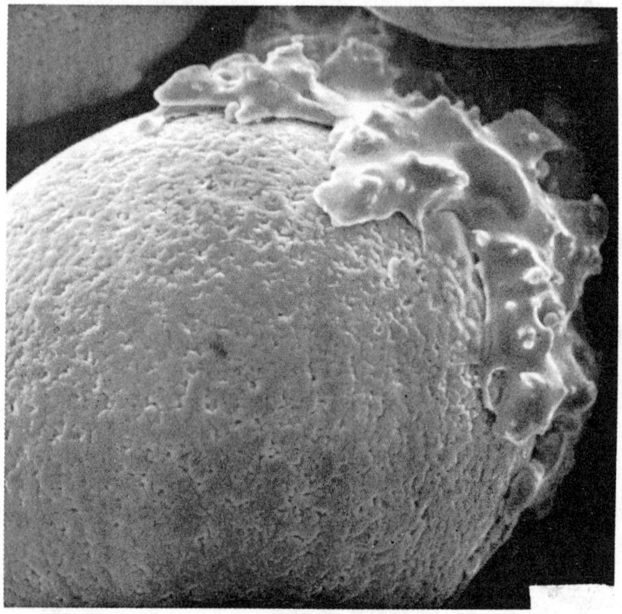

(*b*)

Fig. 4—Evidence for splat quenching of: (*a*) fully molten droplet, and (*b*) partially solidified, "mushy" droplet, on previously solidified particles.

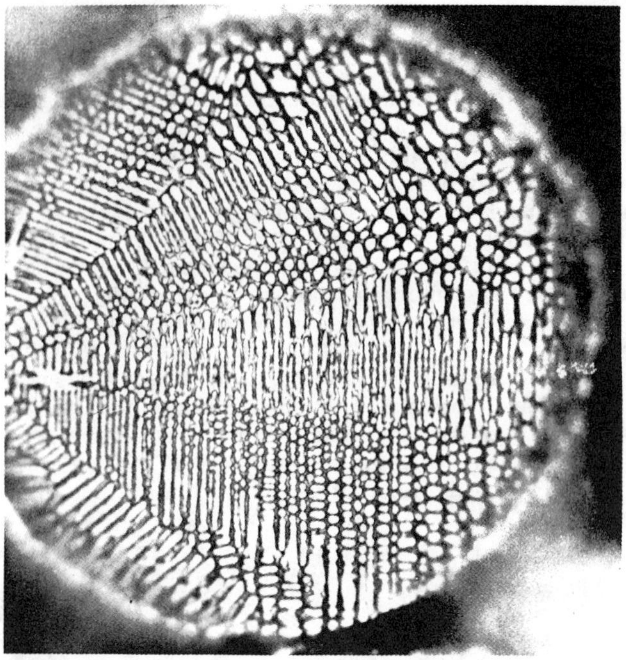

(*a*)

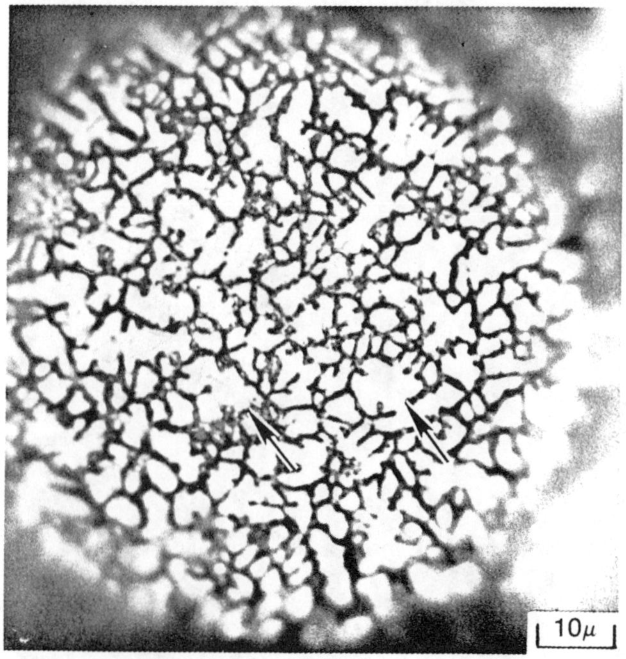

(*b*)

Fig. 5—Representative optical photomicrographs showing: (*a*) columnar-grained dendritic structure in HS-43 powder, and (*b*) fine grained equiaxed structure in LS-26 powder, with evidence for one "cored" dendrite particle per grain.

showed indications of phase decomposition. Even though no phase separation was detected in the very thin splat quenched material, some evidence was obtained for a columnar grain structure. In complete contrast, LS-26 splats had very irregular shapes, and contained many small particles, frequently with dendritic morphologies, embedded in an otherwise featureless matrix, Fig. 7(*b*); thicker sections showed visible structure in the matrix, but on a very fine scale.

Another feature of interest was the obvious difference in etching behavior of the dendrites in the two batches of powder. In the coarse grained dendritic structure characteristic of HS-43 powder, the center of each dendrite was somewhat darker in appearance than its periphery, Fig. 5(*a*). On the other hand, in the fine grained LS-26 powder, the individual dendrites exhibited a layered structure of alternating light and dark contrast, Fig. 5(*b*). Occasionally, a small dot was visible at the center of such a layered dendritic structure. This was taken to be the point of initial nucleation of the growing dendrite. Despite the presence of such "cored" structures in the dendrites of both LS-26 and HS-43 powders, no evidence was found for significant concentration variations within the dendrites,

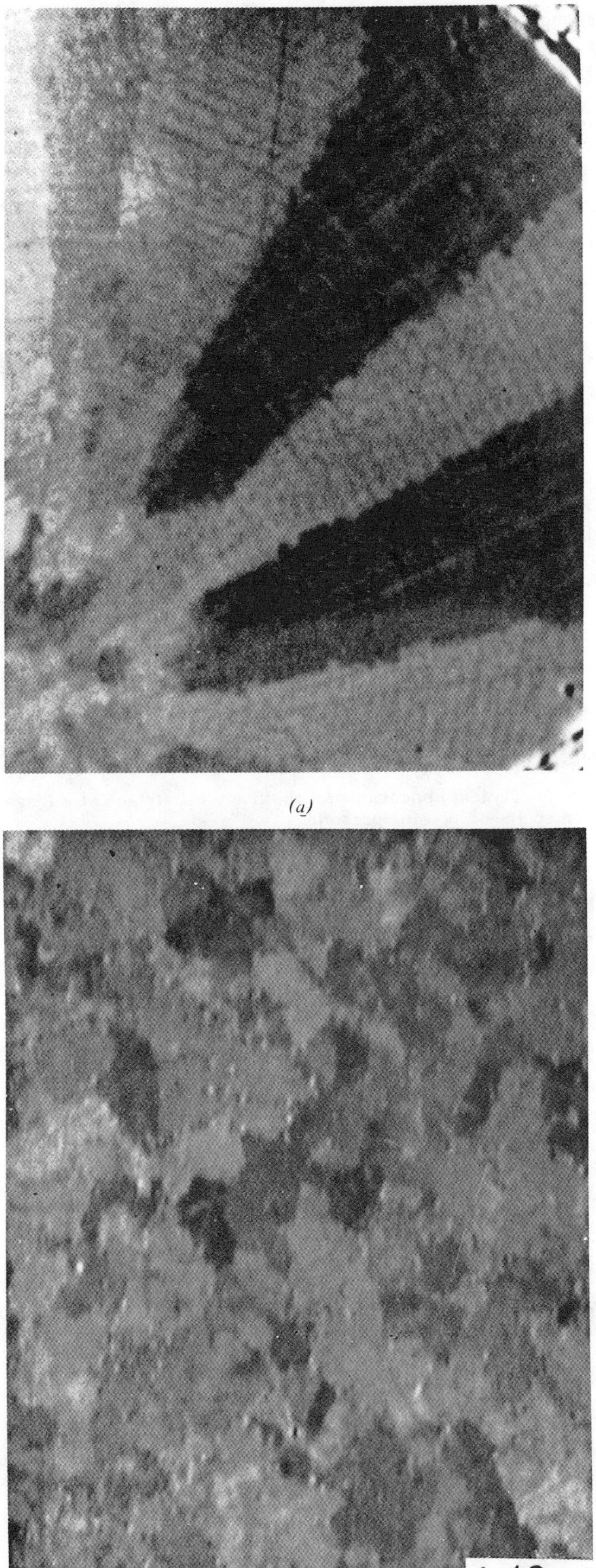

(*a*)

(*b*)

Fig. 6—Electron channeling contrast showing: (*a*) typical columnar grain structure of HS-43 powder, and (*b*) typical microcrystalline structure of LS-26 powder.

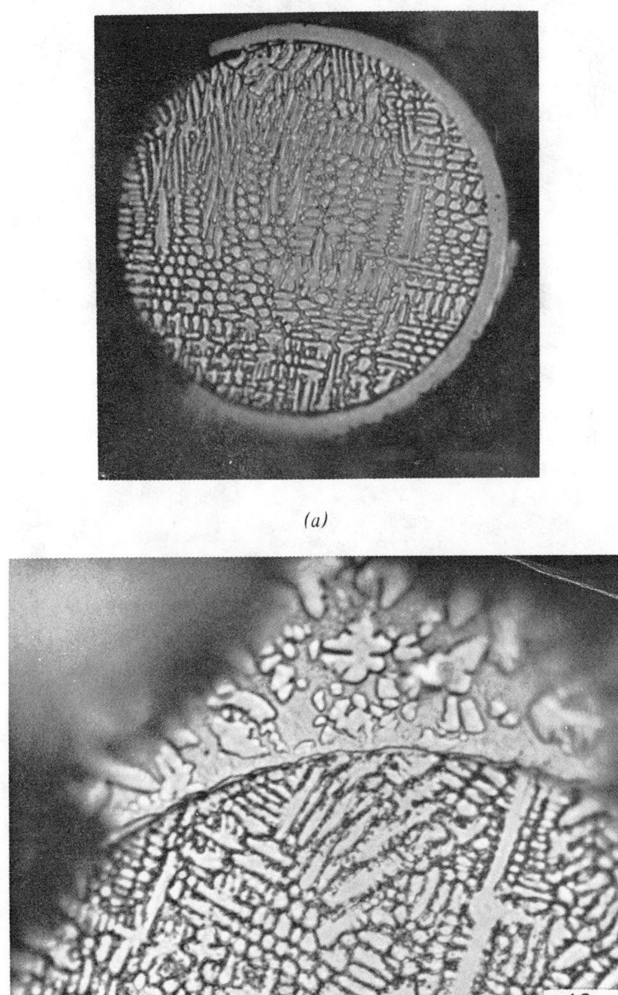

(*a*)

(*b*)

Fig. 7—Optical photomicrographs showing: (*a*) uniformly thin, featureless splat in HS-43 powder, and (*b*) irregularly shaped splat containing visible dendrite particles in LS-26 powder.

perhaps because of insufficient instrumental sensitivity. On crossing into the interdendritic regions, however, a sharp increase in Ti and Al concentrations was detected. In a few cases, the increase in Ti was correlated with the presence of Ti-rich MC carbide phase.

Evidence for elastic collisions between particles was found in both batches of powders. Ordinarily, the effect of elastic collisions was to leave small depressions on the surfaces of the colliding particles, but occasionally the impact was severe enough to cause extensive cracking, Fig. 8. Fracture along grain boundaries was the typical failure mode, although some interdendritic cracking, *i.e.* cracking between primary dendrites within a grain, was also observed. Columnar dendritic structures, such as that shown in Fig. 5(*a*) appeared to be particularly prone to such cracking, and several examples of actual particle fragmentation were seen, Fig. 9. Apparently, in

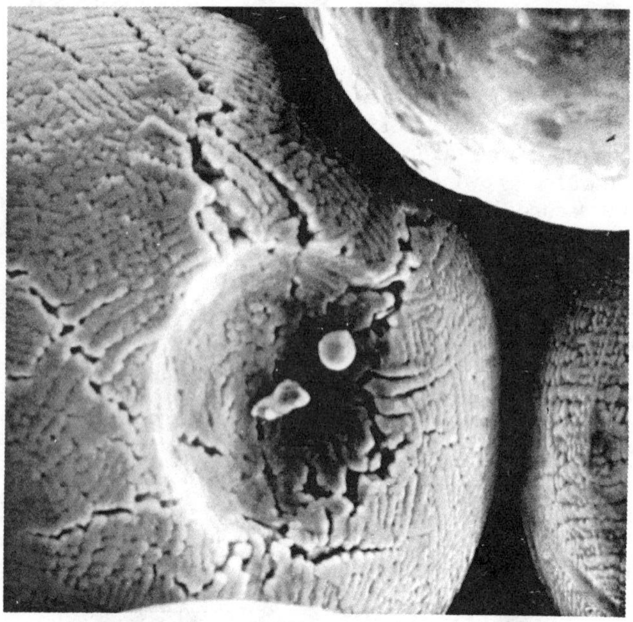

(*a*)

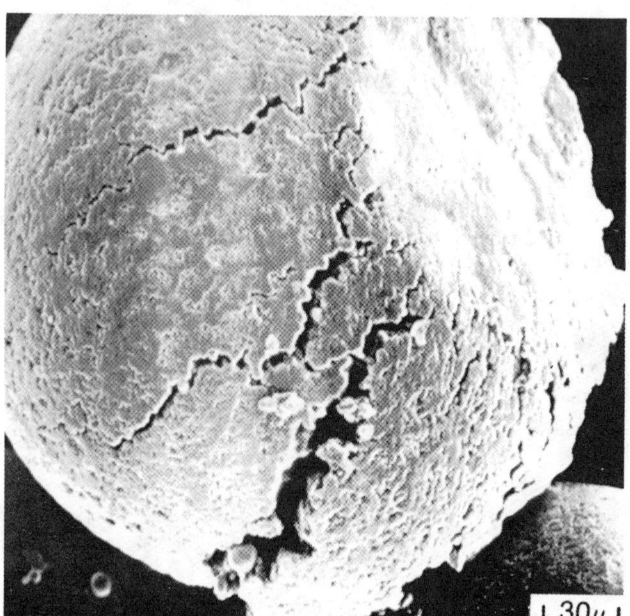

(*b*)

Fig. 8—Evidence for elastic collisions between solid particles, involving grain boundary cracking: (*a*) HS-43 powder particle, and (*b*) LS-26 powder particle.

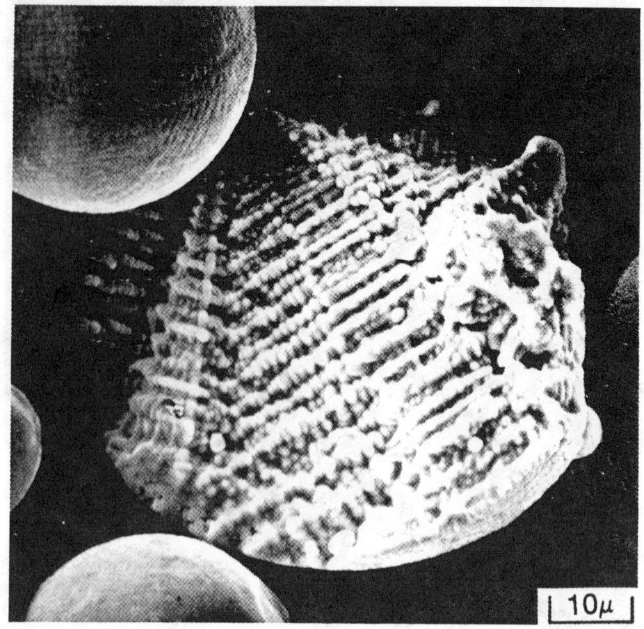

Fig. 9—Typical appearance of the fracture surface of a fragmented HS-43 powder particle.

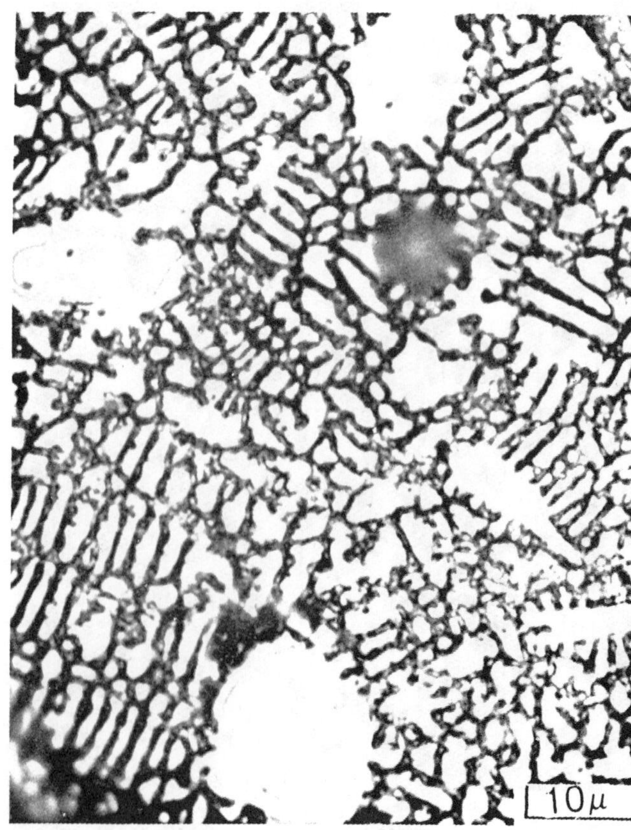

Fig. 10—Bimodal distribution of "cored" dendrite particles in LS-26 powder.

such cases, crack propagation is facilitated by the regular manner in which the columnar grains extend from one side of the particle to the opposite side. Presumably, grain boundary cracking in both batches of powders is most favored when the material is in the "hot short" condition, *i.e.* shortly before or just after completion of solidification.

Although the microstructure of the majority of particles in both HS-43 and LS-26 powders was as described above, it must be emphasized that in each case a smaller fraction of the particles also had structures that were comparable with that found in the other batch. Thus, in HS-43 powder, a small fraction of the particles, in all size ranges, possessed a fine, equiaxed grain structure similar to that found in LS-26 powder, and vice versa. A small fraction of particles exhibited microstructures that did not fit the broad classification into two distinct types, as described above. One of the more interesting features was the fairly frequent occurrence of mixed dendritic and microcrystalline structures. Many examples were also found of powder particles that contained unusually large isolated "cored" dendrites, Fig. 10.

DISCUSSION

In the LS-26 powder, the large majority of the particles possess a fine grained equiaxed, or microcrystalline structure, with typically one dendrite particle per grain. This is easily explained in terms of dendrite prenucleation on the disc of the atomizer. A high rate of heat extraction by the disc of the atomizer is the special feature of the process used to produce this particular batch of powder. Moreover, as noted earlier, calculations indicate that the melt temperature must have fallen below the liquidus temperature, at least by the time the melt reaches the periphery of the disc. Accordingly, it can be argued that the edge of the disc develops a "mushy" layer, supported by a solid layer of skull. Under steady state conditions of atomization, it is envisaged that this mushy layer is continuously being formed and eroded away in a state of dynamic equilibrium. Thus, a continuous spray of "mushy" droplets is generated around the rim of the disc. Since a major fraction of the particles generated will be in the mushy, prenucleated condition, *i.e.* containing small fragments of predendrites, the resulting material after completion of solidification will be predominantly microcrystalline in character as, in fact, observed. Some coarse grained, fully dendritic material will also be produced, because a small portion of the superheated melt will tend to ride over the top of the mushy layer without necessarily picking up predendrities, or, in fact, if it does so it is conceivable that the predendrites will undergo resolution in the superheated melt. The size, number and distribution of predendrites will vary from particle to particle, but not necessarily with particle size, which is in accord with the observed variability in grain size of the microcrystalline material, and its apparent insensitivity to particle size. Particles containing very large predendrites can also be understood in terms of a longer residence time for dendrite growth in the mushy layer, just prior to atomization. Interestingly enough, the model also provides some basis for explaining the curious "cored" structures seen in the heavily etched dendrites. It is proposed that this effect is merely a reflection of a discontinuous growth process for individual predendrite particles, which is to be expected, since the local environment surrounding a dendrite particle must always be changing, at least slightly, both with respect to composition and temperature on the surface of the atomizer disc.

In the HS-43 powder, the major fraction of the particles exhibit a coarse grained dendritic structure, with only a small fraction of the particles having a microcrystalline structure. This situation is the inverse of that found in the LS-26 powder, but nevertheless, the same line of reasoning can be used to explain it. Thus, it can be argued that in the HS-43 case the preponderance of particles with coarse grained structures is a consequence of a major portion of the atomized melt having retained the initial melt superheat, by virtue of the lower rate of heat extraction by the atomizer disc used in this batch. The much smaller fraction of the atomized material in the microcrystalline condition, on the other hand, may be attributed to erosion of a mushy layer on the rim of the disc, albeit in this case much attenuated in its ability to seed particles with predendrites relative to the LS-26 case. When fully molten, superheated particles undergo rapid solidification in the coolant gas stream, the scale of the resulting microstructure should be strongly influenced by particle size. This is because the effectiveness of convective heat transfer in generating high cooling rates increases with decreasing particle size ($T \sim d^{-1.7}$). The observed refinement in dendritic structure with decreasing particle size is consistent with the predicted trend in cooling rate.

This picture of the essential difference in the nature of the atomization process in the two cases draws strong support from the splat quenching observations. In the case of the HS-43 splats, Fig. 7(*a*), evidently the particles must have been in the fully molten superheated condition at the time of collision with the substrate particles. The force of the impact, coupled with the high fluidity of the superheated melt, ensures that the melt spreads out evenly over the substrate particle to form a uniformly thin splat quenched layer. Intimate contact between the melt and the clean, chilled substrate particle gives rise to very efficient conduction heat transfer, with corresponding high cooling, or melt quenching rates. The average cooling rate clearly will depend sensitively on the thickness of the splatted layer—the thinner the layer the faster the cooling rate. Thick splats show visible phase decomposition in the form of a fine distribution of MC carbide particles. Thin splats are generally quite featureless in appearance, except for the indication in heavily etched material for the presence of a fine columnar-grained structure. Such variations in structure with thickness of splat seem reasonable, since the tendency to form an extended solid solution phase, *i.e.* complete suppression of phase decomposition, should increase with the cooling rate. In the case of the LS-26 splats, the evidence is clearly in favor of the particles having a mushy character at the time of collision with the substrate particles. Perhaps the most compelling evidence is provided by the direct observation of predendrite particles in an otherwise featureless matrix, Fig. 7(*b*). Such a microstructure is just what would be expected from splat quenching of mushy material, since only the residual melt in the particles should be affected by fast cooling. As in the HS-43 case, the residual melt forms an extended solid solution phase, with varying degrees of phase decomposition, depending on the cooling rate.

In most of the splat quenched LS-26 material examined, the large majority of the predendrites have a distinctly blocky morphology, rather than the more familiar tree-like appearance of dendrites. Moreover, it appears that all the predendrites, whatever their morphology, are discrete and remain separated from one another in the semisolid material. Such a structure is reminiscent of that produced by the processing technique known as rheocasting,[2] in which ingots of semisolid material containing a high volume fraction (~50 pct) of spheroidal dendrites are produced by subjecting the melt to a high rate of shear during cooling through the liquidus temperature. This raises the question as to the possibility of using the centrifugal atomizer, operating under appropriately low melt superheat conditions, as a method for generating bulk quantities of mushy material. Test experiments, using a probe placed in the path of the atomized spray, indicate that *in-situ* collection and consolidation of the material is feasible. Further attempts to develop and to exploit this new approach to mushy casting are contemplated.

In metallic systems, undercooling of the melt prior to solidification exerts a profound influence on dendrite morphology and segregation. This has been shown for a variety of high purity and commercial nickel- and iron-base alloys.[3,4] At low undercoolings, the dendrites assume the characteristic fern-like morphology. At moderate undercoolings, 60 to 170 K, the dendrites are composed of cylindrical arms elongated along the three equivalent ⟨100⟩ directions. Under extreme conditions of undercooling, > 170 K, the dendritic structures have a distinctly spheroidal morphology. At the same time, with increasing undercooling, the degree of solute segregation progressively diminishes. Of particular significance is the occurrence, at some critical degree of undercooling, of solidification without segregation. This corresponds with the condition that C_p (T solidus $-$ T nucleation) $\geq H$, where C_p = average specific heat of solid and liquid and H = latent heat of fusion. Under less severe conditions of undercooling, only the dendrite cores experience solidification without segregation.[6-8] Due to the recalescence effect, the temperature of the material surrounding the dendrite cores rises above the solidus temperature, and then undergoes subsequent cooling accompanied by normal segregation behavior. Referring to Fig. 5(*a*), it appears that the main features exhibited by the dendritic structure in HS-43 are consistent with a moderate undercooling effect, for two reasons. First, the dendrite arms are elongated in the three equivalent ⟨100⟩ directions. Second, the dendrite cores exhibit a different etching behavior compared with the surrounding matrix material, which may be interpreted as evidence for dendritic growth without segregation, followed by interdendritic solidification in a conventional segregation mode. Unfortunately, it was not possible using existing microanalytical capabilities to determine the segregation profile within this very fine-scale dendritic structure; this must await the availability of the high spatial resolution capabilities of STEM. Referring to Fig. 5(*b*), it seems reasonable to ascribe the observed microcrystalline structure to homogeneous nucleation under extreme conditions of undercooling. However, for the reasons discussed above this does not appear to be the correct interpretation of the observations. On the contrary, all the evidence points to a prenucleation mode of solidification as a result of dendrite erosion occurring in a mushy zone of dynamic solidification on the disc of the rotary atomizer. Recently, a completely different type of microcrystalline material, both in general appearance and in scale, has been observed in certain superalloys (notably those low in carbon) processed under high melt superheat conditions, such as used in the processing of HS-43. All the evidence to date indicates that this is a genuine undercooled microcrystalline structure formed by homogeneous nucleation. This topic will be the subject of a forthcoming paper.

ACKNOWLEDGMENTS

Part of this work was sponsored by the Advanced Research Projects Agency under ARPA Order 3152. The monitoring agency is the Air Force Materials Laboratories, Wright-Patterson Air Force Base, Ohio.

REFERENCES

1. A. R. Cox, J. B. Moore, and E. C. Van Reuth: *Proc. Third International Symposium on Superalloys*, p. 45, Seven Springs, PA, Claitor's Publishing Division, Baton Rouge, LA, 1976.
2. R. Mehrabian and M. C. Flemings: *Proc. Materials Science Symposium*, p. 98, ASM, Detroit, 1974.
3. T. Z. Kattamis and M. C. Flemings: *Trans. TMS-AIME*, 1965, vol. 233, p. 992.
4. T. Z. Kattamis and R. Mehrabian: *J. Vac. Sci. Technology*, 1974, vol. 11, p. 1118.
5. T. Z. Kattamis and R. Mehrabian: *J. Mater. Sci.*, 1974, vol. 9, p. 1040.
6. T. Z. Kattamis, W. F. Bower, and R. Mehrabian: *J. Cryst. Growth*, 1973, vol. 19, p. 229.
7. S. P. Iyer and W. V. Youdelis: *J. Inst. Metals*, 1973, vol. 101, p. 218.
8. H. Biloni and B. Chalmers: *Trans. TMS-AIME*, 1965, vol. 233, p. 373.

Principles of Fabrication

SHELDON KAVESH
Allied Chemical Corporation
Morristown, New Jersey 07960

Metallic glasses and the techniques by which they are fabricated are the remarkable results of two lines of modern inquiry into ancient arts. On the one hand, glassy metals are members of the venerable family of metastable alloys. Their discovery by Duwez stemmed from investigations of metastable alloys pursued to extreme quenching conditions. On the other hand, metallic glasses are fabricated into useful wire and ribbon forms by direct melt processing techniques, an outgrowth of efforts to bypass the age-old limitations of rolling and die drawing.

In this paper, a close examination will be made of two published and possibly characteristic techniques for fabrication of glassy alloys—chill block casting of ribbon and free jet spinning of wire. The task at hand will be deduction and description of the principles of fabrication. The following broad principles of glass formation will be accepted as axiomatic:

1 The fabrication technique must cool the glass forming melt at a rate greater than its "critical quench rate".
2 The fabrication technique must permit cooling of the glass forming alloy to a final state at or below its recrystallization temperature or, more severely, at or below its glass transition temperature, T_g.

The first condition is of course a statement that the glassy state is achieved, and the second is a statement that the product is recovered in that state.

Chill Block Casting of Glassy Alloys

The term "chill block casting" describes a process in which a free jet of molten material is impinged onto a moving chilled substrate.

A ribbon is formed by attenuation and solidification of the jet by the moving substrate. The process is characterized by high lineal speeds, high volumetric throughput and high quench rates for metallic materials (10^5 to 10^8°C/sec). This particular combination of characteristics has prompted several adaptations for fabrication of metallic glasses.

The first description of a chill block casting technique appears to have been by Strange and Pim (1) in 1908. Modifications and improvements of the essential technique have been reported by Staples (2), Nagy (3), Merle (4) and, in more recent years, by Pond (5) and co-workers (6-9). Adaptations for preparation of metallic glasses have been described by Duwez (10), Bedell (11), and Chen and Miller (12).

The technological status of chill block casting was reviewed by Mobley (13). In a related area, Jones (14) has reviewed the process physics of splat cooling. Liebermann and Graham (15) have presented a study of the effects of chill block casting conditions on ribbon dimensions.

A number of basic and applied questions remain unresolved. How does the ribbon form? What are the energetic, thermal and hydrodynamic constraints on processing? What are the fundamental relationships of ribbon dimensions to casting conditions? What are the nature and effect of dynamic instabilities which may be present?

Description of the System

Figure 1 illustrates one embodiment of the chill block casting technique as described by Mobley (13). A molten alloy is extruded through an orifice by means of gas pressure to form a molten jet. The jet impinges on a chilled moving substrate (e.g., a rotating disc) and is attenuated and solidified to form a ribbon. The geometry of the impingement zone is illustrated schematically in Fig. 2. The jet of radius a, cm, may be inclined at an angle θ to the tangent plane of the chill surface. The jet, on impacting the chill surface, spreads into a puddle of width w, cm. The chill surface moving at velocity V, cm/sec, creates a ribbon of thickness t, cm, whose width is the same as the width of the puddle.

Possible Modes of Ribbon Formation

It is helpful to list for further consideration some of the possible modes of ribbon formation and likely determinants of ribbon dimensions.

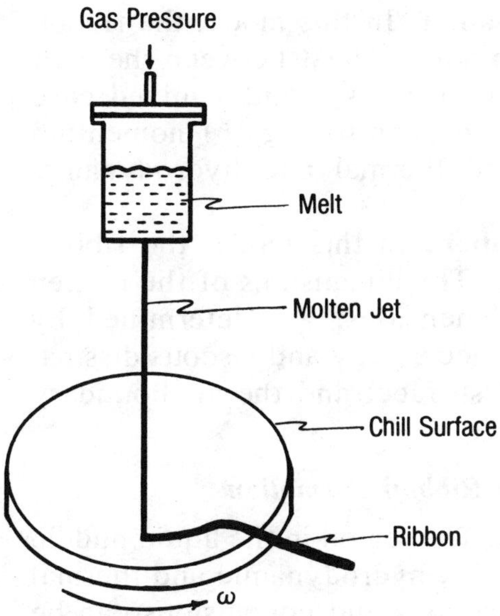

Fig. 1. Schematic of chill block casting (after Mobley)

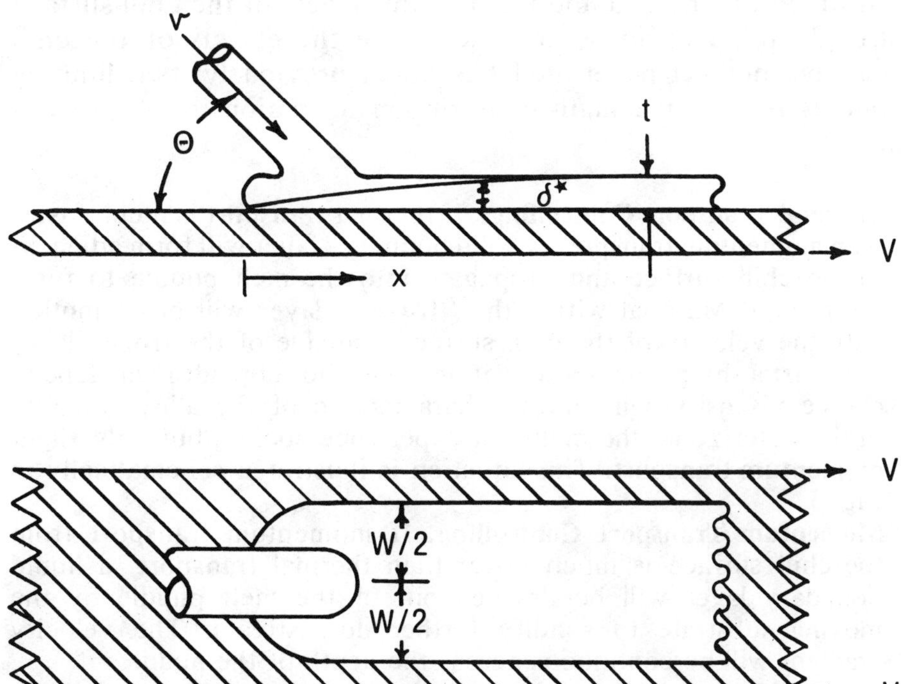

Fig. 2. Ribbon formation by chill block casting

Transport Dominated Ribbon Formation. In this mode, the ribbon is formed by transport of momentum and/or heat between the melt puddle and the chill surface. Two limiting cases and a mixed case can be distinguished: (*a*) thermal transport controlling, (*b*) momentum transport controlling and (*c*) coupled thermal and hydrodynamic regime.

Energy Equilibrium Ribbon Formation. In this mode, the ribbon is formed by wetting of the substrate. The dimensions of the molten puddle, and therefore the ribbon dimensions, are determined by equilibrium of the kinetic energy, surface energy and viscous dissipation of the jet, the puddle, the chill surface, and the air boundary layer in motion with the chill surface.

Consideration of Transport Dominated Ribbon Formation

Determination of Dominant Process. The region in the liquid puddle at the foot of the jet is clearly a complex hydrodynamic and thermal regime. However, important simplifications and conclusions can be obtained upon application of the boundary layer concepts of Schlichting (16).

The liquid puddle is visualized as consisting of a "boundary layer" zone in which the thermal and momentum effects of the chill surface are strongly felt and an outer zone where the effects of the chill surface have not yet penetrated. As noted previously, two limiting possibilities for the mechanism of ribbon formation can be distinguished:

1 **Thermal Transport Controlling.** If thermal transport is much faster than momentum transport, a solid boundary layer will form adjacent to the chill surface and propagate into the melt puddle to form the ribbon. Material within the "frozen" layer will be in motion with the velocity of the chill surface. Outside of the frozen layer will exist a sharp velocity transition zone whose breadth will depend on the viscosity-temperature characteristic of the alloy. Finally, in the outer zone, the melt will experience cooling but only slight momentum transport. This situation is illustrated schematically in Fig. 3.

2 **Momentum Transport Controlling.** If momentum transport from the chill surface is much faster than thermal transport, a liquid boundary layer will be dragged out of the melt puddle by the moving substrate to solidify farther downstream. The velocity gradient will be continuous across the depth of the puddle.

Determination of which of the limiting cases is likely to exist is based on the transport properties of the melt. Schlichting (16) has

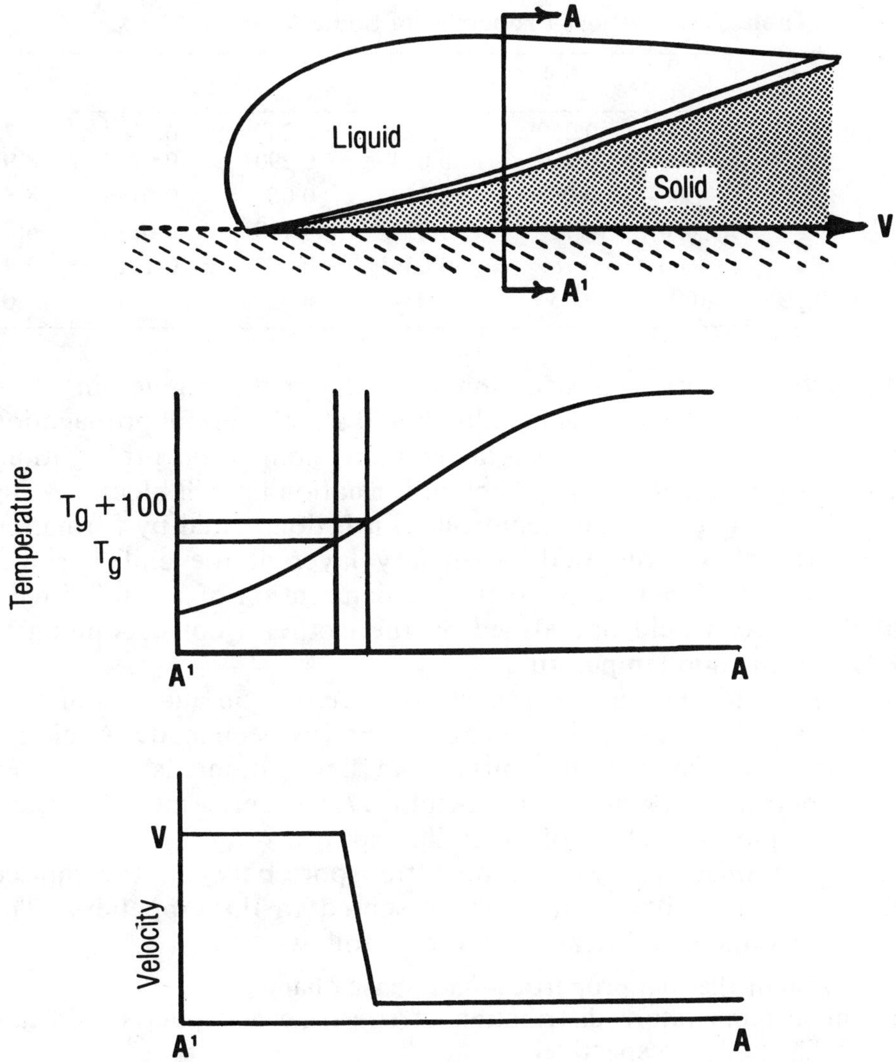

Fig. 3. Schematic temperature and velocity profiles through puddle

performed the analysis and shows that depth of penetration of thermal and momentum effects in a fluid near a solid boundary is given by

$$\frac{\delta_T}{\delta_M} \sim (\mathrm{Pr})^{-1/2} \qquad [1]$$

where Pr is the Prandtl number of the fluid, $\mathrm{Pr} = C_p \mu / k$, dimensionless; δ_T is the thermal boundary layer thickness, cm; and δ_m is the momentum boundary layer thickness, cm.

Table 1. Transport Properties of Some Metallic Melts

Metal	T, °C	C_p, cal/g-°C	μ, poise	k, cal/sec-cm-°C	Pr	$Pr^{-1/2}$
Aluminum	600	0.259	0.029	0.247	0.0304	5.7
	1000	0.259	0.014	0.290	0.0125	8.9
Tin.	300	0.058	0.019	0.08	0.0138	8.5
Zinc	450	0.12	0.032	0.138	0.0278	6.0
Iron	1600	0.189	0.056	0.075	0.141	2.7
$Fe_{40}Ni_{40}P_{14}B_6$. .	1000	0.13	~0.03	0.05	0.078	3.6

A short list of properties for some metallic melts is given in Table 1. It will be seen that for the metallic liquid state, thermal propagation is roughly three to nine times faster than momentum propagation. It is therefore expected that, if ribbon formation in chill block casting of glassy metals is transport controlled, it is dominated by formation and growth of a "solidified" boundary layer at the chill surface. In the context of metallic glasses, the penetration of a "solidified" boundary layer would be defined by the isotherm corresponding to the glass transition temperature.

Description of Thermal Transport. Apparently, an analysis of heat transport in chill block casting has not previously been made. Available studies of solidification in semi-infinite liquid domains have been used for purposes of guidance (13). Ruhl (17) has performed numerical studies of splat quenching of crystalline alloys.

An approximate analysis of heat transport between the molten puddle and the chill substrate is presented in the Appendix. The assumptions and simplifications were as follows:

1 Constant thermal properties—no phase change.
2 Initial temperature distribution of substrate and puddle constant at T_0 and T_m, respectively.
3 Negligible cooling at top surface of puddle.
4 Bottom temperature of chill block maintained at T_0 by circulation of coolant.
5 Negligible viscous heating in transition zone between solidified boundary layer and liquid puddle.
6 Constant thickness of puddle.
7 One-dimensional heat flow—edge effects were not considered.
8 Negligible contact resistance at interface between puddle and chill block.

Assumptions 1 to 5 are reasonably supported by data and calculations. Assumption 6, though not strictly true, is believed to be a justifiable simplification since the effect of puddle thickness is greatest near the end of solidification, where puddle thickness is well defined. Assumptions 7 and 8 are not justified. Work is under way to relax these restrictions.

The results of the present analysis are given by equations A1 and A2 in the Appendix and are summarized in Fig. 4 and 5. Several aspects of these results are worth comment. First, it should be noted that the appropriate figure of merit for the material of the chill substrate is the product of its thermal conductivity, density and heat capacity ($k\rho C_p$) rather than its thermal diffusivity ($k/\rho C_p$). This point has been indicated by others but appears not to have become generally known. On the other hand, copper has a higher ($k\rho C_p$) product than other metals, and indeed copper has been the prototypical chill substrate.

Secondly, it will be noted that the shape of δ^* vs time in Fig. 4 is concave upward. In this respect the finite puddle behaves rather differently from a semi-infinite liquid. At very short times, the slope of log δ^* vs log Θ is 1/2, as it is for the semi-infinite liquid. However, as the heat content of the finite puddle is exhausted, the boundary

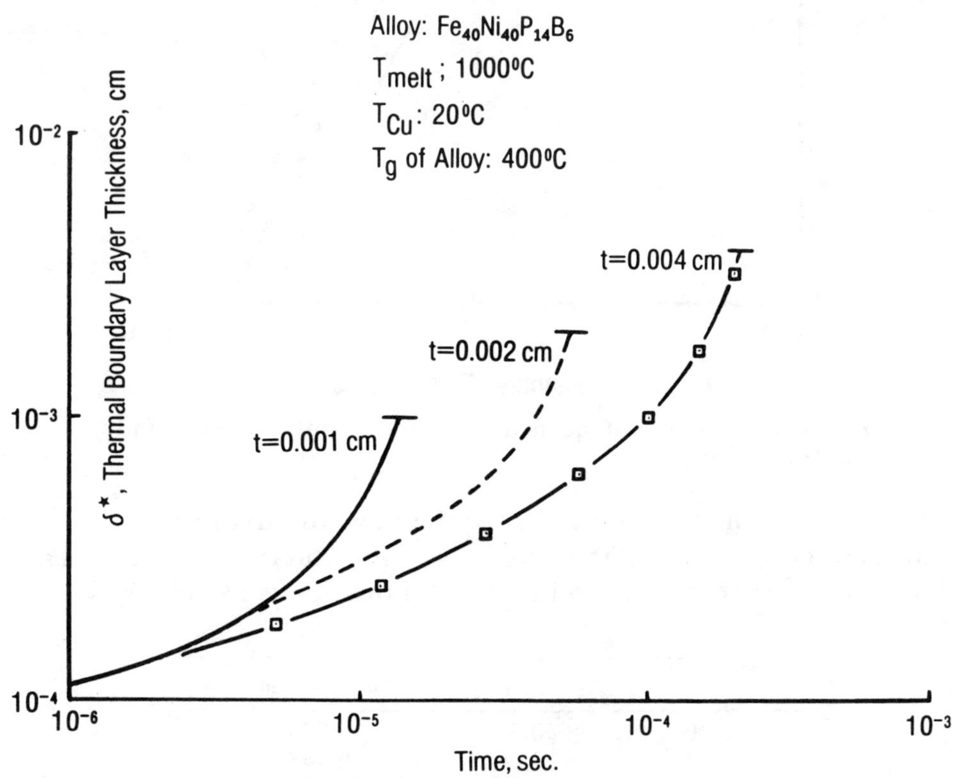

Fig. 4. "Solidified" boundary layer thickness vs time

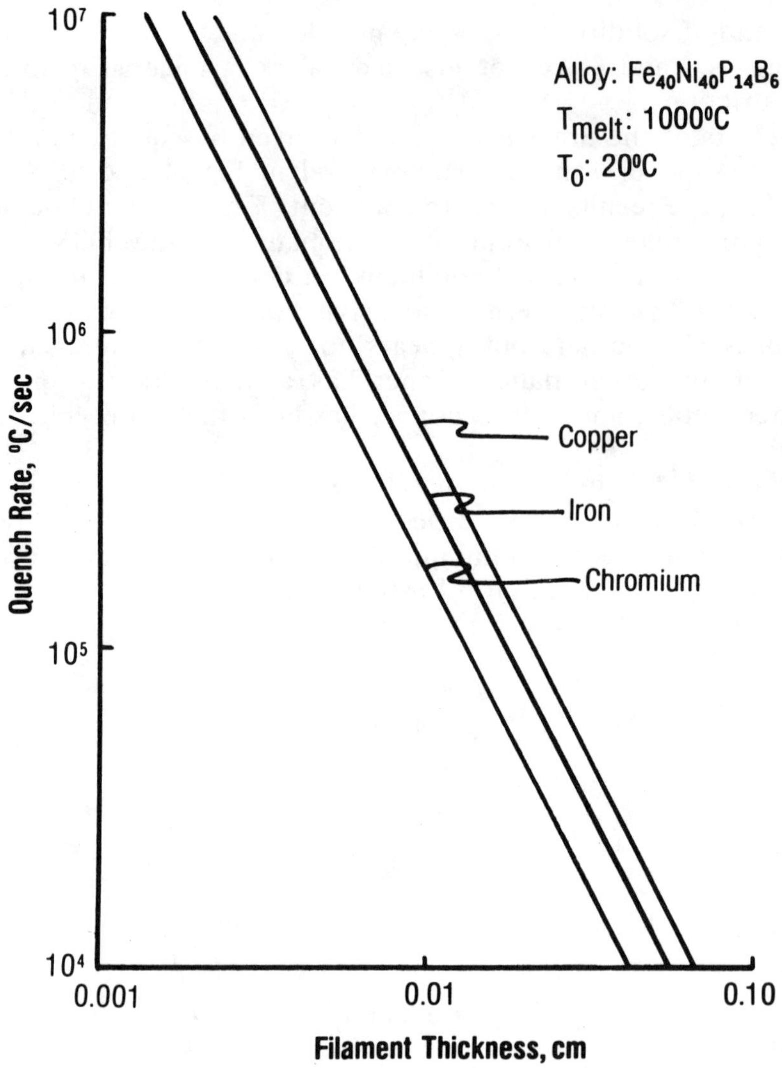

Fig. 5. Relationship of quench rate of filament thickness and substrate character

layer propagates more rapidly; hence the upward curvature.

An average (secant) slope over the time interval between 1% solidification ($\delta^*/t = 0.01$) and total solidification is as follows:

t, cm	m	$n = 1/2 - m$
0.001	0.711	0.775
0.002	0.664	0.748
0.004	0.637	0.734

m is the exponent in the expression

$$\delta^* = c(\Theta)^m \qquad [2]$$

This result and the value of n will be used later.

Relationship of Ribbon Dimensions to Casting Conditions. Sufficient information now exists to derive the functional relationships between casting conditions and ribbon dimensions on the premise of ribbon formation by propagation of a thermal boundary layer. The volumetric rate of flow of molten alloy in the jet and in the ribbon is

$$Q = \pi a^2 v = w\bar{t}V \qquad [3]$$

where $\bar{t}$ is the thickness of the ribbon averaged across its width:

$$\bar{t} = \int_0^{w/2} t\,dy \qquad [4]$$

The average thickness, $\bar{t}$, is calculable from the previous description of the growth of the thermal boundary layer. At the point of total solidification of the puddle,

$$\bar{t} = \bar{\delta} = \int_0^{w/2} \delta^* dy \qquad [5]$$

Recalling Eq 2 for δ^*,

$$\bar{t} = \int_0^{w/2} c(\Theta)^m dy \qquad [6]$$

The solidification time, Θ, varies as x, the distance from the nose of the puddle to the solidification point, and inversely as the speed of the chill surface:

$$\Theta = \frac{x}{V} \qquad [7]$$

For the sake of definiteness it is assumed that the shape of the puddle is elliptical and given by

$$\frac{x^2}{a^2} + \frac{y^2}{b^2} = 1$$

$$x = \frac{a}{b}(b^2 - y^2)^{1/2} \qquad [8]$$

where b = w/2 is the minor axis of the ellipse and the half width of the ribbon. Thus, by Eq 6, 7 and 8,

$$\bar{t} = c\int_0^{w/2}\left(\frac{x}{V}\right)^m dy$$

$$= c\left(\frac{a}{bV}\right)^m \int_0^b (b^2 - y^2)^{m/2} dy \qquad [9]$$

Expansion of the integrand by the binomial theorem and evaluation of the integral yields

$$\bar{t} = c\left(\frac{a}{V}\right)^m b\left[1 - \frac{m}{6} + \frac{m(m-2)}{40} + \ldots\right] \qquad [10]$$

Observations of the shape of the puddle indicate that

$$ab \cong \text{const.} \qquad [11]$$

From Eq 10 and 11 and consolidation of constants,

$$\bar{t} = c'b^{1-m}V^{-m} = c'\left(\frac{w}{2}\right)^{1-m} V^{-m} \qquad [12]$$

Inserting Eq 12 into Eg 13 gives

$$c''w^{2-m}V^{1-m} = Q \qquad [13]$$

Solving for Eq 13 for w gives

$$w = c''\frac{\frac{1}{Q^{2-m}}}{\frac{1-m}{V^{2-m}}} \qquad [14]$$

To simplify, let

$$n = \frac{1}{2-m} \qquad [15]$$

Then

$$w = c''\frac{Q^n}{V^{1-n}} \qquad [16]$$

And, from Eq 3,

$$\bar{t} = \frac{1}{c''}\frac{Q^{1-n}}{V^n} \qquad [17]$$

Equations 16 and 17 are the functional relationships of ribbon dimensions to casting conditions sought for under the premises.

To review the foregoing, $n = 1/2 - m$, where m is the slope of the solidification curve (Eq 2): n is approximately 0.75.

The factor c'' includes the coefficient of the solidification expression and depends also on the precise shape of the puddle. Under constant thermal conditions, c'' should depend principally upon θ, the angle between the jet and the chill surface.

While it will not be detailed here, a momentum balance on the jet and the puddle and the further assumption that the "backward flow" component is entrained in the growing boundary layer yields a first approximation of the dependence of c'' upon θ:

$$c'' \sim (1 - \cos\theta)^n \qquad [18]$$

Consideration of Energy Equilibrium Ribbon Formation

Before attempting a mechanical and surface energy balance on the system, a few remarks are in order. To begin with, the system has several components and is defined as follows:

1 A section of the molten jet whose volume is the same as the volume of the liquid puddle.
2 A section of the chill surface whose area is the same as that of the liquid puddle.
3 The mass of air in motion with the chill surface which impacts on the puddle in time Θ.

The energy balance will evaluate differences in the state of the system at an initial time when the elements of the system are separated and after a time Θ when they have interacted.

An implicit assumption will be that the chill surface contributes negligible kinetic energy to the liquid puddle. (Here the liquid puddle is to be clearly distinguished from its solidified boundary layer.) The justification for this treatment is the previous conclusion that in liquid metals momentum transport lags thermal transport. Hence, outside of the solidified boundary layer and a very narrow velocity transition region, the liquid puddle does not "feel" the motion of the chill surface.

Energy Balance on Jet, Puddle, Chill Surface and Air Boundary Layer. It is sought to evaluate the terms in a mechanical and surface energy balance:

$$\Delta(\text{K.E.}) + \Delta(\text{S.E.}) + \text{dissipation} = 0$$
$$E_D(\text{dissipation}) = -[\Delta(\text{K.E.}) + \Delta(\text{S.E.})] \qquad [19]$$

Kinetic Energy Change of Jet. Let A be defined as the cross-sectional area of the liquid puddle and $\bar{t}$ as its average thickness. Then the length of the molten jet which has the same volume as the puddle is

$$l = \frac{A\bar{t}}{\pi a^2} \qquad [20]$$

The free surface area of this jet segment is

$$A_J = 2\pi a l$$
$$= \frac{2A\bar{t}}{a} \qquad [21]$$

In the initial state the jet segment has a velocity v inclined at an angle θ to the chill surface. In the final state the jet segment has become the liquid puddle.

The component of the kinetic energy of the jet parallel to the chill surface is assumed to be conserved in the puddle:

$$\therefore \Delta(K.E.)_{jet} = -\frac{1}{2} m(v \sin \theta)^2$$
$$= -\frac{1}{2} \rho A\bar{t}(v \sin \theta)^2 \qquad [22]$$

Kinetic Energy Change of Air Boundary Layer. It can be shown that the air boundary layer, at distances from the chill surface comparable to the puddle thickness, is at essentially constant velocity with the chill surface. The time required for the jet segment of interest to displace the liquid in the puddle is

$$\Theta = \frac{A\bar{t}}{\pi a^2 v} \qquad [23]$$

The mass of air which impacts the puddle in time Θ is

$$\rho_a \frac{VAw\bar{t}^2}{\pi a^2 v} \qquad [24]$$

In the initial state, this mass of air moves at a velocity V. In the final state, the air has been brought to rest at the face of the puddle.

$$\therefore \Delta(K.E.)_{air} = -1/2\, \rho_a \frac{VAw\bar{t}^2}{\pi a^2 v} V^2 \qquad [25]$$

Surface Energy Change of Liquid and Chill Surface. In the initial state, the liquid metal in the jet is in contact with air at its lateral surface. In the final state, the liquid metal is in the puddle in contact with a solidified layer of the same composition. In the initial state, the metal of the chill surface is in contact with air. In the final state, the chill surface is in contact with the solidified layer of the alloy. Therefore

$$\Delta(\text{S.E.}) = \sigma_{LV}(A - A_J) + A(\sigma_{SS} + \sigma_{SL} - \sigma_{SV}) \qquad [26]$$

The terms σ_{SL} and σ_{SS} may be neglected in comparison with σ_{SV} (18). Substituting for A_J from Eq 21:

$$\Delta(\text{S.E.}) = \sigma_{LV} A \left(1 - 2\frac{\bar{t}}{a}\right) - A\sigma_{SV} \qquad [27]$$

Over-all Mechanical and Surface Energy Balance. Finally, combining terms, Eq 19 becomes

$$\frac{E_D}{A} = \frac{1}{2}\rho\bar{t}\,(v \sin\theta)^2 + \frac{1}{2}\rho_a \frac{w\bar{t}^{\,2}V^3}{\pi a^2 v} + \sigma_{SV} - \sigma_{LV}\left(1 - 2\frac{\bar{t}}{a}\right) \qquad [28]$$

Comments on Energy Balance. Upon substituting numerical values in the energy balance, some interesting points emerge.

Under many conditions the term σ_{SV} dominates. If the substrate were not wetted, the energy dissipation in such circumstances would be indicated to be negative. Since this is an impossibility, it can be concluded that wetting of the substrate by the molten alloy is generally essential to chill block casting.

Since the kinetic energy term is quadratic in jet velocity and therefore fourth power in jet radius (at constant flow), dissipation processes in the puddle are expected to be quite sensitive to jet radius.

Relationship of Ribbon Dimensions to Casting Conditions. As noted previously, implicit in the energy balance is the notion that the liquid puddle does not "feel" the chill surface velocity. It follows that the surface energy and, therefore, the shape of the puddle are independent of V. Since the puddle wets the substrate, the width of the ribbon is the same as the width of the puddle. The expectation is therefore that w will be independent of V. Further, since $w\bar{t}V = Q$ and w is independent of V, it follows that

$$\bar{t} \sim \frac{1}{V}$$

These proportionalities may be contrasted with Eq 16 and 17.

Ribbon Dimensions vs Casting Conditions—Comparison of Theory With Experiment

Data Analysis. Liebermann and Graham (15) have reported experimental measurements of ribbon dimensions as a function of casting conditions. Similar measurements have been made in our laboratory (Table 2). Both sets of measurements will be discussed in relation to the models considered. By way of introduction, note that the experimenter may make at least three kinds of dimensional measurements:

1 Ribbon width, w
2 Ribbon thickness, t
3 Ribbon cross-sectional area, A

Please note also that

$$wt \neq A$$

unless the ribbon cross section is rectangular, in which case

$$w\bar{t} = A$$

Liebermann and Graham presented the following correlation for t:

$$t = \frac{v_r}{(0.354 - 0.510\phi)v_r^{-6.18+9.18\phi}} + \frac{12.6_P}{v_r^{1.35}}$$

(The reader is referred to the original paper for nomenclature.) While the ribbon width measurements could not be correlated, it was reported that width increased with increasing wheel speed at constant pressure (volumetric flow rate). This conclusion was based on identification of w with A/t. The authors have kindly provided us with their optical measurements of w.

The data of Liebermann and Graham and the data in Table 2 were subjected to regression analysis to determine best fits to the following sets of expressions:

Two constant	*Three constant*
$w = c_1 \frac{Q^n}{V^{1-n}}$	$w = c_3 \frac{Q^a}{V^b}$
$\bar{t} = c_2 \frac{Q^{1-n}}{V^n}$	$\bar{t} = c_4 \frac{Q^c}{V^d}$

The results were as follows:

Model	Data source	"Best fit" c	n	$1-n$	C.O.V., coefficient of variation, %
Two Constant					
$w = c_1 \frac{Q^n}{V^{1-n}}$	Ref 15	0.371	0.87	0.13	13.7
	Table 2	0.625	0.83	0.17	6.9
$\bar{t} = c_2 \frac{Q^{1-n}}{V^n}$	Ref 15	2.65	0.13	0.87	20.3
	Table 2	1.85	0.17	0.83	7.9
Three Constant					
$w = c_3 \frac{Q^a}{V^b}$	Ref 15	0.259	0.61	0.11	18.1
	Table 2	0.332	0.79	0.09	6.5
$\bar{t} = c_4 \frac{Q^a}{V^b}$	Ref 15	3.86	0.39	0.89	16.0
	Table 2	3.01	0.21	0.91	6.6

The data of Liebermann and Graham and the data reported here are in substantial agreement. Note that the three constant regressions show little better fit over the two constant formulations suggested by Eq 16 and 17.

Some details of the agreement between the experimental and calculated values are shown in Fig. 6 to 10. It is clear from the data that ribbon width varies as about $(V)^{-0.2}$.

Comparison of Theory With Experiment. It will be recalled that the predictions of the models were as follows:

1 Thermal boundary layer development:

$$w = c'' \frac{Q^{0.75}}{V^{0.25}} \qquad \bar{t} = \frac{1}{c''} \frac{Q^{0.25}}{V^{0.75}}$$

2 Energy equilibrium:

$$w \text{ independent of } V \qquad \bar{t} \sim \frac{1}{V}$$

The data are not in perfect agreement with either model and lie between the two with respect to the dependence of w and t on V. It appears that the thermal boundary layer model does provide a more complete framework for correlation and prediction. The model

Table 2. Experimental Measurements

Alloy: $Fe_{40}Ni_{40}P_{14}B_6$; T_{melt}: 1000°C; T_{cu}: 20°C; θ: 90°

V, surface velocity, cm/sec	Q = AV, melt flow rate, cm^3/sec	Ribbon: A, sectional area, cm^2	Ribbon: w, width, cm
1150	0.0775	6.74×10^{-5}	0.0210
1190	0.0781	6.57×10^{-5}	0.0231
2200	0.0810	3.68×10^{-5}	0.0246
2210	0.0799	3.61×10^{-5}	0.0205
3300	0.0764	2.31×10^{-5}	0.0208
3310	0.0783	2.36×10^{-5}	0.0192
1220	0.1010	8.26×10^{-5}	0.0267
1230	0.1030	8.34×10^{-5}	0.0269
2230	0.0985	4.41×10^{-5}	0.0236
2280	0.0997	4.37×10^{-5}	0.0269
3350	0.0955	2.85×10^{-5}	0.0223
3350	0.0966	2.88×10^{-5}	0.0229
1230	0.1286	1.05×10^{-4}	0.0337
1260	0.1292	1.02×10^{-4}	0.0336
2250	0.1187	5.27×10^{-5}	0.0305
2300	0.1242	5.40×10^{-5}	0.0331
3350	0.1204	3.59×10^{-5}	0.0346
3350	0.1219	3.64×10^{-5}	0.0310
1210	0.1505	1.24×10^{-4}	0.0431
2250	0.1362	6.05×10^{-5}	0.0385
3350	0.1550	4.63×10^{-5}	0.0385
1600	0.4427	2.78×10^{-4}	0.0866
2230	0.4646	2.08×10^{-4}	0.0915
2230	0.4521	2.02×10^{-4}	0.0841
3350	0.4666	1.39×10^{-4}	0.0805
3350	0.4347	1.30×10^{-4}	0.0793
1600	0.5037	3.16×10^{-4}	0.0963
1640	0.5032	3.06×10^{-4}	0.1067
2220	0.5025	2.27×10^{-4}	0.0920
2270	0.5007	2.21×10^{-4}	0.0927
3350	0.5173	1.54×10^{-4}	0.0860
3350	0.5100	1.52×10^{-4}	0.0950

appears to be successful in predicting that Q, the volumetric flow rate, is the fundamental casting parameter rather than orifice size or jet velocity individually. The predicted exponent of the Q dependence is somewhat small. The effect of a more rigorous thermal analysis will be of interest, and a more complete experimental test involving variation of superheat or thermal parameters should be enlightening.

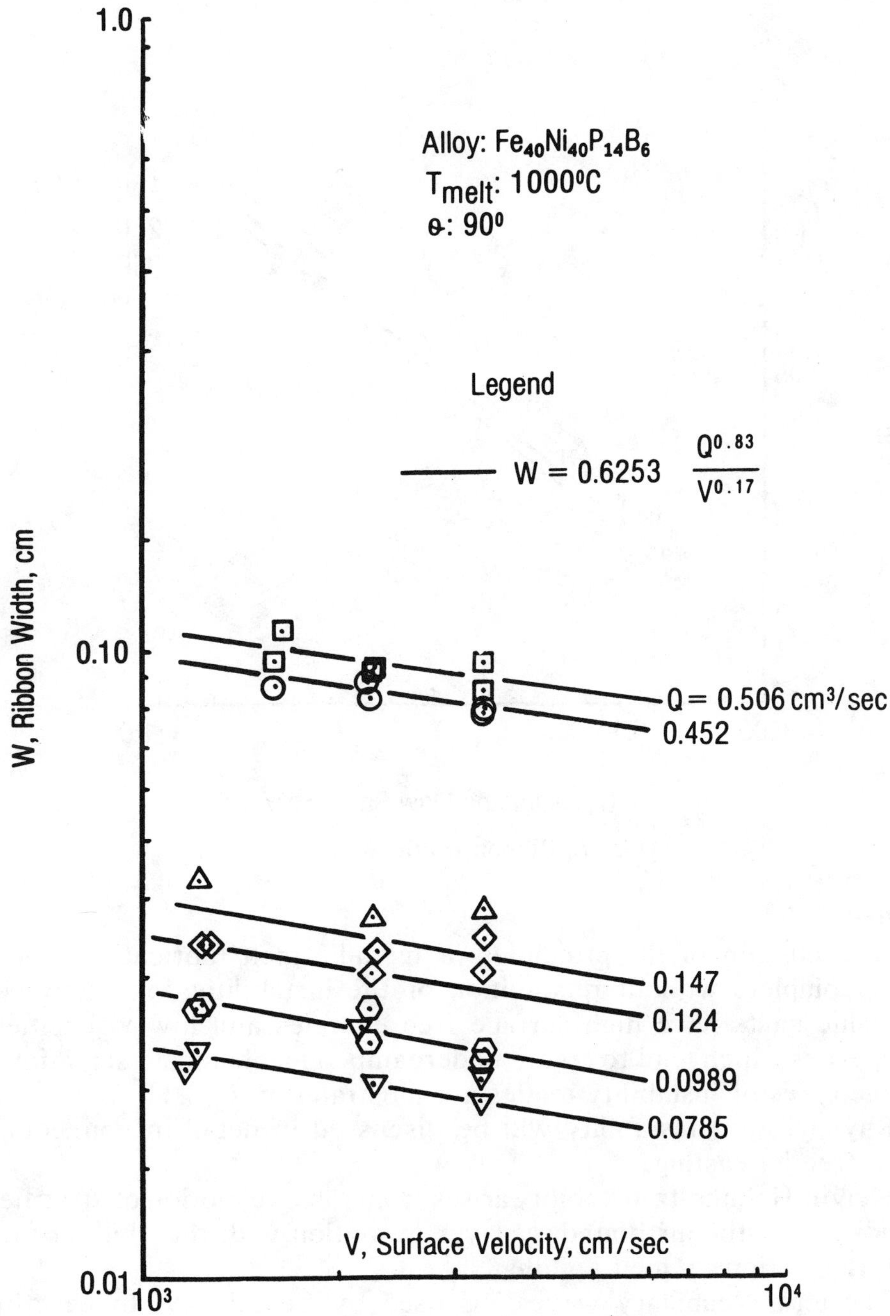

Fig. 6. Ribbon width vs velocity

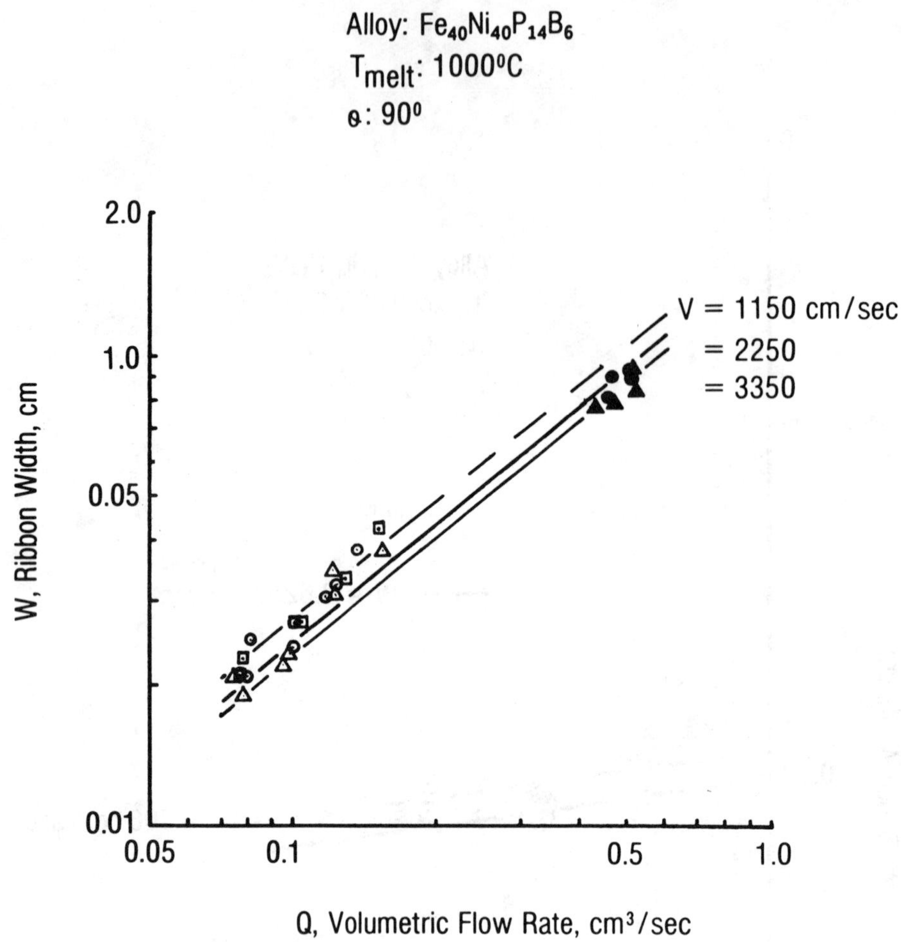

Fig. 7. Ribbon width vs velocity

Instabilities

A discussion of the principles of metallic melt fabrication would be incomplete without recognition of the instabilities encountered. Metallic melts have high surface free energies and low viscosities, properties which tend to create underdamped mechanical oscillations. Some types of instability modes are illustrated in Fig. 11.

Rayleigh (jet) instability will be discussed in detail in connection with free jet casting.

Kelvin-Helmholtz instability arises from relative motion of stratified fluids—e.g., the air boundary layer in motion with the chill surface and the stationary melt puddle.

The term "capillary waves" is used by Levich (19) to describe wave motion which penetrates only to a depth of the order of one

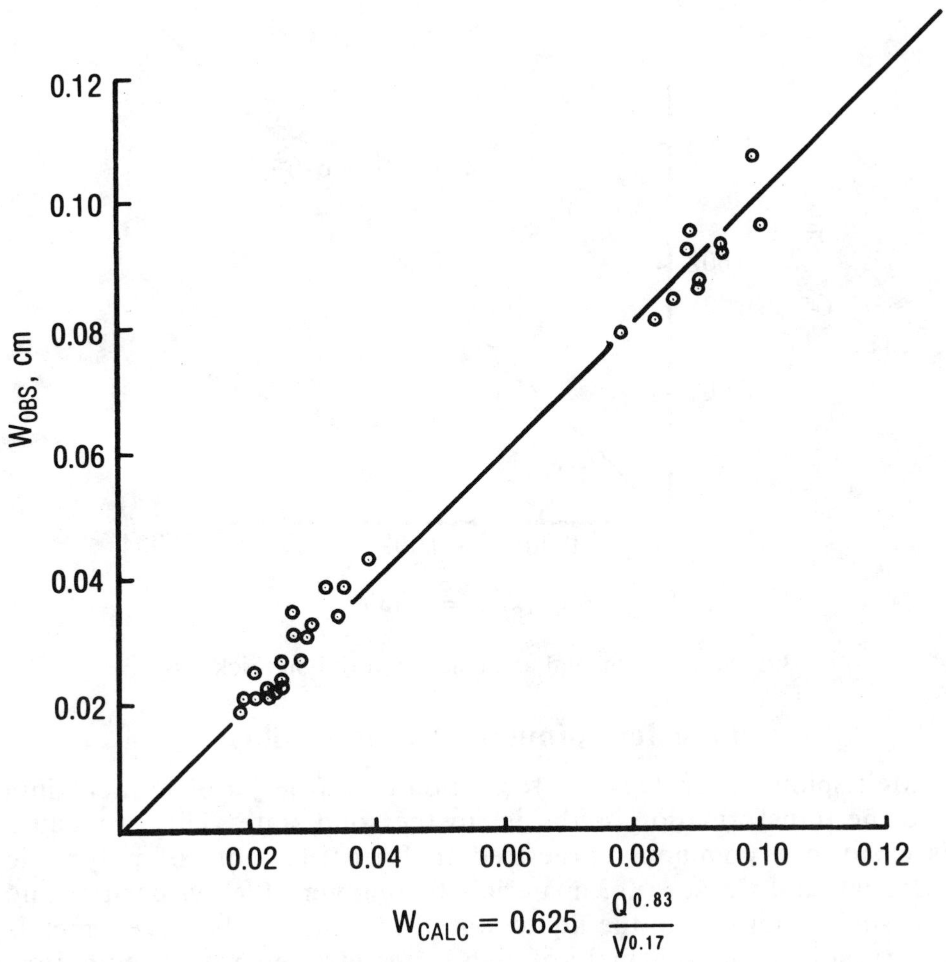

Fig. 8. Experimental and calculated ribbon width

wavelength. Such waves may arise from random perturbations of any source. In the case at hand, they may arise from turbulence within the melt puddle or from air boundary layer separation (eddying) on its surface.

The Marangoni instability is a bulk flow arising from local variations of surface tension. See Brimacombe and Weinberg (20) for an example in liquid metals. The Marangoni instability may be driven by concentration or temperature fluctuations.

It has recently been conjectured that the maximum chill casting velocity may be limited only by the strength of the substrate material. Recognition and resolution of problems which may arise from fluid instabilities may be helpful in reaching that status.

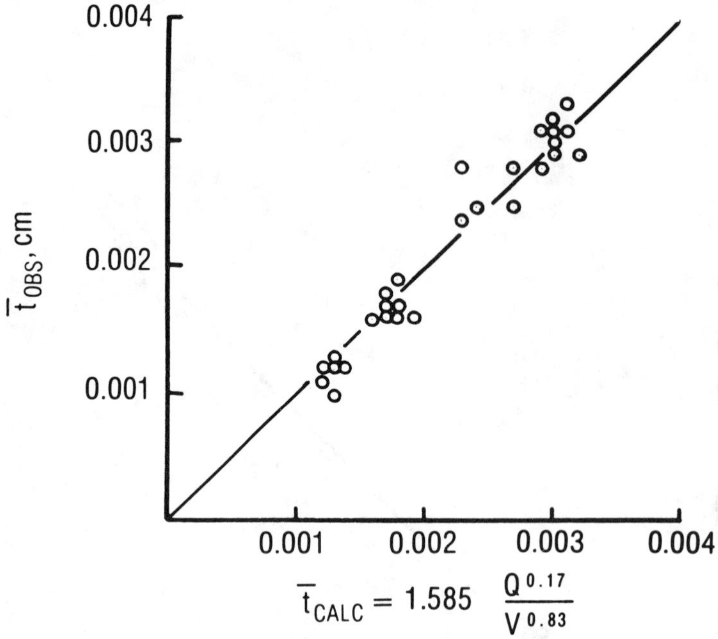

Fig. 9. Experimental and calculated ribbon thickness

Free Jet Spinning of Glassy Alloys

Melt spinning involves the formation of a free jet of molten fluid and the transformation of the jet to the solid state. This procedure is easily and commonly practiced to form filaments of polymeric materials and glasses (i.e., materials having very high viscosities and low surface tension in the liquid state). In contrast, however, metals have relatively inviscid melts of high surface free energy. A cylindrical jet of such a material is inherently unstable. Its surface becomes increasingly perturbed as it issues from the nozzle until at some distance the jet breaks up into droplets. It is evident that if a process is to be capable of producing continuous filaments, a favorable balance between the kinetics of jet solidification and of jet breakup is required.

Several processes for the free jet melt spinning of metal filaments have previously been described. In one group of processes the required balance between jet solidification and jet breakup is obtained by stabilization of the jet by chemical (21-26) or electrostatic (27,28) means. The jet is extruded into a gaseous atmosphere capable of chemical reaction with one or more components of the jet. Stabilization occurs by formation of a solid sheath or skin on the liquid jet. Alternatively, electrostatic charges have been used to stabilize the jet.

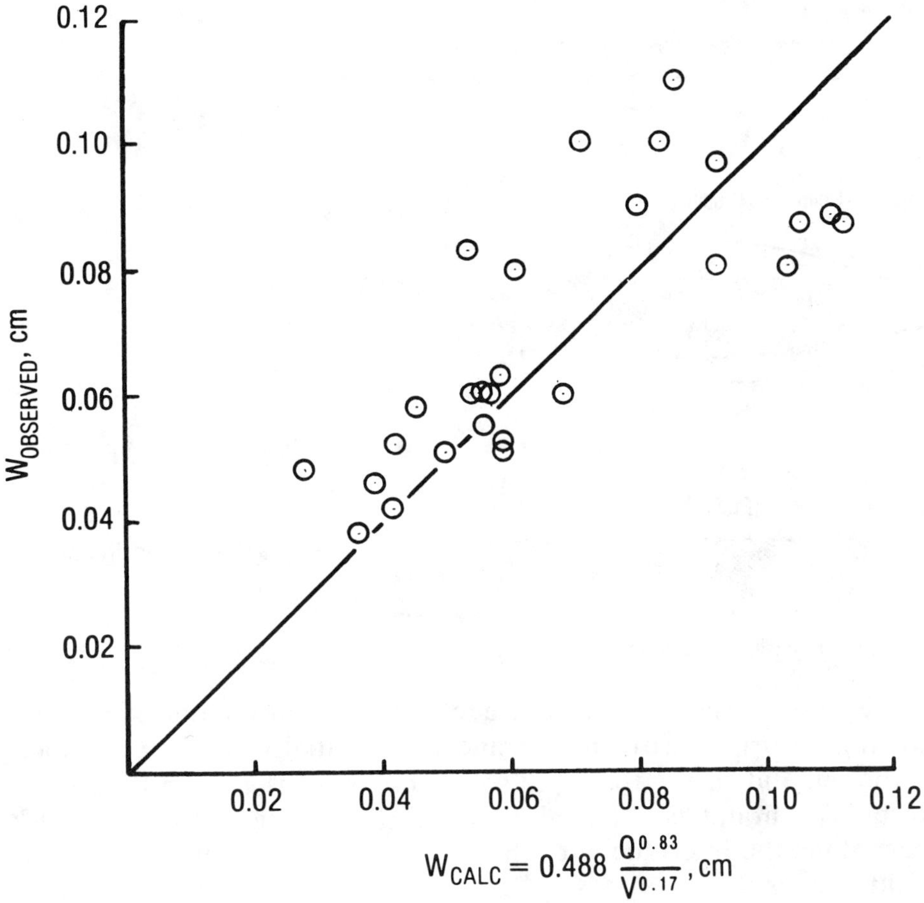

Fig. 10. Fit of model to data of Liebermann and Graham—ribbon width

Several techniques have also been described for hastening the solidification of the jet. Pond (23) described cooling of the molten jet by a mist of a vaporizing liquid dispersed in a gas. Schile (27,28) employed cooling by a gas-solid dispersion. Engelke *et al.* (29) immersed the nozzle from which the molten jet issued in a ''liquid mold stream'' maintained at a slightly lower temperature than that of the melt. However, the solidification rate in each of these processes remained fairly slow. Chemical or electrostatic stabilization of the jet was still required.

Process Description (39)

A process has been devised for the spinning of metallic glasses from their melts. It consists of the steps of formation of a free jet of molten material in a gaseous or evacuated environment, traversal

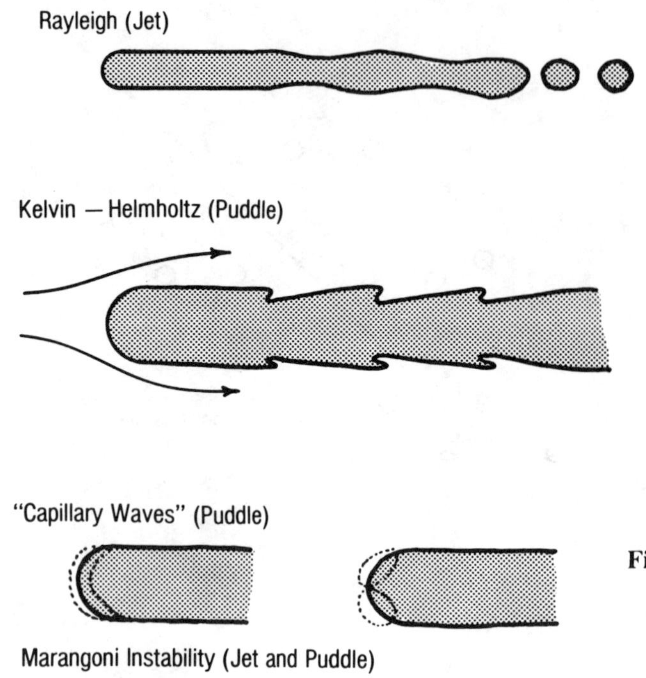

Fig. 11. Casting instabilities

of the free jet through an interface into a liquid medium, the liquid flowing cocurrent with the jet and at essentially the same velocity as the jet, and finally the rapid solidification of the jet in filamentary form. The liquid is selected to cause a high heat flux regime to exist along the interface with the molten jet (e.g., a nucleate surface boiling regime).

The process and a laboratory spinning apparatus are illustrated in Fig. 12. The material to be melt spun is charged to an insulated ceramic crucible. The crucible has one or more bottom openings (spinning orifices) whose diameters are the order of size of the desired filaments (0.002 to 0.020 cm). The crucible lies within a chamber connected to an inert gas pressure source. Energy to melt the metal charge is provided by a 450 KHz induction power supply connected to a coil wound concentrically about the melting chamber. The melting chamber is suspended immediately above a reservoir containing the fluid quench medium. Water at 0 to 20°C is a satisfactory medium for quenching materials whose melting points are less than about 700°C (e.g., aluminum, zinc, lead, tin, bismuth, cadmium, etc.). For metallic glass formers whose melting points are above 700°C to about 1300°C, a refrigerated brine quenchant is satisfactory.

The level of the fluid medium is controlled 0.25 to 0.50 cm below the bottom plate of the melting chamber. Within the fluid reservoir,

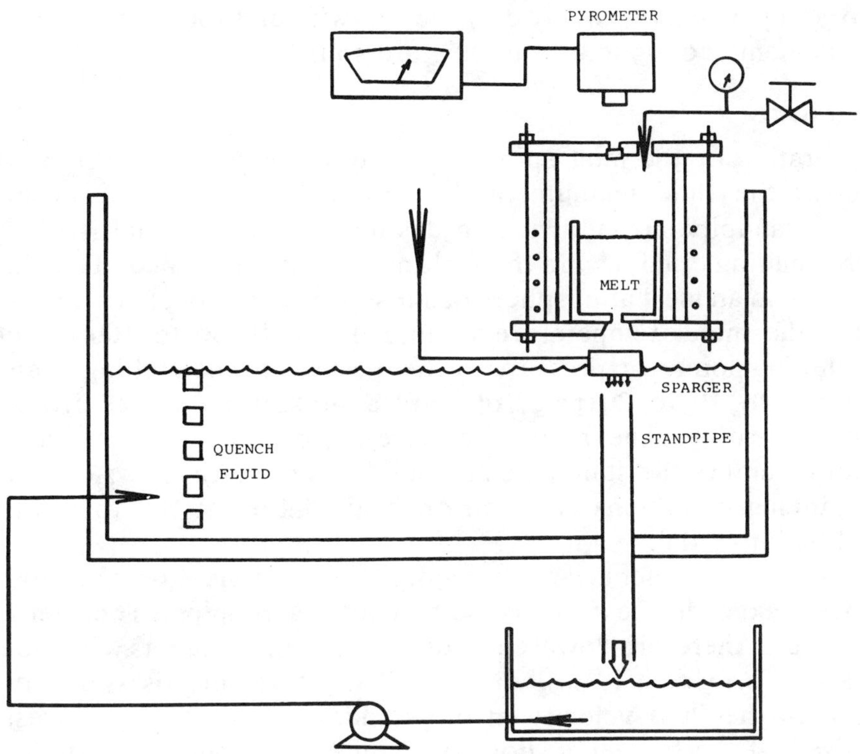

Fig. 12. Laboratory apparatus for spinning metallic glasses from their melts

disposed vertically below the spinning orifices, is a quartz standpipe. The fluid quench medium enters the reservoir at one end, passes horizontally through a calming section, flows vertically down through the standpipe into a catch basin and finally, by means of a pump, is returned to the reservoir.

The top of the standpipe is typically within 2 cm of the fluid surface. A vortex baffle consisting of a bar of rectangular cross section is placed to one side of the standpipe and extending to the fluid surface. A sparger ring is placed immediately above the standpipe immersed in the fluid medium. Fine jets of the quench fluid may be sprayed from the sparger ring into the mouth of the standpipe. The velocity of the fluid medium in the standpipe is determined by the diameter and length of the standpipe, the viscosity and density of the fluid medium, the height of fluid surface above the standpipe, and the velocity and volume of flow sprayed into the standpipe from the sparger ring. A typical fluid velocity using a 1.4 cm I.D. standpipe of 40 cm length, an aqueous quench medium of 1.0 CPS viscosity and 1.0 g/cc density, a fluid level 2 cm above the standpipe, and zero spray velocity is 200 cm/sec. A winder (not shown) collects the filaments continuously as they are spun.

Operation

Operation of the melt spinning process is simple and direct. The level of the fluid medium in the reservoir and the fluid velocity in the standpipe are adjusted to desired levels. Material is charged to the melting crucible and the melting chamber is sealed. The charge is melted in an inert atmosphere at substantially atmospheric pressure. When the metal temperature is approximately 50 to 100°C above the melting point, the pressure of the inert gas in the melting chamber is raised to 10 to 20 psig, or until a molten jet issues from the spinning orifice. The molten jet is ejected down into the quench medium, enters the standpipe and solidifies as it travels. Adjustment of spinning conditions is based on examination of the form of the solidified material.

If sinuous filaments are formed, it is an indication that the jet velocity exceeds the fluid velocity in the standpipe. The inert gas pressure is, therefore, lowered. If discontinuous filaments with tapered ends are formed, it is an indication that jet velocity is substantially less than the fluid velocity in the standpipe and inert gas pressure is raised. If the filaments show axisymmetrical nodes, melt temperature is reduced. Temperature and pressure are adjusted until smooth continuous filaments are being wound. Under these conditions it is found that the jet velocity closely matches the fluid velocity in the standpipe.

Process Analysis

The reasons why the process does in fact work well have been analyzed and are documented in the following sections.

Stability of Liquid Metal Jets Theory. Investigation of the stability of cylindrical jets dates to the experiments of Savart (30) in 1833, and the analyses of Plateau (31) in 1873 and Rayleigh (32) in 1879. If two nodes on the liquid cylinder are farther apart than the circumference of the cylinder, then the pressure is less at the node than at the constriction. Fluid will tend to flow from the constriction to the nodes and the nodes will amplify. If the nodes are less than the circumferential distance apart, the pressure is less at the constriction than at the nodes and the nodes will diminish. In the actual dynamic case, the pressure field within the cylinder is coupled with the flow field as expressed by the Navier-Stokes equations.

The most general analysis of jet stability was by Meister (33,34).

Meister considered both axisymmetrical (varicose) instability and unsymmetrical (sinuous) instability. It will be sufficient for present needs to restrict attention to axisymmetrical breakup. The Meister analysis was in two parts. First, the question of stability was posed for a stationary cylindrical liquid filament, surrounded by another fluid axisymmetrically perturbed at time zero. The cylinder undergoes no axial motion relative to its enviroment. Next was posed the question: "If the filament is in motion, how far do the surface waves travel before the jet breaks up?". Following Rayleigh, the expression for the radius of the jet was presumed to be given by a constant term, a, plus a perturbation term, ϵ, which is exponential in time and periodic in z.

$$r = a + \epsilon(t,z) \qquad \epsilon = \epsilon_0 \exp(\alpha t + ikz)$$

The stability question hinges on the maximum value, α_{max}, of the perturbation growth rate. If α_{max} is negative, an initial perturbation is damped. If α_{max} is positive, an initial perturbation amplifies until for some time t* and position z*

$$r = a + \epsilon(t^*,z^*) = 0$$

(i.e., the filament is pinched off). The dominant wavelength $\gamma = 2\pi/k$ appearing on the surface of the jet is the one for which the growth rate is maximum.

The Meister relationships are lengthy and will not be reproduced here. Predictions of these relations for liquid metal tests in liquid media will be compared with experiments farther on. The equations may be considerably simplified when density and viscosity terms for the continuous phase are negligible. The resulting relationship, Eq 29, applicable to a liquid jet in a gaseous medium, was first derived by Weber (36):

$$\frac{L}{D} = (N_{We}{}^{1/2} + 3\frac{N_{We}}{N_{Re}} \ln \frac{a}{\epsilon_0} \qquad [29]$$

Jet Stability—Experimental. The breakup of liquid metal jets was studied by means of high speed photography using an 8 microsecond General Radio flash source. To investigate the question of jet stability, free of effects of solidification and /or chemical reaction, mercury at room temperature was used as the model fluid. Molten tin in air provided a contrasting chemically reacting system.

Mercury Jet Stability in Air. Figure 13 shows the results of measurements of the stable lengths of mercury jets in air when using capillary nozzles at jet velocities of 55 to 475 cm/sec. The data

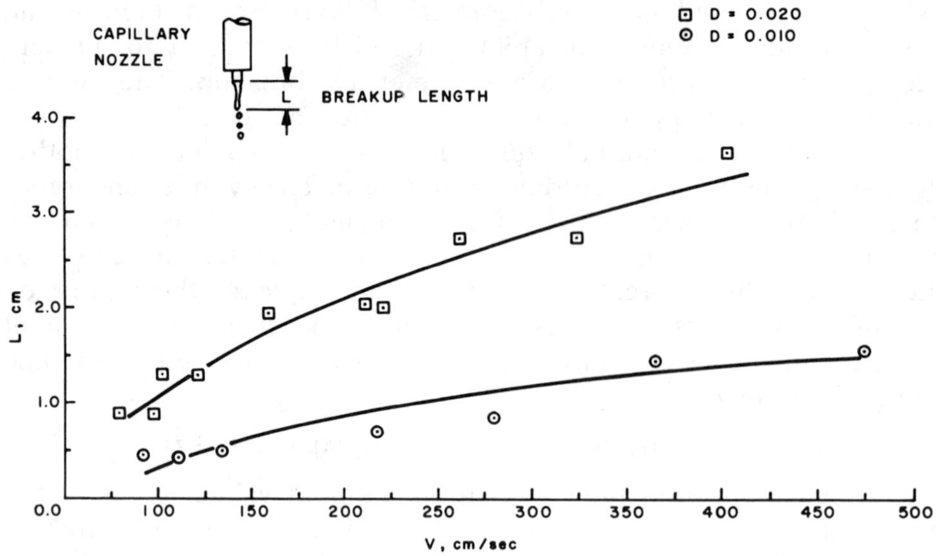

Fig. 13. Mercury jet stability in air

of Fig. 13 have been replotted in Fig. 14 in terms of the parameters of Eq 29. As can be seen, the Weber expression is in fair agreement with the data. The slope of the experimental curve is less than the theoretical value of one. A least squares fit yields the following expression:

$$\frac{L}{D} = 7\left[N_{We}^{1/2} + \frac{3N_{We}}{N_{Re}}\right]^{0.92} \qquad [30]$$

Mercury Jet Stability in Stagnant Liquid Media. Figure 15 shows the results of measurements of the stable lengths of mercury jets injected into water, glycerine and mixtures of water and glycerine of intermediate viscosity. The continuous phase was uncirculated except as motion was induced by the jet. Figure 16 presents the theoretical predictions for these systems.

Experimentally observed jet lengths at jet velocities of 100 to 500 cm/sec exceeded predictions for all systems. Agreement was best for water although even here observed jet lengths at 250 cm/sec were about twice as great as expected. Stable jet length was not observably affected by the viscosity of the continuous medium within the experimental error.

Stability of Mercury Jets in Melt Spinning Configuration. Figure 17 shows the lengths to breakup of mercury jets injected through an air-water interface as in melt spinning. The measurements of Fig.

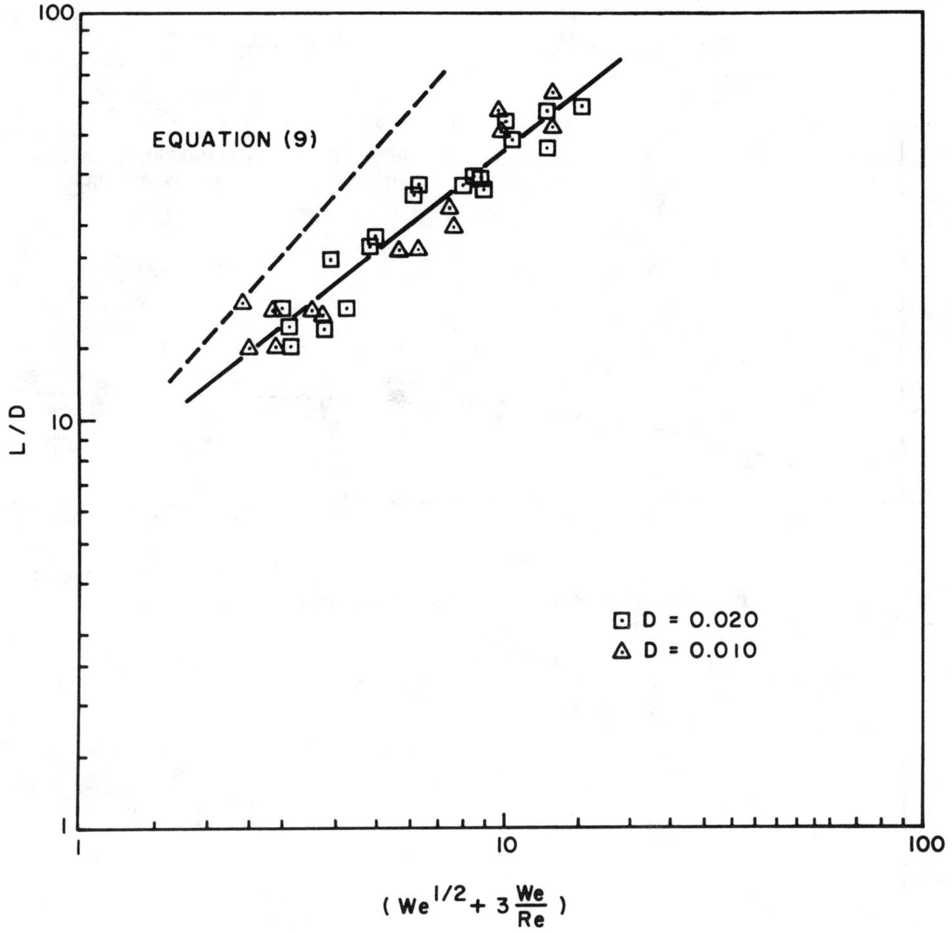

Fig. 14. Mercury jet stability in air

17 were made first with the water stagnant, then with water flowing cocurrent with the jet into a standpipe at 180 cm/sec. A vortex baffle was placed at the mouth of the standpipe as illustrated. The data of Fig. 17 show several interesting features. First, the total length of the mercury jet in air and in the flowing water may be compared with measurements where the jet was in one phase or the other and did not traverse an interface. When this is done it is found that the effect of traversing the interface into flowing water was to reduce the lifetime of the mercury jets by 0.0, 5.9, 6.5 and 23% for the several conditions examined. These results may be compared with observations (23) that " . . . [a molten jet] tends to blow apart upon striking the surface of [a] liquid." The effect of motion of the water cocurrent with the jet was to increase jet

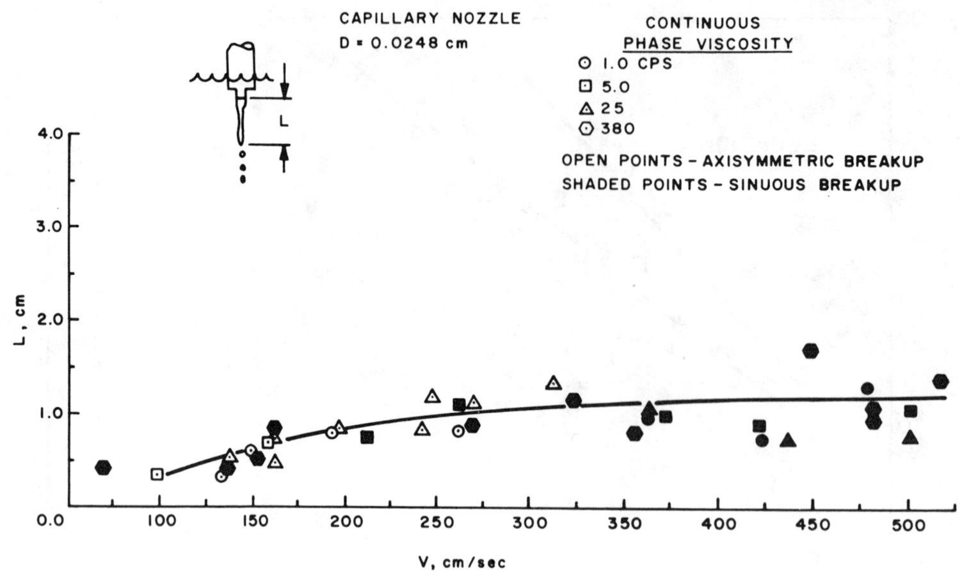

Fig. 15. Mercury jet stability in liquid media

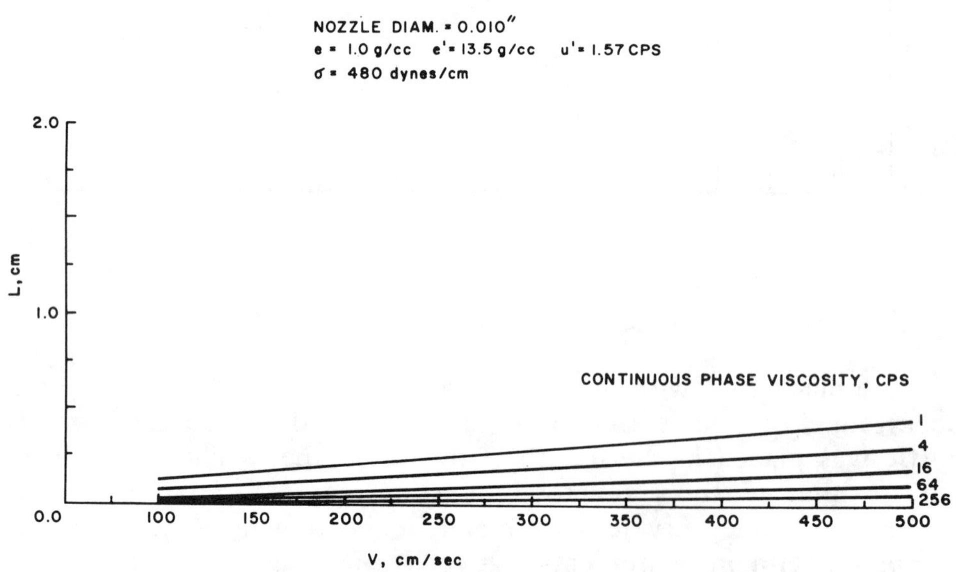

Fig. 16. Predicted stability of mercury jets in liquid media

length to breakup by 0.2 to 0.4 cm, or approximately 30%.

Stability of a Skin Forming Jet—Molten Tin in Air. The breakup of jets of molten tin in air was studied by a high speed flash photographic technique. The behavior of the tin jets showed some parallels to, and also some highly significant departures from, jets

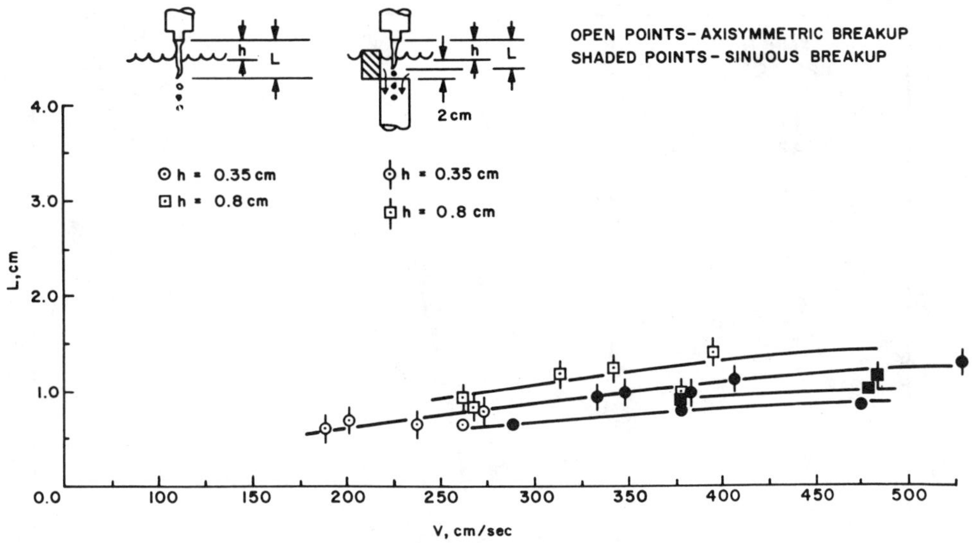

Fig. 17. Mercury jet stability on passing through air-water interface

of liquid mercury. Jet lengths were approximately two orders of magnitude greater than would be calculated for varicose breakup (Eq 29) using the reported value of the interfacial tension of tin in air (33). Breakoff lengths increased sharply with increased jet velocity and were greater at 359 to 442°C than at 234 to 258°C. The lower end of the experimental temperature range (234°C) was only a few degrees above the melting point of tin (232°C). Cooling and solidification of the surface of the jet could have contributed to the marked stability of the jet at that temperature. However, analysis of the jet cooling process shows that, at the higher jet temperatures (359°C, 442°C), surface solidification could not have occurred before the jet disrupted. It seems evident that the operative factor at the higher temperatures must have been the formation of an oxide skin on the jet. Such a skin is observed when tin is melted in an open crucible.

Alber and Smith (24) suggest that the stabilizing effect of an oxide skin results from a lessening of surface tension. However, the surface tension of tin in air has been reported as 662 dynes/cm, and the surface tension of tin in vacuum as 579 dynes/cm (33). It seems more likely that an oxide skin stabilizes a jet by its solid-like character with respect to compressive deformation.

The significance of Fig. 18 is clear: the presence of an oxide skin which makes processes for spinning in gaseous atmospheres workable. Heat transfer in gaseous atmospheres is too slow to effect solidification

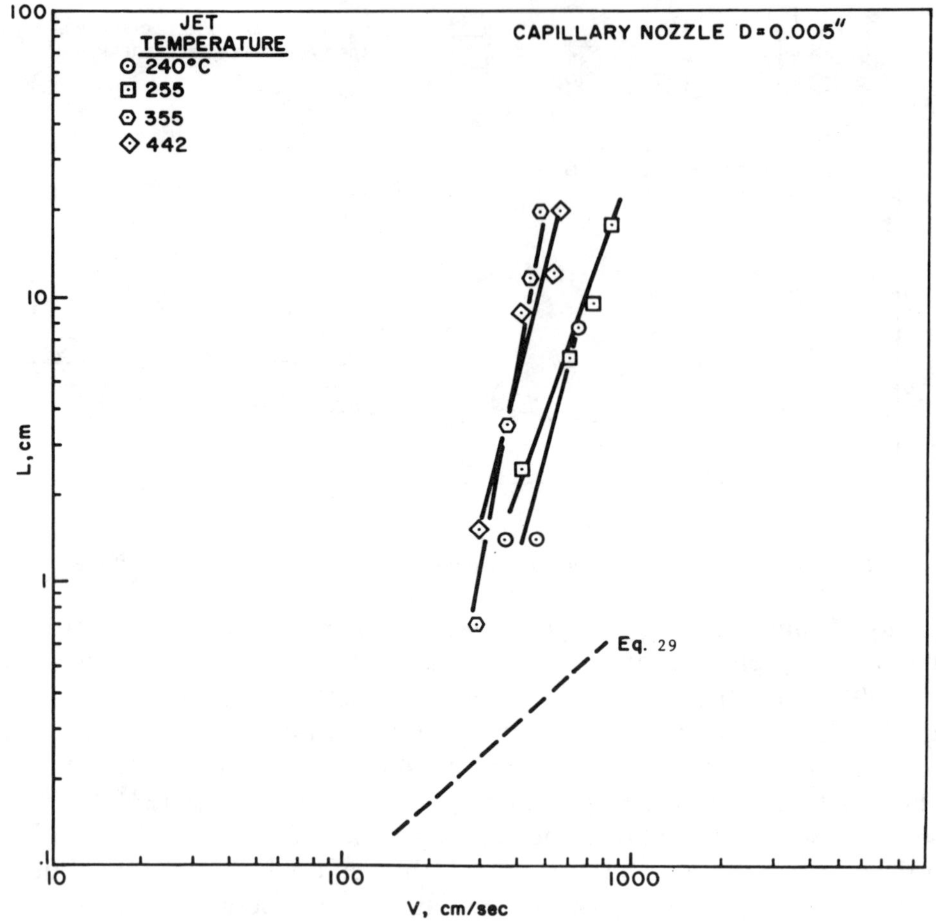

Fig. 18. Stability of molten tin jet in air

before a jet would become disrupted in the absence of a stabilizing skin. Dunn *et al.* (26) note that oxide skin formation is a special example of film stabilization of inviscid jets by reaction or decomposition of the atmospheres into which the jets are injected.

Heat Transmission. Evidence on the nature and rate of heat transmission between a molten jet and a fluid quench medium is composed of the following elements:

1 High-speed photography of molten jets and preheated steel wires entering water.
2 Measurements of heat transmission between an electrically heated 0.004″ diameter platinum wire and flowing water.
3 Deduction of surface coefficient of heat transfer from quenching of glass forming alloys.

Each of these procedures is limited in some respect, but taken together they provide a comprehensive picture.

Photography of Molten Jets and Preheated Steel Wires Entering Water. The nature of the heat transfer regime along molten metal jets and preheated stainless steel wires (type 316) entering cold water was determined using high-speed photography. The observations were as follows:

Photographs of a 0.006″ diameter, stainless steel wire preheated to 800°C entering water at 15°C showed no vapor film around the wire and few active boiling nuclei. Thermal gradients visible in the water around the wire suggest that the heat transmission mechanism was predominantly forced convection. Wire velocities were 200 to 500 cm/sec. A nucleate boiling regime existed along the interface between a 0.006″ stainless steel wire initially at 800 to 1150°C and water at 15°C.

Heat Transmission From an Electrically Heated Platinum Wire to Flowing Water. The test section consisted of a 5 cm length of 0.004″ diameter platinum wire immersed in water flowing within a 1.85″ diameter Pyrex tube. D.C. power input to the wire was measured with a 0.5% precision ammeter and a 0.1% precision digital voltmeter. The temperature of the wire was measured by its change in resistivity. Figure 19 is a plot of the surface coefficient of heat transfer h, as related to the temperature difference between the wire and water and to the water flow velocity. Transition from forced convection to nucleate boiling occurred at the breaks in the experimental curves.

The data show that a heat transfer coefficient of at least 10^4 Btu/hr-sq ft-°F will be obtained on spinning molten jets into water if the heat transfer regime is forced convection or nucleate boiling.

Thermal Analysis of Jet Cooling. Models of the jet cooling and solidification process have been derived and considered for several sets of premises. The models, premises and solutions will be presented elsewhere. In outline, the cases analyzed were as follows:

1 Cooling of a cylindrical jet; no phase change, constant thermal properties.
   - (a) Radial heat conduction through the jet; negligible axial conduction [solution given by Carslaw and Jaeger (38)].
   - (b) Radial and axial conduction both significant.

2 Cooling and phase change of a cylindrical jet.
   - (a) Surface heat transfer limiting—uniform radial temperature through the jet; constant thermal properties.
   - (b) Radial heat conduction only; temperature-dependent properties.

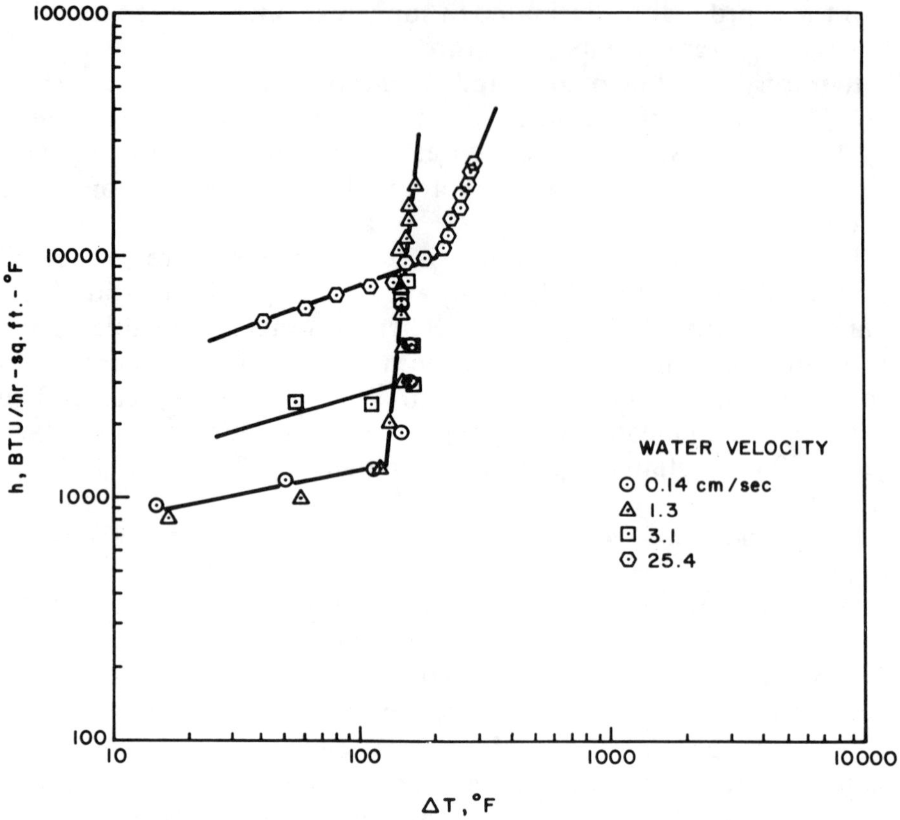

Fig. 19. Heat transmission between electrically heated 0.004″ diameter platinum wire and water

(c) Axial and radial heat conduction; temperature-dependent properties.

Predictions of the models were compared with one another to determine the simplest set of premises which yield the necessary information and accuracy.

Solidification Without Latent Heat Effects—Glass Forming Alloys. As a first approximation to the description of the quenching of glass forming alloys, latent heat effects were taken as zero and thermal diffusivity assumed constant. Under these premises it was found that axial conduction had a negligible effect on the temperature profile of a glass forming jet. Furthermore, the temperature gradient from the inside to the surface of a jet of 0.015 cm radius was predicted to be less than 5°C at a jet temperature of 1000°C.

Key parameters determining the quench rate of a glass forming jet are the surface coefficient of heat transfer, h, and the jet radius.

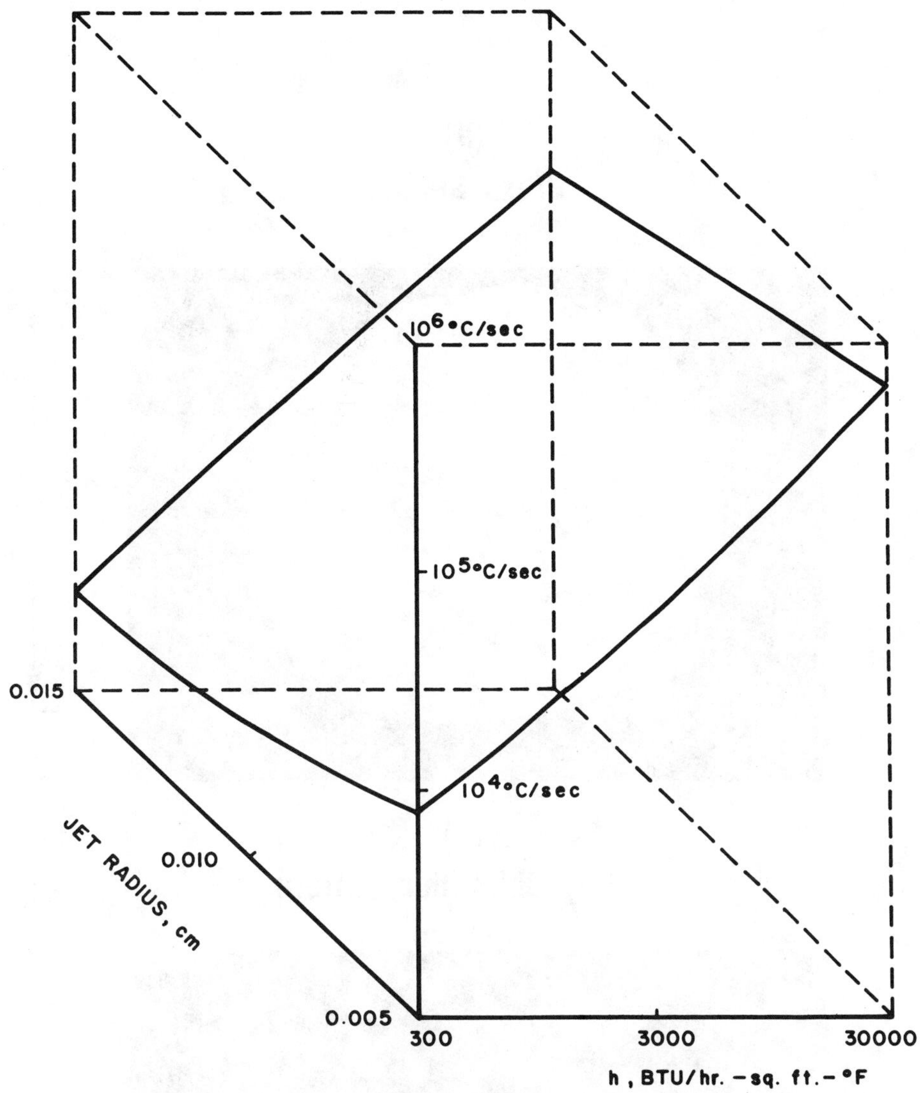

Fig. 20. Predicted quench rate surface for glass forming alloy

Figure 20 presents predicted quench rate contours. In consequence of the flat jet temperature profiles, average quench rate from the melting point (about 950°C) to below T_g (about 400°C) is nearly linearly related to h and the jet radius.

In a film boiling regime (h = 300 Btu/hr-sq ft-°F) quench rate is expected to be approximately 10^3 to 10^4°C/sec for jets of 0.005 to 0.015 cm radius. In a nucleate boiling regime quench rates of 10^5 to 10^6°C/sec may be obtained. Quench rate is expected to depend on jet velocity only as *h* depends on jet velocity.

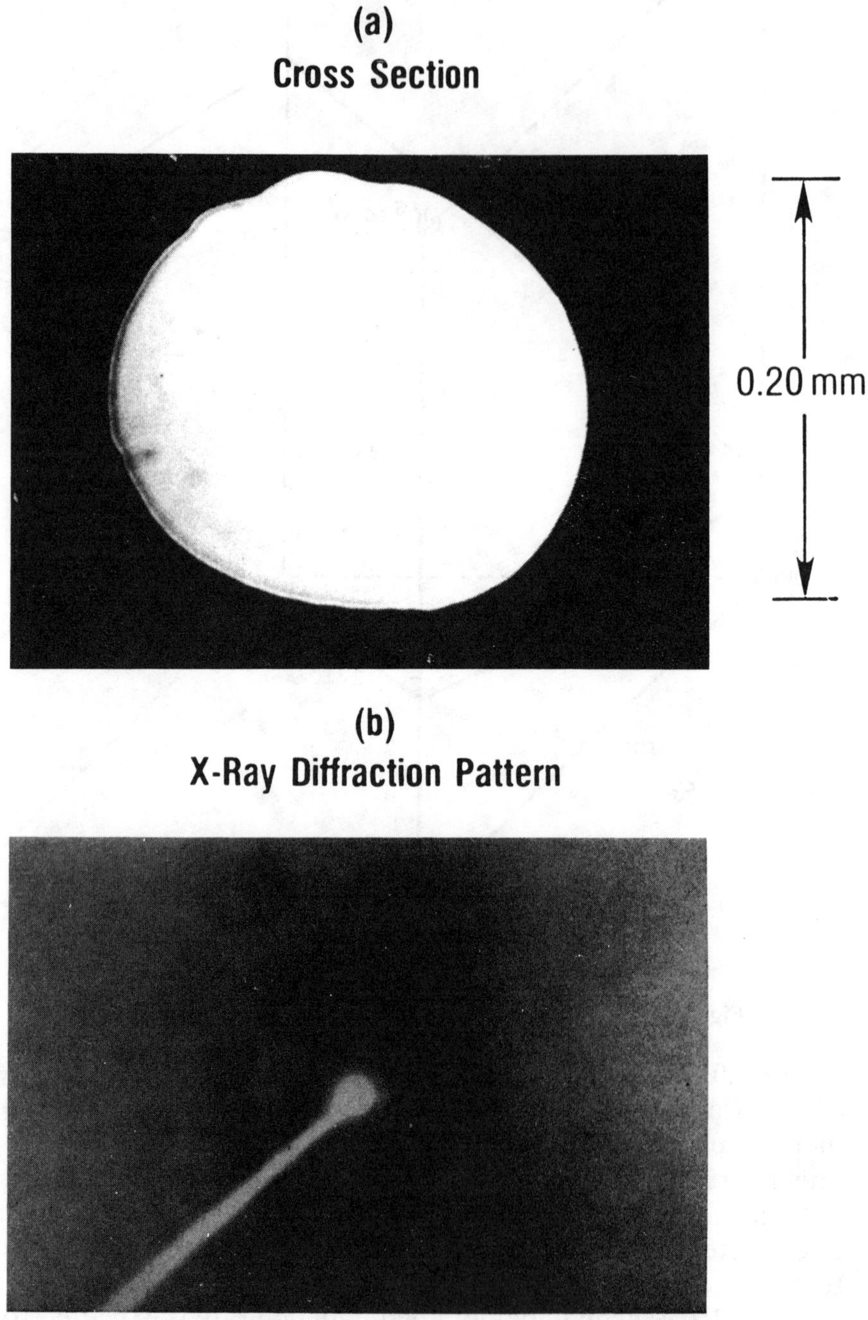

Fig. 21. Texture of Metglas® glassy wire

Metallic Glass Wire. Figure 21 illustrates an axial cross section of a wire of Metglas® (Fe-Ni-Cr-P-B) glassy alloy and its x-ray diffraction pattern. The cross section shows no evidence of grain boundaries. The x-ray pattern shows only a broad amorphous halo. Crystallization has been totally suppressed by rapid quenching, and grain size (if any) is less than 50 Å. Tensile strength of the melt spun wire exceeded 300 ksi. Corrosion resistance in sulfuric acid and sulfuric acid-chloride environments was equivalent to that of type 316 stainless steel.

References

(1) E. H. Strange and C. A. Pim, U.S.P. 905,758, December 1, 1908
(2) W. G. Staples, U.S.P. 989,075, April 11, 1911
(3) S. A. Nagy, U.S.P. 1,291,603, January 14, 1919
(4) T. M. Merle, U.S.P. 2,129,702, September 13, 1938
(5) R. B. Pond, U.S.P. 2,825,108, March 4, 1958
(6) W. Wade, U.S.P. 2,886,866, May 19, 1959
(7) T. P. Gibbons, U.S.P. 2,899,728, August 18, 1959
(8) W. Wade, R. B. Pond, and T. P. Gibbons, U.S.P. 2,904,859, September 22, 1959
(9) R. B. Pond, U.S.P. 2,910,744, November 3, 1959
(10) P. E. Duwez, U.S.P. 3,297,436, January 10, 1967
(11) J. R. Bedell, U.S.P. 3,862,658, January 28, 1975
(12) H. S. Chen and C. E. Miller, *Mat. Res. Bull.*, *11*, 49 (1976)
(13) C. E. Mobley, "Metallurgy and Applications of Rapid Solidification", Battelle Seattle Research Center, June 28-30, 1971
C. E. Motley, R. B. Pond and R. E. Maringer, "New Trends in Materials Processing", American Society for Metals, Metals Park, Ohio, 1976
(14) H. Jones, *Rep. Prog. Phys.*, *36*, 1425 (1973)
(15) H. H. Liebermann and C. D. Graham, Jr., Joint Magnetism and Magnetic Materials-Intermag Conference, A.I.P./I.E.E.E., June 15, 1976, Pittsburgh, Pa.; private communication of optical width measurements, July 21, 1976
(16) H. Schlichting, "Boundary Layer Theory", Chapter XIV, McGraw-Hill, New York, 1960; *Ibid*, Eq. 14.34, p. 301
(17) R. C. Ruhl, *Mat. Sci. Eng.*, *1*, 313 (1967)
(18) S. Z. Beer, "Liquid Metals Chemistry and Physics", Chapter 4-VIII, Marcel Dekker, New York, 1972
(19) V. G. Levich, "Physiochemical Hydrodynamics", Chapter XI, Prentice Hall, Englewood Cliffs, N.J., 1962
(20) O. K. Brimacombe and F. Weinberg, *Met. Trans.*, *3*, 2298 (1972)
(21) R. B. Pond, U.S.P. 2,879,566, March 31, 1959
(22) R. B. Pond, U.S.P. 2,976,590, March 28, 1959
(23) R. B. Pond, U.S.P. 3,602,291, August 31, 1971
(24) N. E. Alber and W. E. Smith, U.S.P. 3,216,076, November 9, 1965; Re. 27,123, May 18, 1971
(25) R. Kafestian, "Melt Spinning of Superalloys", AFML-TR-68-397, January 1969

(26) S. A. Dunn, L. F. Rakestraw, and R. E. Cunningham, U.S.P. 3,658,979, April 25, 1972
(27) R. D. Schile, U.S.P. 3,461,943, August 19, 1969
(28) R. D. Schile, U.S.P. 3,543,831, December 1, 1970
(29) J. L. Engelke, P. C. Johnson, and B. A. Stein, U.S.P. 3,347,959, October 17, 1967
(30) F. Savart, *Ann. Chim. Phys., 53,* 368 (1833)
(31) M. Plateau, *Statique des Liquides,* Paris (1873)
(32) Lord Rayleigh, *Proc. London Math. Soc., 10,* 4 (1879); *Proc. Royal Soc., 29,* 71 (1879); *Phil. Mag., 34,* 145 (1892)
(33) B. J. Meister, *A.I.Ch.E. J., 13,* 682 (1967)
(34) B. J. Meister, *A.I.Ch.E. J., 15,* 689 (1969)
(35) S. Tomotika, *Proc. Royal Soc. (London), A150,* 322 (1935)
(36) C. Z. Weber, *Anew. Math. Mech., 11,* 136 (1931)
(37) R. C. Weast, ed., "Handbook of Chemistry and Physics", Chemical Rubber Co., Cleveland, 1970
(38) H. S. Carslaw and J. C. Jaeger, "Conduction of Heat in Solids", Oxford at the Clarendon Press, London, 1959
(39) S. Kavesh, U.S.P. 3,845,805, Nov. 5, 1974

Appendix: Heat Transport in Chill Block Casting

Statement of the Problem

The melt puddle, Phase 1, at $-a < x < 0$ is initially at a uniform temperature T_m. The substrate, Phase 2, at $0 < x < l$ is initially at a uniform temperature T_0. The top surface of the puddle is uncooled. The bottom surface of the substrate is maintained at temperature T_0 by circulation of coolant.

Assumptions and Simplifications

1 Constant thermal properties—no phase change.
2 Negligible viscous heating in transition zone between solidified boundary layer and liquid puddle.
3 Constant thickness of puddle.
4 One-dimensional heat flow—edge effects were not considered.
5 Negligible contact resistance at interface between puddle and chill block.

Energy Equations (see Nomenclature at end of Appendix)

$$\frac{\partial Y_i}{\partial \Theta} = \alpha_i \frac{\partial^2 Y_i}{\partial x^2} \qquad i = 1,2$$

Initial Conditions

$$Y_1(x,0) = 1 \qquad Y_2(x,0) = 0$$

***Boundary Conditions* ($\Theta > 0$)**

$$Y_2\Big|_{x=1} = 0$$

$$Y_1\Big|_{x=0} = Y_2\Big|_{x=0}$$

$$\frac{\partial Y_1}{\partial x}\Big|_{x=-a} = 0$$

$$K_1\frac{\partial Y_1}{\partial x}\Big|_{x=0} = K_2\frac{\partial Y_2}{\partial x}\Big|_{x=0}$$

Analysis

Laplace transform methods

Transforms

$$\bar{Y}_1 = \frac{1}{p}\left\{1 - \frac{1}{\sigma+1}\,\frac{e^{q_1x} + e^{-q_1(x+2a)} + e^{-q_1(x+2a+2\nu l)} + e^{q_1(x-2\nu l)}}{1+\gamma\left[e^{-2q_1a} + e^{-2q_1\nu l} + \frac{1}{\gamma}e^{-2q_1(a+\nu l)}\right]}\right\}$$

$$\bar{Y}_2 = -\frac{\sigma}{\sigma+1}\,\frac{1}{p}\left\{\frac{-e^{-q_1\nu x} + e^{q_1(\nu x-2\nu l)} + e^{-q_1(\nu x+2a)} - e^{q_1(x-2\nu l-2a)}}{1+\gamma\left[e^{-2q_1a} + e^{-2q_1\nu l} + \frac{1}{\gamma}e^{-2q_1(a+\nu l)}\right]}\right\}$$

Temperatures

$$Y_1 = 1 - \frac{1}{\sigma+1}\sum_{n=0}^{\infty}(-1)^n\sum_{k=0}^{n}\binom{n}{k}\sum_{j=0}^{k}\binom{k}{j}(\gamma)^{n-j}\sum_{i=1}^{4}\operatorname{erfc} Z_i(x,\Theta) \qquad \text{[A1]}$$

$$Y_2 = \frac{-\sigma}{\sigma+1}\sum_{n=0}^{\infty}(-1)^n\sum_{k=0}^{n}\binom{n}{k}\sum_{j=0}^{k}\binom{k}{j}(\gamma)^{n-j}\sum_{i=1}^{4}\operatorname{erfc} W_i(x,\Theta) \qquad \text{[A2]}$$

Quench Rate

$$\text{Quench rate} = \frac{T_m - T|_{x=-a}}{\Theta}$$

Nomenclature for Thermal Analysis

a	Thickness of puddle, cm
i,j,n,k	Integers

K_i	Thermal conductivity of ith phase, cal/cm-°C-sec.
l	Thickness of substrate, cm
p	Laplace transform variable
Y_i	Dimensionless temperature of ith phase

$$Y_i = \frac{T_i - T_0}{T_m - T_0}$$

α_i	Thermal diffusivity of ith phase, cm^2/sec
ν	Defined by $\nu = \sqrt{\alpha_1/\alpha_2}$
σ	Defined by $\sigma = K_1/K_2 \sqrt{\alpha_2/\alpha_1}$
Θ	Time, sec.
γ	Defined by $\gamma = 1 - \sigma/1 + \sigma$
W_i and Z_i	Defined by:

$$W_1 = -\frac{\nu x + 2a(n-k) + 2\nu l(k-j) + 2j(a+\nu l)}{2\sqrt{\alpha_1 \Theta}}$$

$$W_2 = \frac{-\nu x + 2a(n-k) + 2\nu l(1+k-j) + 2j(a+\nu l)}{2\sqrt{\alpha_1 \Theta}}$$

$$W_3 = \frac{\nu x + 2a(1+n-k) + 2\nu l(1+k-j) + 2j(a+\nu l)}{2\sqrt{\alpha_1 \Theta}}$$

$$W_4 = -\frac{-\nu x + 2a(1+n-k) + 2\nu l(1+k-j) + 2j(a+\nu l)}{2\sqrt{\alpha_1 \Theta}}$$

$$Z_1 = \frac{-x + 2a(n-k) + 2\nu l(k-j) + 2j(a+\nu l)}{2\sqrt{\alpha_1 \Theta}}$$

$$Z_2 = \frac{x + 2a(1+n-k) + 2\nu l(1+k-j) + 2j(a+\nu l)}{2\sqrt{\alpha_1 \Theta}}$$

$$Z_3 = \frac{x + 2a(1+n-k) + 2\nu l(1+k-j) + 2j(a+\nu l)}{2\sqrt{\alpha_1 \Theta}}$$

$\alpha_1 \Theta$

$$Z_4 = \frac{-x + 2a(n-k) + 2\nu l(1+k-j) + 2j(a+\nu l)}{2\sqrt{\alpha_1 \Theta}}$$

RAPID CASTING OF METALLIC WIRES AND RIBBONS

by

G. Haour and H. Bode
Battelle, 1227 Geneva.

SYNOPSIS

The purpose of this paper is to report on the fabrication and properties of metallic fibres having a microcrystalline microstructure. Such material belongs to the general class of glassy metals, which present a unique combination of tensile, magnetic and corrosion-resistant properties. This results from their liquid-like microstructure obtained by rapid solidification with cooling rates of the order of 10^6 °C per second.

Wire fabrication techniques will be reviewed. These include : Taylor wire casting, melt spinning, melt drag. In the technique of melt extraction, used in the present work, a disk spins in contact with the metal surface of a melt, thus removing by extraction metal which solidifies into a wire or a ribbon. A short film will be presented. This film was made using high speed cinematography to study the process of formation of wires extracted from the melt.

Fibres have been rapidly solidified in this way using iron-carbon-phosphorus and iron-carbon-phosphorus-chromium alloys. These fibres showed an X-ray diffraction pattern indicating a microcrystalline (or glassy) structure. Their properties will be discussed, with special reference to corrosion resistance, as compared to stainless steel.

INTRODUCTION

There has been a long standing commercial interest in processes that could make it possible to produce metallic wires and tapes direct from the melt. The short route from melt to product is all the more attractive in these days of energy conservation, as it avoids the intermediate steps of ingot casting, rolling and working to final size. New momentum in this field has recently emerged, since some of the techniques designed for the production of tapes and wires can be used for the rapid solidification of the melt into glassy or microcrystalline material. Such material requires cooling rates from liquid to solid in the range of 10^4 to 10^6 °C per second.

The glassy or microcrystalline material thus obtained has a liquid like microstructure and a very uniform chemical composition on a microscopic scale. This results in a unique combination of mechanical, magnetic and corrosion resistance properties (1). This paper will first review some of the techniques that can be used for the production of wires and ribbons by rapid solidification. Results will then be given concerning the production of wires by the technique of melt extraction. Iron-based wires obtained by means of this technique have been characterized by their microstructure and corrosion resistance.

"Melt to Wire" fabrication techniques of interest in this paper are: melt extraction, Taylor wire casting, melt spinning, and melt drag. The two latter techniques are also well suited to the manufacture of ribbons or tapes. These techniques, schematically illustrated in Figure 1, are described briefly below, while the melt extraction technique, which was used in the present work, will therefore be the object of a separate discussion in a following section.

In the Taylor wire process, a metallic rod contained in a glass tube is melted by induction for example, as shown in Figure 1. The glass is softened by contact with the molten metal; it can thus be drawn, while acting as a continuous mould crucible during the solidification of the metal, ensuring a regular surface and diameter of the wire. The composite product obtained, consisting of a metallic wire in a glass sheath, is collected on a rotating drum at speeds of about five metres per second. Typical wire diameters obtained in this way are 100 microns or below. Steel wires as thin as 25 microns have been produced at Battelle and in other laboratories (2). A critical aspect of this technique concerns the availability of sheath material having a sufficient chemical inertness towards the molten metal used, as well as a softening temperature consistent with the melting point of that metal. In addition, the necessity of removing the sheath by a relatively simple treatment might place further limitations upon the choice of the glass material to be used.

Melt spinning denotes a continuous casting process in which a liquid metal jet, formed by extrusion of molten metal through a small orifice, is solidified as wire or ribbon. As shown in Fig 1, the solidification of the molten stream takes place either in free flight or against a rotating chill. In free flight spinning, the diameter of the fibre produced is controlled by the size of the orifice; the main advantage of this technique is the simplicity of the principle involved, although the liquid jet stability requires strict control of the operating parameters. For this, relatively high extrusion velocities (several metres per second) are necessary.

In chill block melt spinning, the liquid metal jet is made to impact on a rotating cold surface, a disk or a drum for example. On impact the jet somewhat spreads into a liquid transient puddle which then solidifies very rapidly. Under certain conditions, this technique makes it possible to produce ribbons, as shown schematically in Figure 1. For this, appropriate orifice(s) and pressures should be used, in order to create a wide enough liquid puddle on a rotating drum. Wires and narrow ribbons have been produced in this way using a variety of alloys. A microcrystalline or glassy state can be obtained in the resulting material, if cooling rates of the order of a million degrees Centigrade per second are achieved, using iron-boron-carbon alloys for example (1,3). Work at Battelle has led to the manufacture of thin, ductile ribbons of low melting alloys having widths of up to 4.5 cm High melting point alloys, which are not amenable to conventional ribbon casting, could also be melt-spun into ribbons, owing to the fact that the rapid solidification conditions used suppressed the formation of brittle intermetallic phases. Ribbons were thus produced at a rate of about fifteen metres per second (4).

The melt drag process involves the contact between a metallic meniscus and a rotating water-cooled drum, as illustrated in Figure 1. This process was invented and developed by Battelle to the pilot-plant scale. Wires and ribbons of aluminium and its alloys, and also of steel and copper, have been produced with this technique, with ribbons having a width of up to 5 cm. Further details on this technique are given in reference 5.

Comparing the processes briefly described above, the following remarks can be made:

- the Taylor process is appropriate for the production of micro wires, 100 microns in diameter or less

- melt spinning can be used to manufacture wires having a diameter in the range of 100 to 500 microns, as well as thin ribbons having a width of up to several centimetres.

- melt drag is suitable for the manufacture of wires and ribbons having sections in excess of about 500 microns

- all the above-mentioned techniques can be operated under conditions of rapid solidification. It is then possible to achieve supersaturated solid solutions by imposing fast cooling rates removed from those in conditions given by the phase diagrams, as the formation of certain inter-metallic phases may be suppressed in this way. Further, if even more rapid cooling rates (of the order of 10^6 °C/second) are imposed, such alloys as Fe-B-C can be produced with a microcrystalline or glassy microstructure, which imparts to the material a unique combination of properties, as briefly mentioned in the previous section. Such features are also associated with the technique of melt extraction discussed below.

THE TECHNIQUE OF MELT EXTRACTION AND EXPERIMENTAL PROCEDURE

In this technique, a water-cooled disk rotates, while being kept in contact with the surface of a metallic melt, as illustrated in Figure 1. In this way, liquid metal extracted from the melt remains in contact with the disk for a very short time, it is then released spontaneously as a wire. This patented technique has been the object of several publications (1, 7, 8). and is being currently used to manufacture short steel fibres to reinforce concrete. In the experiment carried out inthe present work, a water-cooled disk was used, having a diameter of 15 cm. The periphery of the disk had a wedge angle of 90 degrees. Rotation speeds of the disk were up to 1200 rpm.

The experimental procedure for a run was as follows. The charge of alloy was melted under a protective atmosphere in an induction furnace Balzers VSG of 100 kW having a capacity of about 10 kg of steel. The bath was then raised by a driving mechanism to meet the rotating disk, extraction begins as soon as contact is established between melt and bath. Experiments have been made to assess optimal operating conditions in order to produce the rapidly-solidified microcrystalline structure sought. Adjustments were made to the speed of rotation of the disk, the area of contact between melt and disk (i.e. depth at which the disk plunged into the bath) and the temperature of the melt, which controls the viscosity of the liquid alloy.

Alloys used in our experiments were based on aluminium, copper, and iron. The present paper reports on tests using iron-based alloys having the composition given below :

Table 1 : Iron-based alloys produced by melt extraction as wires in the present work.

Alloy	Composition atomic %	weight %
Fe - P - C	80Fe-13P-7C	90Fe-8.1P-1.7C
Fe - P - C - Cr	70Fe-13P-7C-10Cr	80.6Fe-[illegible].1P-1.7C-10.6Cr

Each of these alloys was prepared in the crucible of the induction furnace. For this, appropriate amounts of carbon, or carbon and chromium, were alloyed to a commercial iron-phosphorus pre-alloy. Owing to this, it is expected that uncontrolled, but small, amounts of impurities were present in the melt. During melt extraction the melt temperature was maintained at 1350°C. Wires obtained were characterized for their microstructure and corrosion resistance. Results are given below.

EXPERIMENTAL RESULTS

Wires extracted from melts of Fe-C-P and Fe-C-P-Cr alloys had diameters ranging between 0.1 and 0.5mm, depending upon the production conditions. Lengths obtained were up to about 10 cm, as the wires broke due to their impact on the walls of the protective chamber of the furnace.

Since the present work is aimed at producing small quantities of samples to be evaluated, no systematic attempt has been made to design and operate a system enabling the continuous production and collection of the wires.

In order to observe the process of the formation of the wires, a film was taken, using high-speed cinematography. This made it possible to observe the phenomena taking place at a speed of about 100th of the speed in the actual process. It could thus be shown that the stream of melt extracted by the rotating disk remains liquid for a fraction of a second before leaving the disk.

If the disk is immersed to a depth of about 1 cm into the melt, turbulences develop on the surface of the melt and a large unstable stream of melt is extracted. Short, thick fibres, 0.5 mm average diameter, though with a somewhat irregular cross-section, are obtained as a result. If, on the other hand, and keeping the same rotation speed, the disk is made to barely touch the surface of the melt, a thin continuous stream of liquid is extracted, resulting in thin wires 0.1mm in diameter with a more uniform cross-section. This difference in diameter is indeed accompanied by variations in cooling rates which in turn alter the microstructure of the wire obtained, as discussed below.

Studies were carried out to assess the crystallinity of the wires produced. For this, X-ray diffraction was used. Both types of as-cast wires, thin and thick, were tested; results are given in Figures 2 and 3 for each type of alloy investigated.

In Figure 2, the two top diffraction patterns correspond to iron-phosphorous carbon wires in the as-cast state. The pattern a), with a unique broad low, diffraction peak, characteristic of a microcrystalline structure, was obtained with thin fibres. This suggests that the fibre solidified with a rapid cooling rate in the range of 10^5 to 10^6 °C per second. On the contrary, the diagram b) in figure 2 shows a substantial amount of crystallinity in the thicker fibre, about 0.5 mm in diameter. For comparison purposes, the wires obtained were recrystallized by annealing (2 hrs at 500°C). The X-ray diffraction patterns obtained, shown at the bottom of Figure 2, contain the well-developed diffraction peaks of the fully recrystallized material.

Similar results were obtained with iron-phosphorous carbon-chromium wires, as shown in Figure 3. In that figure, a similar difference can be seen between X-ray diffraction pattern of the thin rapidly solidified wires and that of the recrystallized material.

Corrosion resistance of the iron-phosphorous-carbon chromium alloys was tested in the microcrystalline and in the recrystallized states. Corresponding current/potential curves are shown in Figure 4. The medium used was a solution of sulphuric acid and sodium chloride. As shown in figure 4, the microcrystalline material displays an excellent passivity well above that of the same material in a crystalline form, which, in particular, presents a sharp pitting peak at about 400mV. In these conditions, the microcrystalline wire presents a corrosion-resistance similar to that of stainless steel with 18Cr-8Ni (by wt%), while containing only 10.6 weight per cent of chromium. This indicates that a very significant improvement in corrosion resistance is obtained by going from crystalline to microcrystalline.

Specific explanations of such an improvement are still being explored. However, it is clear that the rapid solidification used in production ensures that the material consists of a supersaturated solid solution having a highly uniform chemical composition on a microscopic scale. In addition, passivation phenomena resulting from the presence of chromium ions seem to be enhanced in the microcrystalline material as compared to the crystalline counterpart.

CONCLUSIONS

Wires have been produced by extraction from the melt, using two iron based alloys. The process of formation of the wires was studied by means of high-speed cinematography. It was thus shown that the extent of the contact between disk and melt essentially controlled the cooling rate from liquid to solid, and, in turn, the microstructure of the material obtained.

Thin, rapidly solidified wires were shown to have a microcrystalline structure and thus, a significantly improved corrosion resistance, as compared to the same alloy in a crystalline form. Applications for this type of material possibly include low-cost substitutes for stainless steel, provided that temperature in service remains well below the recrystallization temperature (about 350°C) of the material.

REFERENCES

1) Third International Conference on rapidly quenched metals, held in Brighton, U.K., 3-7 July 1978 (Proceedings to be published).

2) Copper microwires spun from the melt, G. Manfre et al (1974) J. Mat. Science 9, page 80.

British patents 1.153.577 (1969), 1.197.972 (1970) and 1.210.920 (1970) filed on behalf of Monsanto Company

3) Production of amorphous alloy ribbons, H.H. Lieberman (1976) IEEE Transition of Magnetics 12, page 921.

4) Battelle-Geneva, unpublished research.

5) Manufacture of metallic wires and ribbons, J.C. Hubert et al., Zeitschrift fur Metallkunde 64 (1973), 835-843

6) US patent 3.838.185 (R.E. Maringer et al, inventors) issued, Sept 24, 1974 and owned by Battelle Development Corporation

7) Casting of metallic filament and fibre, R.E. Maringer and C.E. Mobley, Journal of Vacuum Science and Technology II (1974) 1067-1071

8) Melt extraction of metallic glass candidate materials, E.W. Collins, C.E. Mobley and R.E. Maringer. Symposium on Spinning Wire from Molten Metal (American Institute of Chemical Engineers, New York 1978) page 102.

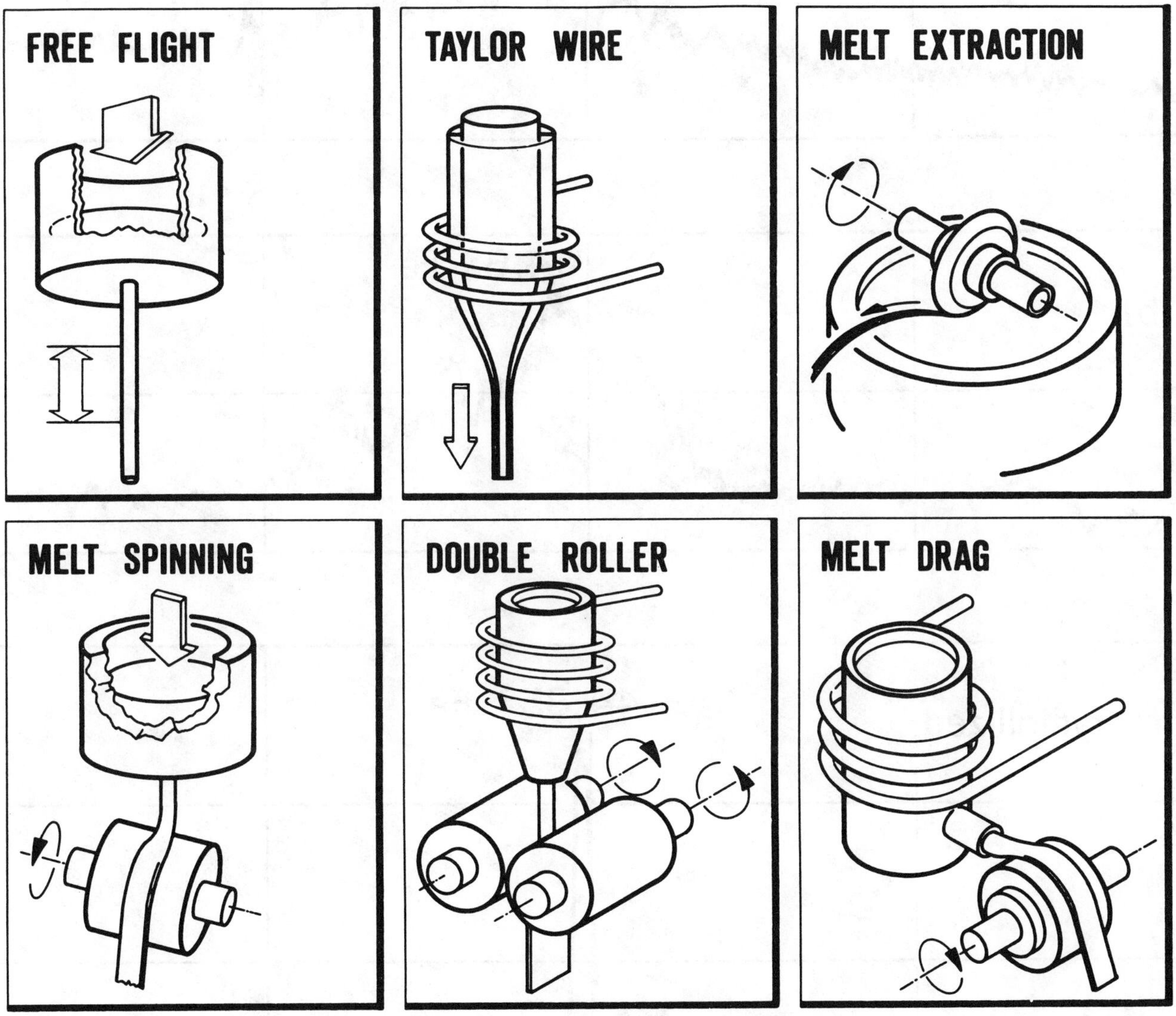

Fig. 1 Techniques for the rapid casting of metallic wires and ribbons.

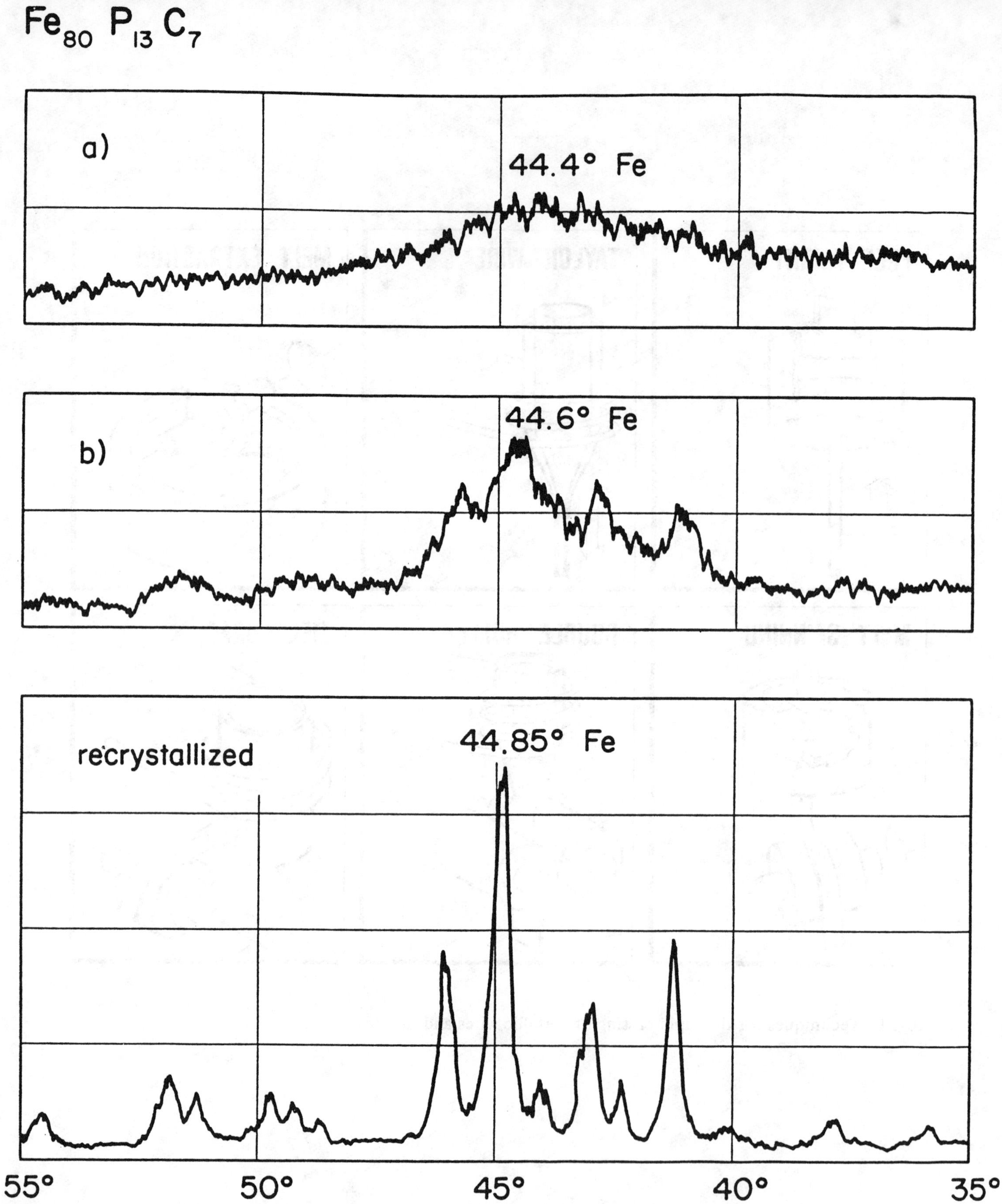

Fig. 2 X-Ray diffraction patterns obtained with melt extracted Fe-C-P wire.
a) as-cast thin microcrystalline fibre.
b) as-cast, thick crystalline fibre.
Bottom: fibre after recrystallization by annealing.

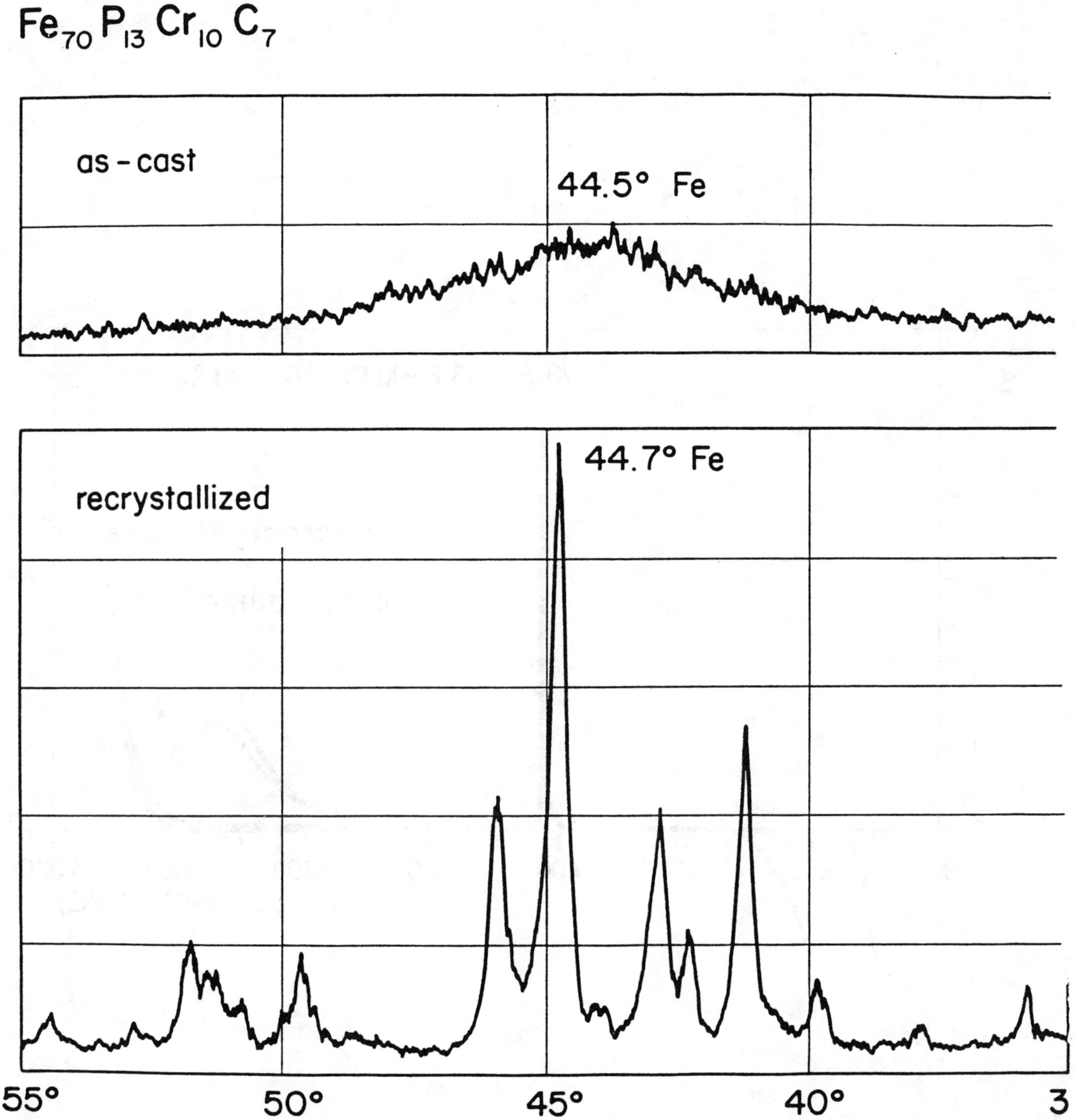

Fig. 3 X-Ray diffraction patterns of melt extracted Fe-C-P-Cr wire.
Top: as-cast, thin,microcrystalline fibre.
Bottom: fibre after recrystallization by annealing.

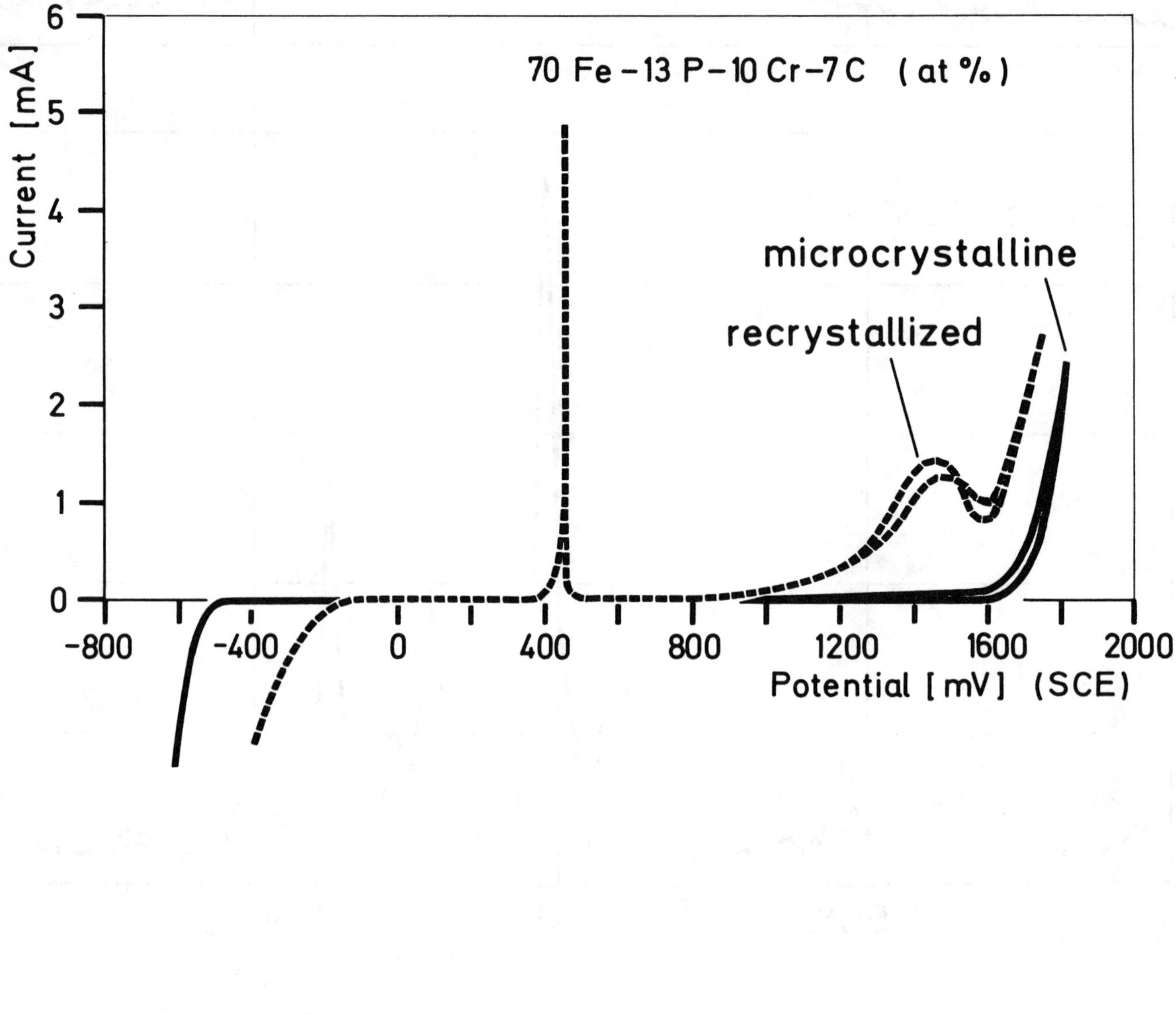

Fig. 4 Corrosion behaviour of microcrystalline and recrystallized Fe-P-C-Cr wire in 1N H_2SO_4 + 0.1 NaCl.

PRODUCTION AND PROCESSING OF RAPIDLY QUENCHED ALUMINUM POWDERS

R. E. Maringer
Battelle-Columbus Laboratories
Columbus, Ohio

Abstract

Aluminum powder metallurgy, using rapidly quenched (RQ) particulates, is showing great promise in providing materials with significantly improved properties. A number of available processes (some still in the laboratory, some in advanced development, and some already developed) appear capable of providing tonnage quantities of RQ particulates should the demand develop. The manufacture of the RQ particulates and their subsequent processing to useful products is discussed.

1. INTRODUCTION

This is an exciting time to be a metallurgist. The developments in rapid quenching over the past two decades have opened up enormous opportunities for research, development, and applications. It now appears possible to design new alloys with enhanced mechanical, physical, and chemical properties. For some properties, the enhancements are incremental – say 10 to 30 percent. For others (specifically corrosion resistance) enhancements can be of orders of magnitude.

The field of rapid quenching alloys is developing across a broad front, with efforts aimed at exploiting ferromagnetism, superconductivity, electrical conductivity, corrosion resistance, resistance to radiation damage, elastic properties, mechanical properties, etc. New patents on processes and alloys are appearing almost weekly. While much of the research is concentrating on the production and property evaluations of amorphous (noncrystalline) alloys, significant research is being directed toward effects of rapid quenching on crystalline alloys. The reasons for this are clear. Rapid quenching produces at least five effects, all of which enhance the ability of the metallurgist to control structure and properties.

These effects, as the quench rate increases, ($\sim 10^2$ to 10^8 C/s), are:

(1) Decreasing grain size
(2) Increasing chemical homogeneity
(3) Increasing supersaturation
(4) Retention of nonequilibrium crystal structures
(5) Increasing number of quenched-in vacancies

As grain size decreases, room-temperature strength increases. At the same time, strength at higher temperatures decreases and ductility at the higher temperatures increases. It is often possible to achieve superplastic properties on RQ materials. Lower pressures and temperatures are often possible in powder consolidation and sintering.

Chemical homogeneity is most often significant in supersaturation. Most of our commercial alloys contain alloy additions in amounts exceeding the equilibrium solid solubility. In the normal process of ingot metallurgy, the additions separate as discrete particles during the casting of the original ingot. They are subsequently broken up by mechanical working or by solution heat treatment. Often, however, the diffusion distances involved in homogenization are too large, and these chemical inhomogeneities survive into the finished product. These then act as stress concentrators or sources of corrosion initiation. It is not surprising that the more uniform materials produced by rapid quenching exhibit enhanced tensile, fatigue, fracture-toughness, and corrosion properties. Furthermore, rapid quenching permits the preparation of alloys well outside of equilibrium limits, which opens up whole new areas for developing precipitation and dispersoid-strengthened materials.

The utilization of nonequilibrium crystal structures (such as austenite or martensite in steels) in present materials technology is reasonably well developed and understood. Rapid quenching may permit us to extend these principles to a wide variety of other alloy systems.

The increased number of quenched-in vacancies provide a significantly enhanced diffusion rate (until the vacancies are used up) and this allows reactions such as precipitation to occur at lower temperatures and shorter times than might otherwise by expected for a given alloy. Thus an alloy such as 7075 aluminum, which is normally aged at 120 C, will age rapidly at room temperature.

Rapid quenching from the liquid state requires that heat be removed from the solidifying solid in very short times. This implies that the RQ product must have at least one very small dimension. Thus thin sheet, flake, ribbon, and powder are the normal RQ products. The heat can be extracted by radiation, convection, or by conduction. Rates of temperature change can vary from tens or hundreds of degrees per second to millions of degrees per second, depending on particle size and/or the mechanics of quenching.

Clearly, rapid quenching introduces new parameters, new properties, and new problems, as well as new opportunities to the metallurgist. Nowhere is this more apparent than in the powder metallurgy of aluminum. This paper describes some production and processing techniques being developed for aluminum and indicates some of the opportunities (and therefore decisions) that will face the aluminum metallurgist in the near future. Meeting these objectives would be a simple job if it required only a listing and description of various techniques for rapid quenching of particulate aluminum. However, the ingenuity of researchers around the world has been amply demonstrated by the number of ways devised to overcome the problems of aluminum. Some of the various methods in use are given below.

2. PROCESS VARIANTS

2.1 A Number of Processes Involve Breaking Up Liquid Streams Into Molten Droplets, With Subsequent Solidification of These Droplets in Flight

2.1.1 Gas Atomization

In gas atomization(1), one or more high-velocity air jets or other gas impinges upon a stream of molten metal (see Figure 1). The stream is broken up into small particles, generally allowed to solidify in flight by convection and radiation. Thus the cooling rates depend strongly upon particle size, with higher rates as particle size decreases.

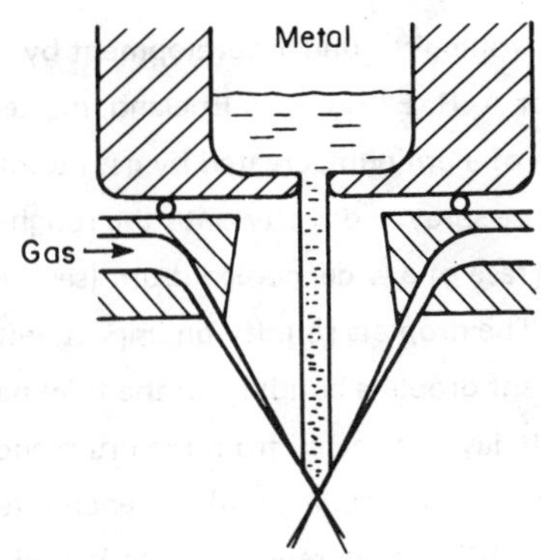

Fig. 1. Gas atomization.

There is a wide distribution in particle size, but generally quench rates in the range of 10^2 to 10^4 C/s are achievable. Such a method represents existing state of the art. Production rates of tens of pounds per minute can be achieved, and the process can be adapted to batch or continuous operations. The powder handles and pours well. With aluminum, however, a special problem exists. When aluminum powders are handled in moist air, hydrates tend to form on their surfaces. These, if not removed, tend to form blisters on subsequent solution annealing. Thus, green compacts of aluminum are generally canned, evacuated, and outgassed at some appropriate temperature (as high as 500 C) prior to consolidation by hot pressing, hot isostatic pressing, extrusion, etc. If the powder is produced and handled in an inert atmosphere, the outgassing step can be greatly simplified, but at added cost.

2.1.2 Ultrasonic Atomization

The ultrasonic atomization(1) is similar to gas atomization except that the impinging gas is excited to a high frequency by an ultrasonic transducer. This causes the molten stream to break up more efficiently, producing finer, more uniform particles and of course, higher average quench rates. This process is currently being developed under the leadership of N.J. Grant at MIT. It is believed that quench rates in excess of 10^6 C/s are possible. Subsequent processing is similar to that described for gas-atomized powder. It must be remembered, however, that finer powders have more surface area, and care must be exerted to insure the safe handling required of all reactive materials in finely divided form. Production rates could be made similar to those achieved in gas atomization.

2.1.3 Centrifugal Atomization

Centrifugal atomization as practiced by Pratt and Whitney in Florida(2), involves the impinging of a stream of molten metal onto the surface of a rapidly spinning cup. The liquid metal is mechanically atomized and thrown off the edges of the spinning cup. Solidification takes place in flight, and can be speeded up by blasting the emerging particles with a stream of helium. The particles are generally spherical, with the average particle size decreasing with increasing disk speeds. Quench rates of the order

Source: The 1980's — Payoff Decade for Advanced Materials, May 1980, 67-74 of 10^4 to 10^6 C/s can be achieved. Production rates and subsequent consolidation procedures should be similar to those for gas atomized powders.

In a variant on this theme, Reynolds Aluminum has developed a method(3) where molten aluminum is poured into a deep rotating cup with holes in the side (see Figure 2).

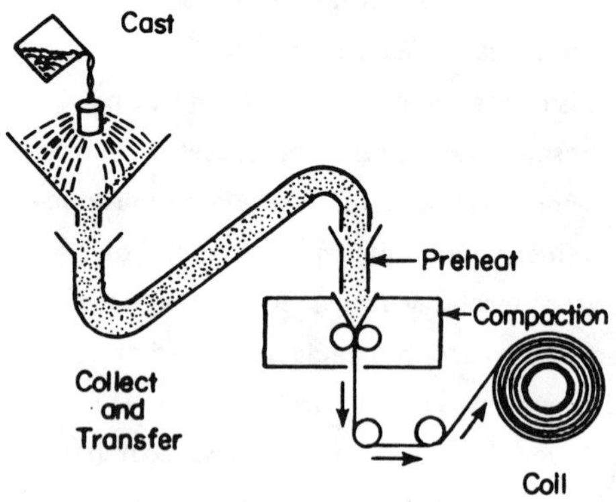

Fig. 2. Reynold's sheet process.

Centrifugal force causes the molten aluminum to flow through the holes. The streams created break up in flight to form football shaped particles perhaps -10 + 70 mesh in size. While these are relatively slowly quenched ($\sim$ 10 to 10^2 C/s), they are still far more rapidly quenched than would be the case in ingot metallurgy. In any event, these particles are subsequently introduced into a rolling mill and consolidated directly as sheet at rates up to 60 meters per minute. Thus, canning and outgassing steps are eliminated, and the process can be made continuous. It has not been demonstrated, to my knowledge, that the same process could be made to operate with finer particulates of more rapidly quenched alloys specifically formulated to achieve improved properties. If so, the economics might be very attractive.

2.2 Substrate Quenching

While most of the methods described above involve forming molten droplets which solidify in flight, it is possible to increase the quench rate appreciably by allowing the molten metal to impinge directly on a cooled substrate. In effect, the spray rolling process discussed below does this to some extent, but with the molten particles adhering to the previously cast material. In general, if the substrate is properly prepared and sufficiently cooled, molten particles will solidify on that surface, but will not adhere there. There are a number of methods of substrate quenching.

2.2.1 Spray Rolling

In spray rolling(4), under development by Professor A. R. E. Singer in England, molten droplets of aluminum, created by gas atomization, are sprayed directly onto the roughened surface of a wide, cooled drum (see Figure 3). The droplets solidify on impact, with subsequent droplets building up the thickness. The built-layer is peeled from the drum and run directly into a rolling mill. Quench rates as high as 10^6 C/s are reported even though the substrate (the previously cast particles) is at a reasonably high temperature. The process can be either batch or continuous. Here all powder handling is eliminated.

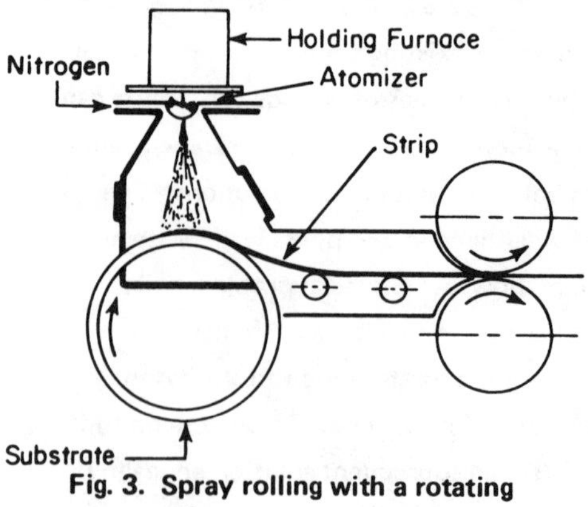

Fig. 3. Spray rolling with a rotating substrate.

2.2.2 Alcoa Splat

At the Alcoa Research Center, gas atomized particles impact on the cooled surface of a rotating drum (see Figure 4). Here they form small splats that leave the drum surface after solidification and are subsequently collected and screened. Production rates can be tens of pounds per minute. Quench rates depend upon the size and velocity of the molten particles, but rates of up to 10^6 C/s appear possible. Obviously some of the molten particles will strike previously cast particles, thus forming multiple-layered particles with a lower average quench rate. The particle, or flake, size is quite variable. Handling and pouring are not as easy as with spherical powders, and the tap density is significantly lower than for spherical powders. On the other hand, such flakes are subjected to higher quench rates than most of the spherical particulates and therefore should permit a wider range of alloy compositions to be processed.

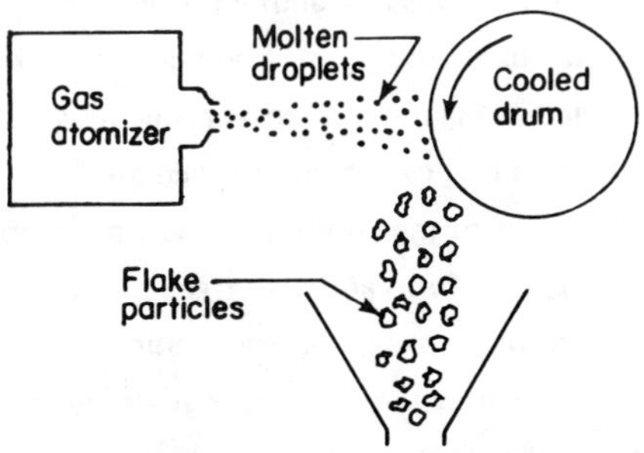

Fig. 4. Alcoa flake process.

2.2.3 Double Roller.

Professor N. J. Grant at MIT is developing a method which involves introducing molten particles into the bite of a double roller (see Figure 5). The particles are quenched on both sides; therefore, they have the advantage of a more uniform structure and an enhanced quench rate. Quench rates in excess of 10^6 C/s are probably feasible. As with the Alcoa process, the flakes produced are irregular and depend strongly on the size of the incoming droplet and on just where it strikes the drum on entering the bite. Advantages and drawbacks are similar to those of the Alcoa splat.

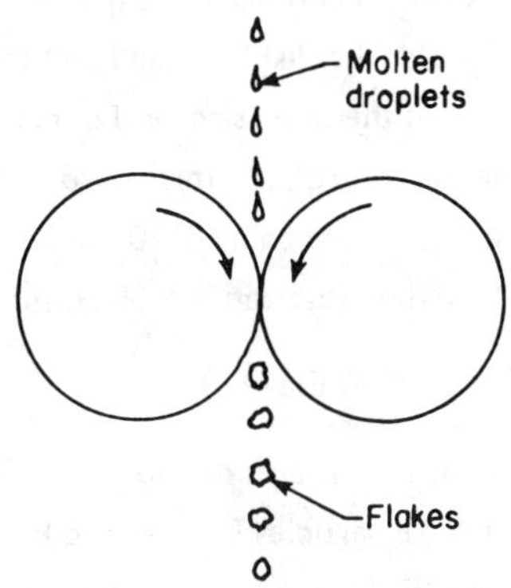

Fig. 5. Double roller flake casting process.

2.2.4 Melt Extraction

In melt extraction[5], the edge of a spinning disk is introduced to the surface of a pool of molten metal (see Figure 6). The metal

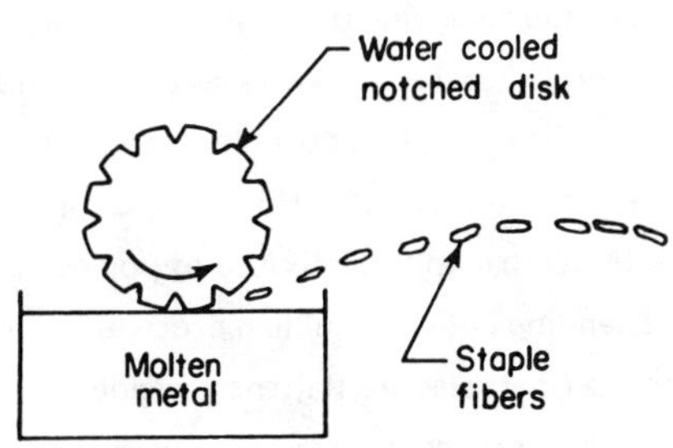

Fig. 6. Melt extraction of staple fibers.

Source: The 1980's — Payoff Decade for Advanced Materials, May 1980, 67-74 solidifies on the disk edge, adheres there briefly, then is thrown free. Notching the disk surface at appropriate intervals allows staple fibers of a specified length to be cast. The fiber thickness (and quench rate) is controlled largely by the rotational speed of the disk. Transmet, Inc., a subsidiary of Battelle, is nearing production of aluminum staple fibers with dimensions of approximately 0.05 x 4 mm. Quench rates are probably in the 10^4 to 10^6 C/s range. Since each particle is cast against a specific segment of the mold (the disk), virtually all the particles are of the same size, and of course the size can be controlled. Initial production rates are expected to be 5-10 pounds per minute and production will be continuous.

2.2.5 Transmet Flake

Transmet, Inc., is also developing a method to cast flake particles (~ 1 mm square x 0.03 mm thick). The process has not yet been disclosed, but is known to produce quench rates at least up to 10^6 C/s. Once again, flake dimensions are uniform and handling problems are thereby reduced. Production is continuous and rates about 10 pounds per minute are anticipated.

2.3 Mechanical Alloying

All of the foregoing discussion has centered around RQ and the advantages derived therefrom. However, it is possible to create such materials without RQ. Novamet, subsidiary of INCO, ball mills a mixture of powders.[6] Depending upon the milling procedures, the powder particles are flattened, welded together, broken up, etc., continuously. After a period of time, the powders are mechanically alloyed and appear to be compositionally homogeneous. By this means it is possible to develop homogeneous powders of mixtures of elemental materials, of mixtures of alloys and nonmetallic particles (for dispersion strengthening), etc. The alloyed powders can then be consolidated using the same methods as used for the RQ particulates.

3. DISCUSSION

There is already ample evidence that the production of metastable aluminum alloys using the RQ technology is a viable path to achieve improved properties. If we assume that such developments will continue, and that a variety of new and improved alloys will flow from research laboratories in the future, then we must begin to consider the problems to be faced in the commercial production of these alloys. Ultimately, the cost of the alloy will be the dominant factor. Of course, design trade-offs will take advantage of the improved properties, but these can absorb only a limited increase in cost. In this context, the decisions required to bring metastable aluminum alloys to the market place begin to come into focus.

The various steps in manufacturing components utilizing the RQ technology are briefly outlined in Figure 7. In effect, one must melt the alloy, cast it into particulate form, transport it to the consolidation step, consolidate it, then form and heat treat the component. In each step we are opening up new ground. For example, there is evidence to suggest that in the melting step, the properties of the product will vary depending upon the melt temperature. In the particulate step, a given alloy may require a quench rate of 10^6 C/s to achieve its best properties while another alloy may require only 10^3 C/s. A specific powder shape, ideal from a quench rate point of view, may be difficult to handle

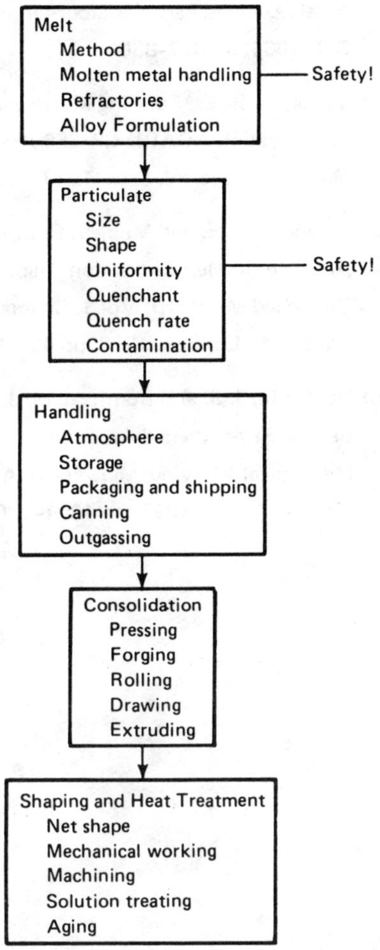

Fig. 7. Some parameters of concern producing and processing aluminum powders.

from safety or packing considerations. Consolidation can be done at conventional temperatures and pressures, but may be worthless if the temperatures are high enough to eliminate the advantages in grain size and homogeneity brought about by rapid quenching. The same is true for heat treatment. RQ alloys will respond differently to heat treatment, and the differences must be taken into account.

Obviously, among the processes outlined earlier, it is possible to by-pass certain of these steps. In the spray rolling process, for example, the powder-handling step is eliminated since the particulates are formed, quenched, and at least partially consolidated in a single continuous step. Similarly, in the Reynolds centrifugal atomization and sheet processing, the particulate does not have to be handled, outgassed, or canned in the conventional way. There is also the potential of continuous extrusion of wire and rod, which eliminates canning and much handling. We have demonstrated at Battelle that we can cast aluminum as thin, RQ sheet which is subsequently coiled and extruded to consolidate it. Here the particulate form is eliminated and handling and outgassing are greatly simplified. The economic advantages of these various routes to a product have not yet been fully explored.

One final point must be made. The RQ technology can benefit metals and alloys other than aluminum. In the laboratory, we have been able to triple the tensile strength of a titanium alloy. In other words, those metals and alloys (including composites) which are today's competition for aluminum alloys are also being improved. The future looks very bright indeed for those who make their living by calculating cost/benefit ratios.

4. CONCLUSIONS

Methods for producing large quantities of RQ aluminum alloy particulates are under development both in the U.S. and abroad. Some are entering the commercial market place while others appear capable of doing so. Methods for subsequent handling, compacting, consolidating, and forming these particulates into useful shapes are also under intense development. All of these developments range from near conventional to highly nonconventional. Processing developments appear, in fact, to be ahead of the development of the special alloys which can take advantage of the RQ technology. It seems clear that, if these new and exciting RQ materials are desired by industry, the basic technology to produce them is there. Cost and competition will decide the issue.

5. REFERENCES

(1) Lawley, A., An Overview of Powder Atomization Processes and Fundamentals Int. J. of Powder Met. and Powder Tech. Vol 13, No. 3, July 1977, pp 169-187.

(2) Cox, A. R., Moore, J. B., and VanReuth, E. C., Rapidly Solidified Powders, Their Production, Properties, and Potential Applications. AGARD CP-256 Advanced Materials Processes, Sept. 1978, Paper 12.

(3) Daugherty, T. S., Aluminum Sheet from Finely Divided Particles, Journal of Metals, Oct. 1964, pp 827-830.

(4) Singer, A. R. E., Forming Metals by Rapid Solidification AGARD CP-256 Advanced Materials Processes, Sept. 1978, Paper 19.

(5) Maringer, R. E. and Mobley, C. E., Advances in Melt Extraction, Rapidly Quenched Metals III, Vol 1, Chameleon Press Ltd., London, 1978, pp 49-56.

(6) Benjamin, J. S. and Bomford, M. J., Dispersion Strengthened Aluminum Made by Mechanical Alloying, Met. Trans. A, Vol 8A, August 1977, pp 1301-1305.

Melting and solidification during dynamic powder compaction of tool steel

D. G. Morris

A detailed study has been carried out of the structures produced by the dynamic compaction of tool steel powder and used to deduce information on the processes occurring during compaction. Some of the compacting powder is seen to melt during compaction while subsequent solidification occurs so rapidly that metastable structures are produced. Rates of solidification may be determined by heat-conduction calculations or from the details of solidification microstructure. In all cases solidification occurs extremely rapidly and structures and properties very similar to those produced by conventional rapid solidification processing of tool steels may be identified. Using such information of quenching rates it is possible to deduce in some detail the temperature variations suffered by the compacting material during the dynamic processing.

MS/0622

 Manuscript received 18 July 1980; in final form 6 October 1980. Dr D. G. Morris is with the Institut CERAC, Ecublens, Switzerland.

Dynamic powder compaction has recently been developed as a technique whereby high pressures produced by the impact of fast moving punches on to powder cause very effective densifications and interparticle bonding.[1-4] Although features of the compaction technique have been described and also some of the applications enumerated, little has been detailed about the physical mechanisms occurring within the materials as compaction occurs. It has previously been stated[2] that during such dynamic compaction the powder particles are thrown against each other so violently that oxide films are broken off, particle surfaces may melt, and a strong metallurgical bond is formed. The present work was carried out to examine in more detail the processes occurring for cases of dynamic compaction where particle surface melting has occurred and subsequent solidification has led to strong bonding.

EXPERIMENTAL PROCEDURE

The material used for the present study* was a water-atomized, type M2 tool steel powder supplied in the annealed (softened) condition by Davy-Loewy Ltd. The powder particle size was typically 150 μm. Plastic punches were propelled using a two-stage gas gun (compressed air in first stage and helium in second) at velocities in the range 1000–2500 m s^{-1}, producing impact pressures in the range 2–8 GPa in the steel powder. Compaction was carried out in a specially designed steel pressure container, from which the specimens were subsequently extracted for examination. (Fuller descriptions of the techniques involved have been outlined elsewhere.[1-3])

The microstructures of the compacted materials were examined optically by scanning electron microscopy using a Cameca instrument and by transmission electron microscopy using a Philips 300 microscope. Thin foils were prepared using a Struers Tenupol twin-jet device at a voltage of 18 V with a 10% perchloric acid–ethanol mixture. Additional structural studies were carried out by X-ray diffraction using a Philips diffractometer with filtered copper radiation and a proportional counter with pulse-height discrimination for detection. Microhardness testing was carried out on polished sections of samples using a Leitz hardness tester with a Vickers diamond indenter. An indenting load of 50 g was used: this small load was necessary in view of the small size of the melted zones to be examined.

RESULTS

As-compacted structure

An optical micrograph showing the structure of a typical high-velocity compaction is given in Fig. 1. There was generally little indication of porosity or of poorly bonded interfaces between powder particles. However, of striking interest were white unetched regions between the mottled powder particles. The softened powder particles were composed of a bcc iron phase and mainly M_6C carbides, as confirmed by X-ray diffraction: the micrometre-sized carbide particles were responsible for the mottling inside the powder particles. Scanning electron microscopy using EDX analysis failed to find any structure within the unetched regions and showed that the chemical composition was the same as the bulk alloy composition.

Figure 2 indicates the structure observed within an unetched region where a solidification cell structure can be distinguished. It is obvious that the white regions in Fig. 1 correspond to material which melted during the dynamic compaction and subsequently solidified to produce the fine cell structure observed. The solidification cell sizes observed depended on the size of the melted zone (recorded as the distance across the thin dimension) and Fig. 3 gives examples of cells observed, illustrating the range of variations studied: the one scanning electron micrograph was recorded on the polished and heavily etched surface of a particularly large melt pool near the edge of an overcompacted specimen. For very small melt-pool sizes corresponding, as will be discussed later, to very high cooling rates, extremely fine structures were seen and there

* Bulk alloy composition: 0·95%C, 3·78%Cr, 6·03%W, 2·14%V, 4·92%Mo, (wt.-%), balance Fe.

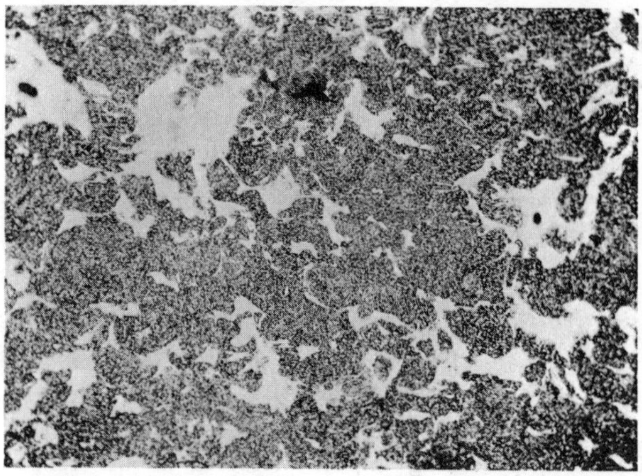

1 Optical micrograph showing typical structure in dynamically compacted tool steel ×350

appeared to be a breakdown of the solidification cell structure giving way to a more homogeneous or uniformly transformed microstructure (Fig. 4). Martensite was occasionally observed within the melted zones (Fig. 5). Such observations were exceptional and it was not normal to find any martensite within a melted zone. Where observed it appeared to lie near the interface between melted and unmelted materials. Diffraction patterns taken from the material of the melted zones showed spotty ring patterns corresponding to the large number of grains and precipitates examined (Fig. 6). Following camera length calibrations using thallium chloride as standard, the diffraction patterns were analysed to identify a bcc iron solid-solution phase and M_2C carbides with lattice parameters of 0·292 (bcc) and 0·290 and 0·468 (M_2C) nm. No signs of other carbides or of an fcc (iron) phase were found. Figure 6*b* shows a diffraction pattern taken from a region containing martensite where some streaking of the diffraction spots may be seen.

Hardness and aging response

Hardness measurements performed on the compacted specimens showed that the melted regions were very much harder (900–1000 HV) than the softened particle interiors (about 500 HV) and that there was no significant variation of hardness with melt-pool size. The hardness values measured on specimens after aging for 30 min at temperatures in the range 100–700°C are shown in Fig. 7 and compared with hardnesses quoted for a solid-state quenched and aged type M2 tool steel.[5] It may be seen that the hardness of the particle interiors was approximately constant, while that of the melted zones varied in a fashion similar to that of solid-state quenched material. Nevertheless, the melted zone's hardness was always considerably greater than that of the solid-state quenched material and the secondary hardening peak was extended to a temperature of about 50 K higher. Similar improvements in properties have previously been observed for splat-quenched tool steels.[6]

Microstructural changes on aging at temperatures of up to 700°C have been examined *in situ* using a hot stage and transmission electron microscopy. This study, to be reported in more detail at a later date, showed that microstructural changes were hardly detected until about 500°C and even in the range 600–700°C took place very slowly (Fig. 8). Extensive precipitation and coarsening of M_2C carbides were observed, but there was only very slight indication of other, probably M_6C, carbides and then only at the highest temperatures (above 600°C).

Microstructures observed after aging at even higher temperatures are shown in Fig. 9. It may be seen that over the temperature range 800–1100°C more extensive precipitation and coarsening has occurred within the melted zones which are now darkly etched. Over the same temperature range, greater precipitate coarsening has occurred in the original powder particle interiors and there was also evidence of the ferrite to austenite transition. Even at such high temperatures, precipitation and coarsening processes were delayed in the melted and quenched material relative to the solid-state quenched material. After heat treatment at 1200°C, the structure of the original powder particle interiors has greatly simplified, since considerable

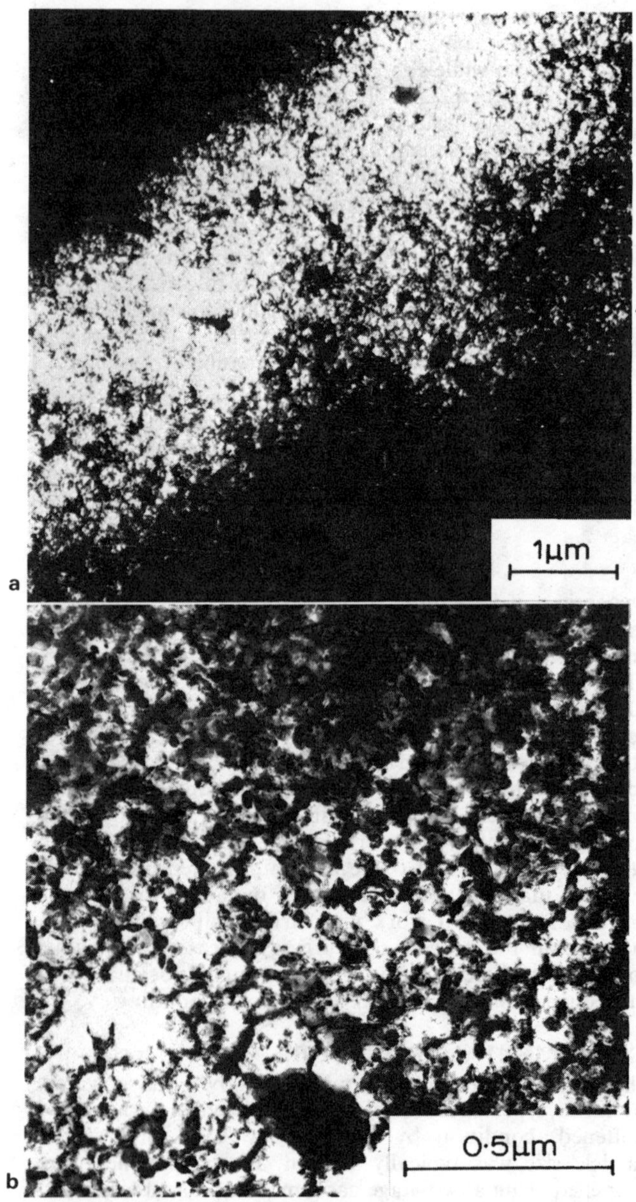

2 Electron micrograph illustrating solidification cell structure observed within a melted zone showing *a* entire melt-zone thickness and *b* detail within melt zone

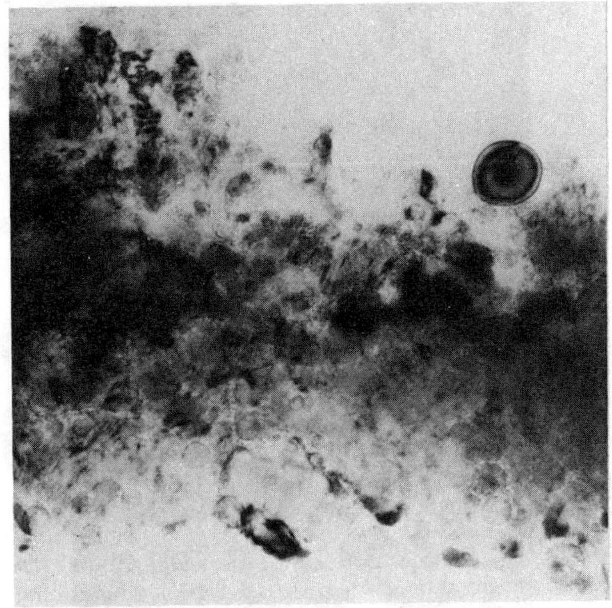

4 Breakdown of solidification cell structure for very small melt pools (solidification rate extremely rapid) ×142 500

austenite grain growth has occurred with the dissolution of some, and coarsening of the remainder, of the carbide precipitates. The sites of the originally melted zones were distinguished as small regions of fine-grained material, although whether this was produced by slower precipitate dissolution and austenite grain growth or by melting of an unusually highly alloyed matrix[7] was not established.

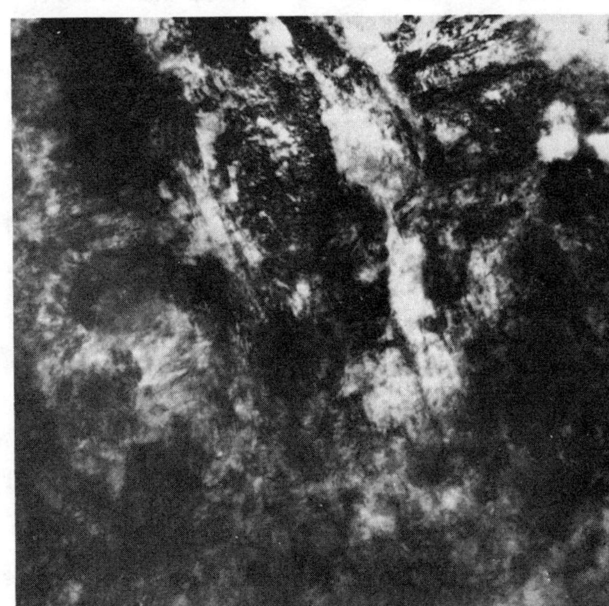

5 Martensite within melted zones ×108 000

3 Scanning and transmission electron micrographs illustrating variation of cell size with melt-pool thickness; melt-pool total widths were about *a* 240 μm, *b* 20 μm, *c* 1·5 μm

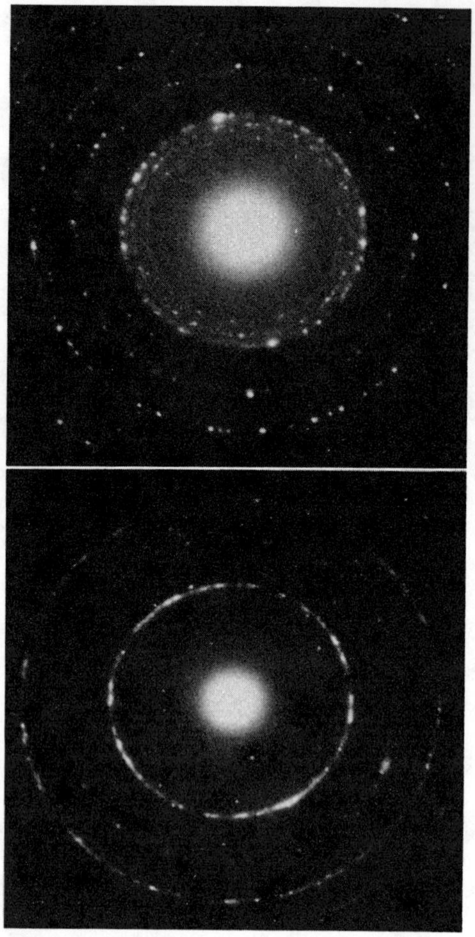

a

b

6 **Diffraction patterns from *a* typical melted zone, *b* martensite-containing region**

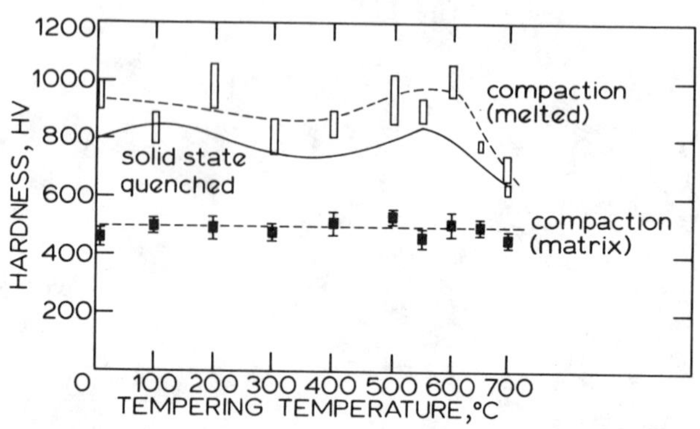

7 **Hardness as function of aging temperature for melted zones and matrix interiors of compacted material and for solid-state quenched steel**

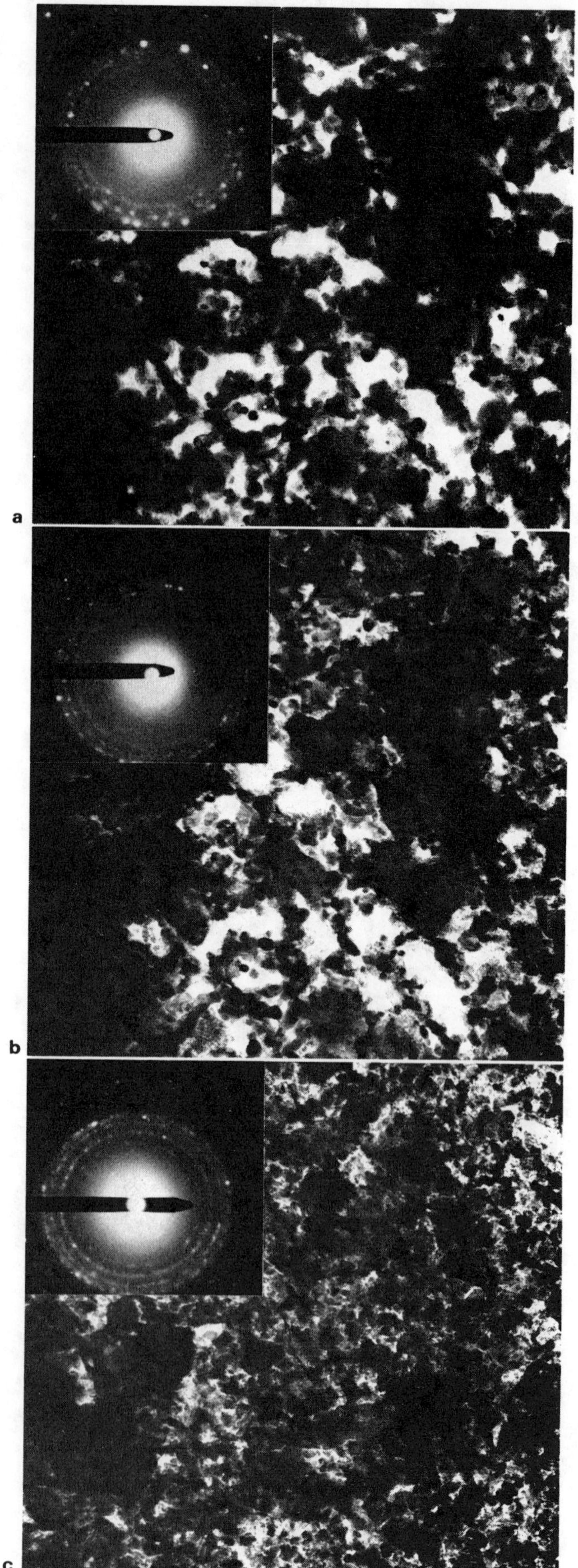

a

b

c

a as compacted; *b* at 500°C; *c* at 650°C

8 **Electron micrographs illustrating changes observed on heating (for about 30 min)** × **46 400**

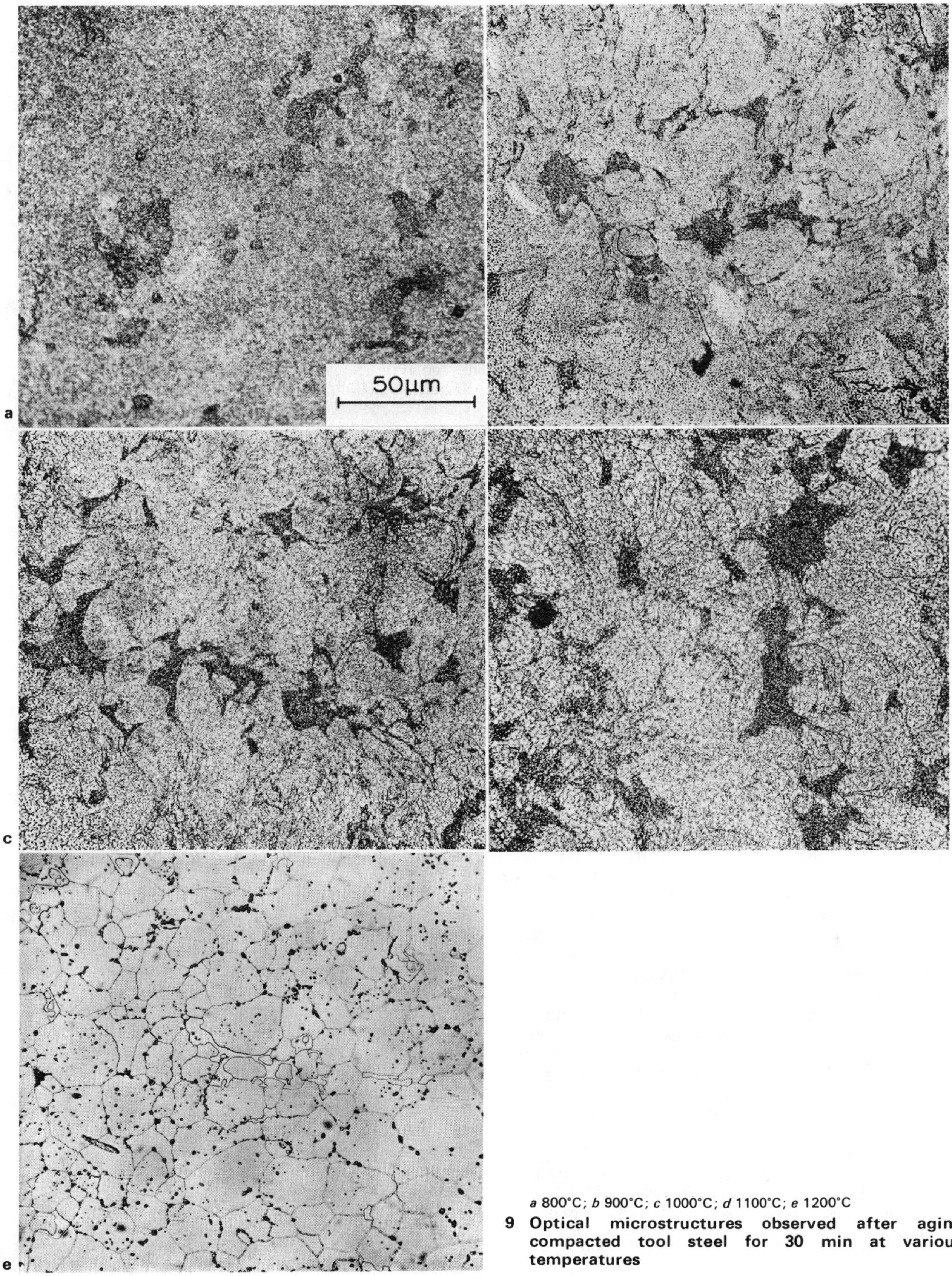

a 800°C; *b* 900°C; *c* 1000°C; *d* 1100°C; *e* 1200°C

9 Optical microstructures observed after aging compacted tool steel for 30 min at various temperatures

Source: Metal Science, March 1981, 116-124

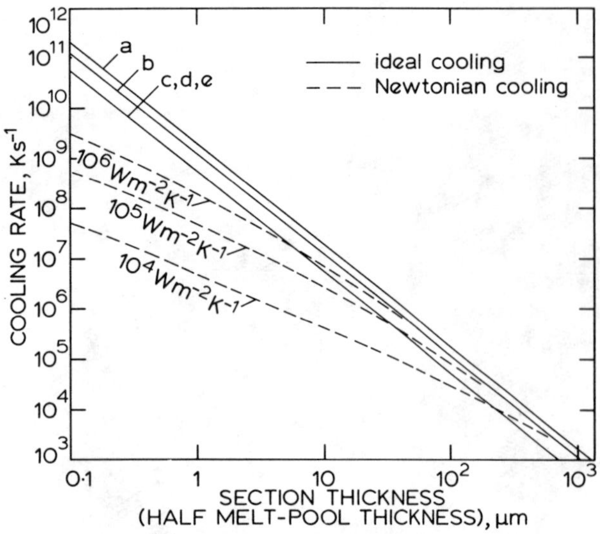

a 2200 K (liquid) and 300 K (chill); *b* 2000 K (liquid) and 300 K (chill); *c* 1900 K (liquid) and 300 K (chill); *d* 2000 K (liquid), 300 K (chill), and freezing temperature depressed to 1600 K; *e* 2000 K (liquid) and 800 K (chill)

10 Calculated ideal and Newtonian cooling rates at freezing point of molten pool in contact with cool chill metal as functions of molten-pool thickness (half thickness for case of a molten slab with chill metal on both sides), at various liquid-metal and chill-metal temperatures: Newtonian cooling rates were calculated for a freezing temperature of 1800 K and substrate temperature of 300 K

DISCUSSION

Quenching rates

Quenching rates have been determined in two ways: by heat-conduction calculations and from the details of the solidification microstructure.

For the heat-conduction calculations, the simple model of a hot, thin, molten slab in contact with a very much thicker and cool heat sink, a model extensively used for analysis of splat-cooling and melt-spinning/extraction situations, has been used. This model should apply reasonably well here, as indicated by the geometry of the melt zone in Fig. 2. The case of heat extraction from one surface with a melt-pool thickness of half the actual value is considered. For the ideal cooling case (where the interface resistance to heat is not noticed) the heat-conduction rates may easily be calculated.[8,9] Cooling rates at the freezing point are shown as a function of melt-pool thickness in Fig. 10 for several values of liquid-metal temperature and chill substrate temperature, making use of reasonable values for the other thermal properties.* For the range of liquid-metal and chill substrate values considered, there is little influence on the cooling rate and this can be considered to depend principally on the melt-pool thickness. For cases where the molten liquid/chill solid interface resistance plays a part, namely for non-ideal or Newtonian cooling, the cooling rates are deduced as indicated also in Fig. 10 for several

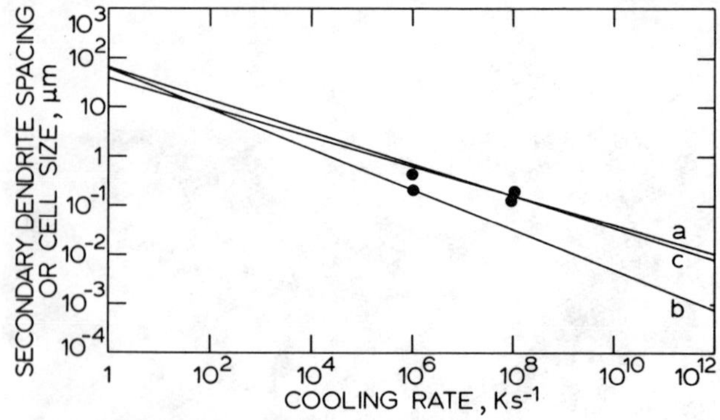

a maraging steel; *b* Fe–18Cr–IC
c Fe–25Ni and (•) Fe–10Mo and M1 steel

11 Cell size–cooling rate plots used to approximately describe behaviour of present M2 tool steel on solidification[12–14]

values of the interface heat-transfer coefficient. Cooling rates in the intermediate range are estimated by interpolation between ideal and Newtonian values. It is difficult to evaluate the important interface coefficient (values in the range 10^5–3×10^6 W m^{-2} K^{-1} have been estimated for melt spinning–extraction and splat-quenching processes[10]) and consequently to estimate the cooling rates valid in the present situation. It may be seen from Fig. 10 that for melt-pool half thicknesses in excess of about 100 μm, the cooling rates can reasonably be estimated; but for thicknesses significantly less than this, the unknown value of the interface heat-transfer coefficient makes even an estimate difficult.

Quenching rates during solidification can be deduced from the solidification structure, specifically from the size of the cell structure.[11] Such a procedure has previously been carried out making use of a calibration cell size–cooling rate plot, often established by extrapolation from slow cooling rates. There is some difficulty in establishing a calibration plot for M2 tool steel and the data shown in Fig. 11, drawn from various sources,[12–14] have been used to describe approximate bounds for the cell size–cooling rate relationships for the present material. By making use of such a calibration plot, it has been possible to estimate the cooling rate of a given melt pool from the solidification cell size (measured as the mean linear intercept on micrographs) and the results are shown in Fig. 12 as a function of the half thickness of the melt pool. The divergence in the possible cell size–cooling rate relationships at small cell size noted in Fig. 11 is represented by the bars which become larger at the smaller melt-pool thicknesses. The following conclusions may be drawn from Fig. 12. The cooling rates for all melt-pool thicknesses can probably be regarded as ideal cooling values, with the true quenching rate corresponding to a value slightly lower than the upper bound of the estimated range. Newtonian cooling with interface heat-transfer coefficients from 10^5 to 10^7 W m^{-2} K^{-1} (and probably even as high as about 10^8 W m^{-2} K^{-1}) would not give a consistent picture of the observed cooling rate–melt-pool thickness variation. In a typical, well compacted material, for which all melt pools are less than 20–30 μm in total thickness, the quenching rates are all high, ranging from values well in excess of 10^6 K s^{-1} to values in excess of 10^{10}–10^{11} K s^{-1}.

Rapidly quenched structures

The structures observed on solidification of the melted regions between powder particles are clearly similar to those

* The values taken for thermal properties were: thermal conductivity and diffusivity of chill metal 50 W m^{-1} K^{-1} and $8{\cdot}6\times10^{-6}$ m^2 s^{-1}; thermal conductivities and diffusivities of hot metal after freezing and hot liquid metal 30 W^{-1} K^{-1} and $8{\cdot}4\times10^{-6}$ m^2 s^{-1}; freezing temperature 1800 K; dimensionless latent heat 0·39.

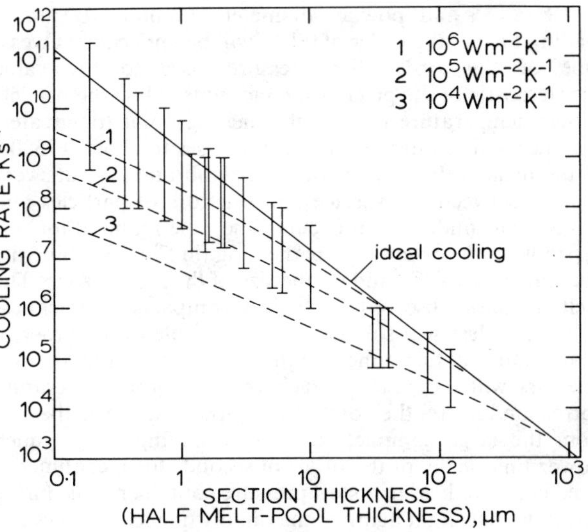

12 Comparison of experimentally determined and calculated cooling rates in tool steel as function of melt pool thickness: experimentally determined cooling rates were established from cell size using calibration plot of Fig. 11

previously observed on rapidly quenching similar steels.[6,7,14–16] In these cases it has been normal practice to observe mixtures of bcc and fcc matrix phases and minor quantities of carbides, as summarized in Table 1. Here, a clear picture of the process is given, with higher quenching rates corresponding to retention of austenite and eventually to retention of the δ-ferrite phase. The sequence from M_6C and $M_{23}C_6$ to M_2C or MC and possibly eventually complete suppression of carbide is also noted. The present observations of retention of the δ-ferrite phase, with occasional observations of martensite in the relatively slowly cooled melt/non-melt interface, are consistent with the very high quenching rates at solidification. The observations of a mild precipitation of the M_2C carbide are also reasonably consistent, particularly when the possibility of some aging having occurred during late compaction is taken into account, which will be considered in the following section. The relatively low-temperature *in situ* aging study has shown that much more extensive carbide precipitation occurs on heating in excess of 500–700°C. The matrix δ-phase present in the as-quenched structure is consequently in a highly supersaturated state.

There is some evidence to suggest that δ-ferrite alone is a relatively soft phase[15,16] and that high hardnesses are principally caused by the austenite to martensite transition or by extensive carbide precipitation (as gives rise to the hardness peak observed near 550–600°C). The high hardnesses observed in the present cases for as-quenched materials are explained by the very high degree of refinement of the δ-ferrite cell structure and by the very finely distributed, incipient M_2C carbide precipitation.

Temperature variations during dynamic powder compaction

The picture developed for dynamic powder compaction is one whereby the work of compression is converted to heat which produces several changes in the material. Work is concentrated at the powder particle surfaces by frictional rubbing and localized deformation and consequently heating occurs preferentially at these surfaces leading to softening, melting, and eventual welding between particles. In addition, some deformation and heating occurs uniformly throughout the powder particle volumes. The maximum amount of melting during dynamic compaction can be deduced from micrographs such as Fig. 1 and this information used to estimate the distribution of deformation or heat at the moment of compaction.

Table 1 Published data on rapid solidification of tool steels

Authors	Quenching rate, K s^{-1}	Matrix phases observed	Precipitate phases observed
Rayment and Cantor[6]	10^5–10^6	Mostly austenite; some bcc phase, not distinguished as δ or α	Some carbide, but type not identified
	10^4	Nearly all austenite; a little bcc phase, probably α (martensite)	Some carbide, but type not identified
	Very slow (solid state)	All bcc phase, namely α (martensite)	M_6C carbide
Sare and Honeycombe[7,14]	10^6–10^8	Mainly bcc phase, probably δ	At slower quenching rates, either M_6C or $M_{23}C_6$ carbides observed At higher quenching rates, sometimes fine MC or M_2C, sometimes no precipitates
Strutt *et al.*[15]	'Slow'*	bcc phase, namely α (martensite)	A little carbide seen, but type not identified
	'Medium'*	Austenite	
	'Fast'*	bcc phase, namely δ, with some austenite	
Kim *et al.*[16]	5×10^5	Mixture of austenite and bcc, namely δ	M_2C and $M_{23}C_6$ carbides seen

* These are qualitative and relative descriptions of quenching rates obtained throughout a laser surface-melting operation. Thus, 'slow' would be regarded as very fast in all but the very rapid solidification context.

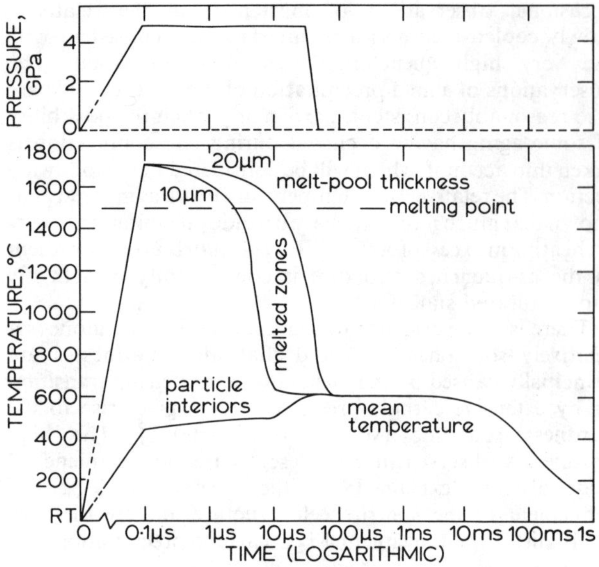

13 Schematic indication of pressure and temperature variations during typical dynamic compaction cycle

By assuming that at the moment of maximum melting the liquid temperature is about that of the melting point and knowing the total amount of energy introduced during the shock, essentially the product of force and distance or $\int PdV$ (Refs. 3, 4), the respective energies imposed in the liquid and solid fractions can easily be deduced. Some of the data obtained are shown in Table 2 and indicate that the energy deposition rate of the particle interiors is about 0·75–0·8 of the value of the specimen as a whole. It should be noted that these specific values apply for the particular powder used and for the compaction conditions operating and under other test conditions will adopt different values. Nevertheless, the values deduced give some idea of the energy distribution during dynamic compaction.

A schematic indication of the pressure and temperature variations occurring throughout a dynamic compaction cycle of the type used here is shown in Fig. 13. The pressure-loading rate is not accurately known but will be related to the time taken for the shock wave to pass through the powder particle. By taking typical figures for the compaction described here, namely a shock speed in excess of 1 Km s^{-1} and powder diameter of about 100 µm, a loading time of the order of 0·1 µs will be applicable. Over a time of this order the pressure rises to the value characteristic of the given test conditions, while the particle surface temperature rises to the melting point (or greater) and the particle interior temperature rises to about 0·75–0·8 of the mean value. Temperature equilibration occurs very rapidly between the surface and interiors of particles and rapid local solidification occurs. The time required for this depends on the extent of superheating (in Fig. 13 arbitrarily illustrated as 200 K) and on the size of the melted zone. The melted zones observed in typical compacts of thickness 20 µm and less will solidify on a time scale of the order of 10 µs. During this time the temperature of the particle interiors will increase to reach that uniform temperature corresponding to the total work input. Heat will be lost from the large compact to its surroundings on a much slower time scale, of the order of seconds for a centimetre-sized compact held in a steel pressure container. It is during this time that slight aging of the rapidly quenched zones can occur, leading here to incipient precipitation of the M_2C carbides.

CONCLUSIONS

1. A detailed study of the microstructural changes occurring during dynamic compaction has enabled a much greater understanding of the processes occurring to be deduced.

2. During the compaction process, preferential mechanical working and heating occurs at the powder particle surfaces, while the particle interiors suffer work at a rate of the order of 0·75–0·8 of the average compaction value.

3. The preferential surface working leads to localized heating, even to the melting point, and such melting followed by solidification appears to be responsible for the formation of effective interparticle welds.

4. Quench rates during solidification of the melted zones have been calculated by heat-conduction analysis and also determined from the solidification cell structure. Quench rates are always very high, ranging in a typical compact from about 10^6 to 10^{10} K s^{-1} or even greater.

5. Such localized heating followed by rapid solidification helps to explain why dynamic powder compaction can successfully be applied to the compaction of thermally unstable, rapidly quenched materials. Surface melting and subsequent solidification leads to welding between the

Table 2 Temperature distribution during dynamic compaction

Compacting pressure,* GPa	Mean compact temperature† T_{mean}, °C	Amount melted‡ a, %	Temperature of particle interiors‖ $T_{interior}$, °C	$\frac{T_{interior}}{T_{mean}}$	$\frac{T_{interior}}{T_{mean}}$ (with superheat)
3·6	420	5	360	0·85	0·83
5·5	650	15	490	0·75	0·72
6	730	10	640	0·87	0·85
6	730	10	640	0·87	0·85
7	840	15	715	0·85	0·82
7	840	20	660	0·78	0·84
8	950	35	625	0·66	0·57
8	950	30	690	0·73	0·66

* Deduced by impedance matching of the Hugoniots of the material involved.[3]
† Taking starting temperature 20°C and calculating temperature rise as work input/specific heat.[3]
‡ Measured by random point counting on micrographs such as Fig. 1.
‖ Calculated from $T_{mean} = a.T_{MPt} + (100-a)\ T_{interior}$, where T_{MPt} is the melting point (1550°C).
§ These values apply when considering the liquid to be superheated to 1700°C.

powder particles, yet this can occur even though the mean temperature rise is very small, thus avoiding structural degradation. Furthermore, the quenching rates of the molten material are sufficiently high to cause resolidification in the metastable state. The quenching rates involved are typically greater than those used in the commercial production of rapidly quenched powder.

6. The dynamic compaction process, by producing in a single sample many microsamples quenched from the liquid state over a very wide range of cooling rates, may lend itself to application as a technique for alloy development studies. By not only indicating whether metastability is possible, but by also indicating the cooling-rate range where metastability becomes important, the selection or development of alloys for rapid-quenching applications may be facilitated.

REFERENCES

1. D. RAYBOULD: Proc. 15th Int. Machine Tool Design and Research Conf., (eds. S. A. Tobias and F. Koenigsberger), 1975, London, Macmillan.
2. D. RAYBOULD, D. G. MORRIS, and G. A. COOPER: *J. Mater. Sci.*, 1979, **14,** 2523.
3. D. G. MORRIS: *Met. Sci.*, 1980, **14,** 215.
4. D. G. MORRIS: 2nd Int. Conf. on 'Rapid solidification processing, principles, and technologies', (eds. R. Mehrabian, B. H. Kear, and M. Cohen); 1980, Baton Rouge, Claitor's Publishing Division.
5. 'Tool steel', 1978, British Independent Steel Producers Association.
6. J. J. RAYMENT and B. CANTOR: *Met. Sci.*, 1978, **12,** 156.
7. I. R. SARE and R. W. K. HONEYCOMBE: *ibid.*, 1979, **13,** 269.
8. H. S. CARSLAW and J. C. JAEGER: 'Conduction of heat in solids', 2 ed.; 1959, Oxford, Clarendon.
9. H. JONES: *Mater. Sci. Eng.*, 1969–70, **5,** 297.
10. H. A. DAVIES: 'Rapidly quenched metals' (ed. B. Cantor); 1978, London, The Metals Society.
11. H. JONES: 'Rapid solidification processing, principles, and technologies', (eds. R. Mehrabian, B. H. Kear, and M. Cohen); 1978, Baton Rouge, Claitor's Publishing Division.
12. R. MEHRABIAN: 'Rapid solidification processing, principles, and technologies', (eds. R. Mehrabian, B. H. Kear, and M. Cohen); 1978, Baton Rouge, Claitor's Publishing Division.
13. W. E. BROWER, R. STRACHAN, and M. C. FLEMINGS: *Cast Metals Res. J.*, 1970, **6,** 176.
14. I. R. SARE and R. W. K. HONEYCOMBE: 'Rapidly quenched metals', (eds. N. J. Grant and B. C. Giessen), 1976, Cambridge, Mass., MIT Press.
15. P. R. STRUTT, H. NOWOTNY, M. TULI, and B. H. KEAR: *Mater. Sci. Eng.*, 1978, **36,** 217.
16. Y.-W. KIM, P. R. STRUTT, and H. NOWOTNY: *Metall. Trans.*, 1979, **10**A, 881.

RAPID SOLIDIFICATION PROCESSING OF SUPERALLOYS USING HIGH POWER LASERS

D. B. Snow, E. M. Breinan and B. H. Kear
United Technologies Research Center
East Hartford, CT 06108

A new form of laser materials processing has been developed (the LAYERGLAZE™ process) whereby bulk parts can be built up from rapidly solidified (~10^4°C/s) materials. This involves the introduction of wire or powder feedstock onto a moving substrate at the point of impingement of a continuous CO_2 laser beam. LAYERGLAZE processing is currently being employed to fabricate model gas turbine disks with very fine microstructures and low defect populations. This application has required the development of high molybdenum, nickel base superalloys which are both laser-weldable, phase stable and strong at ≤760°C. Further development of the LAYERGLAZE technique has produced the concept of general "directed energy processing", which combines rapidly solidified materials fabrication with alloy composition gradients; as well as in-process nondestructive testing, defect repair and thermal treatment.

INTRODUCTION

In 1976, at the 'Third International Symposium on Superalloys', Breinan, et al, described a new method for the surface treatment of materials by controlled laser surface melting, followed by rapid self-quenching (1). Cooling rates up to 10^6°C/sec were readily attainable in appropriately thin sections. The range of microstructures achieved by this so-called

Source: Superalloys 1980, 189-203, © 1980 American Society for Metals

LASERGLAZE™ process included amorphous metallic solids, extended solid solution phases, ultra-fine eutectics, and refined dendritic structures.

Much work has now been done to determine the utility of LASERGLAZE processing as a means to improve the corrosion, erosion and wear properties of metal surfaces. Some good examples of improvements in all three categories have been obtained by different research groups. On the other hand, much less work has been reported on LASERGLAZE processing in conjunction with surface alloying. Two distinct approaches have been evaluated at UTRC: (1) preplacement of alloying material on the workpiece surface prior to laser melting; and (2) continuous delivery of alloying material to the laser-material interaction zone (the LAYERGLAZE™ process).

The preplacement method has been exploited as a hardfacing treatment; e.g., to harden critical regions of wear in a part. However, the continuous-feed method holds greater promise, both for surface alloying and as a means to produce bulk rapidly solidified structures in near-net shapes.

A major R&D program is now being undertaken by UTRC to exploit the continuous-feed method for the purpose of fabricating a high-performance gas turbine disk. This work has progressed to the point that sound parts, free of gross imperfections, can now be produced, using either prealloyed wire or powder feed. Some success has also been achieved in controlling the shape of the disk during fabrication. In addition, several precipitation-strengthened nickel-base superalloys have been developed which exhibit as-fabricated properties comparable to current disk alloys.

This paper constitutes a progress report on the continuous powder-feed method as applied to disk fabrication. Future directions for R&D in the area will also be discussed, including such considerations as combining incremental solidification processing with incremental thermo-mechanical treatment, as well as continuous inspection procedures.

INCREMENTAL LASER MELT PROCESSING

Although the LASERGLAZE™ process has been found to be capable of producing rapidly chilled microstructures, its applications are limited by the small section thickness required to achieve high cooling rates. Typically, an order of magnitude decrease in the melted and resolidified section thickness results in an increase in the average cooling rate of two orders of magnitude. The LASERGLAZE process is thus suited only for treatment of thin layers. In order to achieve the fabrication of rapidly-solidified thick sections (bulk parts), the LAYERGLAZE™ concept was evolved. LAYERGLAZE is essentially the sequential buildup of bulk material with controlled composition and microstructure by simultaneous material addition and LASERGLAZE processing. Although the deposition of each subsequent layer produces a high-temperature thermal transient in the layer beneath, the time duration of this transient is extremely brief. As a result, although the process will not permit the formation of bulk amorphous structures, it does allow fabrication of relatively homogeneous crystalline structures with little or no phase transformation between layers.

A schematic representation of the LAYERGLAZE process is shown in Fig. 1. Feedstock can be in the form of either wire or powder. Wire is convenient to feed but often difficult to obtain in the required alloy compositions. Powder is relatively easy to procure, and can be conveniently fed into the interaction zone. The laser beam melts both the feedstock material and some of the substrate layer, which produces good interlayer bonding, and eptiaxial solidification from layer to layer. There is a critical relationship between the location of the feedstock impingement point and the laser-heated zone. Since the substrate is rotating, feedstock impingement must occur slightly ahead of the laser beam for stable, steady-state material deposition. The feedstock is then carried into the laser beam impingement area by the rotation of the substrate. The part being fabricated is internally water-cooled during material deposition in order to prevent heat buildup, and thus maintain high cooling rates.

The LAYERGLAZE technique for incremental laser melt processing has been used to fabricate a 13.2 cm diam. x 3.2 cm thick cylindrical blank, to be subsequently machined into an ~12.7 cm diam. scale model gas turbine disk. Another model disk will be constructed in late 1980 for spin testing early in 1981. The initial trial disk fabrication produced the disk preform pictured in Fig. 2. This preform was powder-feed LAYERGLAZE fabricated from two nickel base superalloys. The core consisted of UTRC 8-12-3 (Ni-3.4Al-17.9Mo-8.4Ta, wt %), which was one of the initial alloys considered for model disk manufacture in this program. The rationale for this choice of composition and additional details of the alloy development effort are included in the following section. The core was fabricated from alloy 8-12-3 powder to a diameter of 8.4 cm, then the disk preform was completed using IN-718 powder. This test served to demonstrate both the capability of making full-sized parts and the ability to utilize different alloys sequentially in the same part by means of incremental fabrication.

The microstructure of this LAYERGLAZE-processed model turbine disk is shown in Fig. 3. The buildup process has created a uniformly layered appearance, with generally good structural integrity and interlayer bonding. Both the transverse cross-section and the disk surface display the radially-aligned, columnar grain structure and epitaxial growth which is characteristic of the LAYERGLAZE process.

In all LAYERGLAZE buildup tests, compressive deformation of the substrate was observed. This was expected, since the process involves the sequential deposition and shrinkage of successive layers. As each layer is deposited and cooled, it is stressed in tension and yields plastically. As each subsequent layer is deposited on it, the stress on the first layer gradually changes from residual tension to residual compression. Detailed calculations which describe these changes may be found in Ref. 2.

Although to date the utilization of LAYERGLAZE processing at UTRC has been concerned with the fabrication of simple axisymmetric shapes, it should be pointed out that this process is not limited to them, and in fact is capable of producing

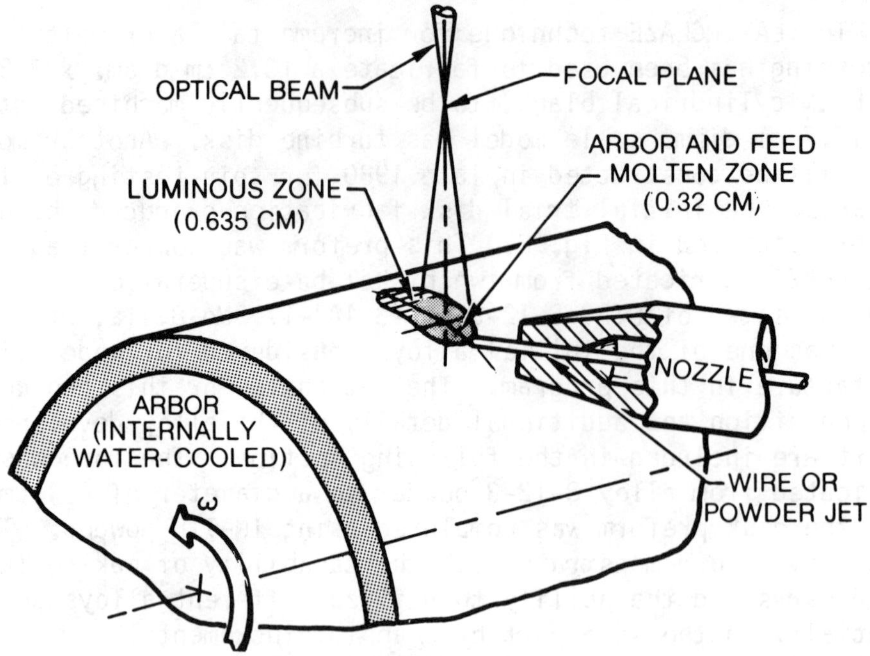

Fig. 1. Schematic Representation of the LAYERGLAZE Process.

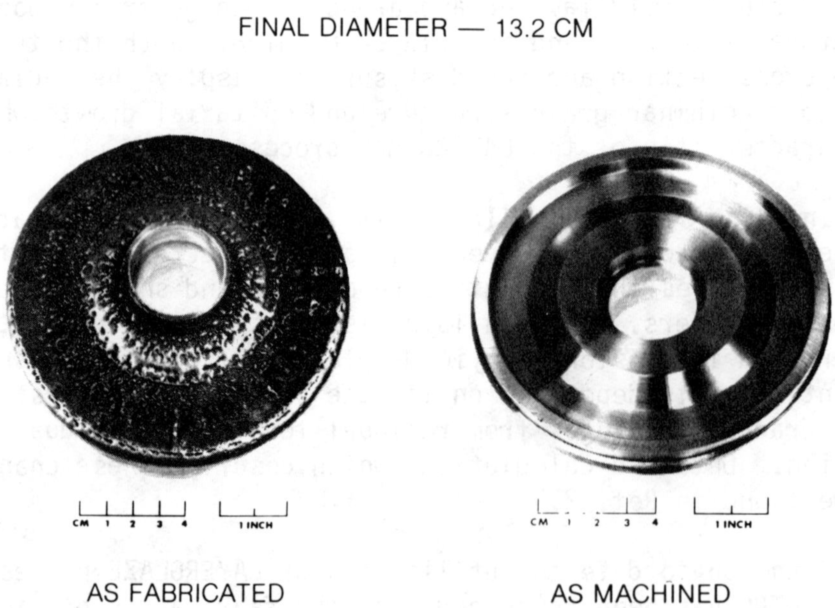

Fig. 2. LAYERGLAZE-Fabricated Turbine Disk Preform.

almost any desired shape. Some of the possibilities are illustrated in Fig. 4. As indicated, the production of more complex shapes requires the use of a numerically controlled work station, which is capable of simultaneous motion about at least three axis. However, even the simple case of translation back and forth on a straight edge has obvious implications for hard-facing of cutting tools.

ALLOY DEVELOPMENT

The development of nickel base superalloys for use as turbine disks fabricated by the LAYERGLAZE process was constrained by three sets of requirements; crack-free laser weldability, beneficial alloy response to rapid solidification ($\sim 10^4$°C/s), and mechanical properties exceeding those of current jet engine disks. The primary consideration in evaluating any alloy was that it should remain free of cracks after laser welding (2). As-chill-cast specimens of ~8 mm thickness were bead-on-plate welded at 5 cm/s with a continuous, CO_2 laser operated at 6 kW in the unstable resonator mode. The beam intensity on each specimen was approximately 1 MW/cm^2. Overlapping pass welds were made under 100% He, with the specimens mounted on a horizontal, rotating wheel. If no post-weld cracks were observed on the specimen surface or on a transverse cross-section through the entire specimen, the alloy was selected for further evaluation.

No conventional nickel base superalloys of adequate strength, including those known to be weldable by other techniques, passed the test described above. Those which cracked when laser processed included advanced disk superalloys of recent development or use, such as AF2-1DA, AF 115, IN 100, etc. Consequently, the main effort of alloy development was directed toward devising and evaluating strong nickel base superalloys which would be compatible with laser materials processing.

Research within United Technologies by Lemkey (3), Pearson (4), and Aigeltinger, et al (5) had suggested that a particular range of alloy compositions based on the Ni-Al-Mo system might be expected to possess the unique properties

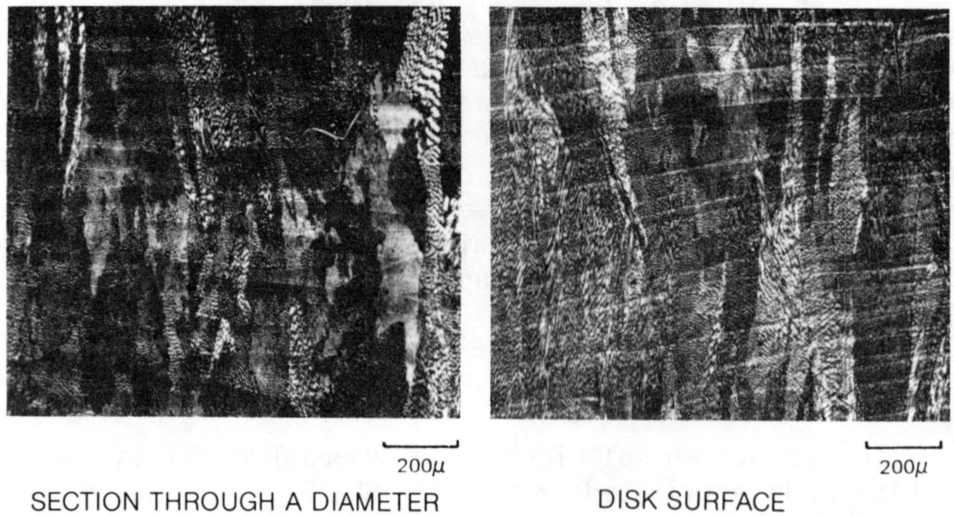

Fig. 3. Microstructure of LAYERGLAZE-Fabricated Turbine Disk.

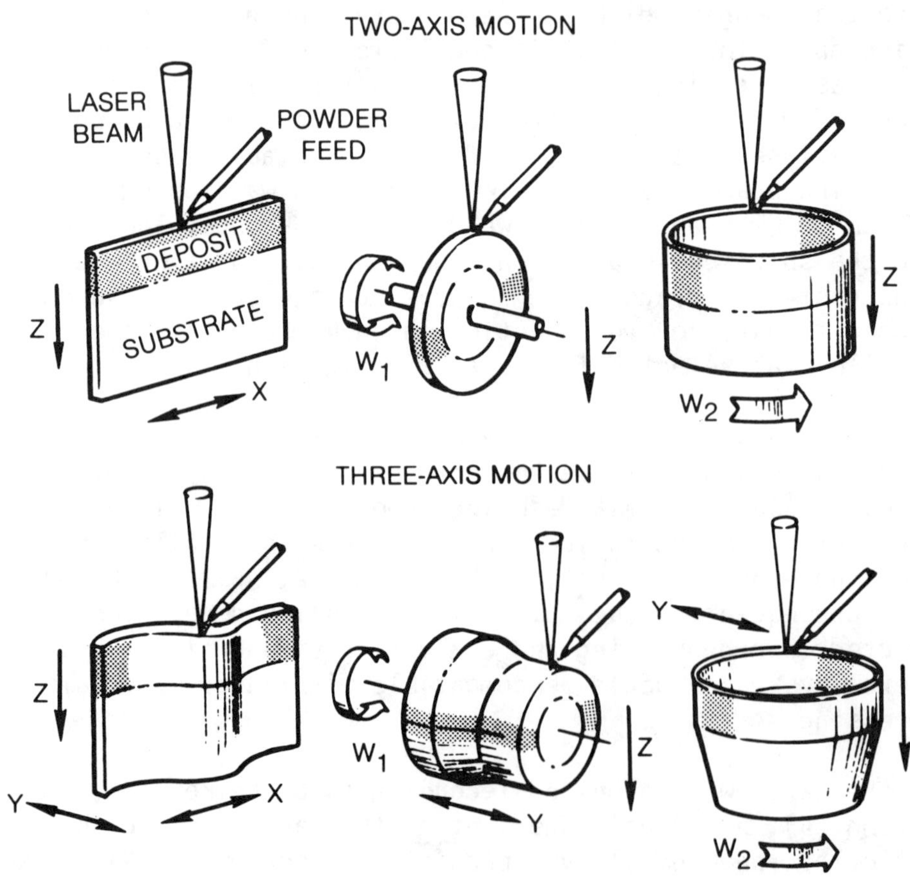

Fig. 4. Examples of LAYERGLAZE Processing for Near-Net-Shape.

required to produce a LAYERGLAZE processed turbine disk. The initial rationale for choosing this group of alloys was based on their observed resistance to laser weld cracking, lower γ' volume fraction for ductility during LAYERGLAZE processing, and potential for age hardening by Ni_3Mo and/or Ni_4Mo intermetallic compounds. Some initial consideration was given compositions intended to avoid the formation of intermetallic δ (NiMo) and to promote the strengthening of γ' by Ta.

A description of the first-generation alloy compositions and properties can be found in Ref. 2. From this group, the first alloy chosen for evaluation by LAYERGLAZE processing was UTRC 8-12-3, Ni-3.4Al-17.9Mo-8.4Ta (wt %). It was argon atomized, sieved to -150 mesh to improve flow characteristics, and utilized to fabricate several model turbine disks by the powder-feed LAYERGLAZE process; one of which is shown in Fig. 2. Extensive observation of the as-LAYERGLAZE processed microstructure of this alloy (Fig. 3) revealed very few inclusions or voids, and none larger than ~4μm. Solidification occurred by dendritic growth in all areas, with little or no observable side branching. The mean secondary dendrite arm spacing was 1.9μm, while the mean primary spacing was 3.8μm. Comparison of these data with similar observations (6) suggests that this microstructure solidified at ~10^4°C/s. Evidence that solidification was primarily epitaxial is found in the continuation of most grains through the striations which mark the successive location of the solid-liquid interface of each added layer.

Examination of as-LAYERGLAZE processed alloy 8-12-3 by transmission electron microscopy (Fig. 5) revealed that the interdendritic areas were delineated by subboundaries and by phases which have been tentatively identified as MoC. There was a high density of uniformly distributed dislocations, predominately on {111}, within the grains and subgrains. The dendrites themselves consisted of a γ matrix which contained a very fine, homogeneous dispersion of both DO_{22}-structure Ni_3Mo and γ' (Fig. 6). No coarse γ' was observed in the interdendritic regions.

As indicated by Table I, the strengths of several first generation, as-LAYERGLAZE- processed alloys were sufficiently high to merit further development. However, examination of

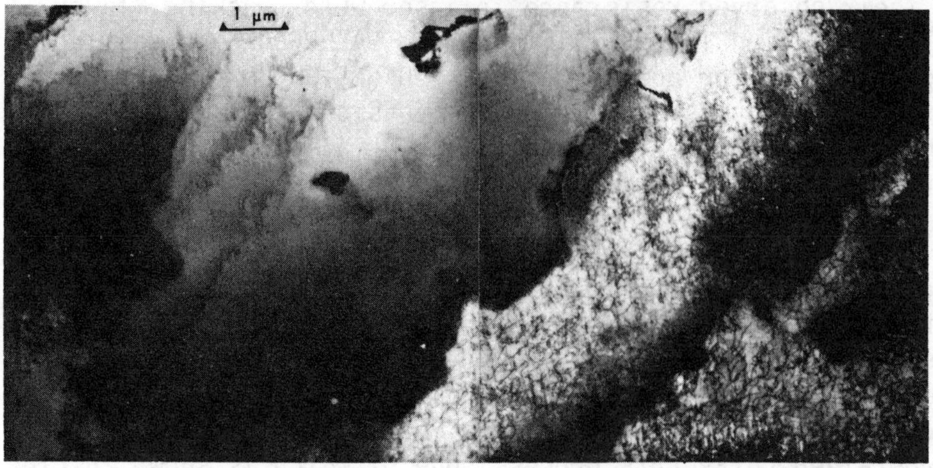

Fig. 5. As-LAYERGLAZE Processed Microstructure, Alloy 8-12-3, (Ni-3.4Al-17.9Mo-8.4Ta), which exhibits high dislocation density, interdendritic subboundaries, and some interdendritic precipitation.

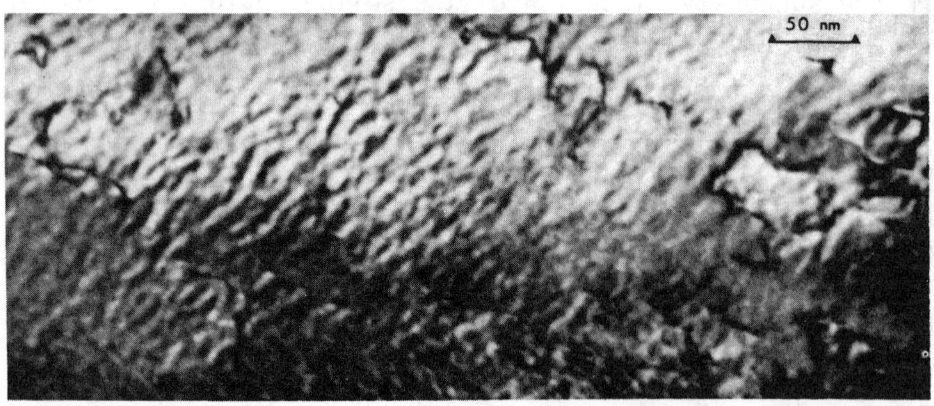

Fig. 6. As-LAYERGLAZE Processed Alloy 8-12-3. Bend contour within dendrite, bright field. γ' and DO_{22} Ni_3Mo in γ matrix.

Table I. Mechanical Properties of LAYERGLAZE Process Model Turbine Disk Alloys

Alloy	Composition (wt %)	Tensile Test Data, As-Fabricated, Radial Orientation Temp (°C)	0.2% YS (MPa)	UTS (MPa)	% E
8-12-3	Ni-3.4Al-17.9Mo-8.4Ta	25	1050	1280	40
		538	965	1100	43
		704	980	1125	32
12-15	Ni-5.4Al-23.8Mo	25	1207	1393	26
		704	1180	1241	13
11-12-0-0-5	Ni-5.0Al-19.4Mo-4.4Cr	25	1100	1407	4

these alloys after annealing at 760°C for 100 hrs revealed extensive discontinuous (cellular) precipitation at both grain boundaries and interdendritic subboundaries. This caused virtually complete embrittlement by the formation of orthorhombic Ni_3Mo in a lamellar structure similar to that observed in binary Ni-Mo alloys (7) (Fig. 7). On the other hand, the dispersed γ' and DO_{22} Ni_3Mo exhibited very little coarsening, while the average microhardness increased from 420 (as-fabricatdd) to 580 VHN (200g load). Other Ni-Al-Mo and Ni-Al-Mo-Ta alloys, which required ≥18 wt % Mo to avoid cracking when laser welded, all exhibited similar cellular phase transformations when annealed at >650°C. Consequently, the program was directed toward a second generation of alloys which involved the partial substitution of Cr for Mo to avoid intermediate temperature phase instability while retaining weldability and strength.

The absence of crack formation in the experimental, high-Mo nickel base superalloys during LAYERGLAZE processing was unusually sensitive to (deliberate) compositional variation, e.g. a change of $\pm$ 1 at % Mo could be critical. The general dependence of crack formation on composition was difficult to predict, and empirical criteria for doing so were inadequate. The mechanism(s) of crack formation in Ni-Al-Mo-X alloys is not obvious, but appeared to be related to high-temperature, post-solidification ductility, and possibly to slip dispersal at grain boundaries by second phases. All alloys which contain >6 wt % Al have cracked when laser welded by means of the procedure described earlier. These problems are clearly related to, but have been confirmed not to be identical with, weld cracking in conventional nickel base superalloys.

The second generation, Ni-Al-Mo-Cr alloys have thus far shown considerable promise with respect to the program goals. Again, the absence of laser weld cracking has been difficult to predict, as incrementally increasing substitution of Cr for Mo does not result in a systematic variation of crack-free laser weldability. Nevertheless, several interesting alloys have appeared, especially alloy 11-12-0-0-10, Ni-5Al-19.5Mo-8.8Cr (wt %). This alloy can be LAYERGLAZE processed without cracking and displays no cellular transformation after annealing at 760°C for 300 hrs. Laser welded specimens annealed under these conditions were age hardened to ~R_C 54 and displayed predominately intergranular fracture surfaces when impact tested. At this time, extensive structure/property characterization of LAYERGLAZE-processed alloy 11-12-0-0-10 is planned, as well as fabrication and spin testing of a model turbine disk LAYERGLAZE-fabricated from this alloy.

DIRECTED ENERGY PROCESSING

Experience with the basic LAYERGLAZE technique has pointed the way to a potential means for controlling structure and properties more closely than has previously been possible by using sequential, in-situ processing techniques on small increments of material. The technique has been termed Directed Energy Processing and is illustrated schematically in Fig. 8. The processing starts at work station #1 with material being added by the laser melting and deposition of feedstock. This step essentially creates the part. At additional work stations, other materials processing operations can be sequentially accomplished on each incremental layer. These include inspection, mechanical deformation, and/or heat treatment. Like deposition, which derives advantages from rapid solidification of a thin section, each of these subsequent operations also derives advantage from being

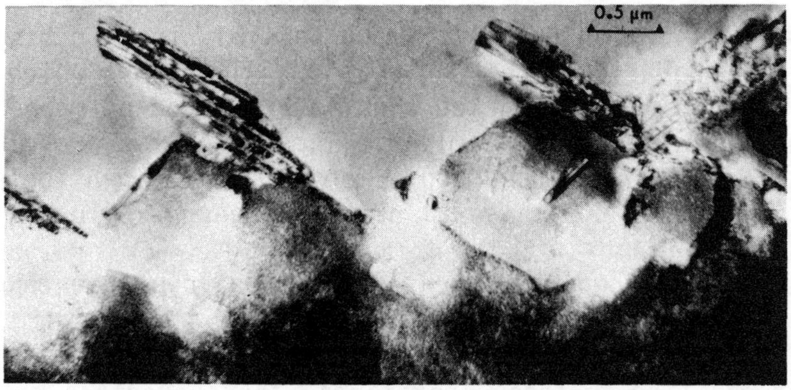

Fig. 7. Alloy 8-12-3, LAYERGLAZE Processed and Annealed at 704°C, 32 hrs. Cellular precipitation nucleated at interdendritic subboundaries and involving orthorhombic Ni_3Mo.

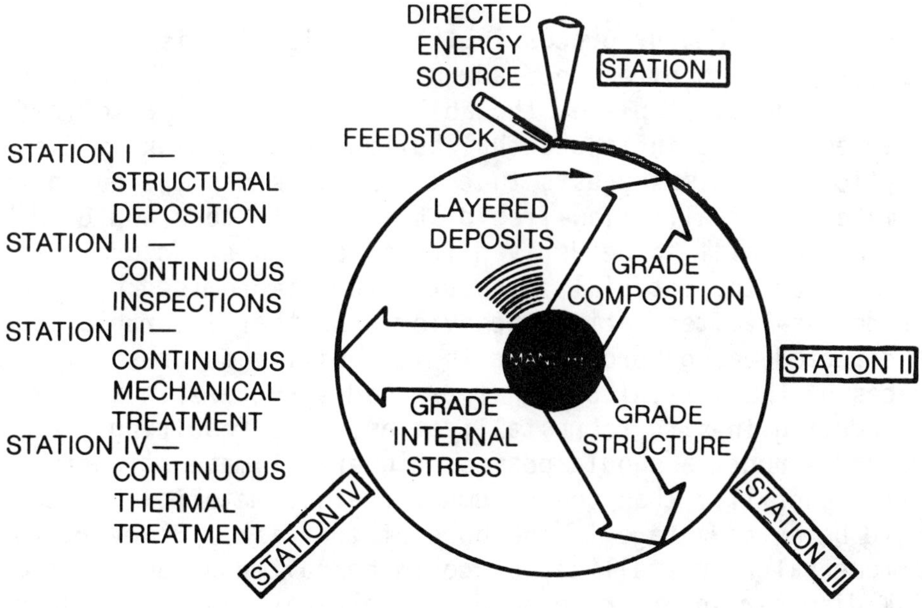

Fig. 8. Schematic Representation of Directed Energy Processing.

applied to a small layer. Incremental inspection can thus be accomplished without destructively testing the disk. Mechanical processing can be applied to small increments without resulting in significant distortion, and thermal treatments or even remelting can be accomplished immediately by interrupting the material feed.

In addition to these operations, the sequential nature of deposition and the ability to change the feedstock continuously will permit the future production of parts more specifically tailored to their applications than is presently possible. The composition, structure, and even the residual stresses in a part can be modified and controlled for optimal properties. While there exist some practical limitations on the degree to which the structure can be tailored using incremental processing, the technique clearly holds the potential for more specific structure/property control than has previously been possible.

Directed energy processing, as applied to disk fabrication, appears to offer some advantages over LAYERGLAZE processing. In the first place, the ability to detect and to remove imperfections during the actual buildup of the disk is clearly beneficial. This suggests that the process has the potential to make a virtually flaw-free disk. This in turn would enable the full benefit to be derived from the high structural strength of the material, without being compromised by concern for defect-limited fatigue behavior. Another favorable aspect of directed energy processing is its ability to erase all traces of the initial columnar grained structure in the disk by inducing in-situ recrystallization. The resulting fine-grained structure should possess higher ultimate strength and better ductility than the columnar-grained material, which would be an advantage in the bore of the disk. However, this incremental recrystallization could be halted at any desired disk diameter in order to retain a columnar-grained structure in the rim of the disk, which should be beneficial because the rim tends to run hot enough for creep to be a critical design factor.

ACKNOWLEDGEMENTS

The authors gratefully acknowledge the helpful technical discussions and experimental contributions of Dr. D. D. Pearson, Dr. E. R. Thompson, and Mr. M. J. Kaufman, as well as the contributions to the development of the laser processing techniques made by Mr. C. O. Brown.

REFERENCES

(1) E. M. Breinan and B. H. Kear, "Surface Treatment of Superalloys by Laser Skin Melting", Superalloys, Metallurgy and Manufacture, Proceedings Third International Symposium, Claitors Publishing Div., Baton Rouge, LA, 1976, p 434.

(2) E. M. Breinan, C. O. Brown, and D. B. Snow, "Program to Investigate Advanced Laser Processing of Materials", Report No. R79-914346-4, Contract N00014-78-C-0387, United Technologies Research Center, August 1979.

(3) F. D. Lemkey, "Development of a Second Generation Ductile/ Ductile $\gamma/\gamma' - \alpha$ Eutectic Alloy", Superalloys: Metallurgy and Manufacture, B. H. Kear, D. R. Muzyka, J. K. Tien, and S. T. Wlodek, eds., Claitor's Publishing Div., Baton Rouge, LA, 1976, p. 321-330.

(4) D. D. Pearson, "Ni-Al-Mo-Ta $\gamma' + \alpha$ Eutectics", Proc. Conf. In-Situ Composites-II, J. L. Walter, M. F. X. Gigliotti, B. F. Oliver and H. Bibring, eds., Ginn Custom Publishing, Lexington, MA, 1979.

(5) E. H. Aigeltinger, S. R. Bates, R. W. Gould, J. J. Hren, and F. N. Rhines, "Phase Equilibria in Rapidly Solidified Nickel Rich Ni-Mo-Al Alloys", Rapid Solidification Processing, Principles and Technologies, R. Mehrabian, B. H. Kear and M. Cohen, eds., Claitor's Publishing Div., Baton Rouge, 1978, p. 291.

(6) R. Mehrabian, "Relationship of Heat Flow to Structure in Rapid Solidification Processing", Rapid Solidification Processing, Principles and Technologies, R. Mehrabian, B. H. Kear and M. Cohen, eds., Claitor's Publishing Div., Baton Rouge, 1978, p. 9.

(7) T. Saburi, K. Komatsu, M. Yamamoto, S. Nenno, and Y. Miyutani, "A New Metastable Phase Ni_2Mo", Trans. TMS-AIME, 245, 1969, p. 2348-49.

SECTION III

Heat Flow

HEAT FLOW LIMITATIONS IN RAPID SOLIDIFICATION PROCESSING

R. Mehrabian, S.C. Hsu, C.G. Levi, and S. Kou

Department of Metallurgy and Mining Engineering
Department of Mechanical and Industrial Engineering
University of Illinois
Urbana-Champaign, Illinois

INTRODUCTION

The term Rapid Solidification Processing, RSP, is equally applicable to the formation of both crystalline and non-crystalline solid phases by quenching of a material from an initial liquid state. During RSP cooling rate in the liquid prior to solidification affects nucleation (undercooling) and growth phenomena in important ways — it influences undercooling in crystalline solidification and is an overriding factor in the formation of non-crystalline structures. On the other hand, the fineness of a crystalline microstructure (e.g. segregate spacing, size of second phase particles, etc.) can usually be correlated to average cooling rate during solidification or time available for coarsening. Thus, a clear distinction must be made between cooling rates in the liquid (or during non-crystalline solidification) and during crystalline solidification; the latter is significantly lower at equivalent rates of external heat extraction due to the heat of fusion.

In what follows some general relationships are presented between cooling rates during crystalline and non-crystalline solidification and process variables in different RSP techniques. Calculations are presented to show the heat flow characteristics and limitations in the three general areas of RSP; atomization and solidification against substrates with and without significant resistance to heat flow at the liquid-substrate interface.

HEAT FLOW DURING ATOMIZATION

During solidification of small spherical alloy droplets, heat flow is controlled by both convection at the surface and by radiation. However, there are no accurately established values for the combined radiative and convective heat transfer coefficient, and direct measurement of the cooling rate or heat flux during solidification of an atomized droplet would be extremely difficult, if not impossible. In gas atomization the convective heat transfer coefficient is overriding and is usually estimated from the following equation:

$$\frac{hD}{k_f} = 2.0 + 0.60 Re^{1/2} Pr^{1/3} \qquad (1)$$

where:

Re = Reynold's number = $vD\rho_f/\mu_f$

Pr = Prandtl number = $C_{pf}\mu_f/k_f$

C_{pf} = specific heat of the gas

D = particle diameter

k_f = conductivity of the gas

h = heat transfer coefficient

v = gas velocity relative to particle

ρ_f = density of the gas

μ_f = viscosity of the gas

An upper limit on achievable heat transfer coefficents can be deduced from equation (1). For example, the calculated heat transfer coefficients during argon gas atomization, with a high relative velocity of one Mach between the gas and the metal droplets, are 5.86×10^3 and 1.1×10^4 $W/m^2.K$ for droplet diameters of 75µm and 25µm, respectively. Using higher conductivity gases and finer particles result in calculated heat transfer coefficients of less than 10^5 $W/m^2.K$.

Indirect estimates of heat transfer coefficients in various atomization processes have also been made by comparison of measured segregate (dendrite arm) spacings in crystalline alloy powders with predetermined relationships between these spacings and average cooling rates during solidification. Table I shows the various heat transfer coefficients during atomization of Maraging 300 steel determined by this method. Note that the heat transfer coefficient for gas atomization is the same order of magnitude as that estimated above from equation (1).

In general, then, a limitation on the achievable heat transfer coefficient at a liquid metal droplet — environment interface can be translated into a limitation on the important dimensionless variable - Biot Number - governing the rate of heat extraction from the droplet. For example, a heat transfer coefficient $h < 10^5$ W/m^2.K translates to a limitation on the range of Biot Numbers of $10^{-2} < Bi < 1.0$ for atomized droplets of liquid aluminum in the size range of 1μm to 1000μm.

$$Bi = \frac{hr_o}{k_\ell} \qquad (2)$$

where h is the heat transfer coefficient at the metal droplet-environment interface, r_o is the radius of the droplet and k_ℓ is the conductivity of the liquid metal.

For smaller heat transfer coefficients, Biot numbers below 0.01, Newtonian cooling expressions are generally considered to be applicable. However, as shown below, Biot numbers should be below 0.001 before temperature gradients in a liquid droplet become negligible. For Bi > 0.001 numerical heat flow models are necessary. A short summary of achievable cooling rates and solidification times, from one such numerical heat flow model, based on the limitations noted above is presented below from a recent paper on solidification of atomized droplets (1).

Figure 1 shows calculated dimensionless temperature distribution in a liquid droplet for various Biot Numbers and an initial superheat of $T - T_g/T_M - T_g = 1.3$ at the instant the droplet surface reaches its melting point. These data show that for Biot Numbers less than ∿0.001 there is no significant temperature gradient in the droplet and the simple Newtonian cooling expressions can be used for crystalline and non-crystalline solidification. On the other hand, the results also indicate that even for small Biot Numbers, in the range of ∿0.01, there may be significant temperature gradients in a metal droplet. For example, in a 20μm diameter droplet of liquid iron where Bi ∿0.01 ($h - 4 \times 10^4$ W/m^2 K) the maximum temperature gradient, at the droplet surface when the surface reaches the melting point, is ∿ 1.5×10^6 K/m and a temperature difference of 7.5K between the surface and the center of the droplet is calculated. Thus, the Newtonian cooling assumption that temperature

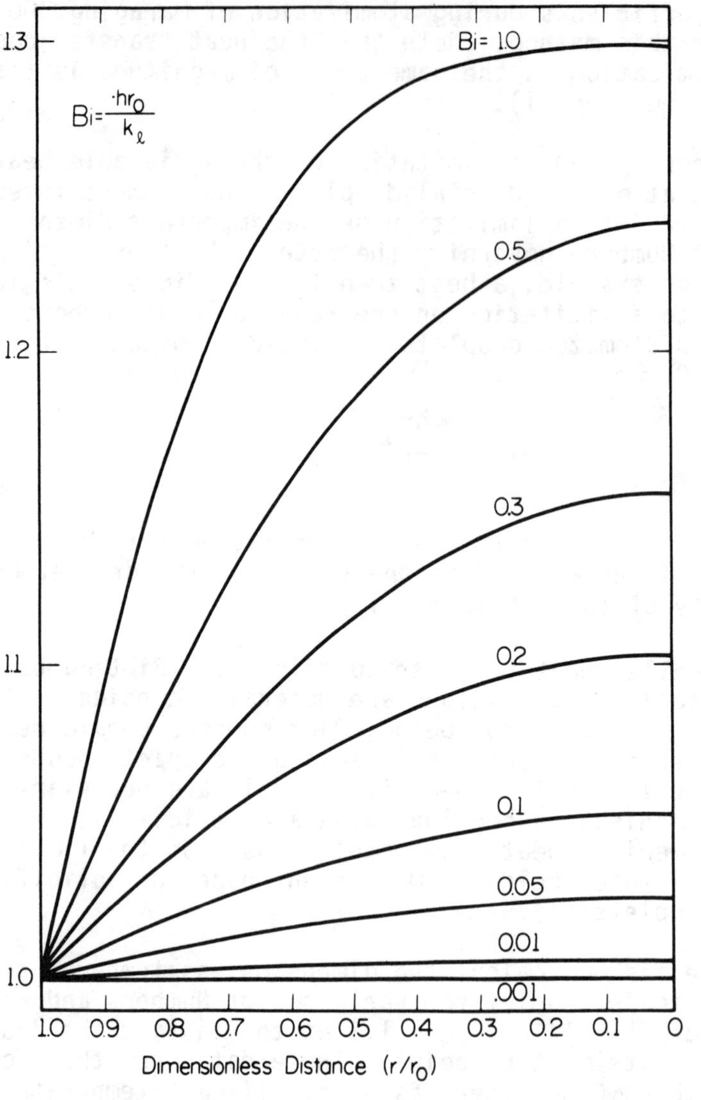

Figure 1. Dimensionless temperature distributions in a liquid droplet when its surface reaches the melting temperature T_M, for different Biot numbers and dimensionless initial superheat $T-T_g/T_M-T_g$ = 1.3 (from Reference 1).

TABLE I

CALCULATION OF HEAT TRANSFER COEFFICIENTS FROM DAS
(From Reference 2)

MARAGING 300 STEEL

$$d = 39.8\ \varepsilon_{Avg.}^{-0.30}$$

ATOMIZATION PROCESS	PARTICLE SIZE, μm	DAS μm	$\varepsilon_{Avg.}$ °K/sec	h (CALCULATED) c.g.s.	h (CALCULATED) S.I.	$\frac{hr_o}{k_\ell}$
ARGON ATOMIZED FINE POWDER	75	∿2	$\sim 2.1 \times 10^4$	∿0.23	9.6×10^3	0.0084
REP	170	∿3	$\sim 5.5 \times 10^3$	∿0.13	5.4×10^3	0.011
STEAM ATOMIZED COARSE POWDER	1000	∿6.5	$\sim 4.2 \times 10^2$	∿0.06	2.5×10^3	0.029
VACUUM ATOMIZED	650	∿6.5	$\sim 4.2 \times 10^2$	∿0.039	1.63×10^3	0.0123

d = secondary dendrite arm spacing, DAS.

ε_{Avg} = average cooling rate during solidification.

$$\text{Biot number} = \frac{hr_o}{k_\ell}$$

differences inside a body are negligible for Bi $\leqslant$ 0.01 may not always be justified in atomization processes.

An important variable effecting undercoolinq prior to crystalline solidification or formation of amorphous structures is the cooling rate in the liquid droplet. A generalized expression relating the dimesionless instantaneous average cooling rate, ε_{avg}, in a liquid metal droplet to the Biot Number and dimensionless surface temperature has been derived (1):

$$\varepsilon_{avg} = 3 \times Bi \, (\theta_{SURFACE} + Ste)$$

where:

$$\varepsilon_{avg} = \frac{r_o^2 \, C_{ps}}{\alpha_\ell \, \Delta H_{s\ell}} \left(\frac{\partial T}{\partial t}\right)_{avg} = -3 \int_0^1 \phi^2 \left(\frac{d\theta}{dFo}\right) d\phi \qquad (3)$$

C_{ps} = specific heat of the solid

$Fo = \frac{\alpha_\ell t}{r_o^2}$ = Fourier number or dimensionless time

$\Delta H_{s\ell}$ = Heat of fusion

$Ste = C_{ps} \, (T_M - T_g)/\Delta H_{s\ell}$ = Stefan number

t = time

T = temperature of the droplet

T_g = temperature of the environment

T_M = melting point of the droplet

α_ℓ = thermal diffusivity of the liquid metal

$\theta = \frac{C_{ps}(T - T_M)}{\Delta H_{s\ell}}$ = dimensionless temperature

$\phi = r/r_o$ = fractional radius

Note that if thermal properties of a non-crystalline solid are assumed to be equal to the liquid the above expression would apply throughout solidification and subsequent cooling of amorphous droplets.

Expression (3) indicates that average cooling rate, $(\partial T/\partial t)_{avg}$, in a liquid metal droplet is directly proportional to the heat transfer coefficient at the droplet-environment interface and inversely proportional to the radius of the droplet. Thus, considering an upper limit for achievable heat transfer coefficients the only other method for increasing cooling rate is to decrease particle size. For example, using $h = 10^5$ W/m^2.K and expression (3) the average cooling rates of 8.3×10^5 K/s and 8.3×10^6 K/s are calculated for iron droplets of 100μm and 10μm, respectively, when their surface reaches the melting point.

Figure 2 shows the effect of process variables on the normalized crystalline solidification time of three different metal droplets. This time is normalized by dividing it by the Newtonian prediction of net solidification time which is:

$$\left.\frac{\alpha_\ell t_f}{r_o^2}\right|_{\text{Newtonian}} = [3 \text{ Bi Ste } (C_{p\ell}/C_{ps})]^{-1} \qquad (4)$$

Crystalline solidification time is the most important variable affecting segregate spacings, inclusion size, etc. It gives a better indication of time available for coarsening phenomena to occur than average cooling rate during solidification. The general trends established in Figure 2 for this normalized solidification time indicate that once heat transfer at the droplet-environment interface is maximized this time can be reduced by decreasing particle size. Furthermore, increasing initial superheat prolongs solidification, and the effect is larger as the Biot Number increases; the sensible heat retained in the liquid portion of the droplet increases resulting in longer times from initiation to completion of solidification. For example, 500μm droplets of iron solidifying with an $h = 5 \times 10^4$ W/m^{2}K (Bi = 0.31) will require 4.8×10^{-3} seconds to complete solidification with no initial superheat and 5.4×10^{-3} seconds with a dimensionless initial superheat of 0.1 (∿150K). Finally the effect of superheat on net solidification times diminishes with decreasing Biot Numbers. For example, for the same heat transfer coefficient, 50μm droplets of iron (Bi = 0.031) would solidify in 2.6×10^{-4} seconds and 2.7×10^{-4} seconds for initial superheats of zero and 150K, respectively.

Figure 3 shows dimensionless solid-liquid interface velocity as a function of Biot Number for various positions of the interface in iron droplets solidifying with concentric isotherms. The data show that dimensionless interface velocity increases with Biot Number and the progress of solidification.

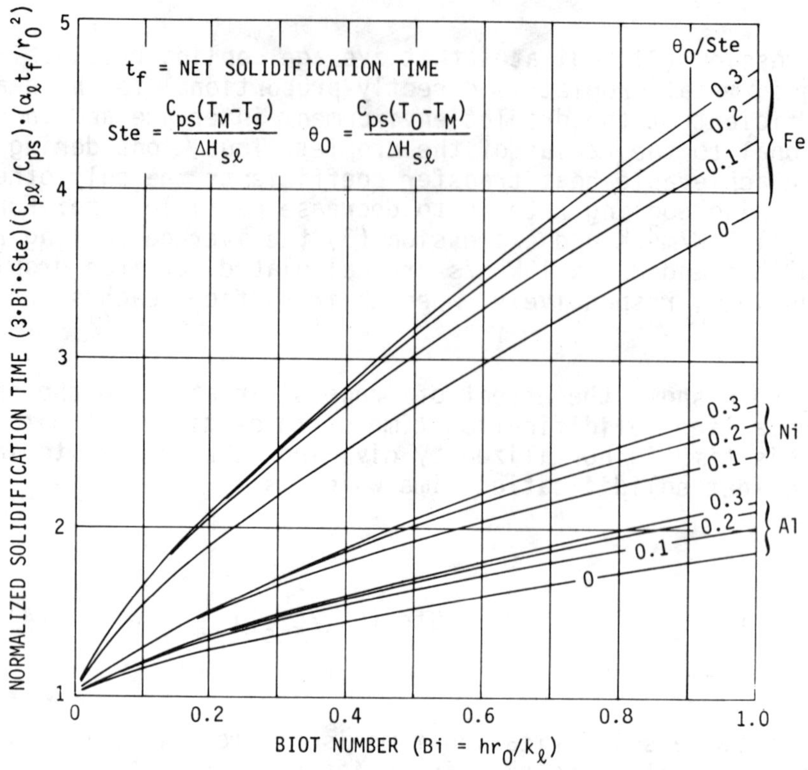

Figure 2. Normalized net solidification time for liquid droplets of aluminum, iron and nickel, as a function of Biot number and dimensionless initial superheat θ_0/Ste. The calculated curve for nickel with no superheat closely follows that of aluminum for θ_0 = 0.3 Ste. T_0 is the initial temperature of the droplets (from Reference 1).

The dimensionless interface velocity for Newtonian solidification is:

$$\frac{r_o}{\alpha_\ell} R \bigg|_{Newtonian} = \left(\frac{1}{\phi^*}\right)^2 \text{ Bi Ste } \frac{C_{p\ell}}{C_{ps}} \qquad (5)$$

where R = solid-liquid interface velocity
ϕ^* = r^*/r_o = dimensionless solid-liquid interface position
r^* = interface position

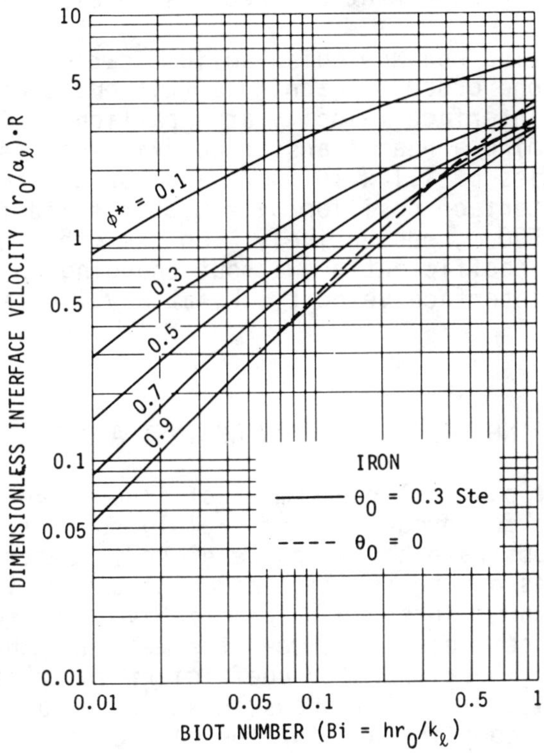

Figure 3. Dimensionless solid-liquid interface velocity during solidification of iron droplets, as a function of Biot number and fractional position of the interface $\phi^*=r^*/r_0$. $\theta_0=C_{ps}(T_0-T_M)/\Delta H_{s\ell}$ (from Reference 1).

For a given dimensionless interface position, ϕ^*, equation (5) gives a straight line with a slope of unity in Figure 3. The ratio of calculated non-Newtonian to Newtonian interface velocities is close to unity in the beginning of solidification for the case of no superheat regardless of Biot number. On the other hand, this ratio decreases with increasing fraction solid and increasing Biot number at a given fraction solid. In general, increasing initial superheat in the non-Newtonian regime results in a corresponding decrease in solid-liquid interface velocities. However, this effect is only noticeable in the initial stages of solidification, and decays rapidly with increasing superheat and decreasing Biot number.

Note that during non-Newtonian solidification increasing the particle size by one order of magnitude (without changing h) reduces the actual interface velocity at a particular fraction solid. For example, assuming a heat transfer coefficient $h = 5 \times 10^4$ W/m^{2}K interface velocities of 1.1×10^{-1} m/s and 5.3×10^{-2} m/s are calculated for an interface position $\phi^* = 0.5$ in solidifying iron droplets of 50μm and 500μm, respectively. On the other hand, increasing the heat transfer coefficient h (without changing r_o) will increase the solid-liquid interface velocity, R, at any location.

HEAT FLOW DURING SOLIDIFICATION AGAINST A METAL SUBSTRATE

In the recent past a large number of innovative batch and continuous techniques for production of rapidly solidified material against a metal substrate have been developed. Determination of exact cooling rates during solidification of crystalline and non-crystalline structures in these processes have required estimates of heat transfer coefficients between the melt and the substrate. Two measurements of heat transfer coefficient in splat cooling have been reported (3,4). Table II shows these values along with measured heat transfer coefficients for the case of a pressurized (13,000 psi) 89.7 MN/m^2 aluminum casting against a steel mold (5) and liquid aluminum die cast against a steel mold at 55 m/s metal flow velocity (6). Again, it appears than an upper limit exists for practically achievable heat transfer coefficients between liquid metals and substrates. It is probably in the range of $h = 10^5$ to 10^6 W/m^2.K.

Computer heat flow calculations carried out for solidification of a non-crystalline aluminum melt are shown in Figures 4 and 5. The data is presented in terms of the dimensionless variables

$$Bi = \frac{hL}{k_\ell} \quad \text{and} \quad Fo = \frac{\alpha_\ell t}{L^2} \qquad (5)$$

where L is casting thickness.

There is no evidence that aluminum has been made amorphous by rapid solidification, however, the numbers generated in these calculations are useful in that they again present the upper limit of

TABLE II

MEASURED HEAT TRANSFER COEFFICIENTS FOR SOLIDIFICATION OF ALUMINUM AGAINST A METAL SUBSTRATE

	TECHNIQUE	THICKNESS OF CASTING	h, $W/m^2.K$
1.	SPLAT ON Ni SUBSTRATE (from reference 3)	1μm	1.1×10^5 to 2.8×10^5
2.	DROP SMASH ON Fe SUBSTRATE (from reference 4)	150μm	$1.7 \times 10^4 - 1.8 \times 10^5$
3.	PRESSURE CAST IN STEEL MOLD (from reference 5)	10^4μm	NO PRESSURE 3.1×10^3 89.7 MPa 3.3×10^4
4.	DIE CAST IN STEEL MOLD (from reference 6)	1.6×10^3μm	7.94×10^4 Metal flow velocity 55 m/s Pressure 175 MPa

achievable cooling rates — heat of fusion did not have to be removed during solidification.

As expected, for small Biot numbers temperature gradients in the melt are negligible, Figure 4 (a) and the linear portion of the curve in Figure 5. The maximum melt thickness for this region of the plot is ∿40μm for a heat transfer coefficient $h = 4.18 \times 10^5$ $W/m^2.K$ (which is higher than those reported in Table II). At Biot numbers larger than ∿10 resistance to heat flow is primarily within the aluminum melt (cooling is essentially ideal) and dimensionless average cooling rate is independent of Biot number — actual average cooling rate $(\partial T/\partial t)_{Avg}$ is inversely proportional to the square of melt thickness. Thus, for Biot numbers in the Newtonian and ideal cooling ranges the actual average cooling rate in the liquid increases by one and two orders of magnitude, respectively, as the melt thickness is decreased by one order of magnitude. Thus, as in the case of metal droplets, the most effective way to increase cooling rate is to decrease melt thickness.

Similar calculations have been reported for crystalline solidification of iron splats against a copper substrate (7). It was found that for Biot numbers less than ∿0.015 and more than ∿30, cooling was Newtonian and ideal, respectively.

Analytical expressions are available for the solid-liquid interface velocity during plane front solidification of a melt against a substrate for the cases of Newtonian and ideal cooling, equations (6) and (7), respectively.

$$R = \frac{h}{\rho_s \Delta H_{s\ell}} (T - T_g) \qquad (6)$$

$$R = \frac{2\gamma^2 \alpha_s}{S} \qquad (7)$$

where R = solid-liquid interface velocity

α_s = thermal diffusivity of the solid forming

ρ_s = density of the solid forming

S = distance solidified

γ = argument of the error function solution of the temperature distribution; it is determined from a characteristic equation which contains metal and substrate thermal constants.

Approximate analytical solutions are available for Biot numbers between these two extremes. However, numerical solutions such as those described above can be readily employed.

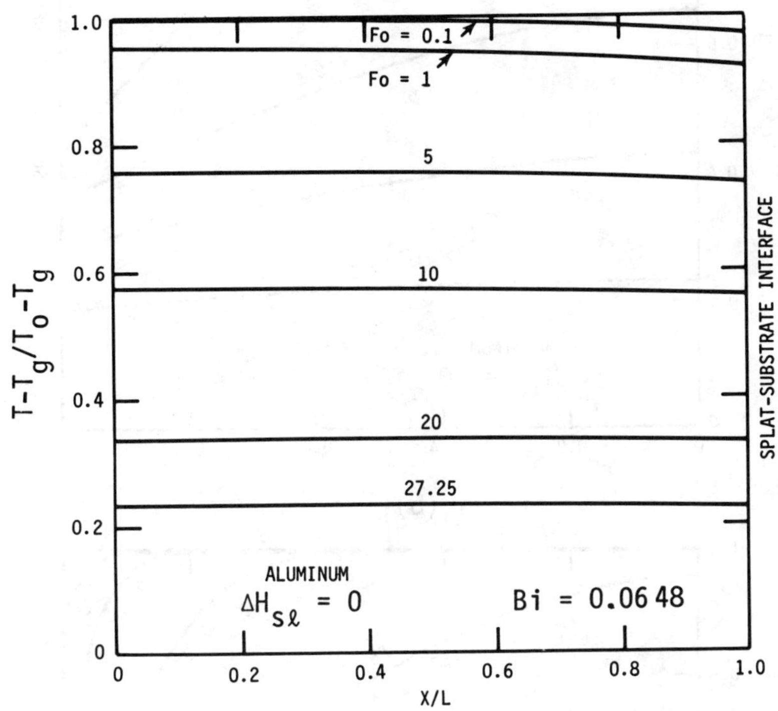

(a)

Figure 4. Calculated temperature distributions during cooling and non-crystalline solidification of splats of aluminum against a copper substrate. T, and T_0 and T_g denote instantaneous and initial aluminum melt and initial copper substrate temperatures, respectively. X/L is fractional distance from the free surface of the aluminum. (from Reference 2).

Several attempts have been made to predict critical cooling rates in the liquid during RSP for the formation of non-crystalline structures (8-10). Heat flow calculations are usually combined with theories of nucleation, growth and transformation kinetics. This area, as well as, rapid solidification of undercooled crystalline structures remains fertile for future investigations.

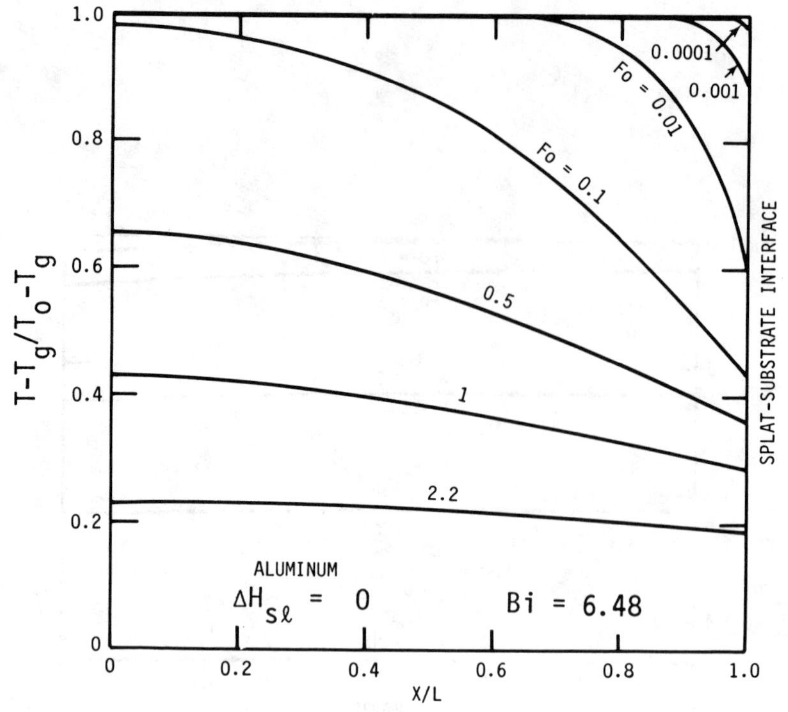

1.0
0.8
0.6
0.4
0.2
0
T-Tg/To-Tg
0.0001
0.001
Fo = 0.01
Fo = 0.1
0.5
1
2.2
ALUMINUM
ΔHsℓ = 0
Bi = 6.48
SPLAT-SUBSTRATE INTERFACE
0
0.2
0.4
0.6
0.8
1.0
X/L

(b)

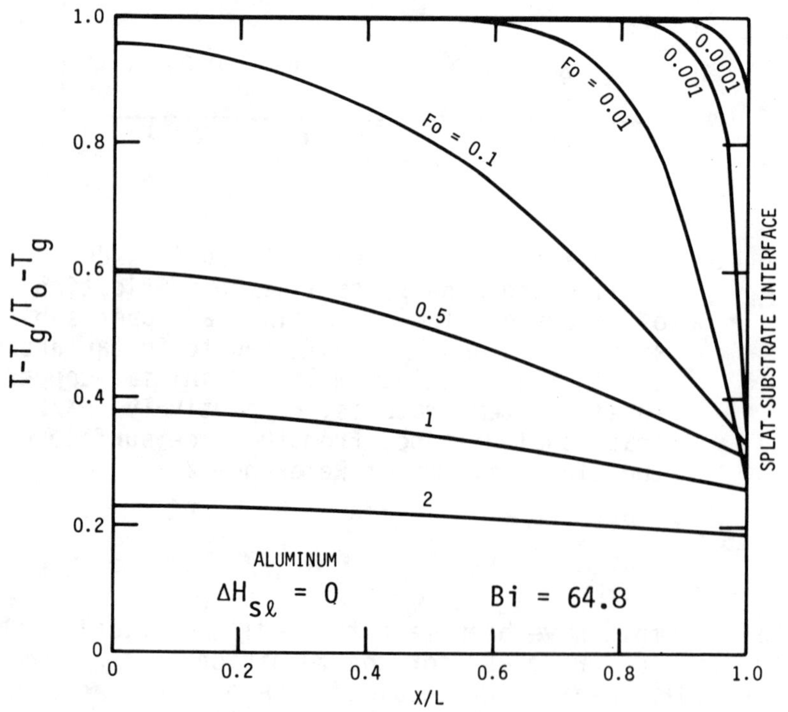

1.0
0.8
0.6
0.4
0.2
0
T-Tg/To-Tg
0.0001
0.001
Fo = 0.01
Fo = 0.1
0.5
1
2
ALUMINUM
ΔHsℓ = 0
Bi = 64.8
SPLAT-SUBSTRATE INTERFACE
0
0.2
0.4
0.6
0.8
1.0
X/L

(c)

Figure 4

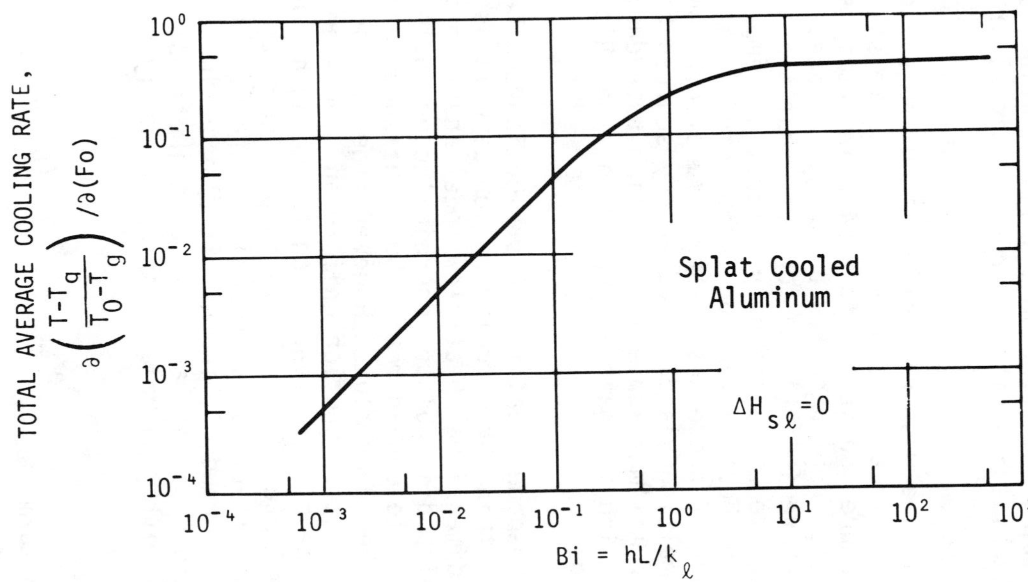

Figure 5. Dimensionless cooling rate averaged over melt thickness and time for temperature to reach half the melting point for non-crystalline solidification of an aluminum melt against a copper substrate. T_0 and T_g are initial melt and substrate temperatures, respectively.

HEAT FLOW DURING MELTING AND SOLIDIFICATION OF A SURFACE LAYER

The recent availability of high power directed energy sources such as the electron beam and the different types of lasers has led to the development of rapid melting and solidification techniques in which a bulk (semi-infinite) substrate in intimate contact with a molten layer acts as the quenching medium. Results of recent one and two-dimensional computer heat flow analyses (11,12) carried out to investigate the effect of high intensity radiation on the important surface layer melting and subsequent solidification variables of three substrate materials: aluminum, iron and nickel are summarized below.

The problem considered was the rapid melting and subsequent solidification of the surface layer of a semi-infinite solid, initially at room temperature, subjected to a high intensity stationary heat flux over a circular region on its bounding surface. Since the melt and the substrate are in intimate contact, heat transfer coefficient between the two tends to infinity. Thus, one important limitation on rapid rate of heat extraction encountered in atomization and splat cooling types of processes has been removed. However, as shown below, surface melting via directed energy sources has its own inherent limitation on maximum achievable cooling rate during subsequent solidification which is again a function of the thickness of the molten surface region.

In general, the absorbed heat flux distribution can be both a function of position within the circular region as well as time. The generalized, two-dimensional heat flow model and numerical solution techniques developed in oblate spheroidal coordinate system (12) can readily take into consideration both the space and the time variation of the absorbed heat flux. However, the numerical results presented below are for step function uniform and Gaussian heat flux distributions within the circular region on the bounding surface of the semi-infinite substrate. The relationship between these two absorbed heat flux distributions is shown in Figure 6. If we assume that the total power absorbed, Q, in the circular region is identical for the uniform and the Gaussian heat flux distributions, as shown in Figure 6, then the following relationship is readily deduced:

$$q_{uniform} = \frac{q_o}{2.313} \qquad (8)$$

where q_o is the absorbed heat flux at the center of the circular region in the Gaussian distribution.

In what follows we will first consider one and two-dimensional heat flow for the case of uniform absorbed heat flux before treating the more complex Gaussian heat flux distribution.

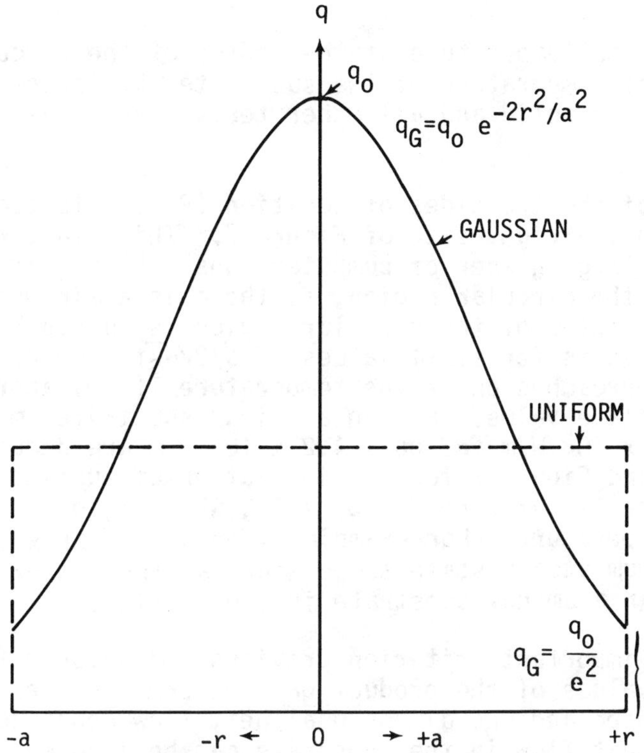

Figure 6. Relationship between uniform and Gaussian absorbed heat fluxes when the total power absorbed over the circular region is identical. q denotes the absorbed heat flux while a is the radius of the circular region. (from Reference 12).

Two important criteria were deduced, from the computer heat flow calculations (12), for the surface melting of a semi-infinite substrate subjected to a uniform heat flux q over a circular region of radius a on its bounding surface. First, it was shown that the product qa should exceed specific values, given by the following expression, if surface melting is to be initiated and the center of the circular region is to reach a given temperature up to the vaporization temperature of the substrate:

$$qa \geq k_{\ell}(T(0,0)-T_M) + \frac{k_s \Delta H_{s\ell}}{C_{p\ell}} + k_s(T_M-T_g) \qquad (9)$$

where T(0,0) is the temperature at the center of the circular region, T_o is the initial temperature of the substrate, k_s is the thermal conductivity of the solid and all other terms have previously been defined.

The ratio of the two sides of equation (9) is plotted on the vertical axis on the right size of Figure 7. This plot, which was obtained from a large number of computer runs, shows that for a given radius of the circular region, a, there is a minimum heat flux required if the center of the circular region is to reach a given temperature. That is for small values of $a/2\sqrt{\alpha_s t}$ the ratio on the vertical axis approaches one - the temperature T(0,0) approaches its maximum steady state value. For an aluminum substrate, minimum values of qa ≃ 2.3 x 10^5 W/m and qa ≃ 4.2 x 10^5 W/m are deduced from expression (9) and Figure 7 for the initiation of surface melting and for a surface temperature T(0,0) = T_v, where T_v denotes the vaporization temperature. For example, when qa ≃ 2.86 x 10^5 the calculated maximum steady state temperature at the center of the molten zone of an aluminum substrate is 1607°K (12).

The second important criterion previously developed (12) was that for large values of the product qa isotherms in the substrate are planar in shape and one-dimensional heat flow conditions prevail - lateral heat flow in the substrate can be ignored. The vertical axis on the left side of Figure 7 is the ratio of the square root of time in one and two-dimensional heat flow. The curve associated with this axis shows that for values of the dimensionless term $a/2\sqrt{\alpha_s t} \gtrsim 2.0$ - the radius of the circular region is larger than the characteristic distance for diffusion of heat-heat flow is essentially one-dimensional. The specific value of qa deduced for a given temperature T(0,0) for this condition to be met is 3.33 times the value calculated from expression (9). As example, for an aluminum substrate, when qa ≳ 1.4 x 10^6 W/m heat flow is essentially in one-dimension for T(0,0) temperatures up to the vaporization temperature. Time for the center of the circular region to reach any temperature above the melting point of the substrate can also be deduced from Figure 7. For example, for an absorbed uniform heat flux of 10^9 W/m^2 and a spot radius a = 380μm, T(0,0) will reach 1700K in 2.1 x 10^{-3} sec. Furthermore, the maximum temperature achieved at the center of the circular region would be 2238K.

Results from one-dimensional heat flow analysis, when the criterion noted above is applicable, are shown in Figures 8 to 11. Calculated melt depths versus total time for several uniform absorbed heat fluxes are shown in Figure 8. The arrows in the figure denote times at which the surface of each material reaches it's vaporization temperature, T_v, and the heat flux, q, is removed. Melting continues a while longer until a maximum melt depth, z_{max}, is reached.

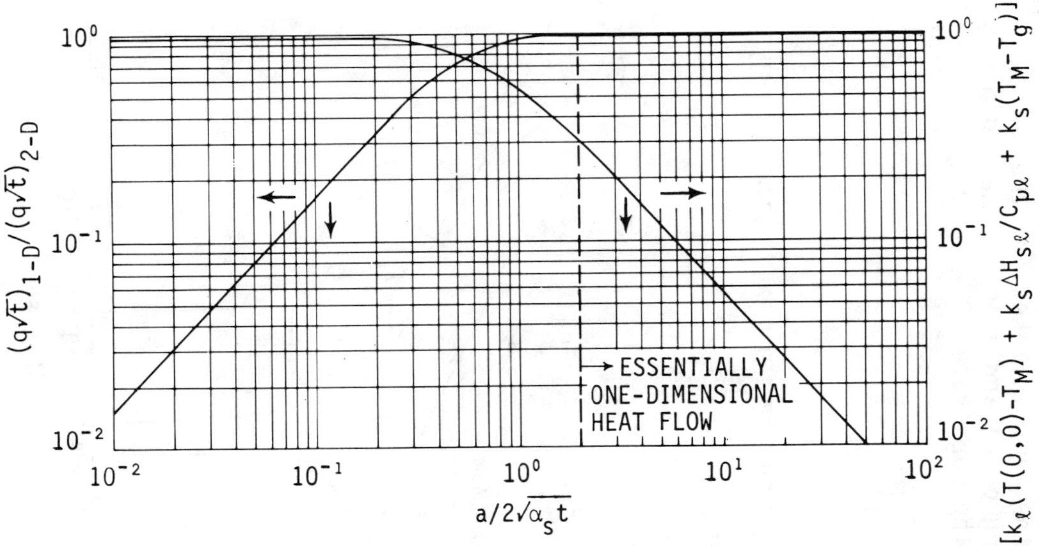

Figure 7. Temperature at the center of the liquid zone of a semi-infinite solid substrate during surface melting as a function of uniform absorbed heat flux, radius of the circular region and time. The vertical axis on the left gives the ratio of square root of time in one and two-dimensional heat flow (see Reference 12).

Note that for an aluminum substrate subjected to uniform absorbed heat fluxes of 5 x 10^8 W/m^2 and 5 x 10^{10} W/m^2 the criterion for one-dimensional heat flow would be met if the radii of the circular regions exceed 2800μm and 28μm, respectively. The corresponding maximum melt depths, z_{max}, from Figure 8 are ∿650μm and 6.5μm, respectively.

Melt depth during melting and subsequent crystalline solidification periods are expressed in dimensionless form, $\rho_s \Delta H_{s\ell} z/qt_m$ in Figure 9. t_m denotes the time it takes for the surface of a material to reach its melting point and z is the melt depth perpendicular to the surface of the substrate.

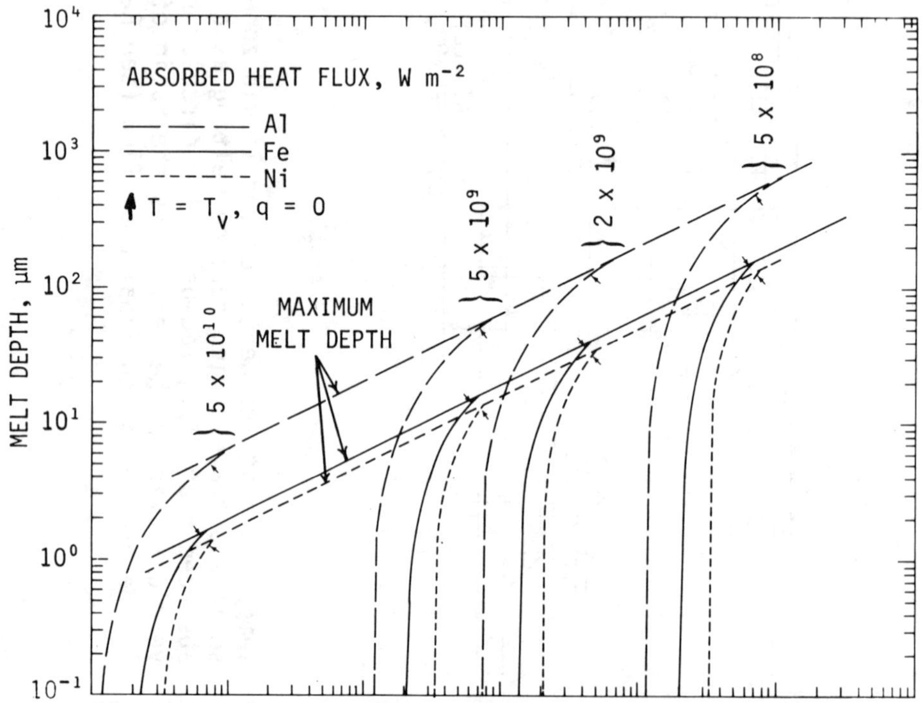

Figure 8. Melt depth versus total time for different uniform absorbed heat fluxes obtained via a numerical technique. Arrows indicate times at which a surface reaches it's vaporization temperature and heat flux is removed (from Reference 11).

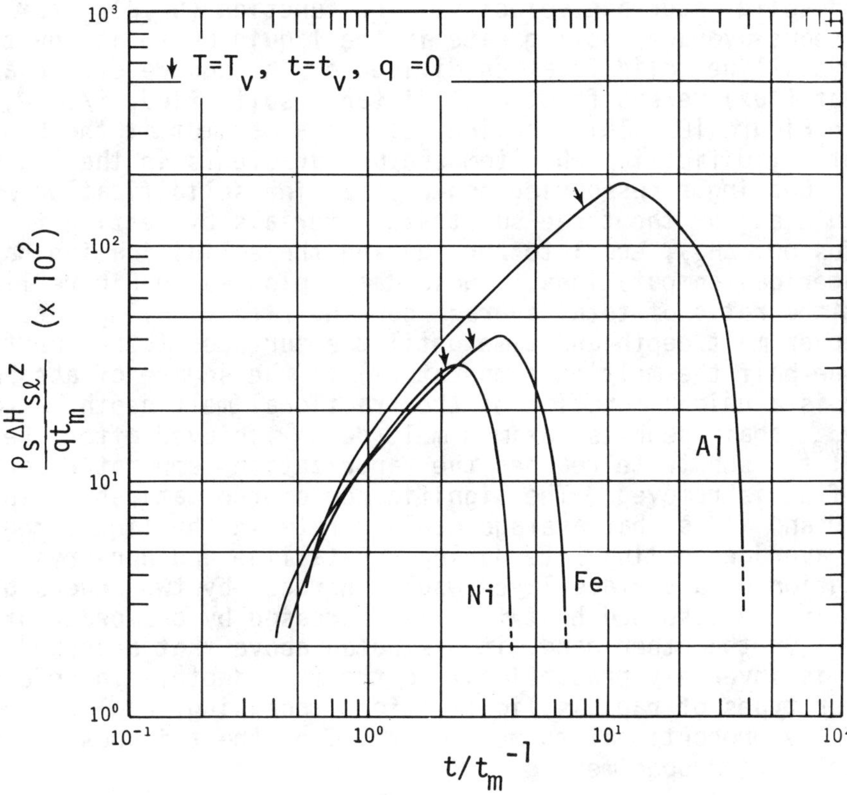

Figure 9. Dimensionless melt depth $\rho_S \Delta H_{S\ell} z/q\, t_m$ during melting and solidification versus dimensionless time, t/t_m (from Reference 11).

Some general trends between the absorbed heat flux, time and melt depth can be readily deduced from the data presented in Figures 8 and 9.

The surface of a substrate reaches a given temperature when $q\sqrt{t}$ is kept constant. The significance of this finding is that at higher power inputs, less time is available for diffusion of heat into the metal substrate - the absorbed heat is concentrated near the surface of the material leading to steeper temperature gradients.

The melt depth is inversely proportional to the absorbed heat flux.

The change in temperature with time in the liquid, heating and cooling rates during melting and subsequent solidification, is given by the product $G_L.R$ where G_L is the temperature gradient and R is the liquid-solid interface velocity. The function $(G_L.R)_{Avg}/q^2$ (instantaneous average cooling rate in the liquid phase at any time during crystalline solidification divided by the square of the absorbed heat flux) versus fractional distance solidified, S/z_{max}, is plotted in Figure 10. This cooling rate is a maximum at the beginning of solidification when temperature gradients in the liquid are high. Cooling rates during non-crystalline solidification were also calculated for the three substrate materials by setting the heat of fusion, $\Delta H_{s\ell}$, equal to zero during the solidification half of the numerical computations. These data, plotted in Figure 11, show that the ratio of total average cooling rate (cooling rate averaged over melt depth and time until the surface of the substrate reaches one-half the melting temperature) to the square of absorbed heat flux is a unique function of the fractional melt depth, z/z_{max} where $z/_{max}$ again denotes maximum melt depth achieved after the surface of the substrate reaches the vaporization temperature and the heat flux is removed. The significance of the data shown in Figures 10 and 11 is that average cooling rate in the liquid region and total average cooling rate during crystalline and non-crystalline solidification of a surface layer would increase by two orders of magnitude if the absorbed heat flux is increased by one order of magnitude. On the other hand, it was noted above that absorbed heat flux is inversely proportional to the melt depth. Therefore, as in other types of rapid solidification processing, cooling rates are inversely proportional to melt depth, i.e. the thickness of the rapidly solidified specimen.

For example, consider an aluminum substrate subjected to uniform absorbed heat fluxes of 5×10^8 W/m^2 and 5×10^{10} W/m^2. As before, the corresponding maximum calculated melt depths are 650μm and 6.5μm, respectively. Now, assume that the heat fluxes are removed once 10% of these melt depths are achieved in each case, $z/z_{max} = 0.1$ in Figure 11, z = 65μm and 0.65μm, respectively. The corresponding total average cooling rates from Figure 11 for the lower and the higher absorbed heat fluxes are $\sim 1.9 \times 10^5$ K/sec and $\sim 1.9 \times 10^9$ K/sec, respectively. For comparison, assume a high heat transfer coefficient of $h = 10^5$ W/m^{2}K between an aluminum splat and a copper substrate. Using the thermophysical data for aluminum given in Reference (11) cooling rates of $\sim 2.3 \times 10^5$ K/sec and $\sim 2.3 \times 10^7$ K/sec are calculated from Figure 5 for splat thicknesses of 65μm and 0.65μm, respectively. Comparison of these values with those calculated above for the rapidly solidified surface layers reveals that the latter permits higher cooling rates as the melt depths are reduced by application of higher heat fluxes. This is due to the fact that in surface melting average cooling rate is inversely proportional to the square of the melt depth (is directly proportional to the square of the absorbed heat flux) while in splat cooling type processes, when solidification is in the Newtonian regime, average cooling rate is only inversely proportional to the melt thickness.

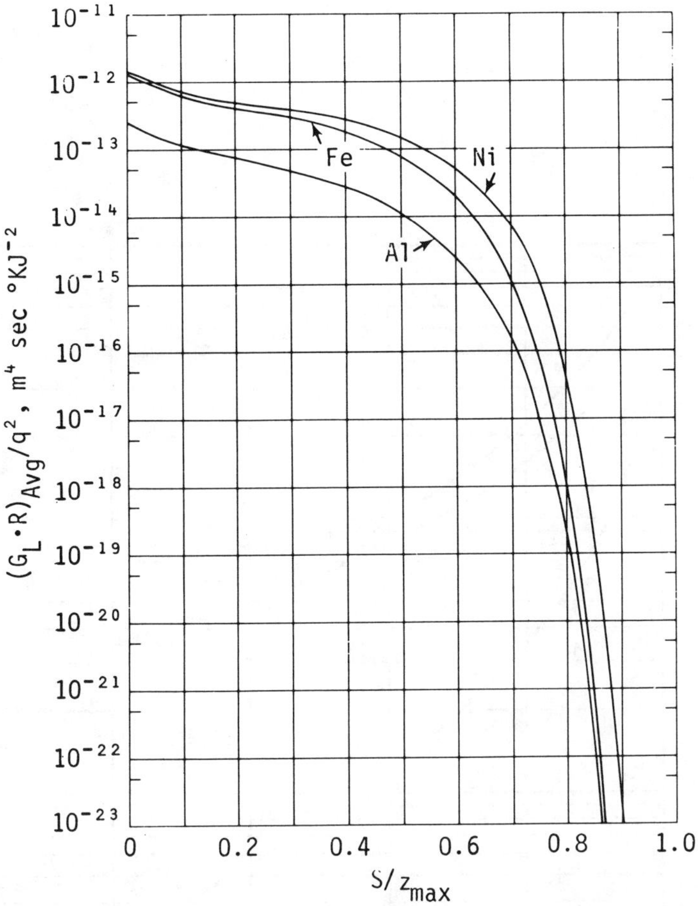

Figure 10. The ratio of instantaneous average cooling rate in the liquid to the square of absorbed heat flux, $(G_L.R)_{Avg}/q^2$ versus fractional distance solidified, S/z_{max} (from Reference 11).

The general trends established in these one-dimensional computer calculations (11) are summarized in Table III below. They show that temperature gradients in the liquid and solid phases and interface velocities are directly proportional to the absorbed heat flux, whereas melt depth is inversely proportional to the absorbed heat flux. Average cooling rates comparable to and exceeding those predicted for splat cooling can be achieved by increasing the heat flux and reducing the dwell time of the incident radiation. An order of magnitude increase in the absorbed heat flux results in a corresponding two orders of magnitude increase in average cooling rates in the liquid during solidification of crystalline and non-crystalline structures.

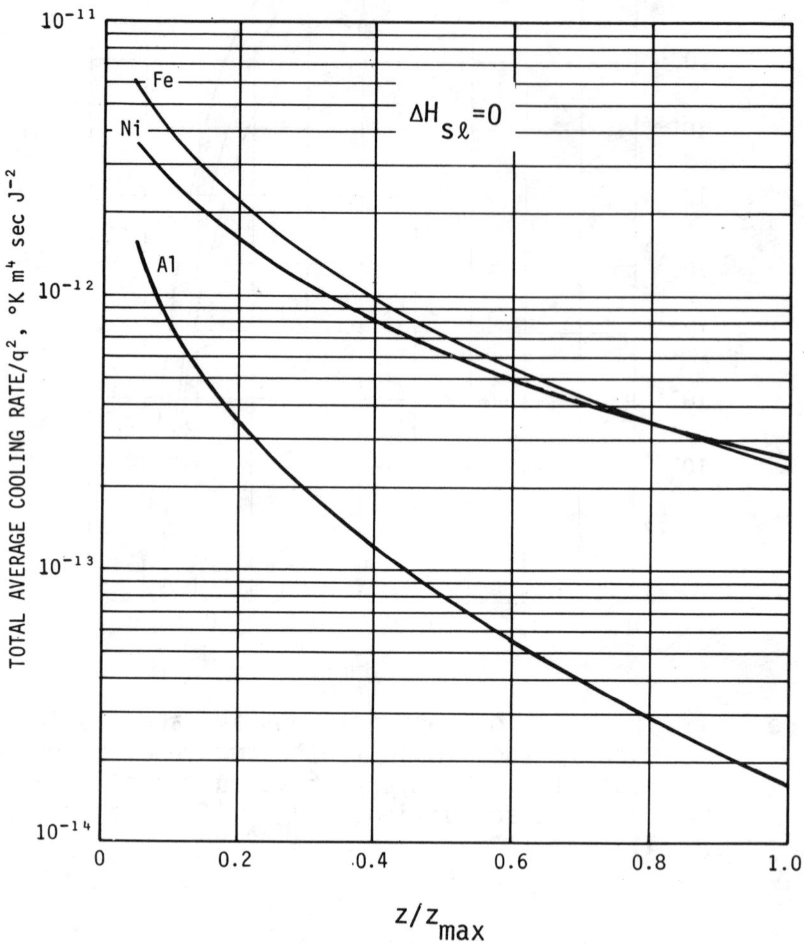

Figure 11. The ratio of total average cooling rate to the square of absorbed heat flux versus fractional distance melted, z/z_{max}, during solidification of a non-crystalline solid (from Reference 11).

TABLE III

One-Dimensional Heat Flow
The Effect of Change in Absorbed Heat Flux on Other Variables

Order of magnitude increase in q results in the following changes:

q	Time	Melt Depth	R	G_L	$G_L \cdot R$	G_L/R
10 ↑	10^2↓	10↓	10↑	10↑	10^2↑	↔

↑ increase

↓ decrease

↔ no change

From Reference 11.

Results from the two-dimensional heat flow calculations (12) for a uniform absorbed heat flux q over a circular region of radius a on the bounding surface of a semi-infinite aluminum substrate are shown in Figures 12 and 13.

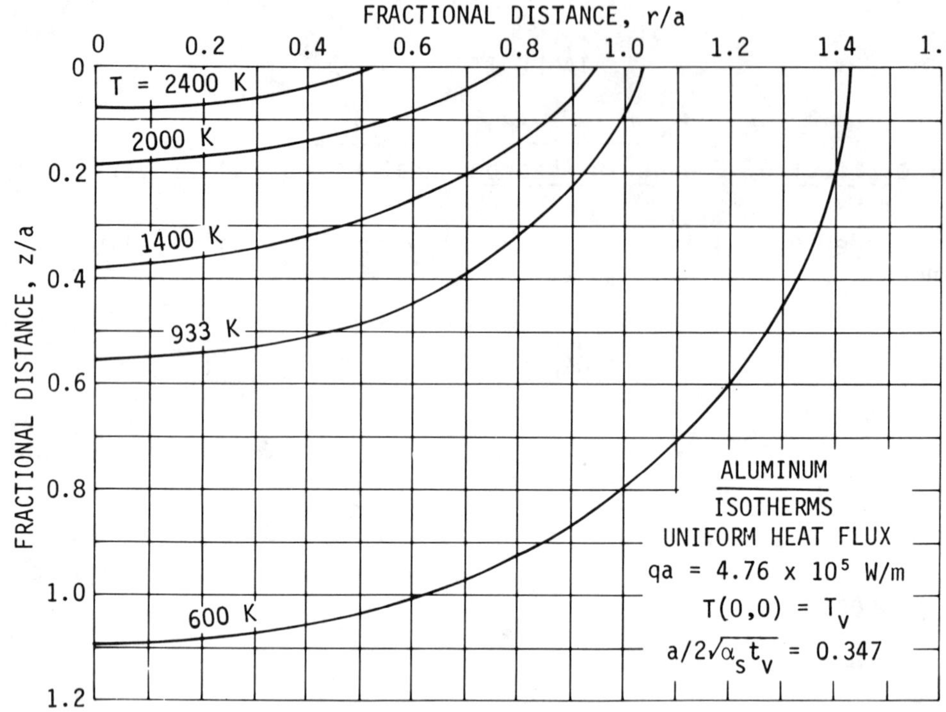

Figure 12. Shape and location of several isotherms, including the liquid-solid interface (T=933K), during melting of an aluminum substrate subjected to a uniform absorbed heat flux q over a circular region of radius a. The isotherms are drawn at the instant when $T(0,0)=T_V=2723K$ (from Reference 12).

An important finding from this work is that for a given value of T(0,0) at the center of the circular region the dimensionless temperature distribution in the substrate material during melting and solidification is identical for any combination q and a as long as qa = constant. Figure 12 shows the shape and location of several isotherms, including the liquid-solid interface, in an aluminum substrate material subjected to a uniform heat flux q over a circular region of radius a where the product $qa=4.76 \times 10^5$ W/m. Note that these are the isotherms at the instant T(0,0) reaches the vaporization temperature of aluminum T_V=2723K and the axis used are made dimensionless dividing them by the radius of the circular region. Identical dimensionless plots were obtained from several computer runs where q and a values were varied over large ranges while the product qa was kept constant.

Figure 13 shows the effect of increasing the product qa on the geometry and the location of the liquid-solid interface. The curve for qa $\simeq$ 1.9 x 10^6 W/m is in the range where heat flow can be assumed to be essentially one-dimensional and the shape of the isotherm verifies this fact. On the other hand, with decreasing values of qa, lateral heat flow becomes significant and the liquid-solid interfaces assume increasingly more convex geometries.

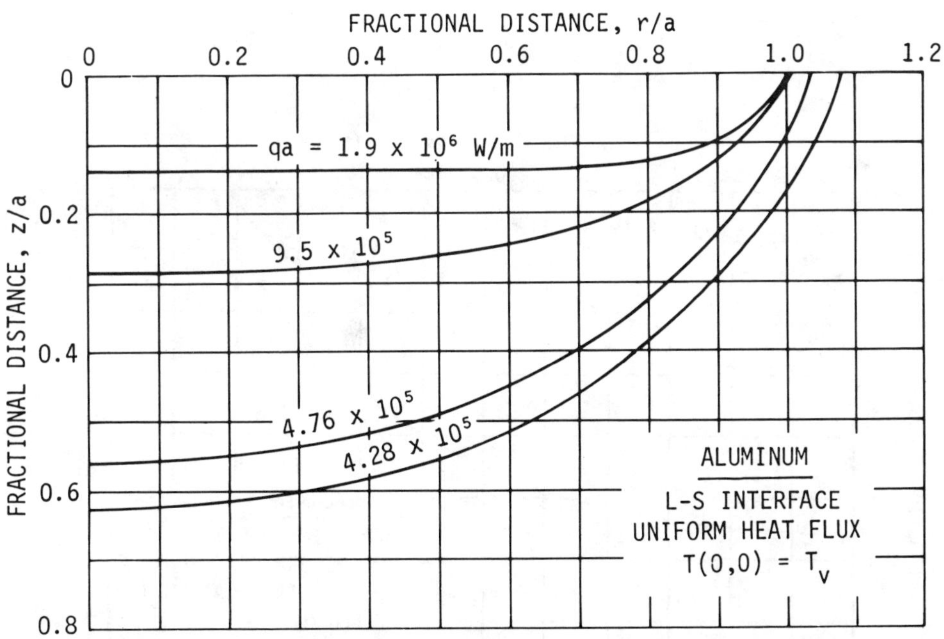

Figure 13. Liquid-solid interfaces during melting of an aluminum substrate subjected to a uniform heat flux q over a circular region of radius a. The interfaces for each constant qa are shown at the instant the center of the circular region reaches the vaporization temperature.(from Reference 12).

Using the relationships between uniform and Gaussian heat flux distributions developed in Figure 6 and expression (8) the following correlations between melt depths and shapes of liquid-solid interfaces were obtained (12). For a given temperature at the center of the circular region T(0,0) and total power absorbed, Q, melt depths in the Gaussian heat flux distributions are ∿2.313 times shallower than that obtained in the case of the uniform absorbed heat flux.

Figure 14 compares the shapes and locations of the liquid-solid interfaces between uniform and Gaussian heat flux distributions when T(0,0) reaches the vaporization temperature of an aluminum substrate. As indicated above, for a given total absorbed heat flux expression (8) holds and the center of the circular region heats up much more rapidly when the heat flux has a Gaussian distribution. The melt depth along the z-axis is shallower, while the melt width is smaller due to the diminishing absorbed heat flux with increasing r in the circular region.

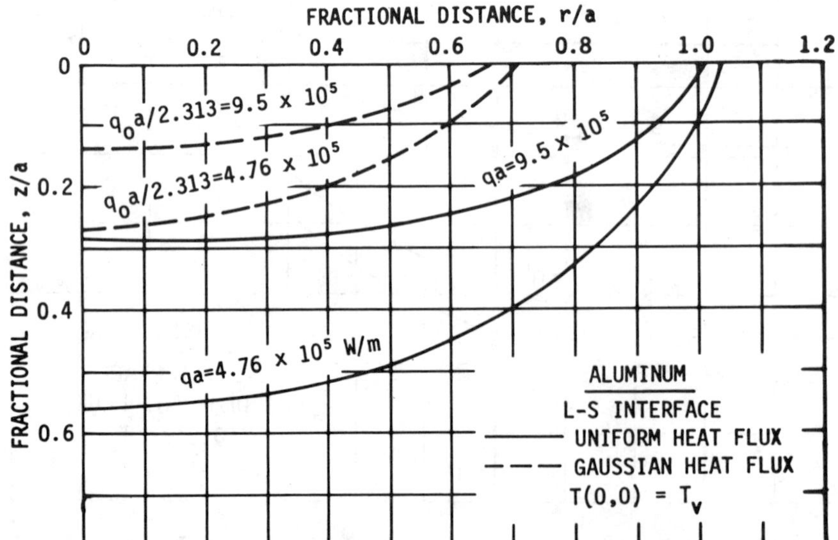

Figure 14. Comparison between the shapes and locations of the liquid-solid interfaces during melting of an aluminum substrate subjected to uniform and Gaussian absorbed heat flux distributions at the instant $T(0,0) = T_V$ (from Reference 12).

Figure 15 shows the general relationship developed from numerical computations between temperature at the center of the circular region and the product of absorbed heat flux and melt depth.for an aluminum substrate. z in this figure denotes melt depth along the z-axis at the center of the circular region r = 0. For example, for

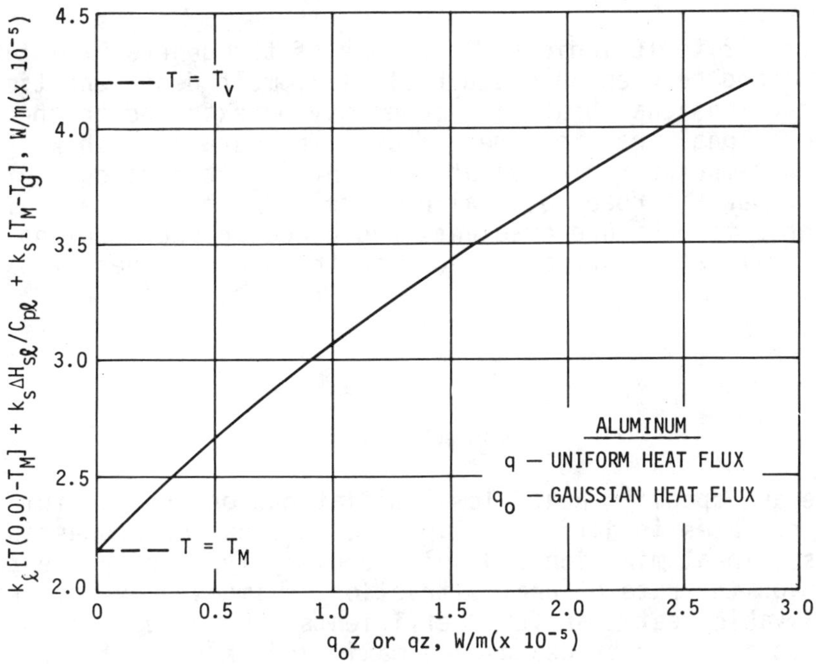

Figure 15. Temperature at the center of the circular region T(0,0) over which a heat flux is absorbed in an aluminum substrate versus the product of the absorbed heat flux and melt depth in the axial direction away from the surface. q_o denotes the maximum heat flux at the center of the Gaussian heat flux distribution. See Figure 6 and equation (8) (from Reference 12).

uniform heat fluxes of 3×10^8 W/m^2 and 2.5×10^9 W/m^2 absorbed over circular regions with radii of ≃ 1590μm and ≃ 190μm, respectively (qa ≃ 4.76×10^5 W/m) the corresponding times and melt depths during melting of an aluminum substrate when $T(0,0) = T_v$ are $t = 5.4 \times 10^{-2}$ sec and $t = 7.7 \times 10^{-4}$sec from Figure 7 and z = 933μm and z = 112μm from Figure 15, respectively. Similar information can also be obtained from these two figures for any T(0,0) temperature between T_M and T_V. While the data plotted in Figure 7 can be used for any substrate material, the curve in Figure 15 is specifically calculated for an aluminum substrate. Similar curves can readily be generated for other substrate materials using the equations and computations procedure outlined in the previous sections. As notea above, the data in Figure 15 also show that for given total powers absorbed, Q, melt depths for the Gaussian heat flux distribution are 2.313 times shallower.

Figures 12 to 15 indicate that some of the general relationships developed between absorbed heat flux, melt depth and time in the one-dimensional heat flow model may be extended to the case of two-dimensional transient heat flow. The data also show that for a given temperature T(0,0) at the center of the circular region and total power absorbed, Q, shallower melt depths and heat affected zones, higher temperature gradients and hence, higher cooling rates during solidification are achieved with the Gaussian heat flux distribution.

SUMMARY

There are specific heat flow limitiations on the maximum achievable cooling rates in different Rapid Solidification Processing, RSP, techniques. In atomization and splat cooling types of processes the limitation on the rate of heat extraction is imposed by the practically achievable heat transfer coefficients. Once heat transfer coefficients in these processes are maximized, $h = 10^5$ to 10^6 W/m^2K, further improvements in cooling rates can only be obtained at the expense of reduced specimen size or thickness. The rapid surface melting and solidification technique, via directed high energy sources, has its own limitation on specimen size and thickness, even though the melt is intimate contact with its own substrate. High cooling rates during solidification are only achieved at the expense of high absorbed heat fluxes, reduced heat affected zones and melt depths. Furthermore, the requirement of high absorbed heat flux imposes restrictions on the spot size (e.g. the radius) of the surface melted region.

REFERENCES

1. C. G. Levi, R. Mehrabian, submitted for publication to Met. Trans.

2. R. Mehrabian; Proceedings of Conference on Rapid Solidification Processing, Principles and Technologies, Nov. 1977, Reston, Virginia, p. 7. R. Mehrabian, B. H. Kear and M. Cohen, Eds., Claitor Publishing Division, 1978.

3. P. Predecki, A. W. Mellendore and N. J. Grant, Trans. Met. Soc. AIME, 1969, Vol. 233, p. 1581.

4. D. R. Harbur, J. W. Anderson and W. J. Maraman, Trans. Met. Soc. AIME, 1969, Vol. 245, p. 1055.

5. S. D. E. Ramati, G. J. Abbaschian, D. G. Backman and R. Mehrabian, Met. Trans. B, 1978, Vol. 9B, p. 279.

6. S. Hong, D. G. Backman and R. Mehrabian, submitted for publication to Met. Trans.

7. R. C. Ruhl, Mater. Sci. Eng., 1967, Vol. 1, p. 313.

8. P. H. Shingu and R. Ozaki, Met. Trans. A, 1975, Vol. 6A, p. 33.

9. D. R. Uhlmann, J. Non-Cryst. Solids, 1976, Vol. 7, p. 337.

10. F. Spaepen and D. Turnbull; Proceedings of Second International Conference on Rapidly Quenched Metals. Edited by N. J. Grant and B. L. Giessen, 1975, M.I.T. Press, Cambridge, Mass., p. 205.

11. S. C. Hsu, S. Chakravorty and R. Mehrabian, Met. Trans. B, 1978, Vol. 9B, p. 221.

12. S. C. Hsu, S. Kou and R. Mehrabian, submitted for publication to Met. Trans.

ACKNOWLEDGEMENT

This research is being sponsored by the Defense Advanced Research Projects Agency and monitored by the Office of Naval Research under contract # N00014-78-C-0275. Technical monitor of the contract is Dr. B. A. MacDonald.

Rapid solidification

R. Mehrabian

Points of progress in current understanding of rapid liquid to solid transformation are reviewed. Emphasis is placed on those aspects of the emerging science that would permit development of guidelines and predictive models for alloy design and process control to achieve desired microstructures and properties. Recently available computer-based phase-stability data compilations could be coupled to predictive models to provide a complete description of phase relations (both equilibrium and metastable) for design of new alloys. Heat-flow models are now available for the rapid solidification of ribbons, surface layers, and atomized droplets. These models predict strict limitations on achievable cooling rates in the various processes. The need to balance emphasis on heat-extraction rate and other factors, such as nucleation rate and undercooling governing the attainment of unique microstructures, is now clearly recognized. The extension of morphological stability theory to high interface velocities has shown that plane-front solidification becomes increasingly stable with increasing interface velocity. While some metastable structures obtained during rapid solidification indicate departures from local equilibrium at the solid/liquid interface, the exact form or magnitude of these departures are not known. Amorphous solidification in the presence of a crystalline phase can now be explained in terms of thermodynamically forbidden regions of a phase diagram for partitionless solidification of a single phase. Microstructural observations made on submicrometre powders are discussed with particular reference to recent theoretical developments in interface stability and heat-flow concepts. Finally, the potential for coupling non-destructive testing for on-line control of parameters during rapid solidification processing is noted. IMR/107

LIST OF SYMBOLS

Dimensioned variables

a = radius of circular region

C = solute concentration

C_0 = initial solute concentration

C_∞ = bulk solute concentration

D_L = liquid diffusion coefficient of solute

g = volume fraction solid

G = temperature gradient

G_C = unperturbed solute gradient, $RC_\infty(k-1)/D_L k$

G_L = temperature gradient in liquid at solid/liquid interface

G_S = temperature gradient in solid at solid/liquid interface

h = heat-transfer coefficient

H = enthalpy content over that of solid at melting temperature H_{SM}

ΔH_M = heat of fusion

k = equilibrium partition coefficient

$\bar{k}$ = average thermal conductivity

k_L = liquid thermal conductivity

k_S = solid thermal conductivity

K_M = kinetic parameter relating velocity and undercooling in linear law, evaluated at melting temperature

L = casting thickness

m_L = slope of liquidus line

q = absorbed heat flux

r = radial coordinate

r_0 = droplet radius

R = solid/liquid interface velocity

S = distance solidified

t = time

t_f = local solidification time

T = temperature

T_C = critical temperature for supersaturation

T_E = eutectic temperature

T_G = temperature of environment or substrate

T_H = hypercooling temperature

T_L = liquidus temperature

T_M = melting temperature

T_N = nucleation temperature

T_S = solidus temperature

Dr R. Mehrabian is Director of the Center for Materials Science, National Bureau of Standards, Department of Commerce, Washington, DC, USA.

T_0 = temperature at which solid and liquid free energies are equal

ΔT = undercooling

$\Delta(T)$ = maximum temperature difference inside droplet during solidification

ΔT_N = undercooling at moment of nucleation

U = velocity of high-energy heat source

x, y, z = Cartesian coordinates

z_{max} = maximum melt depth

α = thermal diffusivity

γ = solid/liquid surface energy

δ = perturbation amplitude

$\dot{\epsilon}$ = cooling rate

ζ = interatomic distance

κ = heat capacity

λ = wavelength

ξ = argument of error function

ω = spatial frequency

ρ = density

σ = time constant in morphological stability analysis

Dimensionless variables

Bi = Biot number, hr_0/k_L

Fo = Fourier number, $\alpha t/r_0^2$

K = kinetic coefficient, $(r_0/\alpha_L)(\Delta H_M/\kappa_L)K_M$

β = Cahn's parameter in kinetic equations

θ = temperature, $\kappa_L(T - T_M)/\Delta H_M$

ψ = enthalpy, $(H - H_{SM})/\Delta H_M$

Subscripts

L = liquid

M = at melting temperature

N = at moment of nucleation

S = solid

The present status of the various aspects of rapid-solidification processing can be found in the proceedings of a number of conferences and workshops held in the recent past.[1-7] A review of this literature shows that while significant advances have recently been made in the engineering aspects of rapid solidification, understanding of the basic phenomena is still in its infancy. The full potential and limitations of this evolving technology, however, will only be realized through the precise explanation of the underlying phenomena. An enhanced understanding of the fundamentals should permit development of predictive models elucidating the interrelationships between 'rapid-solidification' processing/structure/property and performance. It is the aim in this article to review both recent advances in understanding of the rapid liquid-to-solid transformations involved, and outline, where appropriate, fruitful directions for future research.

The two major ways in which rapid solidification processing provides improved structures and hence improved properties are:

1. Production of refined structures and more homogeneous materials, e.g. through the production of fine dendrites and eutectics, or through the reduction and elimination of intercellular and interdendritic alloy segregation.
2. Production of new alloy compositions, microstructures, and phases, e.g. through extended solid solubility, new phase reaction sequences, and the formation of metallic-glass microstructures.

In the past few years, a number of generic research programmes have focused on the development of the necessary scientific framework, data, and measurement techniques for rapid solidification. Major objectives have been the development of guidelines for alloy design based on kinetic and thermodynamic solidification theory, and for prediction and control of rapid-solidification processes and structures. In particular, heat-flow limitations, segregation effects, and rules governing the formation of equilibrium and non-equilibrium phases, including metallic glasses have been investigated. The results and the guidelines that have been developed as a result of these investigations are described below.

PHASE DIAGRAMS

Advances in the development of predictive models for phase diagrams are essential in the development of new alloys specifically designed to exploit the beneficial effects of rapid-solidification processes on microstructure. Improved predictive techniques have recently been employed, including computer-graphics display of phase-stability information to estimate rapid solidification effects. This effort is part of a broader joint programme between the National Bureau of Standards and the American Society for Metals aimed at developing a critically evaluated phase-diagram data bank and on-line computer-graphic system.

The computer-based phase-stability data compilation programme has four basic and interrelated components. These include compilation and evaluation, development of computerized bibliographic files, thermodynamic optimization, and development of computer software. In addition to data on the equilibrium phase boundaries, non-equilibrium structures and transformations are also considered. The aim is to provide a complete description of the phase relations (both equilibrium and metastable) of the alloy system.

Figure 1 shows the Ag—Cu phase diagram from this study.[8] The diagram includes metastable extensions of two-phase equilibria (solid lines), spinodal curve (dashed line), and T_0 curve for partitionless solidification (short—long dashed lines). Experimental data (closed circles) are also included. The Ag—Cu alloy system contains several interesting features for the study of rapid solidification. Both equilibrium and metastable phase boundaries as well as additional phase-diagram features have been calculated from free energy functions for the liquid and solid phases. These free energy functions were determined by fitting the equilibrium regions of the phase diagram for which excellent experimental data exist. Implicit in this approach is the assumption that, while phases may fail to nucleate or equilibrium at interfaces may break down, phases have well behaved free energies represented by the thermodynamic state variables. For example, the short—long dashed curve is the T_0 curve for the transition from liquid to crystalline solid. At temperatures below this curve liquid can transform to solid without change in composition (partitionless solidification). In Ag—Cu, this curve is only about 100 K below the liquidus and so it is not surprising that by rapid solidification a complete series of fcc solid solutions has been previously obtained by Linde.[9]

A second interesting feature of the metastable diagram in Fig. 1 is the existence of retrograde solidus curves for both the Ag-rich and Cu-rich phases. The retrogrades are points on the metastable solvi with infinite slope. The existence of these retrogrades are important for they show that over a very large range of compositions, enhanced solubility must be caused by a solute-trapping mechanism rather than by second-phase nucleation avoidance.

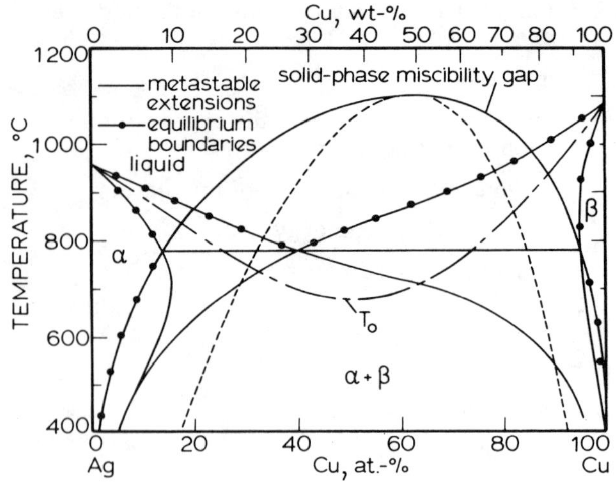

1 Au—Cu phase diagram[8]

HEAT FLOW

The development of predictive models for rapid-solidification processes requires the coupling of realistic heat-flow models to the emerging theories for rapid-solidification processing. Heat-flow models are now available for splat cooling, surface melting with directed high-energy sources, and solidification of atomized liquid droplets. There are limitations in attainable cooling rates in these processes imposed by attainable heat-transfer coefficients and/or useful melt thickness and powder size. These limitations are discussed herein along with a short summary of the various heat-flow models developed in the past few years.

It is generally agreed that the term rapid-solidification processing is equally applicable to the formation of both crystalline and non-crystalline solid phases by quenching of a material from an initial liquid state. On the other hand, a cursory review of the published literature on the subject shows that confusion arises when rapid solidification is defined in terms of high cooling rates. For example, it is possible to obtain rapidly solidified structures during recalescence of significantly undercooled melts. In general, achievable undercooling is a function of available sites for heterogeneous nucleation and time ('incubation period'). The latter is inversely proportional to cooling rate in the liquid as a specimen is rapidly cooled from above to below its melting temperature. Clearly, therefore, a measure of rapid-solidification processing could be the cooling rate of an alloy at its liquidus temperature. On the other hand, the fineness of a crystalline microstructure (e.g. segregate spacing, size of second-phase particles, etc.) can usually be correlated to average cooling rate during solidification or the time available for coarsening. Thus, a clear distinction must be made between cooling rates in the liquid (or during non-crystalline solidification) and during crystalline solidification: the latter is significantly lower at equivalent rates of external heat extraction due to the heat of fusion. Finally, many structural manifestations of rapid soldification are best correlated to solid/liquid interface velocity. Examples include formation of amorphous structures in the presence of a crystalline substrate and solute trapping at high interface velocities.

Heat flow during solidification against a metal substrate

In the recent past a large number of innovative batch and continuous techniques for production of rapidly solidified material using heat extraction by a metal substrate have been developed. Determination of exact cooling rates during solidification of crystalline and non-crystalline structures in these processes has required estimates of heat-transfer coefficients between the melt and the substrate. Two measurements of heat-transfer coefficient in splat cooling have been reported.[10,11] Table 1 gives these values along with measured heat-transfer coefficients for the case of a pressurized 89.7 MNm^{-2} aluminium casting against a steel

Table 1 Measured heat-transfer coefficients for solidification of aluminium against a metal substrate

Technique	Ref.	Thickness of casting, μm	Conditions	h, $Wm^{-2}K^{-1}$
Splat on Ni substrate	10	1	...	$1.1–2.8 \times 10^5$
Drop smashed on Fe substrate	11	150	...	$0.17–1.8 \times 10^5$
Pressure cast in steel mould	12	10^4	No pressure	3.1×10^3
			89.7 MPa	3.3×10^4
Die cast in steel mould	13	1.6×10^3	Metal-flow velocity 55 ms^{-1}, 175 MPa	7.94×10^4

mould[12] and liquid aluminium die cast against a steel mould at 55 ms^{-1} metal-flow velocity.[13] It appears that an upper limit exists for practically achievable heat-transfer coefficients between liquid metals and substrates. It is probably in the range of $h = 10^5–10^6$ $Wm^{-2}K^{-1}$.

The results of computer heat-flow calculations carried out to estimate cooling rates in a pure aluminium melt are shown in Fig. 2. The data are presented in terms of the dimensionless variables

$$Bi = \frac{hL}{k_L} \quad \text{and} \quad Fo = \frac{\alpha_L t}{L^2} \quad . \quad . \quad . \quad (1)$$

where L is the casting thickness, k_L is the melt conductivity, t is time, α_L is the thermal diffusivity of the melt, Bi is the Biot number, and Fo is the Fourier number.

As expected, for small Biot numbers temperature gradients in the melt are negligible — the linear portion of the curve in Fig. 2. At Biot numbers larger than ~10 resistance to heat flow is primarily within the aluminium melt (cooling is essentially ideal) and the dimensionless average cooling rate is independent of Biot number — actual average cooling rate $(\partial T/\partial t)_{ave}$ is inversely proportional to the square of melt thickness. That is, for Biot numbers in the Newtonian and ideal cooling ranges the actual average cooling rate in the

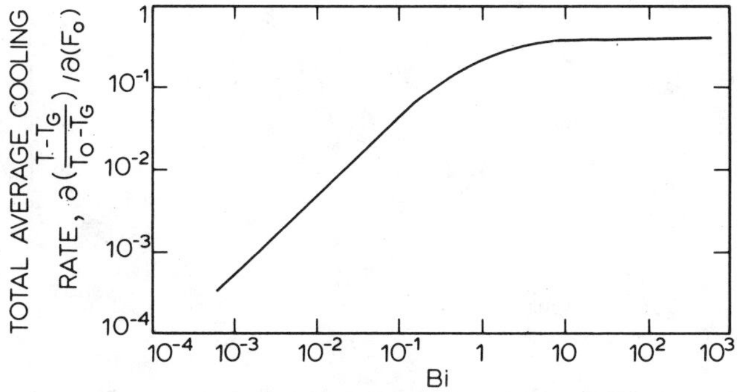

2 Dimensionless cooling rate averaged over melt thickness and time for temperature to reach half the melting point for an Al melt spat cooled against a Cu substrate

liquid increases by one and two orders of magnitude, respectively, as the melt thickness is decreased by one order of magnitude. Thus, the most effective way to increase cooling rate is to decrease melt thickness.

Similar calculations have been reported for crystalline solidification of iron splats against a copper substrate.[14] It was found that for Biot numbers less than ~0.015 and more than ~30, cooling was Newtonian and ideal, respectively.

Analytical expressions are available for the solid/liquid interface velocity during plane-front solidification of a melt against a substrate for the cases of Newtonian and ideal cooling, equations (2) and (3), respectively.

$$R = (h/\rho_S \Delta H_M)/(T - T_G) \quad . \quad . \quad . \quad . \quad (2)$$

$$R = 2\xi^2 \alpha_S/S \quad . \quad . \quad . \quad . \quad . \quad . \quad . \quad (3)$$

where R is the solid/liquid interface velocity, T_G is the initial substrate temperature, α_S is the thermal diffusivity of the solid forming, ρ_S is the density of the solid forming, S is the distance solidified, and ξ is the argument of the error-function solution of the temperature distribution — it is determined from a characteristic equation which contains metal and substrate thermal constants.

Equations (2) and (3) can be used to estimate maximum attainable solid/liquid interface velocities in various alloys. For example, an interface velocity of ~0.06 ms^{-1} is calculated for an aluminium ribbon or splat solidifying against a metal substrate under Newtonian heat-flow conditions. Assumptions of this calculation include: no significant undercooling and a heat-transfer coefficient $h = 10^5$ $Wm^{-2}K^{-1}$. For the typical physical-property values of aluminium[15] the Newtonian condition, Biot number less than ~10^{-2}, will be satisfied for a melt thickness of up to ~50 μm.

Approximate analytical solutions are available for Biot numbers between the two extremes of equations (1) and (2). However, numerical solutions such as those described above can be readily employed.

Several attempts have been made to predict critical cooling rates in the liquid during rapid solidification for the formation of non-crystalline structures.[16–18] Heat-flow calculations are usually combined with theories of nucleation, growth, and transformation kinetics. This area, as well as rapid solidification of undercooled crystalline structures remains fertile for future investigations.

Heat flow during melting and solidification of a surface layer

The recent availability of high-power directed energy sources such as the electron beam and the different types of lasers has led to the development of rapid melting and solidification techniques in which a bulk (semi-infinite) substrate in intimate contact with a molten layer acts as the quenching medium. Results of recent one- and two-dimensional computer heat-flow analyses for stationary and moving heat sources[19–22] carried out to investigate the effect of high-intensity radiation on the important variables during surface-layer melting and subsequent solidification are summarized below.

The problem considered was the rapid melting and subsequent solidification of the surface layer of semi-infinite solid pure metals[19,20] and alloys,[22] initially at room temperature, subjected to a high-intensity stationary heat flux over a circular region on its bounding surface. Since the melt and the substrate are in intimate contact, the heat-transfer coefficient between the two tends to infinity. Thus, one important limitation on rapid rate of heat extraction encountered in atomization and splat-cooling types of processes has been removed. However, surface melting via directed energy sources has its own inherent limitation on the maximum achievable cooling rate which is again a function of the thickness of the molten surface region.

In general, the absorbed heat-flux distribution can be both a function of position within the circular region as well as time. The generalized, two-dimensional heat-flow model and numerical-solution techniques developed in stationary and moving oblate spheroidal coordinate systems[20–22] can readily take into consideration both the space and the time variation of the absorbed heat flux. The one- and two-dimensional stationary heat-flow problem are treated below followed by an example of the temperature distributions when the heat source is moved with respect to the substrate.

Two important criteria were deduced, from the computer heat-flow calculations,[20,22] for the surface melting of a semi-infinite pure or alloy substrate subjected to a uniform heat flux q over a circular region of radius a on its bounding surface. First, it was shown that the product qa should exceed specific values if surface melting is to be initiated and the centre of the circular region is to reach a given temperature up to the vaporization temperature of the substrate. The second important criterion was that for large values of the product qa isotherms in the substrate are planar in shape, and one-dimensional heat-flow conditions prevail, i.e. lateral heat flow in the substrate can be ignored. As an example, for an aluminium substrate, when $qa \geq 1.4 \times 10^6$ Wm^{-1} heat flow is essentially in one dimension.

Results from one-dimensional heat-flow analysis, when the second criterion noted above is applicable, are shown in Fig. 3. These are the calculated melt depths v. total time for several uniform absorbed heat fluxes. The arrows in the figure denote times at which the surface of each material reaches its vaporization temperature T_V and the heat flux q is removed. Melting continues until a maximum melt depth, z_{max}, is reached. Note that for an aluminium substrate subjected to uniform absorbed heat fluxes of 5×10^8 and 5×10^{10} Wm^{-2} the criterion for one-dimensional heat flow would be met if the radii of the circular regions exceed 2800 and 28 μm, respectively. The corresponding maximum melt depths z_{max} from Fig. 3 are ~650 and 6.5 μm, respectively.

The general trends established in these one-dimensional computer calculations[19] are summarized in Table 2. They show that temperature gradients in the liquid and solid phases and interface velocities are directly proportional to the absorbed heat flux, whereas melt depth is inversely proportional to the absorbed heat flux. Average cooling rates comparable to and exceeding those predicted for splat cooling can be achieved by increasing the heat flux and reducing the dwell time of the incident radiation. An order of magnitude increase in the absorbed heat flux results in a

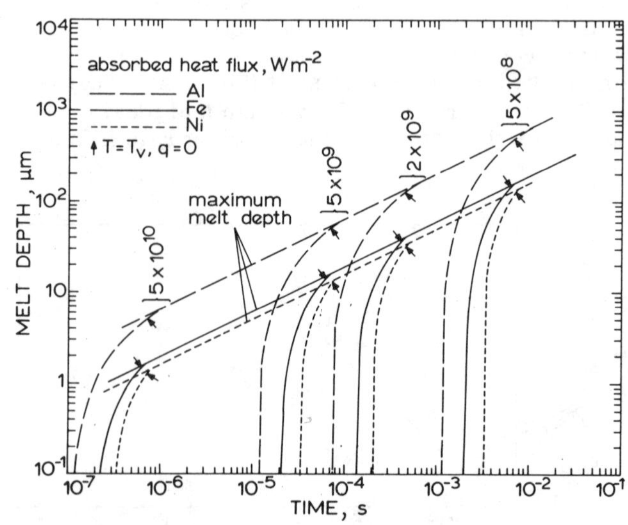

3 Melt depth v. total time for different uniform absorbed heat fluxes obtained via a numerical technique; arrows indicate times at which a surface reaches its vaporization temperature and heat flux is removed[19]

Table 2 One-dimensional heat flow: effect of change in absorbed heat flux on other variables

Order of magnitude increase in *q* results in the following changes:

q	Time	Melt depth	R	G_L	G_LR	G_L/R
10↑	10^2↓	10↓	10↑	10↑	10^2	*

↑increase, ↓decrease, and * no change.

corresponding two orders of magnitude increase in average cooling rates in the liquid during solidification of crystalline and non-crystalline structures.

Numerical calculations in the one-dimensional model[19] show that the solid/liquid interface velocity remains almost constant during 90% of the solidification sequence after an initial transient during which time the interface accelerates from an initial velocity of zero. For aluminium, the calculated interface velocity to absorbed heat-flux ratio for 90% of the solidification sequence is $R/q \simeq 7 \times 10^{-11}$ $J^{-1}m^3$. This translates into interface velocities of $R \simeq 0.035$ and $\simeq 3.5$ ms^{-1}, respectively, for the absorbed heat fluxes of 5×10^8 and 5×10^{10} Wm^{-2} noted above (z_{max} $\simeq 650$ and 6.5 μm, respectively).

Examples of results from the two-dimensional heat-flow calculations[20,22] for a stationary absorbed heat flux q over a circular region of radius a on the bounding surface of semi-infinite pure aluminium and Al–4.5Cu alloy substrates are shown in Figs. 4 and 5, respectively. An important finding from this work is that for a given value of $T(0, 0)$ at the centre of the circular region the dimensionless temperature distributions in the substrate materials during melting and solidification are identical for any combination q and a as long as qa is constant. Figure 4 shows the effect of increasing the product qa on the geometry and the location of the liquid/solid interface. The curve for $qa \simeq 1.9 \times 10^6$ Wm^{-1} is in the range where heat flow can be assumed to be essentially one dimensional and the shape of the isotherm verifies this fact. On the other hand, with decreasing values of qa, lateral heat flow becomes significant and the liquid/solid interfaces assume increasingly more convex geometries.

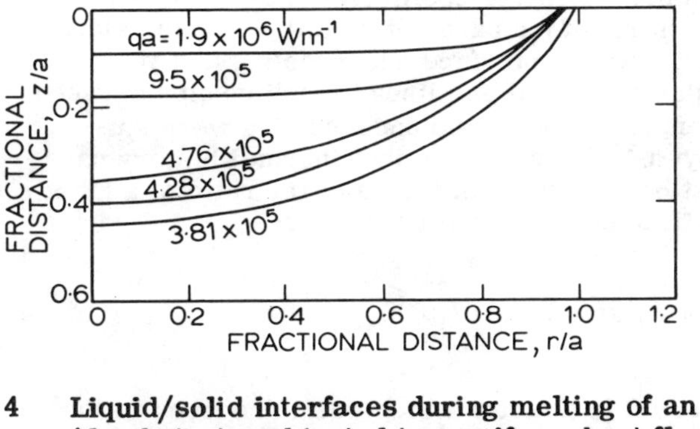

4 Liquid/solid interfaces during melting of an Al substrate subjected to a uniform heat flux *q* over a circular region of radius *a*; interfaces for each value of *qa* are shown at instant centre of circular region reaches 2100 K (Ref. 20)

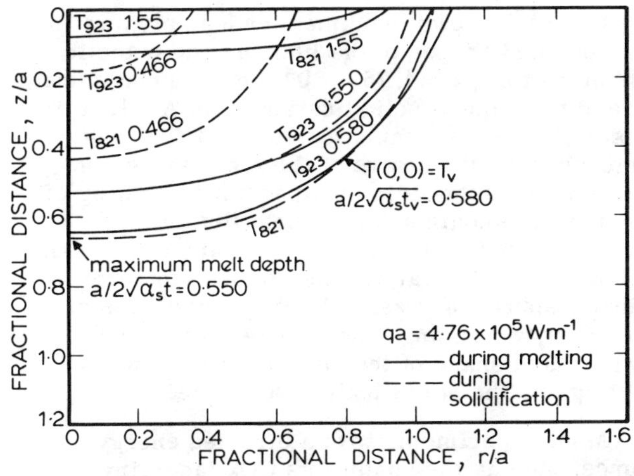

5 Liquidus and solidus (eutectic) interfaces during melting and subsequent solidification of Al–4.5Cu alloy substrate subjected to uniform absorbed heat flux *q* over a circular region of radius *a*; curve marked $T(0, 0) = T_V$ shows location of interface at time heat flux was removed[22]

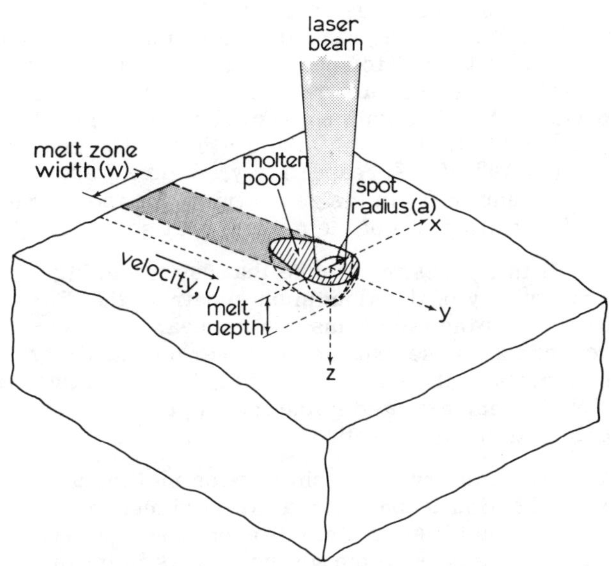

6 Schematic illustration of laser beam–substrate geometry during rapid surface melting and solidification

Figure 5 shows the dimensionless plots of the locations and shapes of the liquidus and eutectic isotherms for $qa = 4.76 \times 10^5$ Wm^{-1} during melting and subsequent solidification of an Al—4.5Cu substrate. In this simulation, the heat flux was removed when the centre of the circular region reached the vaporization temperature. Melting of the substrate continues until heat flux in the 'mushy' zone and the solid at the solidus isotherm (821 K) become equal, i.e. the solidus isotherm velocity approaches zero. Decreasing values of $a/2\sqrt{(\alpha_s t)}$, increasing time, denote successive position and shape of the mushy zone during the melting and the solidification sequence.

Most experiments with a directed energy source, such as the continuous CO_2 laser, involve scanning of the source over the surface of the substrate. Recently, Kou *et al.*[21] considered the rapid melting and solidification of a semi-infinite substrate subjected to a high-intensity heat flux over a circular region on its bounding surface, moving with a constant velocity U in the positive y-direction in a Cartesian coordinate system, Fig. 6. An important finding of the investigation was that the three variables, absorbed heat flux q, the radius of the circular region a, and the velocity of the moving heat flux U, could be combined into two independent variables. That is, the dimensionless temperature distribution in the liquid-metal pool and the solid substrate remain the same as long as the products qa and Ua or U/q are kept constant while the individual values of the three variables are varied.

Examples of the shape and location of several isotherms, including the liquid/solid interface, for given values of the products qa and $Ua/2\alpha$ are shown in Fig. 7. This figure shows composite top and side views of the isotherms at steady state for a stationary and a moving heat source.

The dimensionless velocity $Ua/2\alpha = 0.75$ would, for example, translate into actual velocities of 0.1 and 1 ms^{-1} for radii of the circular region of ~1260 and ~126 μm, respectively. The corresponding absorbed uniform heat fluxes that result in $qa = 6.4 \times 10^5$ Wm^{-1} are $q \sim 5.1 \times 10^8$ and $\sim 5.1 \times 10^9$ Wm^{-2}, respectively. Figure 7 shows significant shifts in the geometry of the isotherms to the trailing end of the moving heat source.

It is interesting to note that due to the high conductivity of the aluminium substrate the distortions in the isotherms are not nearly as pronounced as those expected in lower-conductivity materials such as iron or nickel. This point was clearly demonstrated in the moving point source calculations of Rosenthal.[23]

In summary, the rapid surface melting and solidification technique, via directed high-energy sources, has its own limitation on specimen size and thickness, even though the melt is in intimate contact with its own substrate. High cooling rates during solidification are achieved only at the expense of high absorbed heat fluxes, reduced heat-affected zones, and melt depths.

Heat flow during atomization

Since atomization is more limited than surface melting or splat cooling in the achievable rates of heat extraction, it is desirable to reach substantial undercoolings before nucleation in order to enhance the rapid-solidification effects. Experimental evidence from several investigations[24–29] has shown that large metastable effects can indeed be obtained under slow-cooling conditions by suppressing the onset of nucleation. On the other hand, it has been contended that external cooling during some atomization processes may be sufficiently high to partially offset the recalescence of the droplet and extend the regime of rapid solidification.

The solidification of undercooled spherical droplets with a discrete melting temperature was recently analysed using both a Newtonian and a non-Newtonian (enthalpy) model.[30] Relationships were established between the atomization parameters, the growth kinetics, the interface velocity and undercooling, and other important solidification variables. A summary of the findings from this study is given below.

The important process parameters in the solidification model are the initial undercooling, the particle size, the combined convective and radiative heat-transfer coefficient and a kinetic relationship between undercooling and interface velocity. Droplet diameters from 0.1 to 1000 μm were considered to cover the typical range of most current laboratory and commercial atomization techniques.

The heat-transfer coefficient during atomization may vary between 10 Wm^{-2}K^{-1} for the pure radiation case to significantly less than 10^6 Wm^{-2}K^{-1} for convection under the most favourable experimental conditions.[15] The limitations on h translate into Biot numbers lower than 1 for most metals and particle sizes of interest. For example, 500 μm droplets of Al, Fe, and Ni result in Biot numbers of 0.28, 0.62, and 0.64, respectively, for a high heat-transfer coefficient of 10^5 Wm^{-2}K^{-1}.[15] The third major parameter is the kinetic relationship between interface undercooling and velocity. Both exponential and linear relationships between interface velocity and undercooling were employed.[30] It was shown that thermal history can best be described in terms of the ratio of a kinetic coefficient K to the Biot number:

$$K/Bi = 2 \times 10^7(\beta/h) \quad . \quad . \quad . \quad . \quad . \quad (4)$$

where β is a factor which was introduced in the continuous-growth model by Cahn *et al.*[31] Kinetic considerations generally limit β within 1 and 10^2. With the limitation on the heat-transfer coefficients during atomization noted above the parameter K/Bi can vary from $\sim 10^2$ to $\sim 10^8$.

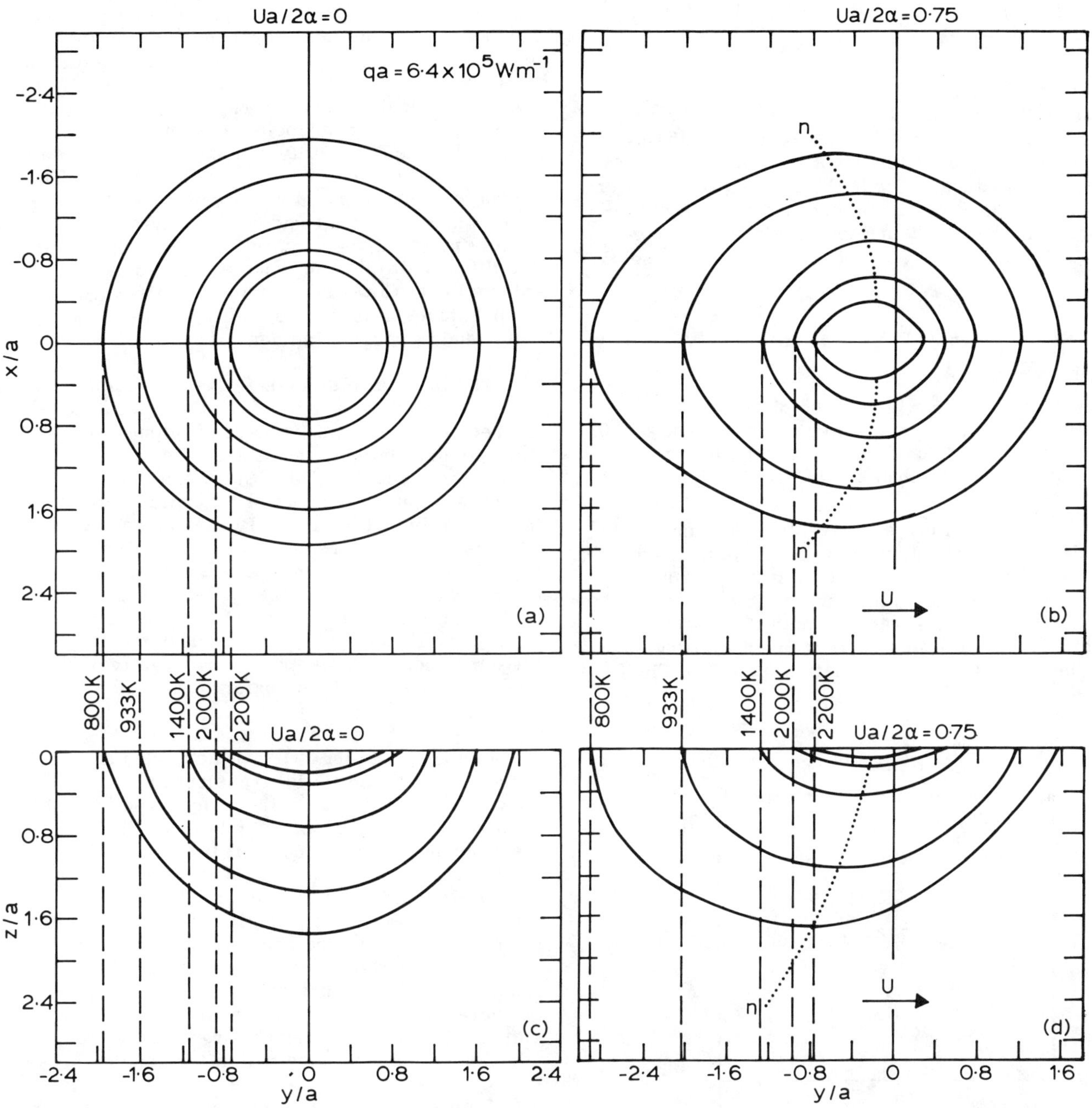

a, *b*, top views, *x*—*y* plane; *c*, *d*, side views, *y*—*z* plane

7 Composite side and top views of isotherms in Al substrate subjected to stationary and moving uniform absorbed heat fluxes over a circular region on its boundary surface; *n*—*n* curve connects points of maximum temperature farthest from *y*-axis[21]

The thermal history of powder solidification can readily be described on the enthalpy–temperature diagram as shown in Fig. 8. This diagram represents graphically the following equation:

$$H - H_{SM} = [\Delta H_M + \kappa_L(T - T_M)](1 - g) + \kappa_S(T - T_M)g \quad . \quad . \quad . \quad . \quad (5)$$

where H is the enthalpy of the system at temperature T and fraction solid g, ΔH_M is the enthalpy of fusion, H_{SM} is the enthalpy of the solid at the melting temperature, and κ is the heat capacity. Alternatively, Fig. 8 can be plotted in terms of dimensionless enthalpy ψ and temperature θ where

$$\psi = (H - H_{SM})/\Delta H_M \quad . \quad . \quad . \quad . \quad . \quad (6)$$

$$\theta = \kappa_L(T - T_M)/\Delta H_M \quad . \quad . \quad . \quad . \quad . \quad (7)$$

For a solid volume ($g = 1$) the dimensionless nodal enthalpy ψ is always negative or zero since $\theta \leq 0$, whereas for a volume of superheated liquid ($g = 0$, $\theta > 0$) ψ is larger than unity.

The system can conceivably follow any 'path' of decreasing enthalpy in the supercooled region ($\psi < 1$) starting at the nucleation temperature ($\theta_N < 0$) on the liquid line, and ending at some point on the solid line. A vertical path represents isothermal solidification, of which a limiting case is the no-undercooling situation (path 1 in Fig. 8).

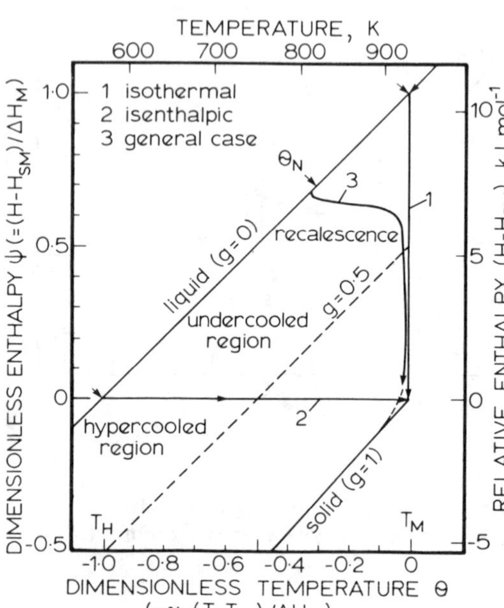

8 Enthalpy–temperature diagram showing possible solidification paths; θ_N is dimensionless temperature at moment of nucleation

A horizontal path represents isenthalpic (or adiabatic) solidification. If nucleation can be suppressed until an amount of enthalpy at least equivalent to the heat of fusion has been removed (i.e. $\psi \leq 0$) the system can undergo complete solidification without further heat extraction (path 2 in Fig. 8). The region of the diagram for which $\psi \leq 0$ is known as 'hypercooled'. In the path such as that described by curve 3, two distinct solidification regimes can be identified. The first one is the recalescence or 'rapid-solidification' stage, where the supercooled droplet absorbs most of the liberated heat of fusion and the heat loss to the surroundings is more or less irrelevant. The second regime is one of a 'slower growth', and develops after droplet undercooling has been largely relieved, thereby limiting the progress of solidification to the conditions imposed by the external heat flow. This regime usually involves temperatures close to the melting point of the metal and is roughly isothermal.

The thermal history reflects the competition between the external rate of heat extraction and the recalescence rate (release of latent heat) which is proportional to the solidification rate (change in fraction solid with time, dg/dt. The latter is in turn a product of the solid/liquid interface velocity and its area. Predominance of the recalescence rate translates into a shallow negative slope for the ψ–θ curve in the rapid-solidification stage. As the heat-extraction process is able to change some of the heat of fusion being transformed into sensible heat, the slope of the curve becomes steeper (more negative) and eventually positive if the external heat flow is dominant.

In addition to the two regimes noted above, there is usually an initial transient in the ψ–θ curve, during which time the overall cooling rate of the droplet decreases rapidly from its value at the moment of nucleation until it changes sign and recalescence begins. At the start of solidification, the solid/liquid interface velocity is finite but its interfacial area is very small, hence, slow solidification rate. Therefore, the rate of heat extraction dominates the rate of recalescence and the droplet temperature drops further below θ_N. This period of decreasing enthalpy after nucleation may be sufficiently long to permit a large number of additional nuclei to form within the droplet.

Figure 9 shows the effect of the initial undercooling on the enthalpy–temperature curve for a constant $K/Bi = 100$, considering both linear and exponential kinetics. It is evident that increasing undercooling before nucleation results in a corresponding increase in the solidification rate, thus reducing the initial transient and the relative contribution of the external heat extraction.

The dependence of the thermal history on the ratio K/Bi for a constant initial temperature $\theta_N = -0.5(\Delta T_N = 182$ K) is shown in Fig. 10.

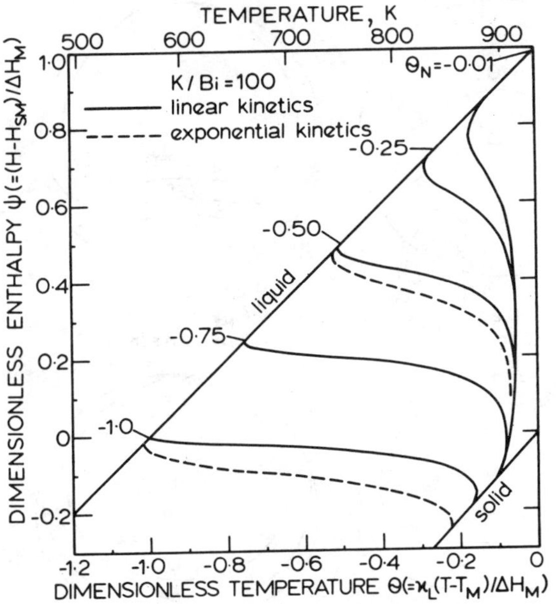

9 **Effect of nucleation temperature on thermal history of undercooled Al droplets calculated from Newtonian model[30]**

Curves for $K/Bi > 10^3$ closely approach the limiting case of adiabatic (isenthalpic) rapid solidification followed by an isothermal slow-growth regime at the melting temperature. As K/Bi decreases (slower kinetics or more efficient cooling), the

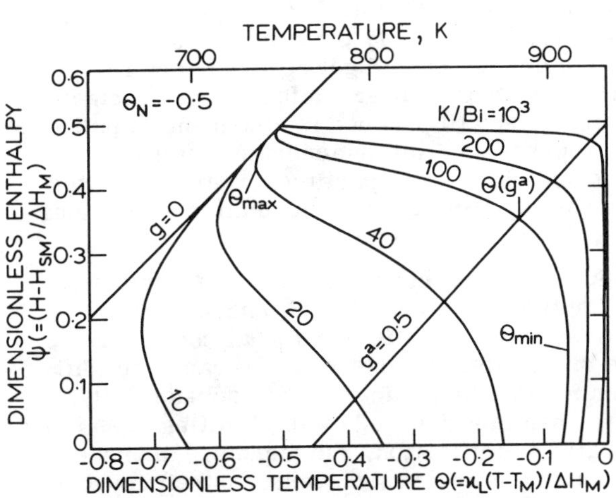

10 **Effect of ratio K/Bi on thermal history of undercooled Al droplets for a fixed nucleation temperature; note that for typical atomization conditions ($K/Bi > 100$) external cooling is not able to slow down recalescence to a significant extent. Linear kinetics and Newtonian conditions were assumed[30]**

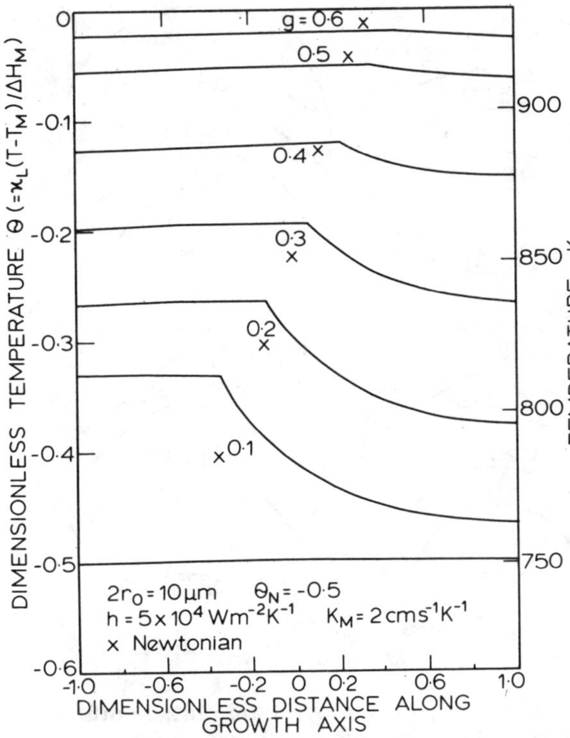

11 **Thermal profiles developed inside Al droplet as solidification progresses according to enthalpy model, for constant particle diameter of 10 μm (Ref. 30)**

curves show an increasing contribution of the heat-extraction rate during recalescence. At the extreme, e.g. $K/Bi = 10$, the regime of rapid solidification is extended to a point where the droplet temperature never goes beyond its initial value θ_N. However, even for a low kinetic coefficient $K_M = 2$ cms^{-1}K^{-1} ($\beta = 1$), the heat-transfer coefficient would have to be about 2×10^6 Wm^{-2}K^{-1} in order to produce this effect. For a higher β such as the value of 100 suggested by Cahn *et al.*,[31] $h = 2 \times 10^8$ Wm^{-2}K^{-1} would be required, which is clearly above the values encountered in atomization.

Figure 11 shows the temperature profiles calculated from the non-Newtonian model along the growth axis, as a function of fraction solid for a fixed diameter droplet of 10 μm. The corresponding droplet temperatures and interface positions calculated from the Newtonian model are given for comparison. As expected from the Biot number ($Bi = 2.8 \times 10^{-3}$) the profiles are Newtonian before nucleation and after recalescence. However, substantial temperature differences $\Delta(T)$ are developed inside the droplet during the period of rapid growth. For example, maximum $\Delta(T)$ values of 7, 52, and 148 K occurring at fractions solid of 0.08, 0.108, and 0.126 were calculated for aluminium droplets 1, 10, and 100 μm dia., respectively, under the conditions of this figure.

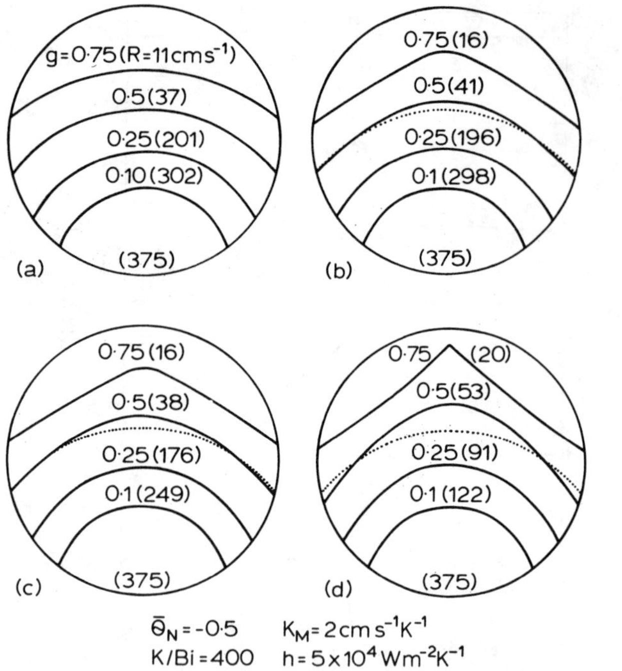

a Newtonian; *b* enthalpy, 1 μm dia.; *c* enthalpy, 10 μm dia.; *d* enthalpy, 100 μm dia.

12 Interface profiles and velocities for Al predicted from enthalpy model for undercooled solidification; deviations from Newtonian velocity limit increase with particle size, but profiles for recalescence stage are quite close in both models for $r_0 < 100$ μm (Ref. 30)

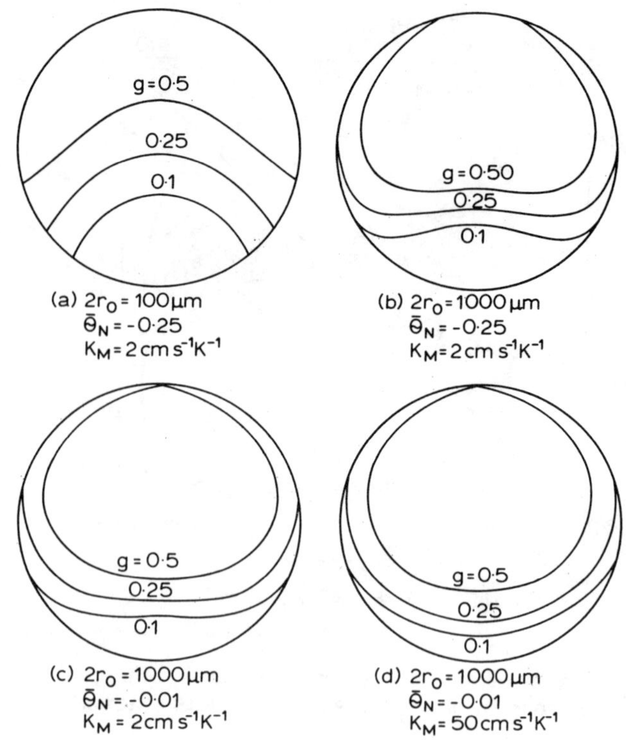

13 Transition from convex to concave interface profiles is produced by increasing Al particle size (*a*, *b*), reducing the initial undercooling (*b*, *c*), and increasing the kinetic coefficient (*c*, *d*). In the limit, the interface tends to the concentric shape assumed in the no-undercooled solidification model[30]

Figure 12 compares the position, shape, and velocity of the solid/liquid interface as solidification progresses, for 1, 10, and 100 μm dia. aluminium droplets and the corresponding Newtonian limit. As expected, the growth velocities increase as the radius decreases, and the values for a 1 μm particle are almost identical to those calculated from the Newtonian model (interface and droplet undercoolings have similar values).

The attainable solid/liquid interface velocities during solidification of droplets is thus a complex function of the initial undercooling, the heat-transfer coefficient at the droplet/environment interface, and the growth kinetics of the interface. Numerical calculations[15] for a 50 μm dia. aluminium droplet, solidifying with no initial undercooling and $h = 5 \times 10^4$ $Wm^{-2}K^{-1}$, give a range of interface velocities, $R \simeq 0.04$–0.13 ms^{-1}, from the beginning of solidification to 0.5 fraction solid. Similar data for an initial dimensionless undercooling of $\bar{\theta}_N = 0.5$ are shown in Fig. 12. In all four cases a high initial solid/liquid interface velocity ($R = 3.75$ ms^{-1}) is calculated. This velocity decreases during recalescence, at different rates depending on the particle size, and reaches a value between 0.37 and 0.5 ms^{-1} at 0.5 fraction solid.

Figure 13 shows the transition from a convex to a concave solid front as θ_N is reduced and r_0 is increased above 100 μm. In addition, increasing K_M will further magnify the concavity. In the limit, then, the droplet should get covered very rapidly with a solid 'shell' after nucleation, and further solidification will proceed in a concentric pattern. This represents a fairly accurate picture of the solidification of large particles where dendrites grow rapidly from several nucleation sites along the surface of the particle.

An important conclusion of the study[24] was that for typical values of K/Bi encountered in atomization ($K/Bi > 100$), external cooling has little or no effect in extending the rapid-solidification regime to a fraction solid substantially larger than that dictated by the limiting case of adiabatic solidification. Furthermore, increasing the number of nucleation events further reduced the role of external cooling during recalescence.

STABILITY AT SOLIDIFICATION INTERFACE AND MICROSEGREGATION

Interface stability during solidification has attracted increased interest recently due to the emergence of rapid solidification as a field of technological interest. A most important contribution has been the extension of morphological stability theory to high interface velocities. It has been shown that plane-front solidification, which would avoid microsegregation effects, becomes increasingly stable with increasing interface velocity during rapid solidification of crystalline alloys.

For most solidification processes at low interface velocities, the classical theory of constitutional supercooling[32] has been used as an adequate guideline in determining the growth conditions that result in unstable interfaces. The theory is basically an application of thermodynamics to determine if the liquid ahead of a moving interface is undercooled with respect to its composition. The undercooled liquid layer, due to rejection of solute by the growing solid, makes the shape of the interface unstable with respect to the formation of cells or dendrites that could protrude into it. The stability parameter emerging from this theory is the ratio of the temperature gradient in the liquid ahead of the moving interface G_L to its velocity R. Greater stability is promoted by increasing G_L/R.

Figure 14 delineates the G_L and R combinations that lead to the three solidification-front morphologies shown. The line between the planar and cellular regions is the constitutional supercooling criterion for a binary alloy:

$$G_L \geq m_L G_c = \frac{[m_L R C_\infty (k-1)]}{D_L k} \quad . \quad . \quad . \quad (8)$$

where C_∞ is the bulk solute concentration, D_L is the liquid diffusion constant, G_c is the unperturbed solute gradient at the interface, m_L is the slope of the liquidus line, and k is the equilibrium distribution coefficient. To the left and above this line, the actual temperature gradient in the liquid is larger than that calculated from equation (8) and plane-front growth, with compositional homogeneity, is morphologically stable. To the right

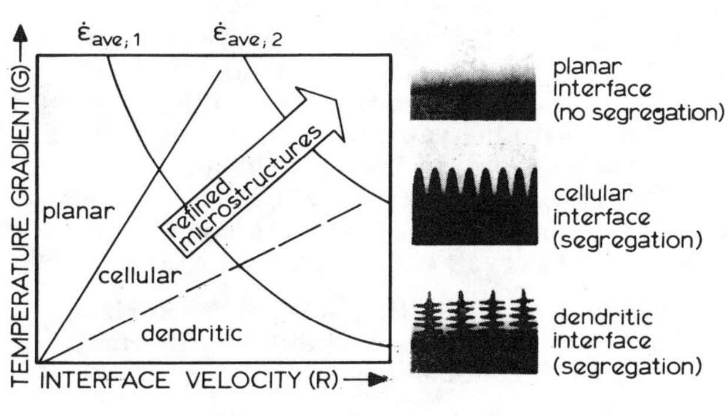

14 Dependence of solidification morphology on temperature gradient *G* and solidification velocity *R*; qualitative effect of increasing average cooling rate $\epsilon_{ave} = GR$ is indicated by sloping arrow

and below this line, the solute-gradient effect dominates (actual temperature gradient is smaller than that given in equation (8)) and chance protuberances can become pronounced leading to cellular structures and microsegregation. For small amounts of constitutional supercooling, the cellular interface is stable. This morphology reduces the constitutional supercooling ahead of the main interface by rejecting solute between the cells. With still larger solute concentrations, or faster interface velocities R relative to the operative temperature gradient, the familiar dendritic-growth morphology is encountered.

For the case of cellular or dendritic solidification, the temperature gradient and interface velocity in Fig. 14 could represent averages for the liquid plus solid mushy region. The product GR is then a measure of the average cooling rate ϵ_{ave} during solidification.[24] This variable is then directly related to local solidification time t_f, which is the time available for coarsening:

$$\epsilon_{ave} = GR = \frac{(T_L - T_E)}{t_f} \quad . \quad . \quad . \quad . \quad . \quad (9)$$

where $T_L - T_E$ is the solidification temperature range.

Therefore, as indicated by the sloping arrow, moving out to larger GR values in Fig. 14, i.e. shorter local solidification times, corresponds to the finer microstructures obtained in rapid-solidification processing.[24] Note that the simple treatment here does not include growth into a supercooled liquid. Glicksman and Huang[33] have developed a series of relationships between the microstructural size scale (segregate spacing) and interface undercooling. They concluded that for small undercoolings, segregate spacing should be a fixed multiple of the critical nucleation radius. Furthermore, as dimensionless supercooling approaches unity, the scale of the morphology should rapidly decrease. They also studied the coarsening process during solidification of a transparent material and noted that the initial microstructure is produced far from equilibrium in order to comply with the transport processes during growth. The final microstructure tends to be substantially different from that produced in the vicinity of the dendrite tip, and can be correlated with the coarsening time, i.e. local solidification time, by a scaling law.

The constitutional supercooling theory predicts cellular or dendritic structures for most solidification experiments. However, for rapid solidification some of the assumptions and simplifications of this theory may not be valid. These include:

1. Surface tension (capillarity) and latent-heat evolution which could stabilize the interface are neglected.

2. The constitutional supercooling theory takes into account a layer of liquid cooler than its local melting point, but it does not solve the kinetic equations for the thermal and diffusional fields to show how interface fluctuations lead to instability.

3. Local equilibrium at the interface: the equilibrium distribution coefficient is assumed to describe quantitatively the amount of solute rejected by the growing crystal.

The theory of morphological stability originally developed by Mullins and Sekerka[34,35] and Sekerka[36] considers the time dependence of a general sinusoidal perturbation of the interface with respect to solute and heat-diffusion fields and surface tension. Extensive reviews of the theory are available in the literature.[36–39] It has been demonstrated that for a large number of conventional solidification processes, the constitutional supercooling and morphological stability approaches lead to similar results. This explains the popularity of the former approach owing to its simplicity. On the other hand, since both surface tension and evolution of latent heat have a stabilizing influence, the morphological stability theory (even if local equilibrium at the interface is assumed) predicts far greater stability during rapid solidification than that predicted by the constitutional supercooling criterion. The two dominant effects are ascribed to the excess surface energy for short-wavelength perturbations, common at high interface velocities, and departure from local equilibrium at the solid/liquid interface.[40]

The first treatments of morphological stability assumed local equilibrium at the solid/liquid interface and considered the transition from planar to non-planar growth during steady-state, constant velocity, unidirectional solidification of a binary alloy. A linear time-dependent stability analysis was carried out[41] in which the perturbation at the solid/liquid interface growing in the z-direction is given by:

$$z = \delta \exp(\sigma t + i\omega_x x + i\omega_y y). \qquad (10)$$

where δ is the perturbation amplitude at time $t = 0$, and ω_x and ω_y are spatial frequencies. The interface is stable if the real part of σ is negative for all perturbations. The value of σ was determined[35,41] by solving the steady-state heat flow and diffusion equations with appropriate boundary conditions. For the stability–instability demarcation ($\sigma = 0$) the equation simplifies to a criterion which identified three factors contributing to the overall stability of the interface. These are: the solute field, the capillary forces, and the thermal field given below:

$$m_L G_C \underbrace{\frac{(\omega^* - R/D_L)}{[\omega^* - (1-k)R/D_L]}}_{\substack{\text{always} > 0 \\ \text{solute field}}} - \underbrace{\frac{T_M \gamma \omega^2}{\Delta H_M \rho_S}}_{\substack{\text{always} < 0 \\ \text{capillary force}}} - \underbrace{\frac{k_S G_S + k_L G_L}{2\bar{k}}}_{\substack{\lesseqgtr 0 \\ \text{thermal field}}} \gtrless 0 \qquad (11)$$

where G_S is the unperturbed temperature gradient in the solid, $\bar{k} = (k_S + k_L)/2$, k_L and k_S are the thermal conductivities of liquid and solid, respectively, T_M is the melting point of the interface in the absence of solute and curvature, ΔH_M is the latent heat of fusion per unit mass, ρ_S is the solid density, γ is the solid/liquid surface energy, and $\omega^* = R/2D_L + [(R/2D_L)^2 + \omega^2]^{1/2}$, where $\omega^2 = \omega_x^2 + \omega_y^2$.

Thus, the interface is stable when the sum of the three terms in equation (11) is negative. The solute field is always destabilizing, but the capillary forces always oppose the growth of a perturbation. For large ω (short wavelengths), this term is large. The thermal field will be stabilizing if the numerator of the third term is positive, and destabilizing if it is negative. Note that while the constitutional supercooling criterion predicts instability for all cases where $G_L < 0$, morphological theory may predict a stabilizing thermal field if G_S is sufficiently positive to make the average gradient term also positive.

Two limiting cases are recognized for equation (11). First, consider growth at slow interface velocities and where $\omega^* \sim \omega \gg R/D_L$, lateral solute diffusion is complete, and capillary forces are small. If, in addition, $R\Delta H_M \rho_S \ll 2k_L G_L$, then equation (11) for the stability criterion becomes:

$$(k_L/\bar{k})G_L \geq m_L G_C \qquad (12)$$

which reduces to the constitutional supercooling criterion for $k_S \approx k_L$. Equation (12) is the solid line in Fig. 15 which shows the calculated critical copper concentrations in dilute Al–Cu alloys (below which the planar interface is stable) as a function of interface velocity R. A constant positive liquid-temperature gradient in front of the interface $G_L = 2 \times 10^4$ Km^{-1} was assumed. The wavelength $\lambda = 2\pi/\omega$ which corresponds to the onset of instability, i.e. to the value of ω for which $C_\infty(\omega)$ is a minimum, is shown in Fig. 16 to be a monotonically decreasing function of interface velocity. For the limiting case noted above (upper left corner of this curve), λ is proportional to $R^{-1/3}$.[40]

For the second limiting case, high interface velocities, $\omega \ll R/D_L$, and a stabilizing thermal field equation (11) becomes:

$$\left.\begin{aligned} &m_L G_C (D_L^2/R^2 k)\omega^2 - T_M \gamma \omega^2 / \Delta H_M \rho_S < 0, \\ &\qquad\qquad \text{for stability} \\ \text{or}& \\ &A = (k T_M \gamma / m_L G_C \Delta H_M \rho_S)(R/D_L)^2 > 1.0 \end{aligned}\right\} \qquad (13)$$

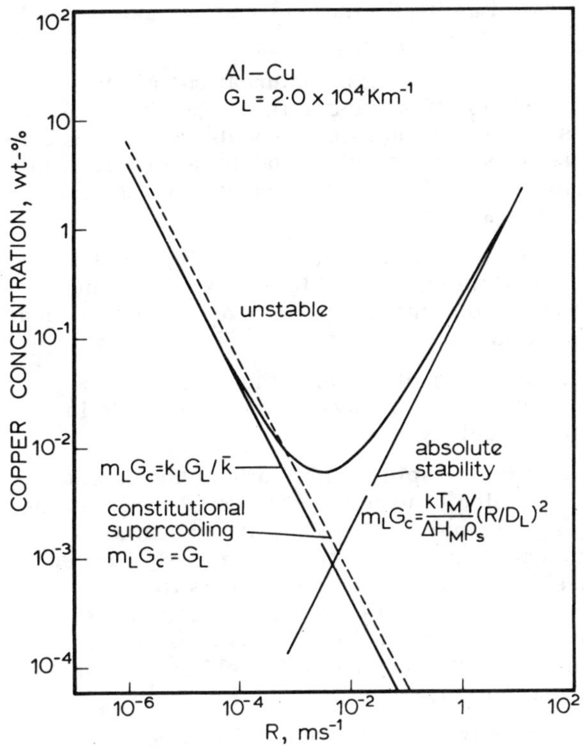

15 Critical concentration of copper above which interface instability occurs as function of interface velocity R in directional solidification of Al–Cu alloys for temperature gradient in liquid of 2.0×10^4 Km^{-1}; curve is based on morphological stability theory while lines correspond to constitutional supercooling, modified constitutional supercooling, and absolute stability[40]

Equation (13) is the absolute stability criterion.[35] It is a good approximation at high interface velocities, right-hand side of Fig. 15, and shows that in this region stability is practically independent of G_L. Furthermore, it has been shown[40] that the wavelength for the onset of instability at such high velocities is proportional to $R^{-3/4}$. This is the lower right-hand side of the curve in Fig. 16 where calculated wavelengths are shown for R up to 10 ms^{-1}. It should be noted that the theory is no longer valid at such high velocities. Since D_L for Cu in Al is ~5×10^{-9} $m^2 s^{-1}$ and the characteristic diffusion boundary layer, D_L/R, is only a few atomic distances, the use of macroscopic transport equations is not justified.

On the other hand, the diffusive velocity of the solute in the liquid can be defined as D_L/ζ, where ζ is the interatomic distance. Then, as the solid/liquid interface velocity approaches this diffusive velocity, the solidification front will begin overtaking the solute atoms, and the assumption of local equilibrium at the solid/liquid interface will no longer be valid. In the limit, when the interface velocity exceeds the diffusive velocity, complete solute trapping should result.

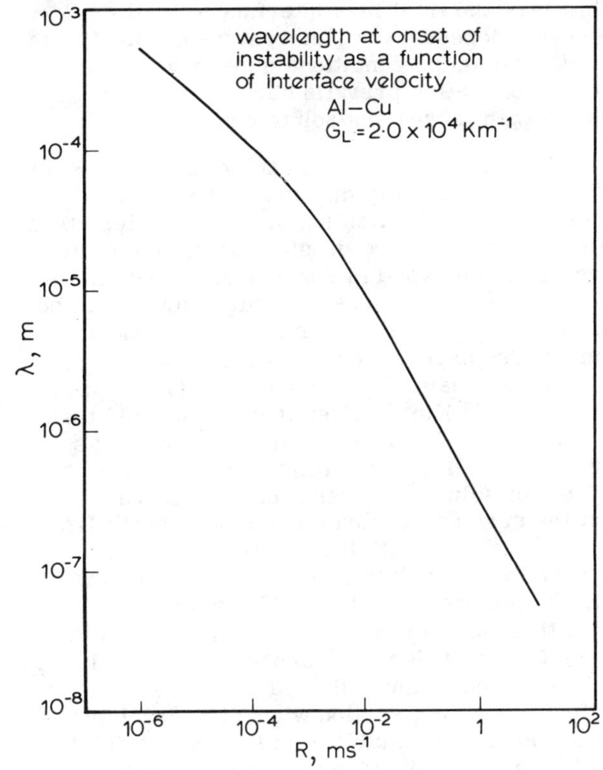

16 Wavelength λ at onset of instability during directional solidification of Al containing Cu as function of interface velocity R (Ref. 40)

Figure 17 maps out the calculated G_L and R combinations that lead to homogeneous plane-front solidifications as a function of copper concentration in Al–Cu alloys. This plot is similar to Fig. 14 except that now a new region of morphological

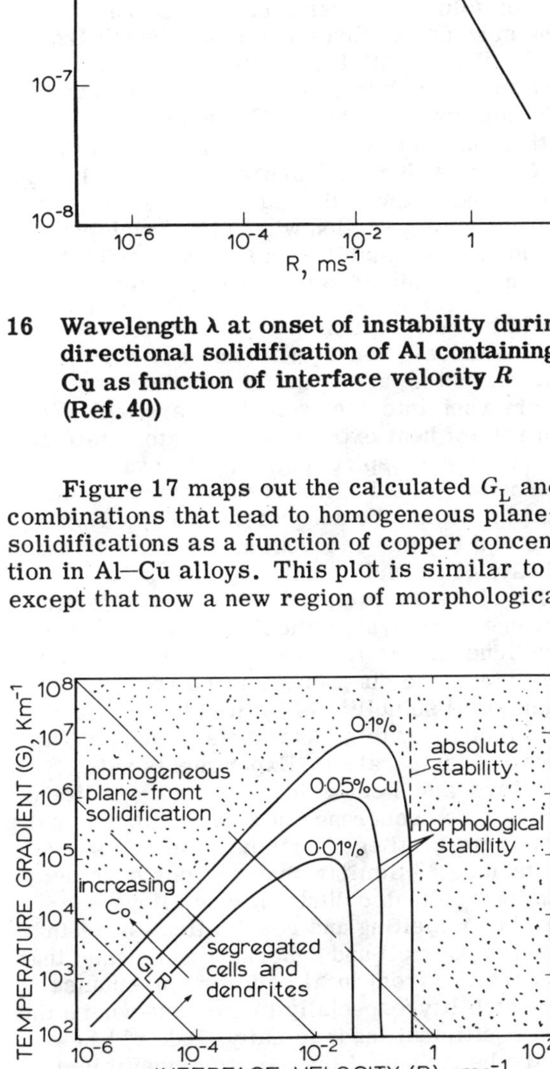

17 Regimes of homogeneous plane-front growth *v.* segregated cellular and dendritic growth as determined by liquid temperature gradient G_L and interface velocity R in Al–Cu alloys; cone-shaped lines separating the two regimes are the locus of morphological stability

stability is shown at high interface velocities which is independent of temperature gradient. As anticipated, the conditions under which plane-front solidification prevails become more restricted with increasing solute content.

Solidification into supercooled melts ($G_L < 0$) has also been recently discussed.[40] Of special interest is the solidification of very fine, less than 1 μm dia., supercooled droplets which are microstructurally analysed in a subsequent section. In a small droplet the largest perturbation has to be less than the droplet size or $\omega \geq \pi/r$. Thus, for a 1 μm dia. droplet of pure aluminium $\omega \geq 6 \times 10^6$ m^{-1} the surface-tension term in equation (11) is larger than 3.6×10^6 Km^{-1}. Even if it is assumed that $G_S = 0$ a stable interface would be expected as long as $G_L \geq -1.2 \times 10^7$ Km^{-1} or $R < 1.0$ ms^{-1}. Addition of solute requires a larger contribution from the surface-tension term, since destabilization is now due to both the negative temperature gradient and the solute-gradient terms in equation (11). Calculations for the Al–Cu system[41] have shown that some minimum composition, less than 1 wt-% Cu, is needed for morphological instability to be observed in these fine particles. This is verified below in a section where the development of morphological instability in fine particles of aluminium containing 4.5 wt-% Cu is noted.

While the assumption of local equilibrium has served to elucidate non-equilibrium solidification-segregation 'paths' at low interface velocities, it falls short when interface velocities are high due to rapid rate of heat extraction as in atomization, surface melting, or splat cooling. For example, the extended solid solubility of metastable structures cannot always be adequately explained when interface equilibrium is invoked. This extension in solubility can result from suppression of nucleation of the second phase below the eutectic temperature. The system thus follows the metastable solidus line as it moves into the two-phase region of the phase diagram. Therefore, the maximum solid solubility is limited to the composition dictated by this metastable solidus at the solidification temperature. Experiments with Zn–Cd alloys, a system with a retrograde solidus, have proven the occurrence of solute trapping and deviation from local equilibrium at the solid/liquid interface.[42] Similar observations have been made on ion-implanted dilute silicon alloys subjected to laser melting and resolidification.[43] In a recent paper, Coriell and Sekerka[40] concluded that most departures from local equilibrium enhance interface stability, especially in the case where the partition coefficient tends to unity. Baker[44] has studied the behaviour of the partition coefficient as a function of interface velocity for various models. In one such model with a variable solute energy at the interface, he showed that given a sufficiently high velocity the partition coefficient will always tend towards unity.

In conclusion, the present contributions of the morphological stability theory to rapid solidification, as well as its limitations, can be summarized as follows:

1. Short wavelength perturbations at high interface velocities are stabilized due to capillary forces. If the thermal field is stabilizing, the interface velocity for absolute stability is a function of the solute content and is independent of the temperature gradient in the liquid.

2. Departures from local equilibrium at the solid/liquid interface tend to have a stabilizing influence on the interface. However, the exact form or magnitude of these departures is not known.

3. In undercooled melts, the temperature gradient in the solid plays an important role in interface stability.

4. In most rapid-solidification experiments, interface velocity varies with undercooling and the changing thermal fields. Thus, the assumption of constant interface velocity has seriously limited the design of conclusive experiments to test the theory. Some attempts have been made to overcome this assumption.[45] These have been for limiting special cases and this field remains ripe for investigation.

5. At very high interface velocities, the solute-enriched layer approaches atomic dimensions and the macroscopic-transport equations used to derive the stability criterion are no longer valid.

6. When the interface velocities approach the diffusive velocity of the solute in the liquid, the assumption of local equilibrium at the solid/liquid interface is no longer valid. Complete solute trapping should occur if the interface velocity significantly exceeds the diffusive velocity.

GLASS FORMATION

Theory has been developed providing boundaries for the region in the phase diagram where partitionless solidification is thermodynamically forbidden.[42,46] In this region, a two-phase structure must be created if crystallization is to occur. It has also been shown[46] that, because diffusion processes are required to arrange the constituents into two-phase eutectic microstructures in this region, crystallization velocities typically will not exceed 5 cms^{-1}, making glass formation very probable during solidification. A short review of these findings is given below.

Solidification of crystalline structures is known to require nucleation and growth processes. On the other hand, attempts to predict the conditions for formation of metallic glasses usually focus only on nucleation, i.e. the degree of difficulty that atoms experience in joining a nucleus – viscosity is high or diffusivity is low. Suppression of nucleation by rapid quenching of the liquid until the alloy reaches its glass-transition temperature is used to predict the glass-forming tendency of alloys. Thus, the glass-forming tendencies of alloys have been classified in terms of their reduced glass temperature (the ratio of the glass-transition temperature to the liquidus temperature of the alloy). As this ratio increases, the critical cooling rate to produce an amorphous structure decreases. A recent investigation[46] has shown that amorphous solidification can be obtained in some eutectic and off-eutectic alloy compositions even in the presence of the crystalline phase, i.e. crystalline solidification is growth limited. This is the case when the alloy is thermodynamically forbidden to form a single phase, and the solidification velocity is sufficiently increased to overcome the diffusion-controlled partitioning process at the interface for the formation of a two-phase structure.

Baker and Cahn[42] have discussed the thermodynamics of non-equilibrium solidification in a binary alloy. With the aid of free energy G *v.* composition C curves, they defined the domain of possible interface compositions for isothermal and steady-state solidification. They concluded that the maximum solid composition C_S that can be formed from a liquid of composition C_L at the temperature of interest is determined by the point at which the liquid and solid free energy curves intersect. This is known as the $C(T_0)$ composition and the locus of all these points forms a curve between the liquidus and the solidus of the equilibrium phase diagram, *see* Fig. 1. Below the T_0 curve partitionless solidification of a liquid to a single-phase metastable solid of the same composition is permitted. In some phase diagrams, the T_0 curve(s) fall to very low temperatures. In this case, the curves represent a thermodynamic bound on the maximum extension of solid solubility and alloys beyond this bound, above the T_0 curve(s), must crystallize as multiphase solids.

The growth of a two (or more) phase solid from the liquid requires the diffusional sorting of the components in the liquid phase as well as the creation of new surface area between the phases.[46] These requirements for coupled growth lead to well known relationships between interface undercooling below the eutectic temperature, and interface velocity. For a given interface velocity, the interface undercooling required for eutectic solidification is much greater than for single-phase crystallization. Owing to this fact and the temperature dependence of the diffusion coefficient (related to the viscosity), a maximum velocity exists above which eutectic solidification cannot occur.[46] Alloys required to crystallize at velocities greater than this maximum, because of the rate of heat extraction, and which cannot crystallize as a single-phase material will form a glass.

It has long been known that alloys with near-eutectic compositions are often good glass formers. An interesting thing to note about compositions near eutectics is that for many alloys they must crystallize as multiphase solids regardless of the cooling rate.[46] This is known by examining the T_0 curves for all possible phases

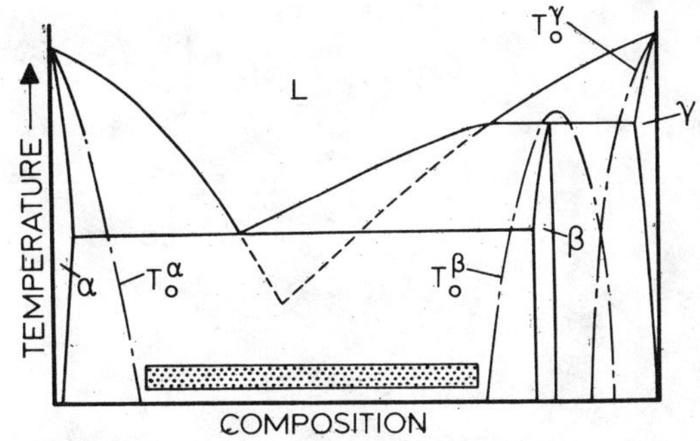

18 Hypothetical phase diagram with a stable and a metastable eutectic; T_0 curves for the three solid phases relative to the liquid are shown; in shaded region of composition between T_0 curves, crystallization to a single-phase crystalline solid is impossible[47]

(stable or metastable). Figure 18 shows a hypothetical phase diagram with a stable or metastable eutectic. If the composition range of an individual phase is very narrow or if the solidus is retrograde,[42] the T_0 curves will not extend far from that phase at temperatures of interest. This produces a rather wide range of alloy compositions between the T_0 curves of the different phases which must crystallize as two-phase solids. Even partitionless (massive) crystallization of liquids in this composition range into a single phase is impossible. Wider composition ranges requiring two-phase crystallization are possible if a particular phase fails to nucleate (for example, β in Fig. 18).

In a recent paper,[46] the microstructure of an easy glass former, Pd–6Cu–17Si (at.-%) alloy, as a function of growth rate up to the critical velocity for glass formation was investigated. As the

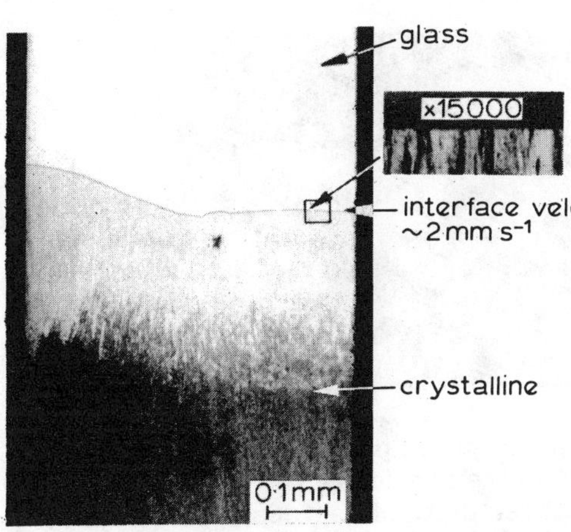

19 Transition from fine eutectic-like growth to glassy alloy observed when solid/liquid interface velocity reached ~2 mms^{-1} in a Pd–Cu–Si alloy[46]

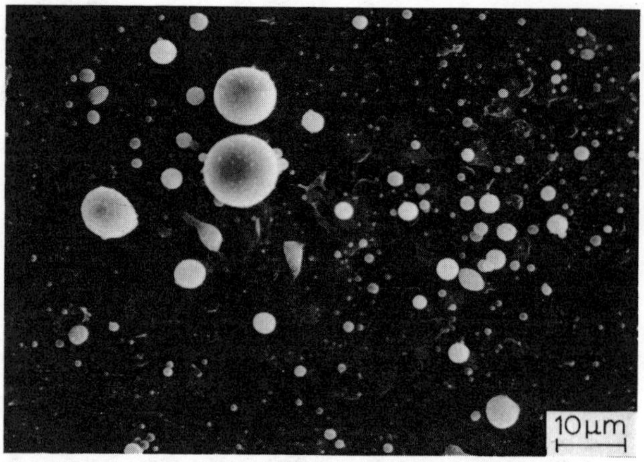

20 **SEM view of Al–4.5Cu powders produced by EHD atomization process**

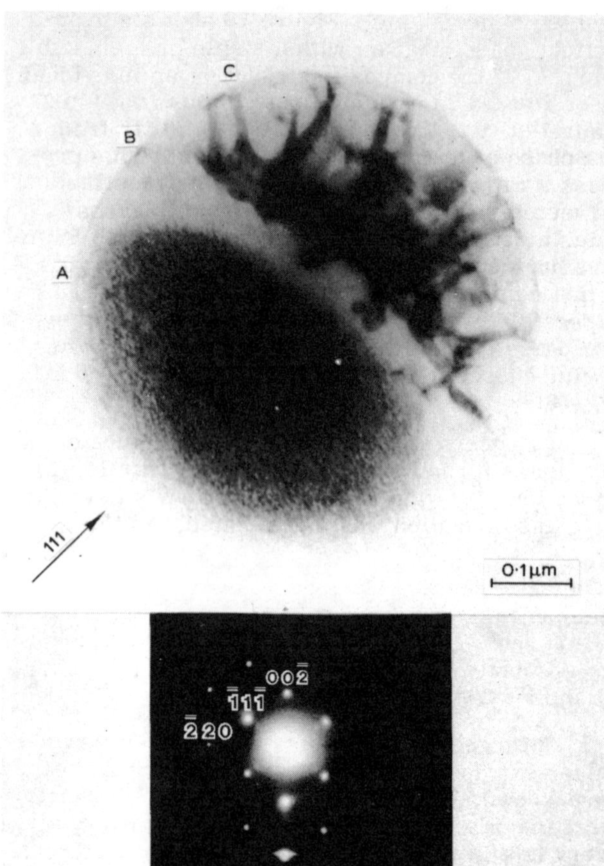

21 **Typical segregation pattern for Al–6Si EHD submicrometre powder (1 μm > $2r_0$ > 0.5 μm) showing three distinct regions; initial rapid solidification produces a supersaturated zone *A*, which later breaks into cells *C* after droplet recalesces; *B* is a transition zone where solute builds up in liquid in front of interface; growth axis is close to a ⟨111⟩ direction[26]**

solid/liquid interface velocity was increased from ~0.25 to ~1 mms^{-1}, the structure of dendrites and interdendritic eutectic became a fine eutectic-like structure. A sharp transition, *see* Fig. 19, from crystalline to amorphous structure was noted at an interface velocity of ~2 mms^{-1}.

MICROSTRUCTURAL OBSERVATIONS

Some of the new theoretical developments have already been qualitatively verified on extremely fine powders of aluminium alloys, less than 1 μm in size, produced in a small laboratory-type atomization apparatus.[26] The submicrometre powders are electron transparent and it is possible to analyse the evolution of a complete microstructure from nucleation until the end of solidification without recourse to thinning procedures. For example, interface breakdown from planar to cellular in the middle of the powders is clearly delineated. Microstructural observations have been coupled with thermodynamic, kinetic, and heat-flow concepts to elucidate the thermal history and solidification mode of the powders and their effects on the microstructure. A summary of these findings[26] is presented below.

The aluminium-alloy powders used in this investigation were prepared by the electrohydrodynamic (EHD) atomization process.[47] Figure 20 shows a typical sample of powders as collected on a plastic substrate. They are nearly spherical and have smooth surfaces, although some appear to have solidified on the tape. Results of the heat-flow studies on powder solidification, presented above, revealed that for typical atomization conditions external cooling plays only a minor role during the recalescence stage of the thermal history. Therefore, the rapid-solidification effects are usually limited by the undercooling achieved before nucleation occurs.

Most of the Al–Si EHD-produced powders conform to a variant of the basic segregation pattern shown in Fig. 21. The microstructure exhibits a region *A* containing a uniform distribution of very fine precipitates or clusters of Si which, according to the electron-diffraction evidence, have an orientation relationship to the primary α-aluminium phase. Dark-field imaging of precipitates slightly coarser than those in the present figure suggest that they are either needles or platelets on the {111} α-aluminium planes. This finding is consistent with the decomposition kinetics of Al–Si solid solutions; the plates evolve from spherical clusters and needles that form in the early stages of precipitation.[48–50] It was concluded that this zone was originally a supersaturated solid solution.

Region *A* is followed by a precipitate-free zone *B*, and the change in microstructure clearly delineates the solidification front. The latter is symmetric about the growth axis, which was consistently found to be close to a ⟨111⟩ direction.

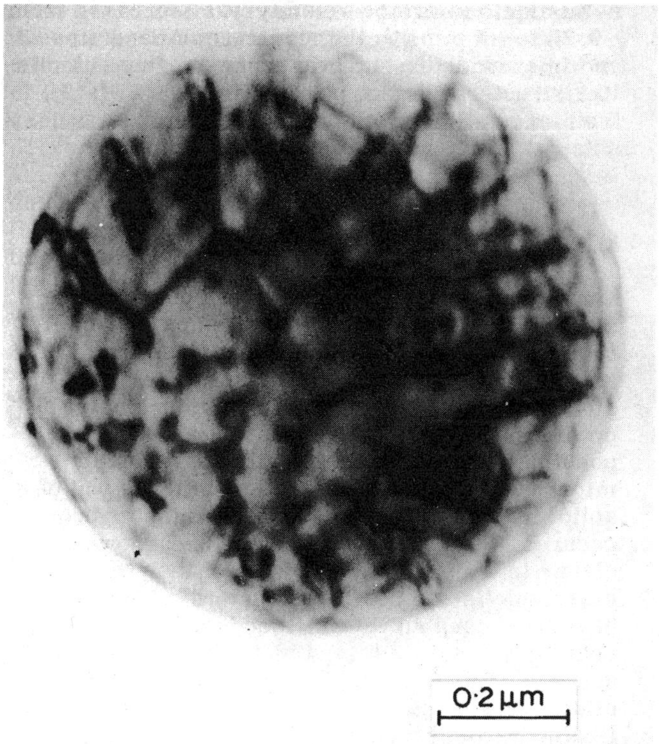

22 **Cellular structures observed in Al—4.5Cu EHD powders**[26]

Table 3 Effect of alloy composition on critical and hypercooling temperatures*[24]

		3%Si T, K	θ	6%Si T, K	θ
Liquidus	T_L	913	0	893	0
Critical	$T_C(T_0)$	848	−0.18	762	−0.36
	$T_C(T_S)$	782	−0.36	631	−0.72
Hypercooling	$T_H(T_0)$	484	−1.18	398	−1.36
	$T_H(T_S)$	418	−1.36	267	−1.72

*Estimated from the equilibrium phase diagram, T_S values are from the metastable solidus, $T_0 = (T_L + T_S)/2$. Dimensionless temperature $\theta = \kappa_L(T - T_L)/\Delta H_M$.

Finally, the particle in Fig. 21 also has a well developed cellular region C, with segregate spacings of the order of 0.1 μm. A similar structure in an Al—4.5Cu powder, shown at a different angle in Fig. 22, reveals the cellular pattern more clearly.

Figure 23 illustrates the general effect of particle size on the extent of the different zones in the powders. As the powder diameter decreases, the cellular structure becomes less well defined (Fig. 23*b*), and eventually disappears, leaving only a few segregates (Fig. 23*c*). At the lower end of the size distribution, the particles appear almost featureless (Fig. 23*d*), showing only the supersaturated region A. The intensity of the silicon spots in electron diffraction was also reduced with diminishing diameter, although the crystallographic relationship noted above prevailed.

These microstructures reveal that substantial undercoolings must have been reached in the liquid droplets before solidification, which is believed to proceed as follows. In general, nucleation occurs at the surface, as evidenced by the growth direction. The subsequent rapid solidification into a highly undercooled liquid produces a homogeneous zone A of extended solubility. As the droplet recalesces, the interface velocity and undercooling are reduced, thereby decreasing the driving force for solute 'trapping'. When the interface temperature surpasses some critical temperature for the supersaturation process, T_C, solute is rejected ahead of the solidification front, and the solid forming (region B) is depleted in solute. The interface is eventually rendered unstable and breaks into cells.

The extent of homogeneous solidification can be rationalized in terms of the heat-flow concepts discussed above. Assuming that the thermophysical properties are reasonably constant, the behaviour of an alloy growing in a plane-front mode without solute partitioning is expected to resemble that of the pure metal.

The onset of segregation probably occurs at a temperature between the solidus T_S and the T_0 curve. For purposes of illustration, the average value of T_L and T_S from the Al—Si phase diagram has been arbitrarily assigned to T_0, and on this basis an upper bound to the hypercooling temperature T_H estimated, *see* Table 3. The lower bound to T_H corresponds to $T_C = T_S$. The values of θ were calculated by substituting T_L for T_M in its definition and hence are proportional to the undercoolings. The values in this table indicate that dimensionless temperatures of the order of −1.18 ($\Delta T_N/T_L$ = 0.47) and −1.36 ($\Delta T_N/T_L$ = 0.55) may be necessary to eliminate segregation in aluminium alloys with 3 and 6 wt.-% Si, respectively.

A large proportion of powders in the lower sizes experienced some segregation but no cellular breakdown. This phenomenon may simply reflect that the onset of segregation ($\Delta T^* < \Delta T_C$) occurred very late in the solidification process. ΔT^* refers to local undercooling at the solid/liquid interface in a general non-Newtonian heat-flow model. On the other hand, interface breakdown may also be prevented by the stabilization of short wavelength perturbations due to surface-tension effects as noted above; the smallest perturbation that can grow is comparable to the particle size. High-purity aluminium submicrometre powders, such as those in Fig. 24, do not exhibit any evidence of cellular structures that could have originated from thermal instability caused by negative temperature

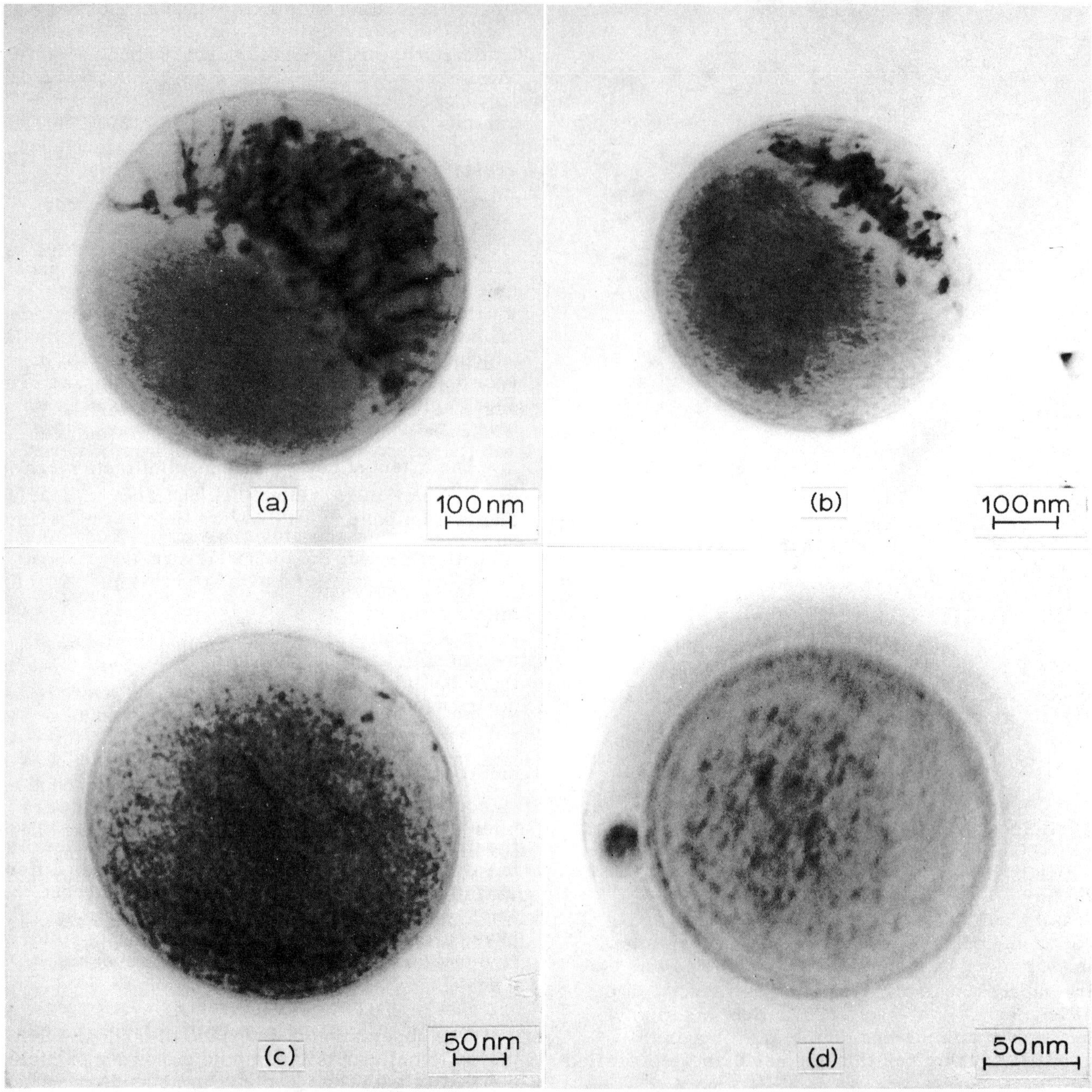

a 570 nm dia.; *b* 470 nm dia.; *c* 220 nm dia.; *d* 170 nm dia.

23 Transition from a partially segregated structure to a homogeneous one as powder particle size decreases in the Al–6Si alloy[26]

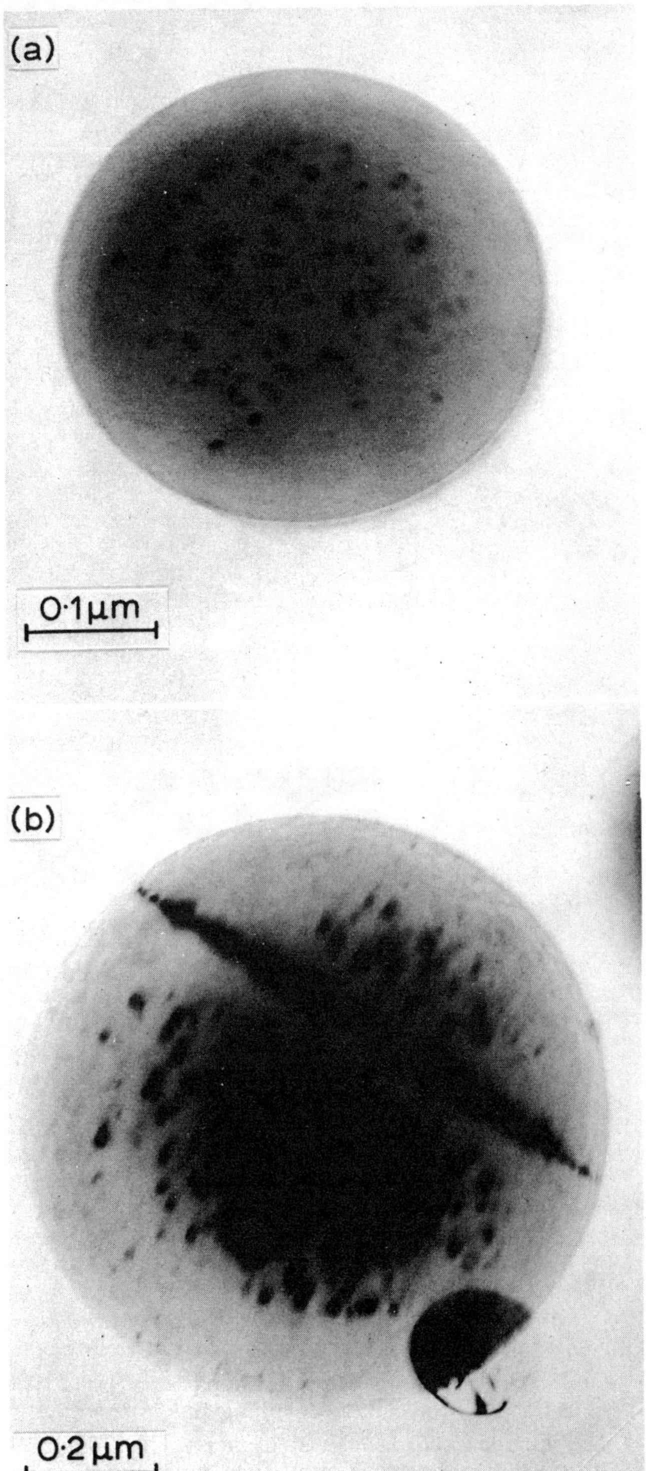

a monocrystalline particle;

b set of twin crystals

24 Typical structures in high-purity Al (99.999%) submicrometre EHD powders[26]

gradients in front of the interface. This observation lends support to the fact that the breakdown observed in the alloys was mainly of a constitutional supercooling nature. It could also be argued that thermal cells in a pure metal would be indistinguishable after solidification, unless they are decorated by impurities or strained areas caused by crystalline misorientation. Nevertheless, pure aluminium particles show some platelike features that can be interpreted either as clusters of impurities or vacancy-condensation loops. The existence of denuded zones near the surface and crystal boundaries suggests that vacancies are involved in the formation of these features.

The microstructural analysis also revealed that most EHD powders are monocrystalline, indicating that nucleation was usually limited to a single event per droplet. A substantial proportion of them, however, consists of two or more crystals which could be either randomly oriented with incoherent boundaries, as in Fig. 25, or twins, as in Fig. 26. It was concluded that each random grain originates from an independent nucleation event, and its boundaries result from impingement with other growing crystals. On the other hand, twinned crystals normally exhibit {111} coherent boundaries and it is unlikely that they were independently nucleated.

In summary then, the availability of submicrometre powders generated by the EHD atomization process has provided a suitable tool for fundamental solidification studies. In addition to having the potential to achieve substantial undercoolings before nucleation, these powders are electron transparent and permit microstructural examination of the complete solidification structure of a small casting without thinning.

The investigation[26] revealed that decreasing the particle size — hence increasing the achievable undercoolings and solid/liquid interface velocities during growth — reduced the extent of segregation. The consequent supersaturation of the primary phase eventually results in homogeneously solidified powders with silicon contents as high as 6 wt-%.

The observation of interface breakdown from planar to cellular in the middle of powder particles coupled with a solute-depleted zone lends credence to the hypothesis that plane-front homogeneous solidification was achieved because of the substantial undercoolings before nucleation and the high interface velocities during recalescence.

Some powders showed segregation but no cellular breakdown. This may be evidence of interface morphological stability owing to surface-tension effects, since the temperature gradient ahead of the interface is expected to be negative throughout solidification and therefore favour the growth of perturbations.

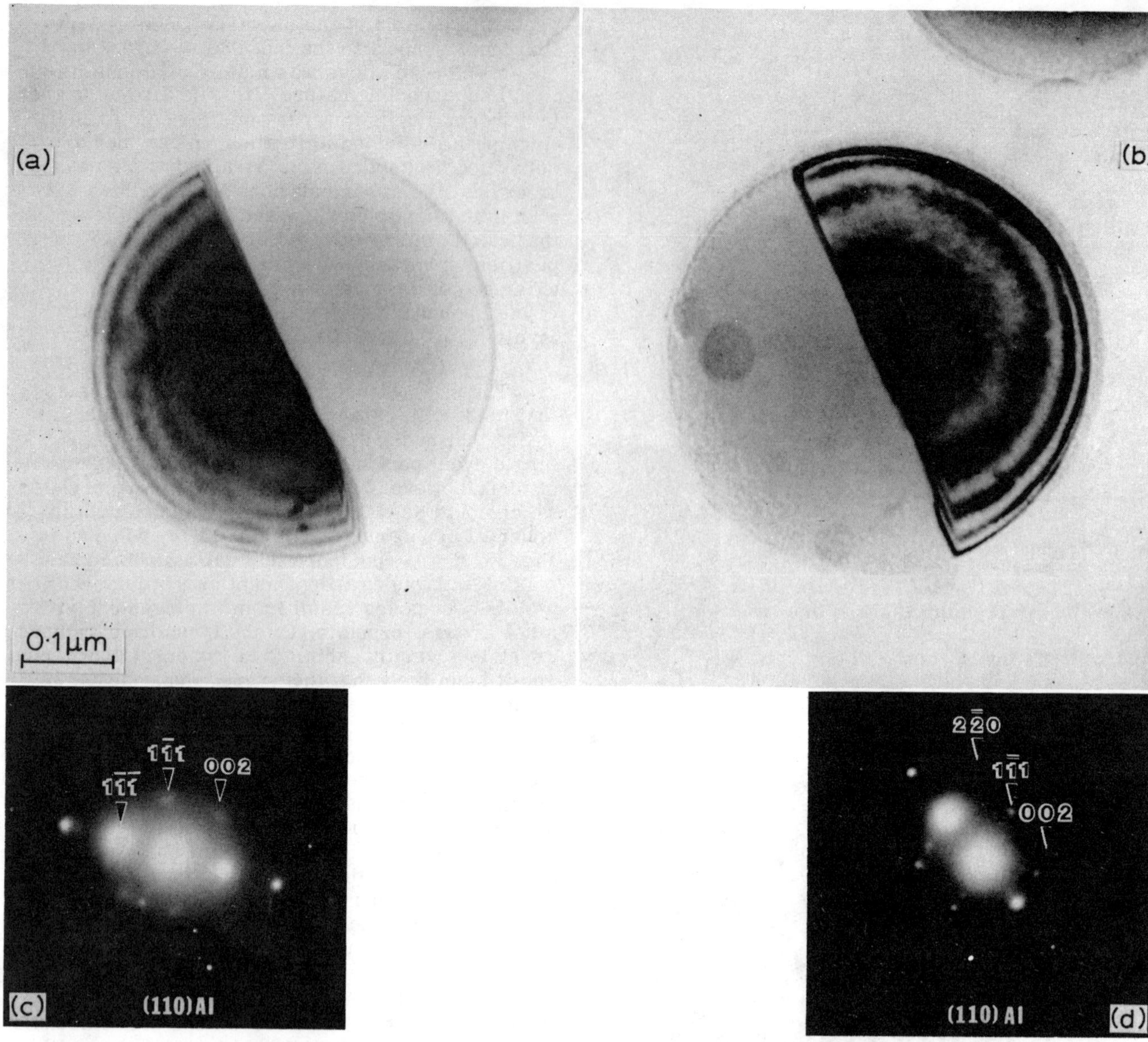

25 Typical polycrystalline powder of Al—3Si alloy where two nucleation events occurred simultaneously in droplet and formed an impingement boundary[26]

NON-DESTRUCTIVE TESTING

A recent investigation[51,52] addresses the need for coupling of rapid-solidification technology to non-destructive testing for on-line control of parameters during processing. Examples of fundamental studies in this area include:

1. The development of quantitative multichannel acoustic emission for monitoring the signals of cracks produced during surface modification by directed high-energy sources.[51]
2. The development of ultrasonic scattering techniques for the characterization of metastable crystalline and glassy microstructures and their subsequent transformations into more stable structures during thermomechanical processing.[52]

An example from the latter study on metallic-glass ribbons of $Pd_{0.775}Cu_{0.06}Si_{0.165}$ alloy is given below. The objective of the investigation was to study the crystallization kinetics of the ribbons (about 50 μm thick) by means of laser-generated and piezoelectrically detected ultrasonic waves whereby the extensional wave velocity, and consequently the Young's moduli, can be determined as a function of crystallization time. The amorphous-to-crystalline kinetics could thus be determined to a high degree of accuracy.

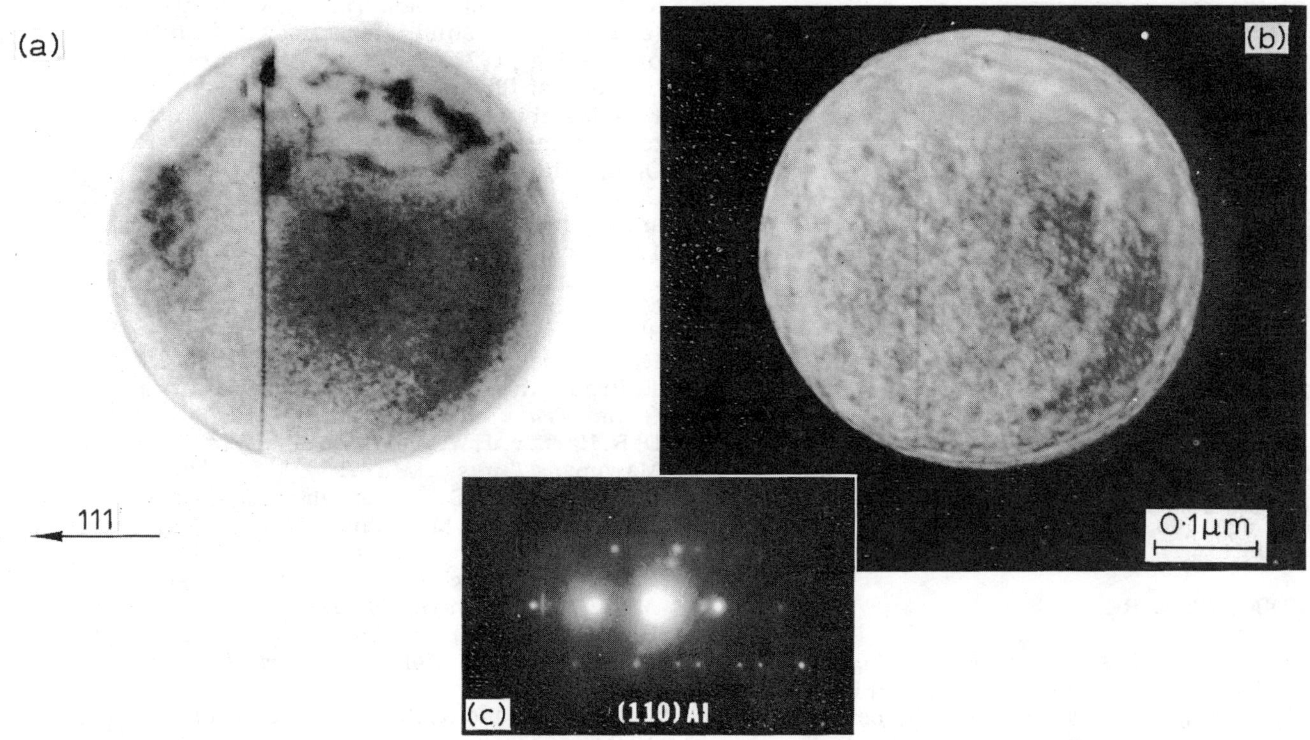

26 TEM of twin crystals in Al–6Si alloy powder; segregation pattern suggests that nucleation occurred at a point on line intersection of powder surface and twin boundary[26]

The inherent difficulty of measuring sound-wave velocities in thin ribbons was overcome by developing a technique whereby a load was applied to the specimen by means of rapid deposition of energy from a single pulse of a Q-switched Nd:YAG laser. A state of unbalanced compressive thermal stress is produced. The stress gradient propagates along the ribbon in the form of an elastic wave. The piezoelectric crystal, at a distance from the spot where the laser radiation was deposited, detects the local displacement. Thus, the transit time of the propagating wave could be determined accurately by means of a transient-pulse recorder.

Results of the investigation[52] indicate that the transition from the amorphouse-to-crystalline state in Pd–Cu–Si is time and temperature dependent, Fig. 27. The kinetics were determined at three isothermal-holding temperatures, and the activation energy for crystallization was found to be 55 kcal mol^{-1}. From the time dependence of the transformation kinetics an insight was gained into the nucleation and growth mechanisms. Corroborative metallographic evidence was found to be in qualitative agreement with the ultrasonic data, Fig. 28, suggesting a linear-growth law.

The investigation[52] also revealed an initial increase of about 7.4% in the elastic modulus, out of 40% for the complete transition from the amorphous to the crystalline state. It appears that this initial increase can be related to structural relaxation of the amorphous phase that may precede the crystallization process.

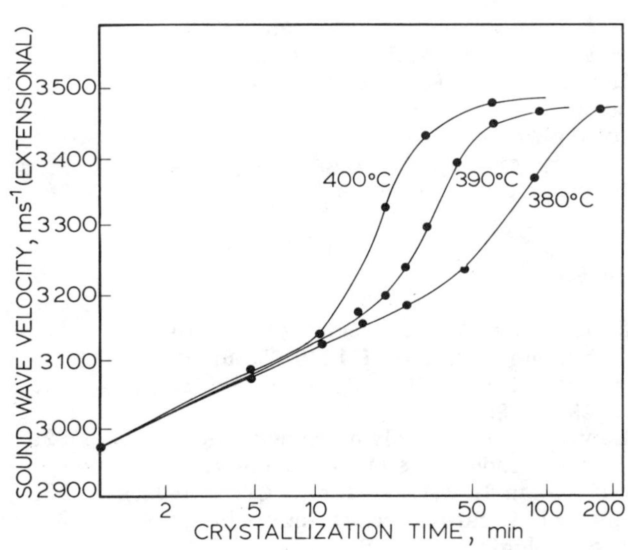

27 Variation of sound wave velocity as function of crystallization time during amorphous-to-crystalline transition in $Pd_{0.775}Cd_{0.06}Si_{0.165}$ (Ref. 52)

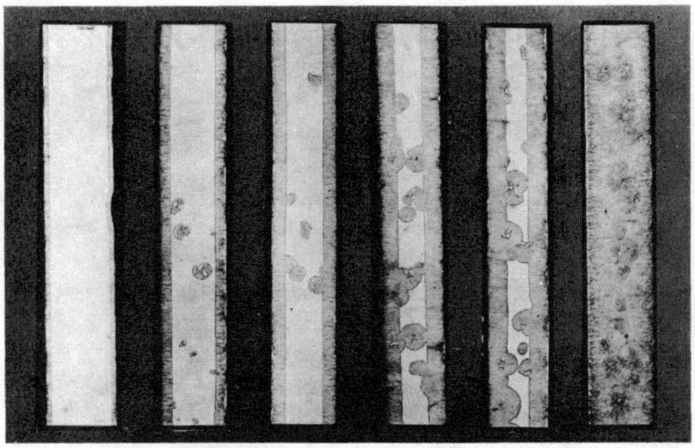

28 Changes in microstructure as function of time during amorphous-to-crystalline transition in $Pd_{0.775}Cu_{0.06}Si_{0.165}$ at 390°C: view from ribbon edge (40 μm) (Ref. 52) × 250

CONCLUDING REMARK

Specific recent advances in the understanding of rapid-solidification processing are covered in this review. No attempt is made to present a complete status of the field, or include progress made in the understanding of basic phenomena taking place in commercially important alloy systems. Emphasis is placed on those evolving principles and guidelines which will govern future developments in the design of new alloy compositions and process-control parameters to achieve refined and more homogenous structures, or new microstructures and phases.

ACKNOWLEDGMENTS

The author is indebted to Drs J.W.Cahn, W.J. Boettinger, S.R.Coriell, M.Rosen, and R.J.Schaefer of the National Bureau of Standards for enlightening discussions, and to Drs E.C.van Reuth and L.A. Jacobson of the Defence Advanced Research Projects Agency for their foresight in supporting most of the generic work required.

REFERENCES

1. N.J.Grant and B.C.Giessen (eds.): 'Rapidly quenched metals'. Part I, 1976, Cambridge, Mass., MIT Press. Part II, *Mater. Sci. Eng.*, 1976, **23**, 81–324.
2. B.Cantor (ed.): 'Rapidly quenched metals III': 1978, London, The Metals Society.
3. R.Mehrabian, B.H.Kear, and M.Cohen (eds.): 'Rapid solidification processing: principles and technologies'; 1978, Baton Rouge, La, Claitor's Publishing Division.
4. R.Mehrabian, B.H.Kear, and M.Cohen (eds.): 'Rapid solidification processing: principles and technologies II'; 1980, Baton Rouge, La, Claitor's Publishing Division.
5. S.D.Ferris, H.J.Leamy, and J.M.Poate (eds.): 'Laser—solid interactions and laser processing'; 1978, New York, American Institute of Physics.
6. E.A.Metzbower (ed.): 'Application of lasers to materials processing'; 1979, Metals Park, Ohio, American Society for Metals.
7. C.W.White and P.S.Peercy (eds.): 'Laser and electron beam processing of materials'; 1980, New York, Academic Press.
8. J.L.Murray: personal communication, Metallurgy Division, National Bureau of Standards, Washington, DC.
9. R.K.Linde: *J. Appl. Phys.*, 1960, **31**, 1136.
10. P.Predecki, A.W.Mellendore, and M.J. Grant: *Trans. AIME*, 1965, **233**, 1581.
11. D.R.Harbur, J.W.Anderson, and W.J. Maraman: *Trans. AIME*, 1969, **245**, 1055.
12. S.D.E.Ramati, G.J.Abbaschian, D.G. Backman, and R.Mehrabian: *Metall. Trans.*, 1978, **9B**, 279.
13. S.Hong, D.G.Backman, and R.Mehrabian: *Metall. Trans.*, 1979, **10B**, 229.
14. R.C.Ruhl: *Mater. Sci. Eng.*, 1967, **1**, 313.
15. C.G.Levi and R.Mehrabian: *Metall. Trans.*, 1980, **11B**, 21.
16. P.H.Shingu and R.Ozaki: *Metall. Trans.*, 1975, **6A**, 33.
17. D.R.Uhlmann: *J. Non-Cryst. Solids*, 1976, **7**, 337.
18. F.Spaepen and D.Turnbull: Ref. 1, p. 205.
19. S.C.Hsu, S.Chakravorty, and R.Mehrabian: *Metall. Trans.*, 1978, **9B**, 221.
20. S.C.Hsu, S.Kou, and R.Mehrabian: *Metall. Trans.*, 1980, **11B**, 29.
21. S.Kou, S.C.Hsu, and R.Mehrabian: *Metall. Trans.*, 1981, **12B**, 33.
22. J.A.Sekhar, S.Kou, and R.Mehrabian: unpublished work.
23. D.Rosenthal: *Weld. J., Res. Suppl.*, 1941, **20**, 2205.
24. R.Mehrabian: Ref. 3, p. 9.
25. J.P.Hirth: *Metall. Trans.*, 1978, **9A**, 401.
26. C.G.Levi and R.Mehrabian: *Metall. Trans.*, 1982, **13A**, 13.
27. T.Z.Kattamis and R.Mehrabian: *J. Vac. Sci. Technol.*, 1974, **11**, (6), 1118.
28. J.H.Perepezko, D.H.Rasmussen, I.E. Anderson, and C.R.Loper, Jr: 'Solidification and casting of metals', 169; 1979, London, The Metals Society.
29. J.H.Perepezko: Ref. 4, p. 56.
30. C.G.Levi and R.Mehrabian: *Metall. Trans.*, 1982, **13A**, 221.
31. J.W.Cahn, W.B.Hillig, and G.W.Sears: *Acta Metall.*, 1964, **12**, 1421.
32. W.A.Tiller, K.A.Jackson, J.W.Rutter, and B.Chalmers: *Acta Metall.*, 1953, **1**, 428.
33. M.E.Glicksman and S.C.Huang: Ref. 4, p. 68.
34. W.W.Mullins and R.F.Sekerka: *J. Appl. Phys.*, 1963, **34**, 323.
35. W.W.Mullins and R.F.Sekerka: *J. Appl. Phys.*, 1964, **35**, 444.
36. R.F.Sekerka: 'Crystal growth: an introduction', 403; 1973, Amsterdam, North-Holland.

37. R.T.Delves: 'Crystal growth', 1, 40; 1974, Oxford, Pergamon Press.
38. D.J.Wollkind: 'Preparation and properties of solid state materials', 4, 111; 1979, New York, Marcel Dekker.
39. J.S.Langer: *Rev. Mod. Phys.*, 1980, **53**, 1.
40. S.R.Coriell and R.F.Sekerka; Ref. 4, p. 35.
41. R.F.Sekerka: 'Crystal growth', 691; 1967, Oxford, Pergamon Press.
42. J.C.Baker and J.W.Cahn: *Acta Metall.*, 1969, **17**, 575.
43. C.W.White, S.R.Wilson, B.R. Appleton, and F.W.Young, Jr: *J. Appl. Phys.*, 1980, **51**, 738.
44. J.C.Baker: PhD thesis, MIT, Cambridge, Mass., 1970.
45. J.S.Wey, K.A.Gautesen, and J.Estrin: *J. Cryst. Growth*, 1973, **19**, 169.
46. W.J.Boettinger, F.S.Biancaniello, G.M. Kalonji, and J.W.Cahn: Ref. 4, p. 50.
47. J.Perel, J.F.Mahoney, P.Duwez, and B.E.Kalensher: Ref. 4, p. 287.
48. K.M.Entwistle, P. Fuller, and I.Brough: *Acta Metall.*, 1978, **26**, 1055.
49. L.F.Mondolfo: 'Aluminum alloys: structure and properties', 373; 1976, London, Butterworths.
50. H.Matyja, K.C.Russel, B.C.Giessen, and N.J.Grant: *Metall. Trans.*, 1975, **6A**, 2249.
51. H.N.G.Wadley *et al.*: Proc. of DARPA/AF Meeting on Quantitative NDE, Aug. 1981, Boulder, Colo.
52. M.Rosen, H.N.G.Wadley, and R.Mehrabian: *Scr. Metall.*, 1981, **15**, 1231.

SECTION IV

Amorphous or Glassy Materials

Overview of Principles and Applications

DONALD E. POLK
BILL C. GIESSEN
Northeastern University
Boston, Massachusetts 02115

The first synthesis of a metallic glass that drew wide attention among materials scientists occurred in 1960 when Klement, Willens and Duwez (1) reported that a liquid Au-Si alloy, when rapidly quenched to liquid nitrogen temperature, would form an amorphous solid. Prior to this, amorphous metals had been prepared by both vapor deposition and electrodeposition methods at least as early as the 1930's, but these studies were not followed up in a continuous line of research. Since 1960, a large number of different alloys have been produced as metallic glasses, and the scientific and technological interest in these materials has burgeoned, as evidenced by previous topical conferences (2-5) and the large number of publications and review papers in this field (6-8).

A starting point in any discussion of metallic glasses should be agreement on the definition of this term. "Metallic", of course, means what it normally does: a metallic glass has a conductivity similar in magnitude to that of solid and liquid metals (though its temperature coefficient may in some cases be negative at lower temperatures) and optical behavior (e.g., sheen) typical of metals. A casual, qualitative inspection would not differentiate between a metallic glass and a common metal. "Glass", in its original sense, denotes an amorphous solid (i.e., a noncrystalline solid lacking long-range periodicity of the atomic arrangement) which has been prepared by cooling the corresponding liquid. In contrast to the first-order crystallization of the liquid, where the solid crystal coexists with and grows into the liquid, the solidification to a glass is homogeneous and occurs over a range of temperature.

However, a definition of a metallic glass based on its origin is too restrictive in that solid amorphous metals of identical composition, and apparently, on the basis of diffraction data, similar structure, can be produced other than from the liquid (e.g., by sputtering). These substances also have other characteristics in common with those produced by liquid quenching and are therefore often considered metallic glasses; however, their identity as such is not universally accepted.

A further problem with any structural definition of amorphous metals results from the practical difficulty encountered in differentiating between a substance which has a continuous random structure and one which is microcrystalline with a grain size of several atomic diameters. The important question of atomic structure is discussed below and in Chapter 4 of this book.

In this chapter we will discuss the more general topic of amorphous metals, defined as noncrystalline metallic solids formed by any method, and we will often use the term "metallic glass" synonymously with "amorphous metal".

Structural Characteristics

As noted above, the lack of crystalline atomic periodicity is the primary distinguishing feature of an amorphous metal. As a consequence, the diffraction patterns from amorphous metals contain only a series of broad maxima, in contrast to the sharp lines observed for crystalline materials. Figure 1 (top) shows the interference functions, I(k), for liquid Ag-Cu and solid amorphous Pd-Si and Ag-Cu, obtained by normalizing the corrected high-angle-scattered intensity (9). Generally, the diffraction patterns of amorphous metals are qualitatively similar to each other and to those of liquid metals, though generally sharper than those of the liquids. The width of the first peak in the diffraction pattern can vary markedly for amorphous metals—e.g., from $\Delta 2\theta = 5.45°$ for $Fe_{75}P_{15}C_{10}$ to $\Delta 2\theta = 11.12°$ for $Pt_{60}Ni_{15}P_{25}$, both using CuK_α radiation (10). This broadening is due both to the variation of the first-neighbor distance inherent in the amorphous structure and to differing atomic sizes within one alloy, as well as to thermal and experimental effects, so that meaningful comparisons between different alloys are necessarily complex.

The diffraction data from amorphous materials provide only statistical information on the atomic structure. The available structural information is contained in the radial distribution function, $4\pi r\rho_0 g(r)$, obtained by Fourier transformation of the interference function. The

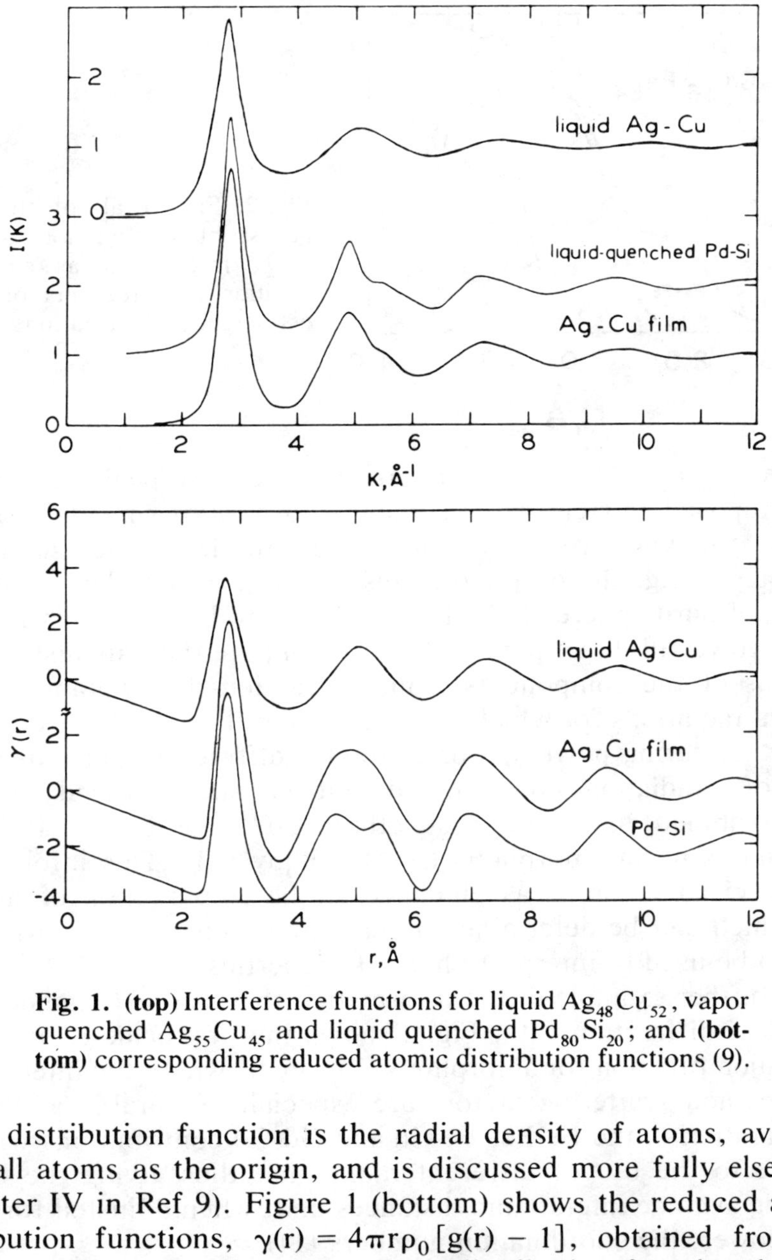

Fig. 1. (top) Interference functions for liquid $Ag_{48}Cu_{52}$, vapor quenched $Ag_{55}Cu_{45}$ and liquid quenched $Pd_{80}Si_{20}$; and **(bottom)** corresponding reduced atomic distribution functions (9).

radial distribution function is the radial density of atoms, averaged over all atoms as the origin, and is discussed more fully elsewhere (Chapter IV in Ref 9). Figure 1 (bottom) shows the reduced atomic distribution functions, $\gamma(r) = 4\pi r \rho_0 [g(r) - 1]$, obtained from the diffraction data of Fig. 1 (top). Such distribution function data are useful in evaluating structural models, as discussed below, and in determining the nature of the short-range order, reflected in the shape of the first peak.

A topic of special interest in this regard is the question of what is the short-range order in the widely studied metal-metalloid glasses

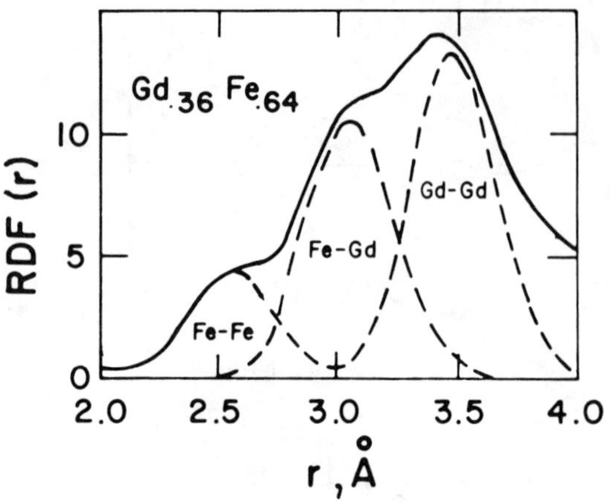

Fig. 2. First peak of the radial distribution function for $Gd_{36}Fe_{64}$, interpreted as a superposition of three sets of nearest-neighbor distributions (11)

such as $Ni_{80}P_{20}$ (11). In one approach, the first peaks calculated from proposed models can be compared to experimental data to appraise the validity of the model and to determine the model parameters—e.g., hard sphere sizes in the case of dense random packing of hard spheres (DRPHS)—which produce the best fit (12). A very powerful direct approach is to compute the pair distribution functions of the components from diffraction data obtained using different radiations for which the elements in the alloy have different relative scattering powers. For example, diffraction patterns taken with MoK_α radiation, nonpolarized neutrons and polarized neutrons were combined to separate out the Co-Co, Co-P and P-P pair distributions in an amorphous $Co_{81}P_{19}$ alloy (13). This information is of special usefulness as guidance in the construction of models from which can be determined actual atomic coordinates, which in turn can be used to interpret physical properties.

As a further example of the information that can be obtained from analysis of diffraction data, Fig. 2 shows the first peak of the radial distribution function for amorphous $Gd_{36}Fe_{64}$, where the interatomic distances and scattering factors are especially favorable and allow the separation of the Fe-Fe, Fe-Gd and Gd-Gd contributions without recourse to the use of different radiations, thus giving the corresponding average interatomic distances and, when adjusted for scattering power, the coordination numbers (11).

Since the atomic configurations present in an amorphous metal cannot be uniquely determined directly as in the case of a crystalline material, investigators have attempted to infer the structure by comparing the experimental data to data calculated from models. Both microcrystalline and "continuous random" structures have received extensive consideration (11); "noncrystallographic cluster" models, intermediate between the other two, have been considered in more detail recently (14).

Microcrystalline models assume the existence of highly ordered regions (i.e., microcrystals) having atomic configurations identical to a small section of an ordinary three-dimensional crystal. These microcrystals are then assembled, often without orientational correlation; generally, the details of the atomic configurations in the boundary layer between microcrystals are not given. [To differentiate the term "microcrystalline" as typically used in discussions of atomic structure from that often used by metallurgists in a different sense—to grain sizes in the μm range—one might wish to term the hypothetical structures with crystallite sizes in the nm (10 Å) range as "nanocrystalline."]

Continuous random models for amorphous metals date back to the concepts of Zachariasen, which have been widely applied to low-coordination, covalently bonded glasses such as SiO_2. Noncrystallographic, densely packed configurations of hard spheres have been produced both physically and with computer algorithms; these DRPHS structures have been proposed as the basis for structural models for metallic glasses (11). An important difference from the microcrystalline models is that the DRPHS structure is statistically homogeneous on a scale of about 10 Å and does not contain the internal boundary regions intrinsic to microcrystalline models.

Noncrystallographic cluster models are similar to microcrystalline models in assuming the existence of discrete, highly ordered regions, except that the microcrystalline models contain symmetry elements, especially five-fold rotational axes, which cannot be the basis of an extended three-dimensional crystal. These concepts have gained impetus from observations that clusters of up to at least 60 atoms have lower energy in noncrystallographic arrangements than in crystalline configurations (15); however, this energy difference appears to be related to surface effects, so that the relevance of this data to bulk material is uncertain. Indications that such clusters (which have a greater proportion of tetrahedral arrangements than DRPHS structures) have higher density than DRPHS structures have increased the degree of scientific interest in such atomic arrangements.

While the diffraction data provide statistical information about both the short-range and long-range order in an amorphous metal, an unambiguous differentiation between a fully amorphous and a microcrystalline metal is still not possible. This is due to the well-known line broadening effect which occurs as the crystals become progressively smaller. In fact, the Scherrer formula predicts that a metal with a grain size of 8 to 16 Å would have a first peak width typical of that observed for "amorphous metals" (10). For such small grain sizes, about one-half of the atoms are on the surface (i.e., in boundary regions), making the meaningfulness of such clusters more uncertain. Though the details are clearly not settled, it would appear that variations of the continuous random or noncrystalline cluster models can account for the experimental observations better than can the microcrystalline models.

Transmission electron microscopy (TEM) is another direct technique for structural analysis of amorphous metals. While truly homogeneous amorphous material is expected to show no contrast effects, one could hope to observe, for example, contrast effects associated with a microcrystalline structure if such a structure were present. The results of high-resolution TEM studies are at the present time ambiguous, however. The observation of contrast effects in the tilt beam dark field mode was originally ascribed to structural clustering, but now appears to be an artifact due to the nonlinearities of the instrument. However, a "salt and pepper" contrast is seen in amorphous Nb-Ni with axial illumination (16); such contrast effects are typical of amorphous metals. The origin of this contrast is still unclear; this effect may also be an artifact.

Thermal Characteristics

The thermal behavior of amorphous metals can be divided into two categories: that associated purely with the amorphous structure and that related to crystallization effects.

Included in the intrinsic behavior is the specific heat behavior shown schematically in Fig. 3. The sudden increase in the specific heat of the glass upon heating, ΔC_p, is the thermal manifestation of the glass transition. This behavior is typical for ordinary (nonmetallic) glasses, though ΔC_p may not be observed upon heating because crystallization may occur before the glass transition is reached. The glass transition temperature, T_g, is generally defined as the point of inflection of the rising C_p. Such behavior has been observed for many amorphous metals; in fact, the first reports of this behavior (17) provided some of the strongest support for the view that these newly produced structures were indeed glasses. Though the dashed section of the curve in Fig. 3 is inaccessible to measurements (because of rapid crystallization at these temperatures) for amorphous metals, it was noted that the measured specific heat of the high-temperature

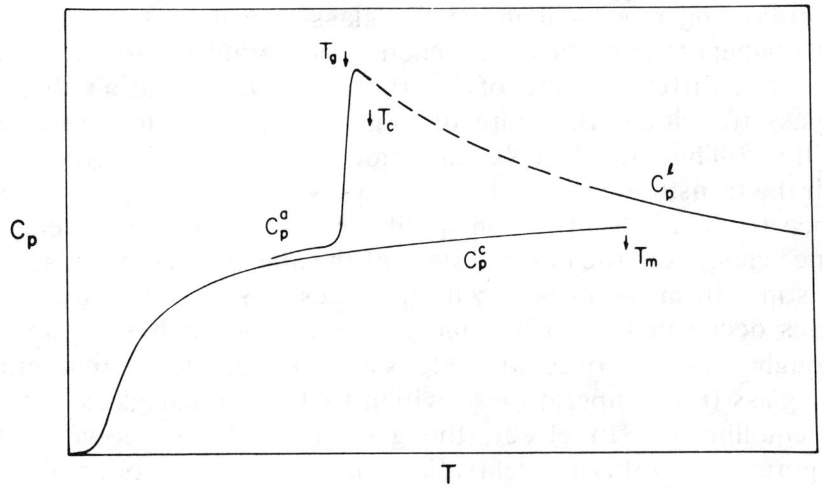

Fig. 3. Schematic representation of the relation between the specific heat of the crystal, $C_p{}^c$; that of the solid amorphous phase, $C_p{}^a$; and that of the liquid, $C_p{}^l$. Here T_m is the crystalline melting point; T_g is the glass transition temperature; and T_c is the temperature at which rapid crystallization occurs upon heating at a selected heating rate.

liquid branch extrapolates smoothly to that measured just above T_g for $Au_{77}Ge_{14}Si_{09}$ (17).

For ordinary glasses, the temperature region near T_g is also the region at which the glass begins to soften mechanically. Again, amorphous metals have also been shown to exhibit such behavior; at temperatures near T_g, a liquid-like (i.e., Newtonian viscous) flow was observed with a viscosity that changed by a factor of 10^4 within 20°C (18).

The observed glass transition effect is a purely kinetic phenomenon. As the liquid is cooled, its viscosity and relaxation time gradually increase to the point where the relaxation time is so large that internal equilibrium can no longer be maintained. Upon further cooling, the specific heat begins to decrease as the equilibrium configurations become inaccessible. This situation is reflected in the nomenclature, as mentioned before: above the glass transition region, the material is a supercooled liquid which is metastable with respect to crystallization, but is in internal equilibrium; below this region, it is a glass which is in a nonequilibrium metastable state; both materials are amorphous.

The temperature of the glass transition is not a constant of the material but is a function of experimental conditions; thus, slower (faster) cooling rates will move the glass transition temperature to lower (higher) temperatures. At each cooling rate the glass will thus freeze in a different state of internal energy, indicating that, like the glass transition, the state of a glass depends on experimental conditions. The "ideal glass transition temperature" is the one at which the transition occurs upon infinitely slow cooling.

Since the glass is in a nonequilibrium state, annealing can also produce changes in the internal state of the amorphous material which are distinct from the crystallization process discussed below. These changes occur increasingly rapidly as one approaches T_g and can be thought of, in first order at least, as lowering the fictive temperature of the glass (the temperature at which that particular glass structure is in equilibrium). In effect, the glass relaxes to a lower energy configuration; exothermic relaxations of this sort have been observed using a differential scanning calorimeter (19). Structural relaxations are discussed in detail in Chapter 6.

Many of the physical properties of amorphous metals—most importantly, the magnetic and mechanical behavior (20)—change upon annealing at temperatures and times that produce no detectable crystallization. This can present serious difficulties in attempting to characterize the physical properties of an amorphous metal, since the nature of the standard preparation processes sometimes leads to samples of uncertain thermal history. While mechanisms such as compositional segregation and stress relaxation may play roles in causing changes in physical properties, structural relaxation effects may also lead to such changes.

Being metastable, amorphous metals will crystallize with the proper combination of temperature and time. The crystallization is generally complex, proceeding in stages that often include intermediate metastable crystalline phases. As an example, Fig. 4 shows the TTT curve for $Pd_{80}Si_{20}$ (21), where four different stages can be observed.

When observed using a calorimeter, crystallization can appear equally complex, involving several distinct exothermic peaks. The total heat of crystallization is generally of the order of 40% of the heat of fusion of the alloy, the remaining enthalpy having been extracted from the sample during quenching of the liquid due to the liquid's much higher specific heat. However, in a multistage crystallization process, the first peak representing the crystallization of the amorphous phase to a metastable crystal may be smaller than the later peaks due to recrystallization and transformation.

The relative thermal stability of amorphous alloys varies widely with composition. Generally, those amorphous alloys which can be

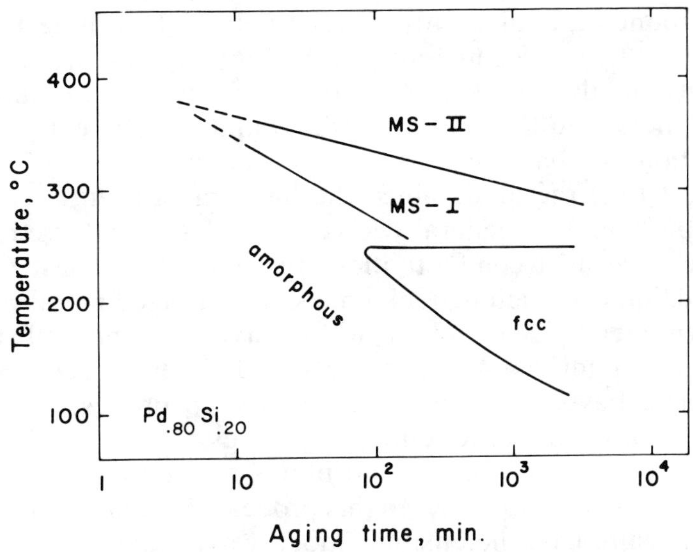

Fig. 4. TTT curve from annealing study of amorphous $Pd_{80}Si_{20}$ (21). MS-I and MS-II correspond to different metastable phases; the stable phase(s) would be obtained by annealing at higher temperatures.

produced from the liquid are the most stable relative to their melting points. These amorphous metals crystallize rapidly at temperatures of about 40 to 60% of their melting points; the temperature for rapid crystallization varies from room temperature to above 1000 K. For alloys made by deposition techniques, the crystallization temperature may vary from several degrees K for nominally pure materials to temperatures similar to those observed for liquid quenched materials.

The mode of crystal growth has also received attention. The very beginning of crystallization at temperatures well below T_g has been studied by TEM for many alloys. Often, isolated crystals embedded in a not noticeably changed amorphous matrix appear and then grow into the matrix. The crystal morphology depends on the alloy composition, with spherical, faceted and dendritic morphologies being observed (22,23). The observation of dendritic crystallization has been cited (22) as further evidence against a microcrystalline structure, for which one would expect to see a more homogeneous mechanism of grain growth.

Preparation by Rapid Liquid Quenching

With the exception of some especially stable compositions, most presently known amorphous metals can be obtained by quenching the corresponding liquids only at relatively high cooling rates (e.g., 10^5 K/sec or higher). This requires that any cooling process provide both a sufficiently high heat transfer coefficient at the interface of the liquid metal and the cooling medium and a sufficiently thin metal cross section so that the heat can be conducted out of the metal in the short time required. Since the heat transferred from the melt to gases and liquids is generally lower than that transferred to solids, in part because heat conductivities are generally highest for solids, most rapid liquid quenching techniques have utilized metal substrates; the molten metal is spread in a thin layer on this substrate and subsequently solidifies. A wide variety of different processes utilizing this principle have been developed, and such processes promise to be relatively low-cost production techniques.

The first preparation of an amorphous metal from the liquid was done by the gun technique (1). In this process, a diaphragm is ruptured by high-pressure gas; the ensuing shock wave travels down the tube to a crucible with a small hole in the bottom. The molten sample is held in the crucible by its surface tension before being driven out of the hole in the form of small droplets by the shock wave; the droplets then impinge on the metal substrate, spreading out and overlapping to form an irregular foil of variable thickness, which typically averages 5 to 25 μm. Some sections (i.e., the edges of some individual droplets) are much thinner (~1000 Å), and thus are transparent to TEM.

The gun technique provides the highest cooling rates of all liquid quenching techniques for the very thin regions of the foil. However, the irregularity of the foil thickness means that the sample contains regions produced with a wide range of cooling rates and thus may contain different structures.

Another group of discontinuous processes can be classified as "hammer and anvil" processes, wherein a molten droplet is quenched to a foil between two metallic plates which are propelled together at high speed. In a very useful variation of this technique, a hammer is used to quench an arc melted droplet against the water cooled metal substrate supporting the droplet, thus avoiding crucible problems and making possible the quenching of alloys with melting points up to 2000°C (24) and above. Regular foils of 20 to 50 μm thick are generally produced, but the quenching may be somewhat unreproducible due to alignment problems.

Because liquid quenched amorphous metals were first made by these methods (i.e., using a low-yield, discontinuous process and producing an irregular, often inhomogeneous product), it had seemed that these materials might remain only laboratory curiosities. A major advance came about as more continuous production processes were developed. These generally produced more reproducible, dimensionally regular samples and thus aided the measurement of physical properties, especially mechanical properties. Further, they enhanced the potential for technological exploitation of such materials.

The twin roll technique (25) was the first potentially continuous process to be applied to metallic glasses. In this technique, molten metal is fed into the nip of two rapidly rotating rollers; the product is in the form of ribbon or sheet. Although this process has the advantages of providing quenching from two sides and good surface contact, it has generally been found hard to control. It is difficult to keep the liquid from either solidifying too early (i.e., before the minimum separation between the rolls occurs), resulting in cold worked strips, or solidifying too late (i.e., leaving the nip partially liquid).

Another, widely used class of processes is known as melt spinning. In these techniques, a stream of molten metal is directed at a rapidly moving substrate; the final product generally is in the form of ribbon 20 to 100 μm thick. The application of melt spinning to production of amorphous metals was first reported in 1969 (26), though processes similar in principle had previously been disclosed in the patent literature. Quenching on the inside or outside of a rapidly rotating cylinder now appears to be a widely utilized technique; production rates of 2000 m/min are readily obtainable.

In another class of processes, called melt extraction, a rapidly rotating quench surface (generally a disc) is used to extract a filament from a stationary melt. In the melt extraction process (27), a spinning disc touches the top surface of a melt held in a crucible; a filament of the rapidly quenched material forms on the edge of the disc. Advantages of this technique are that orifice problems sometimes encountered in melt spinning can be avoided and that a variety of filamentary cross-sectional shapes can be produced. To date, however, the technique has been used mostly to produce rapidly quenched crystalline metals, and it is not clear how useful it is for preparing amorphous metals since the quench rate may be lower than that attained in melt spinning (28).

In the pendant drop process, the disc solidifies a filament from a molten drop held at the tip of a rod (29). Obviously, this totally removes the crucible problem which may still exist with the melt extraction process, especially for alloys having very high melting points.

In a recently developed technique referred to as "laser glazing," a laser beam is used to melt a thin layer on the surface of the sample (30). Relative motion of the sample and the beam removes the molten section from within the beam; the liquid is then quenched by the solid material beneath it. When used to produce amorphous metals, this process will probably involve processing a layer of glass forming composition which is placed on top of another alloy, since alloys known to form glasses at the cooling rates characteristic of this process do not generally have advantageous properties in the crystalline state. In this case, a two-step process would be necessary—for example, plasma spraying to deposit the surface layer followed by "laser glazing."

In addition, a process for forming metallic glass wire has recently been developed; this process, and some characteristics of melt spinning, are discussed in Chapter 2 of this book.

Preparation by Other Techniques

A variety of atomic deposition techniques have been utilized to form amorphous metals (8). These techniques are very powerful tools in the study of amorphous metals since they have higher effective cooling rates than liquid quench processes and thus present the potential for retention of phases with considerably higher free energy excess than the equilibrium phase(s). Hence, many compositions can be produced thereby with amorphous structures which cannot be obtained by liquid quenching.

Sputtering has been an especially valuable technique for the production of amorphous metals. Many variations on the processing conditions are possible for a given composition, and these parameters then control the structure and properties of the film which is obtained (8). These include the sputtering rate, the nature (d.c. or r.f.) and magnitude of the substrate bias, the over-all geometry of the system, and the substrate temperature. Of equal importance is the composition and pressure of the atmosphere supporting the plasma; significant amounts of gases, including the Ar or Ne generally used as the primary bombarding ions as well as more reactive gases present as impurities or as "alloying" additions, may be present in the film and will affect its properties. For such films, "intrinsic" magnetic, electronic and thermal properties are in fact difficult to obtain, because it is difficult to separate the effects on properties of small amounts of impurities from more macroscopic effects (e.g., internal stresses).

The recent development of high-rate sputtering (31) makes the use of sputtering more feasible for other than thin film applications;

deposition rates of the order of 1 μm/min are readily attainable.

Vacuum evaporation, wherein thermally vaporized atoms are condensed on a substrate, is a relatively simple means of producing amorphous metals (8). As with sputtering, the deposition rate, atmospheric composition and pressure, and substrate temperature are important parameters in determining the character of the metal film. The control of alloy composition in the deposit is difficult when only one molten alloy source is used, due to differing vapor pressures; thus, an independently controlled source for each of the alloy components provides the best compositional control. When the substrate is held at very low temperature, vacuum evaporation is the most powerful technique in respect to the ability to produce an amorphous phase. Even nominally pure metals such as Co and Cr have been produced as thin films with amorphous structures by this technique (32), though they may be stabilized at least in part by the inclusion of small amounts of elements from the residual gases present even in ultrahigh vacuum; crystallization temperatures for such high-purity samples generally are low—e.g., ~50 K for Cr (32).

Deposition from baths, both electrolytic (33) and electroless (34), has also been often used to produce amorphous metal alloys. However, the method suffers from the limitation resulting from the need of designing a bath which can provide alloys of the desired composition; to this date, most such alloys have been combinations of Fe, Ni and/or Co with P, B or S. Above a certain metalloid level, amorphous metals are readily obtained; electrodeposited Ni-S alloys reported in 1930 by Brill (35) were probably amorphous.

Cooling Rates

With all of the liquid quenching techniques, the magnitude of the cooling rate is of crucial importance in determining whether or not a specific amorphous metal can be produced, as discussed in a subsequent section. There have been both theoretical and experimental attempts to evaluate the cooling rates achieved in various processes (7); however, the short time periods, small dimensions and possible instabilities inherent in most quenching processes make it difficult to determine an accurate value of the actual cooling rate.

At first sight, the calculation, from heat flow relations, of the expected cooling rate for a thin molten layer spread on a conducting substrate would appear to provide the desired values, given the initial temperatures and thermal properties of the substrate and sample and the thickness of the molten layer. However, the calculation becomes much more complex if the interface between the molten layer and the substrate is considered; the interface can inhibit the heat flow because of factors such as the presence of oxide layers and imperfect contact (e.g., the entrapment of pockets of gas between the molten layer and the substrate). In principle, these effects can be approximated by including the heat transfer coefficient, h, of the interface in the calculation. However, a problem is that for rapid quenching processes one can generally determine the heat transfer coefficient only from an experimentally observed cooling rate—i.e., the quantity which one seeks to calculate.

Alternatively, cooling rates can be estimated directly by measuring the dendrite arm spacing or eutectic interlamellar spacing in rapidly quenched crystalline samples of appropriate compositions (7), but the calibration of the relationship between these quantities and the cooling rates of interest for the production of amorphous metals is still uncertain. However, while not necessarily providing correct absolute cooling rate values, such measurements do provide an effective means of comparing different quenching techniques, different process parameters or different quenching facilities using a supposedly identical process.

For a molten layer placed on a metal substrate, extensive numerical calculations exploring the cooling rate as a function of heat transfer coefficient and splat thickness—the two most important process variables in a given liquid quenching process—have been reported (7,36). The following discussion is based on these investigations.

In one limiting case, the interface provides no barrier to thermal conduction (i.e., there is perfect contact); in this case, the heat transfer coefficient is infinite, and this condition is called ideal cooling. For a 10-μm-thick layer of liquid iron on copper, a cooling rate of $\sim 10^8$ K/sec is predicted for cooling to one-half of the initial temperature. The cooling rate decreases markedly, of course, as the temperature of the layer approaches that of the substrate. For ideal cooling, the average cooling rate to half temperature scales inversely with the square of the layer thickness.

In the other extreme, h is quite small; cooling is then completely interface controlled, and this condition is called Newtonian cooling. For the same conditions selected above for ideal cooling, but with $h = 1$ cal/cm$^2 \cdot$K$\cdot$sec, the cooling rate to half temperature is of the order of 10^6 K/sec—two orders of magnitude lower. For Newtonian cooling, cooling rates are inversely proportional to the layer thickness.

The transition regime is known as intermediate cooling and is

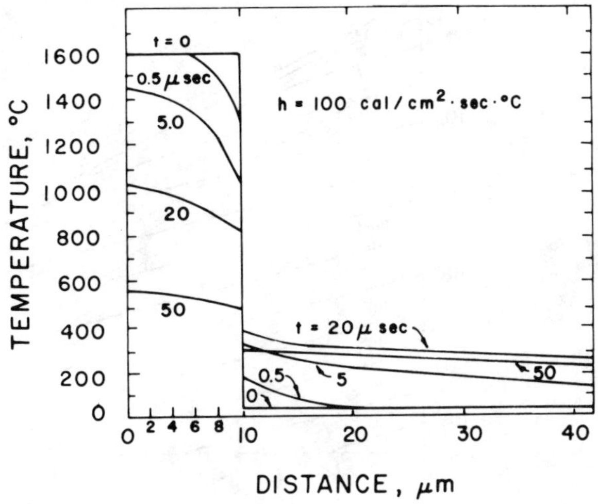

Fig. 5. Temperature profile as a function of time for a 10-μm-thick iron layer, initially at 1600°C, placed onto a copper block which is initially at room temperature (36)

generally the condition encountered in practice. Figure 5 (after Ref 36) shows the temperature profile as a function of time for the same conditions given above when $h = 100$ cal/cm$^2 \cdot$K$\cdot$sec. Here, the cooling rate is reduced only by a factor of about 2 from ideal cooling. As with ideal cooling, significant heating of the substrate occurs; this can be of importance in the case of quenching to a glass which exhibits crystallization or annealing effects at relatively low temperatures.

In general, the type of cooling which occurs is determined by the Nusselt number, $N = h \cdot s/K$, where s is the layer thickness and K is the thermal conductivity of the splat. Ideal cooling occurs for $N \rightarrow \infty$. Because N is proportional to the product of h and s, it becomes more difficult to achieve ideal cooling as s decreases.

Calculation of the cooling rate encountered in practice is further hampered by the fact that actual processes involve a transient period during which the liquid is being spread to its final thickness; for high quench rates, significant cooling occurs during this time period.

Figure 6 (after Ref 7) illustrates the relationship between splat thickness and cooling rate for different values of h, expressed in W/m$^2 \cdot$K [10^5 W/m$^2 \cdot$K = 2.4 cal/cm$^2 \cdot$K$\cdot$sec; line (a) is for ideal cooling; line (b) is an approximation to the upper limit for Newtonian cooling.] The data points are from direct measurements using the thermoelectric effect or optical pyrometry. It is seen that intermediate cooling generally is encountered under typical experimental conditions and that, for those thicknesses most typical of melt spinning processes (~20 to 40 μm), a cooling rate of about 10^6 to 10^7 K/sec is to be expected.

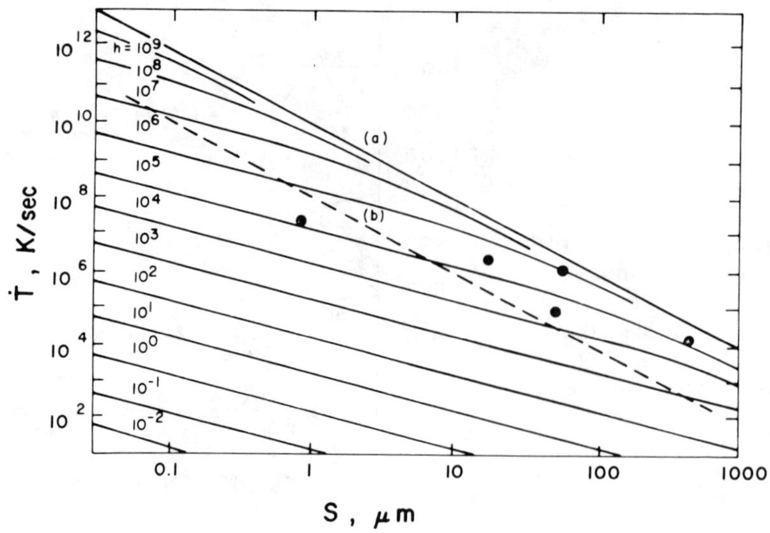

Fig. 6. Cooling rate as a function of splat thickness, s, and heat transfer coefficient, h. Data points summarize experimental data (7)

Criteria for Easy Glass Formation and Glass Forming Compositions

Because readily prepared metallic glasses (i.e., glasses prepared at cooling rates $\dot{T} < 10^7$ K/sec) are the prime topic of this book, we will focus on these glasses here, omitting those requiring ultrahigh cooling rates or atomic deposition. First we will give a brief summary of present theories of glass formation and their implications for glass forming tendency (GFT), followed by a listing of classifications and compositions of amorphous metals and, last, a listing of alloy chemical factors affecting GFT. Some of these topics have been reviewed recently (37,38).

Cohen and Turnbull (39) proposed that every liquid would form a glass upon cooling if crystallization into one or more phases did not intervene. However, experience suggests that crystallization will always occur in equilibrium solidification, and that there are no glass formers in an absolute sense but only for certain experimental conditions where the critical values are determined by the kinetic parameters of the system. We mention here also that Turnbull expressed the conjecture that there are some alloy compositions of which the amorphous state has a lower enthalpy than any crystalline *single* phase and that a glass must form if the system is constrained against crystallization into the equilibrium phase mixture (40).

Thermodynamic conditions that favor the liquid relative to the crystalline solid(s) and kinetic conditions that inhibit crystallization both enhance GFT. This sets the stage for a search for conditions favorable to glass formation and retention. These are of three, interrelated types: thermodynamic, kinetic and processing factors; they can be expressed in terms of several parameters.

Some relevant thermodynamic quantities are the melting temperature, T_m (generally the liquidus temperature); the heat of vaporization, ΔH_v; and the free energies of all phases actually and potentially participating in the solidification process, such as the melt and stable or metastable elemental and alloy phases (9). Kinetic parameters are the melt viscosity, η; the glass transition temperature, T_g; and the homogeneous nucleation rate, I_n (38,41). Of course, these parameters are dependent on energetic characteristics of the system such as bond strength, directionality of bonding, etc. Processing parameters under complete or partial experimental control are the cooling rate, $\dot{T}$ (which is, as discussed above, a function of parameters such as foil thickness, s); the supercooling temperature interval, $\Delta T_s = T_m - T$; and the heterogeneous nucleation rate (which is a function of melt purity and quenching surface preparation).

A simple parameter that has been found especially useful for expressing GFT is the reduced glass transition temperature, $T_{gr} = T_g/T_m$, the ratio of the glass transition temperature to the melting temperature. High values of T_{gr} obviously favor glass formation; values of $T_{gr} > 0.45$ have been observed for presently known metallic glasses (38). The highest observed values of T_{gr} are near 0.65 (38). Several expressions that have been formulated for the reduced melting temperature (39,42) are listed below as part of the discussion of alloy chemical factors of GFT. The critical cooling rate $\dot{T}_c$ is an important processing parameter. It has been empirically related to T_{gr} (38); although there is wide scatter of data, a roughly monotonic increase of ln $\dot{T}_c$ with decreasing T_{gr} is seen. Theoretically, $\dot{T}_c$ is linked to T_{gr} by the application of Uhlmann's equation discussed below.

Systems That Readily Form Glasses

Although classifications of compositions that readily form glasses on a crystal chemical basis have not yet progressed very far, one can outline some major categories of such alloy systems (Table 1). A list of glass forming compositions has been presented by Takayama (37), and a periodic table representation of glass forming elements has been published by Masumoto and Maddin (21).

Table 1. Alloy Systems That Form Metallic Glasses, Based on Chemical Classification of Constituents

Group(a)	Representative systems	Typical composition, at.%	Preparation method(b)
Major Systems			
T^2 (or noble) metal + metalloid (X)	Pd-Si, Co-P, Fe-P-C, Ni-P-B	15–25 X	LQ, ED
T^1 metal + T^2 metal (or Cu)	Zr-Cu	30–65 Cu	LQ, SP
	Y-Cu, Ti-Ni	30–40 T^2,Cu	LQ
	Nb-Ni, Ta-Ni	40–65 Ni	LQ, SP
Miscellaneous Systems			
A metal + B metal	Mg-Zn	25–35 Zn	LQ
T^1 metal + A metal	(Ti-Zr)-Be	20–60 Be	LQ
Actinide + T^1 metal	U-V, U-Cr	20–40 T^1	LQ
"Thermally Stable" Amorphous System Not Formed by Liquid Quenching			
T^2 metal + rare earth metal	Co-Gd(-Mo)	~20 Gd	SP

(a) A metal = Li, Mg groups; T^1 metal = early transition metal (Sc, Ti, V groups); T^2 metal = late transition metal (Mn, Fe, Co, Ni groups); B metal = Cu, Zn, Al groups; metalloid = B, C, Si, Ge, P. (b) ED = electrodeposition; LQ = liquid quenching; SP = sputtering.

Based on the currently known number of their representatives, there are two major groups of systems that readily form glasses ("Major Systems" in Table 1); groups less well delineated and of lesser importance are summarized under "Miscellaneous Systems" in that table. Further, a type of "thermally stable" amorphous system is listed that requires an atomic deposition process for its preparation but that is of technical importance.

The first group is referred to as metal-metalloid systems, $T^2{}_{1-x}X_x$, where T^2 is Mn, Fe, Co, Ni, Pd, Au or Pt and X is B, C, Si, Ge or P, generally with compositions in which x = 0.15 to 0.25. Here both T and X can be combinations of elements from each group; this often increases GFT for reasons discussed below. A good example of this effect is $Fe_{80}P_{13}C_{07}$, which is an alloy with substantially improved ease of formation compared to binary Fe-C or Fe-P (43). Limited amounts of aluminum (and possibly other B-metal elements), as well as early transition metals, can be added to systems of this type.

Systems that have been widely studied are Fe-based—e.g., Fe-B (44), Fig. 7(a)—and Pd-Si (45,46), Fig. 7(b) (47). It is seen that each falls into the region of a narrow, deep eutectic, at approximately the same composition for each system, giving rise to models explaining their occurrence both on structural and electronic grounds (48,49).

The second grouping in Table 1 is referred to as intertransition metal systems, $T^1{}_{1-x}T^2{}_x$, where T^1 is an early transition metal, such as Fe, Co, Ni, Rh or Pd, and Cu; here x = 0.3 to 0.65, depending on the system. The T^1-T^2 systems appear to fall into three subgroups. In some systems, such as Zr-Cu (Fig. 7c), a wide glass forming composition range is accessible by liquid quenching (50). In the second subgroup, glasses form preferentially at T^1-rich compositions; an example is the Ti-Ni system, where glasses can be formed between 30 and 40 at.% Ni (51). In the third subgroup, glasses are retained more readily at T^2-rich compositions, as for Nb-Ni shown in Fig. 7(d) (16); this is even more pronounced for Ta-Ni (52). It is possible that there are no fundamental differences among these three subgroups and that the compositional differences in GFT result only from phase diagram features such as high melting temperatures for alloy compositions of low GFT. There are, however, characteristic crystal chemical differences among the systems representing these types: in the Ti-Ni system there is a fairly stable CsCl-type phase; occurrence of a phase with this simple type of structure generally inhibits glass formation. By contrast, the Nb-Ni diagram (Fig. 7d) is characterized by a Frank-Kasper phase μ of complex structure near the equiatomic

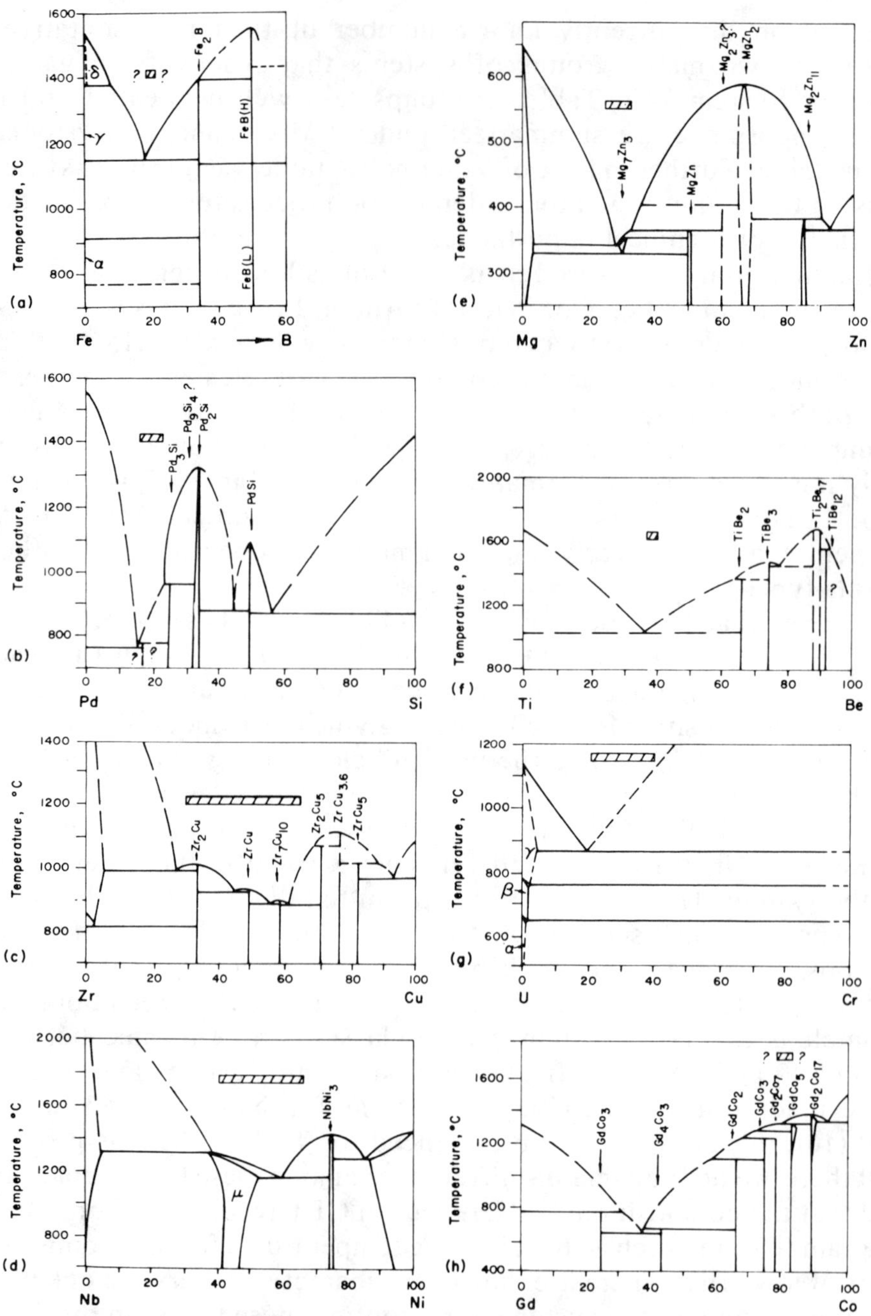

Fig. 7. (a) to (g) Constitution diagrams (47) representative of binary alloy systems which readily form glasses; composition ranges of glasses obtained by rapid liquid quenching are indicated. **(h)** Constitution diagram (47) of Gd-Co system, showing composition range of amorphous alloys formed by sputtering.

composition (see below) with low melting point, while at higher Nb (or Ta) contents the steep rise of the liquidus will rapidly reduce T_g/T_m. By contrast, Zr-Cu combines low melting points over a broad composition range with the presence of complex intermetallic phases.

There are several other T^1-T^2 alloy systems, as well as systems formed by combining a rare earth or actinide metal with a T^2 metal, that have not yet been explored and that can be expected to yield glasses. Thus, in a survey of Zr systems (53) additional glasses were found; superconducting amorphous Zr-Rh alloys were recently reported (54), and some Y-rich Y-T^2 glasses have been prepared (55).

The glass forming systems listed in Table 1 as miscellaneous systems were reported only recently. In the Mg-Zn system (Fig. 7e), glasses were obtained in alloys with 25 to 35 at.% Zn—i.e., bracketing the deep eutectic (56). This was the first amorphous metal, prepared from the melt, that consisted only of "simple" metals (i.e., did not contain either transition metals or metalloids); further, the constituents are isovalent.

A technically interesting series of amorphous alloys combining Ti, Zr, or Ti-Zr with Be (Fig. 7f) is commercially available (57); besides their remarkable mechanical properties (see below), these alloys are constitutionally of interest because the region of glass formation widens from only 4 at.% Be in the binary system Ti-Be to ~40 at.% Be (at constant Ti/Zr ratio) in the ternary system.

Melt spun amorphous U alloys were recently described (58); the T^1 metals V and Cr were added at levels of 20 to 40 at.% to produce glasses (Fig. 7g). The ability to form glasses despite the small melting point depression at the eutectic (especially for U-V) indicates a propensity of U and possibly other actinides for glass formation which may be related to the participation of f-electrons in their bonding.

As a final category, we list Co-Gd, a technically interesting alloy formed by sputtering, as an example of alloys which are "thermally stable" well above room temperature but have not been formed by liquid quenching (Fig. 7h). The high melting points (relative to the cohesive energy) make it unlikely that these alloys could be prepared as glasses from the melt. Binary compositions of interest for magnetic bubble domain memory devices center around 80 at.% Co; Mo may be added at levels between 10 and 20 at.% to modify the magnetic properties (59).

In concluding this section we observe that one can change the present, chemical classification of alloy glasses into one based on glass forming composition ranges; based on the extent of that range, nearly all systems belong approximately to one of the following groupings: 15 to 25%, 25 to 35% and 30 to 70%. The significance of these ranges is not yet clear; they must be interpreted in terms of the criteria outlined in the next section.

Alloy Chemical Factors Affecting Glass Formation

As discussed above, glass formation at a given cooling rate requires that T_{gr} exceed some critical value. The effect of kinetic parameters is well represented by a model proposed by Uhlmann (60), which is based on the theories of homogeneous nucleation and crystal growth and permits the construction of TTT curves for each glass forming composition, using the melt viscosity, η, and a site fraction, f, as partly adjustable parameters. This construction permits calculation of $\dot{T}_c$ required to avoid formation of a measurable amount of crystalline alloy in the melt during cooling (38,61). $\dot{T}_c$ depends on T_{gr} via η; because the kinetic theory underlying this approach is discussed in detail in Chapters 5 and 6, we will focus instead in the following on T_{gr} and consider the alloy chemical factors affecting T_g, T_m and glass formation in general.

As mentioned in an earlier section, T_g varies somewhat but not drastically with experimental conditions. It appears to vary smoothly with composition in binary systems; however, as T_g was found to decrease with increasing Zr content in Cu-rich Zr-Cu glasses despite a concomitant increase of the cohesive energy of these alloys, its change with composition cannot always be monotonic (62). Fundamentally, T_g has been related to a potential barrier $\Delta\mu^*$ for cooperative atomic transitions and hence to the slope of the repulsive potential (37). Davies has related T_g to the cohesive energy (38); however, while this useful relation can serve as a first approximation, it shows too much scatter to stand without modification. Although a comprehensive theory of T_g in terms of more fundamental parameters (e.g., elastic or anelastic constants) still has not been presented, T_g probably is not primarily responsible for the wide variations of GFT with composition (63).

Upon focusing then on low T_m values, theories of glass formation become identical with theories of the formation of stable liquid alloys; however, unlike the theory of stable crystalline alloys, this field has not yet really advanced beyond enlightened empiricism (64). Based on the empirical evidence discussed above, a number of alloy characteristics have been proposed which appear to be more or less associated with GFT. These characteristics are listed in Table 2; they are not independent, but will be seen to be highly correlated.

The primary, phenomenological requirement of a low "relative" T_m for a glass forming alloy has led to the introduction of several expressions: (*a*) A reduced melting temperature, $\tau_m = kT_m / \Delta H_v$; here, ΔH_v, the heat of vaporization, can be understood as a measure of the resistance of the system against the atomic motions required for equilibration (39). GFT increases with decreasing τ_m. (*b*) The reduced melting point depression $\Delta T_{mr} = (\bar{T}_m - T_m)/\bar{T}_m$, where $\bar{T}_m$ is the weighted average of the melting points of the components. GFT should increase with increasing ΔT_{mr}. (*c*) The reduced melting point depression $\Delta T_{mr}{}^{id} = (T_m{}^{id} - T_m)/T_m{}^{id}$, which measures the depression of T_m from $T_m{}^{id}$, the hypothetical liquidus temperature given by ideal solution theory (65); again, GFT and $\Delta T_{mr}{}^{id}$ should scale together.

It is noted that, in some systems, very low melting points occur only in a very narrow composition range—e.g., for Pd-Si (Fig. 7b) or Au-Si. It was thought initially that equilibrium solidification at these low-melting compositions of easy glass formation required the nucleation and growth of phases differing in composition from that of the homogeneous eutectic liquid, providing a constraint against equilibration and enhancing GFT; however, it has since been realized that even in systems with very low-melting eutectics—e.g., Pd-Si (66), Au-Si (67) and Mg-Zn (47)—there are stable or metastable intermetallic compounds at or within a few percent of the eutectic composition. As for the glass forming T^1-T^2 compounds such as Zr_7Cu_{10}, there can thus be no thermodynamic stabilization of the melt with respect to a single-phase solid (40), and the high GFT at these compositions is essentially a kinetic effect.

In T-T metal systems such as Zr-Cu, on the other hand, GFT is high over a broad composition range, and it has been shown (41) that it is not possible to assess GFT in such a system using the departure from an ideal solution (65), since the calculated liquidus curves fall below the actual melting points of glass forming intermetallics in the Zr-Cu system.

Several atomic properties of the constituent elements of glass forming systems have been connected with the GFT. Size difference appears to be of major importance; most alloys readily forming glasses have nominal component size ratios [as measured by $r_1/r_2 = (\bar{V}_1/\bar{V}_2)^{1/3}$, taking molar volumes in the metallic state (68)] appreciably different from unity (9,37).

For the liquid quenched glasses in Table 1, r_1/r_2 (where r_1 is

Table 2. Alloy Chemical Criteria for Easy Glass Formation

Property	Characteristic parameter	Examples	References
Phase Diagram Characteristics (Low Liquidus Temperatures T_m)			
Narrow low T_m region (single deep eutectic)	$\Delta T_{mr} \sim 0.41$(a)	Pd-Si, Mg-Zn	39, 45, 56
Low T_m over extended composition range	$\Delta T_{mr} \sim 0.33$(a)	Nb-Ni, Zr-Cu	16, 50
Atomic Properties of Constituents			
Size difference	$r_1/r_2 = 0.79$–1.41	Fe-B, Pd-Si, Zr-Cu	9, 37, 50
Valence difference	$\Delta n = 4$–5 el/atom	Pd-Si, Nb-Ni	...
Electronegativity difference	irregular	...	37
Position in periodic table	...	Ni-P, Zr-Cu	...
Compositional Properties			
Stoichiometry and compositional ordering	A_4B, A_2B	Pd-Si, Mg-Zn	69, 72, 75, 78
Valence electron concentration (Nagel-Tauc theory)	VEC~1 el/atom VEC~5-7 el/atom	Pd-Si Nb-Ni	79 81
Complex equilibrium phases	Frank-Kasper phases, Fe_3C-type phases	Nb-Ni, Mg-Zn, Pd-Si	9, 70, 82
"Confusion principle"	...	Fe-P-C, Fe-Ni-P-B	43

(a) $T_{mr} = (\bar{T}_m - T_m)/\bar{T}_m$ is the normalized depression of the melting point from the weighted average.

the radius of the majority atom species) ranges from 0.79 (for Cu-rich Zr-Cu) to 1.41 (for Y-Cu), with the values for most glasses lying between 1.15 and 1.25. For some metal-metalloid glasses, however, nominal size ratios near unity occur—e.g., for Pd-Si (9). A better measure of the size ratio is its value in a crystalline phase of similar composition (69) or, when available, its value in the amorphous alloy; this may differ from the ratio calculated from the elemental values due to charge transfer and/or differing coordination numbers for the components in the glass or intermetallic phase. A recent analysis of the partial distribution functions of $Pd_{80}Si_{20}$ using x-ray and neutron diffraction (70) yielded Pd-Si distances of 2.35 Å and Pd-Pd distances of 2.90 Å, indicating a substantial size difference of the components in the glass; crystalline Pd_3Si shows a similar size relation (71).

Thus, it appears that an appreciable size difference ($r_1/r_2 < 0.88$ or > 1.12) between the components in the glassy alloy is a necessary requirement for ready glass formation; no counterexample is known among the several dozen readily formed glasses examined to date. Very recent data on A-metal + A-metal glasses (72) suggest that the size factor may be the only factor active in some metallic glasses, somewhat analogous to its role in some crystalline Laves phases. It had been suggested (69) that, at least in the case of metal-metalloid glasses such as Pd-Si, such size differences make possible a random packed structure with a packing density much higher than that possible for dense random packing of spheres of one size (73); this high packing density, together with the proposed compositional ordering (69), can result in an amorphous structure which is relatively stable to the fluctuations necessary for crystallization to occur.

Valence differences may also be important in promoting glass formation. In T^2-X and in T^1-T^2 diagrams, fairly large differences ($\Delta n = 3$ to 5 and 5 to 6, respectively) between the valences of the components are observed. These large differences may simply be coincidental with the size differences discussed above, but they are probably of significance by themselves. Thus, for T^2-X alloys, the conventional picture of charge transfer taking place on alloying requires the transfer of metalloid s/p electrons into d-orbitals of the transition metals, as confirmed for the corresponding glasses by a reduction in the magnetic moment of the metal atom (74); for T^1-T^2 metals, d-electron transfer from T^2 antibonding orbitals to T^1 bonding orbitals is assumed to take place. In any case, the valence differences are not necessary for glass formation, as shown by the example of Mg-Zn, in which both components are divalent.

The role of electronegativity differences (37) is more uncertain.

The differences between the Gordy electronegativities χ of glass components (68) are generally small ($-0.3 < \chi < 0.2$) and vary irregularly; thus $\chi_{Pd} > \chi_{Si}$ but $\chi_{Ni} < \chi_P$. As for crystalline alloys, the electrochemical factor derived from the values for the elements may often not be meaningful for small differences of χ.

The qualitative criterion related to position in the periodic table may be stated thus: Elements promoting glass formation with a given element tend to be located in different sections in the periodic table—i.e., in sections such that the differences of relevant properties between the components are sufficiently large but not too large. It is not known whether this criterion could be adequately quantified—e.g., analogously to a Darken-Gurry plot (68); however, it is sometimes found to be qualitatively useful in searching for a new amorphous metal.

The composition dependence of GFT suggests that composition-dependent factors are effective in most systems. We list first compositional ordering which has been most frequently considered for T^2-X glasses. Noting the deep eutectics present in some of these systems, Hume-Rothery and Anderson had already attempted to connect melt stability with ideal, quasistoichiometric compositions such as A_6B, A_3B, etc. on the basis of an assumed icosahedral coordination of liquid atoms around a reference atom (75). Chemical ordering has been an important consideration in some structural models (69). Chen and Park have ascribed such ordering to covalently directed T^2-X bonds (76), while Turnbull has associated the relative stability of binary alloy glasses with the softness of the repulsive potential between pairs of atoms of different species, which would favor the liquid and hence the glass by lowering the energy of crystallization (77). Recently, the short-range order of the liquid, as expressed by the absence of minority atom contacts in T^2-X glasses, has been treated theoretically using a semiquantitative pseudopotential argument (78). It is pointed out that such ordering, which results in minimization of the number of first-nearest-neighbor contacts between like atoms, is also a common characteristic of crystalline intermetallic phases.

Favorable values of the valence electron concentration (VEC) have been correlated with the GFT by Nagel and Tauc (49), somewhat in analogy to the corresponding argument for crystalline Hume-Rothery phases. In their approach, GFT is at a maximum at the composition where $2k_F$, the "spanning vector" of the Fermi surface, equals q_p, the maximum of the structure factor, determined by the atomic structure of the glass. When this is the case, the density of states D(E) is at a minimum at the Fermi energy and the amorphous state at this composition has a reduced energy. Presence of a kinetic barrier against fluctuations leading to crystallization has also been inferred (79). Reduced $D(E_F)$ values, as required by this theory, have been observed by ultraviolet photoelectron spectroscopy (UPS) for Pd-Si glasses (80) and were also recently found for amorphous Nb-Ni alloys (81). Systematic UPS measurements on glasses at compositions of high and low GFT would be required to confirm this theory; however, the occurrence of metallic glasses in preferred VEC ranges—e.g., for T^1-T^2 glasses—suggests an electronic contribution to the GFT. Weaknesses of this theory are that it relies on the rigid band model and that it does not express the large qualitative differences in GFT between elements of the same valence.

Another indication of a valence electron effect on GFT comes from a survey of the crystalline equilibrium phases found at the compositions of many metallic glasses. These phases tend to be of complex structure, involving ranges of coordination numbers and interatomic distances, such as Frank-Kasper phases of the σ, μ or M types (82), the Fe_3C (cementite) type, or others (9,82). Thus, at or near the glass forming $Mg_{70}Zn_{30}$ eutectic composition there is a complex orthorhombic phase with about 150 atoms per unit cell (83). The occurrence of Frank-Kasper phases is known to depend strongly on the average VEC; VEC values of 6 to 7 are favorable for both Frank-Kasper phases and metallic glass formation.

Although it is an argument clearly containing a kinetic component, we list here the "confusion principle" often taken recourse to by designers of metallic glass compositions. According to this concept, GFT is enhanced at compositions of low, temperature invariant melting point where the equilibrium alloy consists of as many crystalline solids as is possible and compatible with alloy design, such as a ternary eutectic or, at least, a two-phase composition in a ternary system involving a binary solid solution. An example is the ternary Fe-P-C glass which consists in pseudo-equilibrium of Fe, Fe_3C, and Fe_3P (considering the metastability of Fe_3C) (43).

Potential Applications

Amorphous metals constitute, in effect, a totally new class of materials with physical properties which are distinctly different from those of the corresponding crystalline alloys. While these properties are thus of basic scientific interest, it has also become clear that selected amorphous alloys possess various properties which make them candidates for technological applications. Indeed, one might hazard to say that a significant fraction of the recent work in this field has come about because of these potential applications; the last several years have witnessed concerted efforts to design both the alloys and the processing techniques to optimize the desired properties, as discussed below.

Because of the extensive coverage of mechanical and magnetic properties in subsequent chapters, we will not discuss them in detail here and instead will limit ourselves to a short discussion of characteristics as they pertain to potential applications; some chemical aspects also will be touched upon.

The mechanical properties of the liquid quenched amorphous metals have been extensively studied, as reviewed in Chapters 7, 8 and 9 of this book. It was recognized from the initial tensile-testing data on Pd-Si alloys that amorphous metals could exhibit exceptionally high strength (84); in addition, they were unexpectedly found to possess high compressive (85) or bending ductility. This behavior was also demonstrated for Fe and Fe-Ni based alloys (86). Further alloy development has led to the $Fe_{80}B_{20}$ alloy, which has ultimate tensile strength (UTS) of 370 kg/mm^2 (44)—the highest yet reported for a metallic glass. An alternative development approach has led to alloys exemplified by $Ti_{50}Be_{40}Zr_{10}$, possessing a UTS of 240 kg/mm^2 and density of 4.1 g/cm^3 (57), thus the highest specific UTS yet reported for a metallic glass. While it has been demonstrated that the yield strength of a metallic glass scales with its elastic modulus (87), further work needs to be done to understand more completely the compositional dependence of strength. A promising start in this direction has been made in a study of the microhardness, fracture stress and crystallization temperature of amorphous alloys based on $Fe_{80}P_{13}C_{07}$, where the iron is partially replaced by a second metal. Figure 8 shows the observed compositional dependence of the microhardness and crystallization temperature (88). It was noted that the concentration dependence of the microhardness and the crystallization temperature had "very similar behavior" with concentration; a correlation between the mechanical and thermal behavior and the average outer electron concentration of the metallic atoms was noted (88). However, the initial slopes do not scale with the number of outer electrons of the metal additive. Our further analysis of the data suggests that a qualitatively better fit is obtained by relating the initial slopes of the property changes to the band broadening term arising in calculations of the cohesive energies of the addition elements.

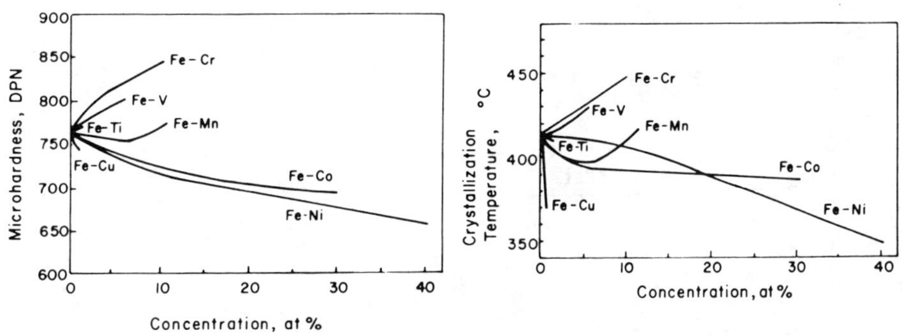

Fig. 8. Microhardness as a function of composition when indicated element replaces Fe in $Fe_{80}P_{13}C_7$ (88)

Since at least one dimension of liquid quenched amorphous metals must be small in order to achieve the necessary cooling rates, filamentary shapes are most readily produced. Considerable effort has been devoted to developing processing techniques which provide as-formed filaments that are sufficiently uniform in properties and cross section. Some of the considerations involved in the formation of rapidly quenched wires and ribbons are discussed in detail in Chapter 2.

A second characteristic of amorphous metals which has attracted major attention is the soft ferromagnetism of Fe-Ni-Co based metallic glasses. The lack of directional anisotropy inherent in an amorphous structure can result in very high permeabilities and low energy losses; properties similar to those exhibited by Permalloy have been obtained (Chapter XI in Ref 89). Thus, these materials may find application where soft magnets are required (e.g., transformers). Once again, efforts are under way to optimize the composition and production process, especially where wider sheet material would be desirable, for this potential application.

Another potential application of amorphous metal films based on their magnetism that is presently under intense study is their use for magnetic bubble devices. It has been demonstrated that Gd-Co films with Mo additions can be made with the desired combination of magnetic properties (59). Generally, such films are prepared by sputtering, and the anisotropy of the film is related to the deposition parameters (59).

A third area of interest involves the corrosion behavior of amorphous metals. The chemical homogeneity, and lack of grain boundaries and line defects such as dislocations, inherent in amorphous metals suggest that superior corrosion resistance might be achievable because of the absence of local electrochemical potential differences. Indeed,

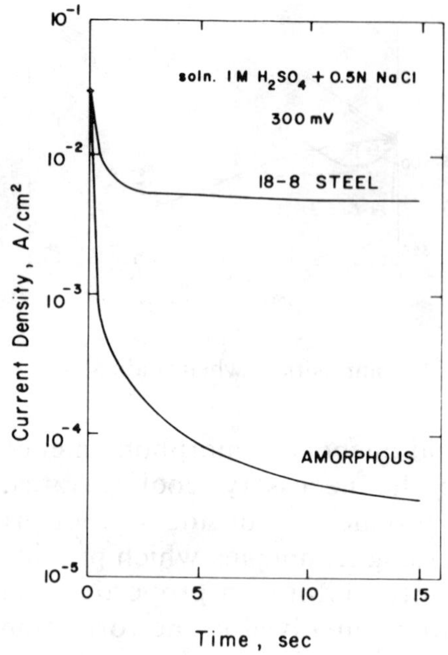

Fig. 9. Current densities for amorphous $Fe_{70}Cr_{10}P_{13}C_{07}$ and crystalline stainless steel as a function of time (90)

very low corrosion rates, compared to those of ordinary stainless steels, have been observed for Cr-containing metallic glasses exposed to standard test solutions (90). Of special interest is the fact that these amorphous alloys are resistant to pitting attack in sulfuric acid solutions containing chloride ions (90). Figure 9 shows the current densities (as a measure of the rate of corrosion attack) for both amorphous $Fe_{70}Cr_{10}P_{13}C_{07}$ and standard 18Cr-8Ni stainless steel in 1M H_2SO_4 + 0.5N NaCl at 300 mV. It can be seen that the rate of attack is 2 orders of magnitude lower for the amorphous metal; in contrast to the amorphous sample, the crystalline stainless steel exhibited pitting corrosion. Other experiments show the interesting result that the *initial* rate of attack (at $t < \sim 2$ sec) is higher for the amorphous alloy, reflecting its higher electrochemical potential (90). ESCA studies indicate that the passive film formed on the amorphous alloy is similar in composition to that found on stainless steels (90); the superior long-time performance of this film must therefore be due to its microscopic homogeneity. Of note is the fact that this superior corrosion resistance is achieved in a material much stronger and harder than crystalline stainless steels—a combination which may be of value in specific applications.

While mechanical strength, ferromagnetic effects and corrosion behavior offer the most promise at this time, some other aspects of the properties of amorphous metals are of potential interest. For example, their already highly disordered structures would appear to be reasonably resistant to radiation damage, suggesting their utilization where conductivity or mechanical properties must remain constant under irradiation.

The hydrogen pressure of an amorphous Pd-Si alloy saturated with hydrogen has been found to be lower than that of the corresponding crystalline alloy (although expectedly higher than that of pure Pd). This suggests an increased hydrogen storage capability of the glass and has been related to the irregular geometry of the voids present in an amorphous structure (91).

Numerous, generally more specialized potential uses of amorphous metals that are based on combinations of the fundamental properties have been envisioned. The combination of very high permeability with high hardness makes metallic glasses useful for tape recorder heads. Amorphous alloys combining superior corrosion resistance, high hardness and high elastic limit may be useful as razor blades. Amorphous alloy coatings produced by laser glazing or electrodeposition may, because of their mechanical, wear and/or corrosion properties, be used as protective surfacings.

A final category of potential applications for amorphous metals is based on current interest in materials with highly controlled microstructures. Thus, it has been suggested that amorphous metals may be highly useful as precursor materials which can be used to get desired crystalline microstructures, not otherwise attainable, through controlled crystallization treatment.

Clearly, however, neither the physical properties nor the potential applications of amorphous metals have yet been fully mapped out.

Acknowledgment

Research in this field was supported by the Office of Naval Research under Contract N14-76-C-820. We thank Mrs. T. Utterback for typographical assistance. Communication No. 28 from the Institute of Chemical Analysis, Applications and Forensic Science.

References

1. W. Klement Jr., R. H. Willens, and P. Duwez, Nature, Vol. 187, 1960, p. 809.
2. "International Conference on Metastable Metallic Alloys" (Brela, Yugoslavia, 1970), Fizika, Vol. 2, Supplement 2, 1970.
3. N. J. Grant and B. C. Giessen, eds., *Proc. Second Internat. Conf. on Rapidly Quenched Metals* (Cambridge, Mass., 1975), Section I, M.I.T. Press, Cambridge, Mass., 1976.

4. N. J. Grant and B. C. Giessen, eds., *Proc. Second Internat. Conf. on Rapidly Quenched Metals* (Cambridge, Mass., 1975), Section II, Mater. Sci. Eng., Vol. 23, 1976, pp. 81–308.
5. D. E. Polk, B. C. Giessen, and F. S. Gardner, "Synopsis of ONR Materials Workshop on State-of-the-Art and Prospects for Magnetic, Electronic and Mechanical Applications of Amorphous Metals" (Boston, Mass., 1975), Mater. Sci. Eng., Vol. 23, 1976, p. 309.
6. H. Jones and C. Suryanarayana, J. Mater. Sci., Vol. 8, 1973, p. 705.
7. H. Jones, Rep. Prog. Phys., Vol. 36, 1973, p. 1425.
8. A. K. Sinha, B. C. Giessen, and D. E. Polk, in *Treatise on Solid State Chemistry*, Vol. 3, N. B. Hannay, ed., Plenum Press, New York, N. Y., 1976, p. 1.
9. B. C. Giessen and C. N. J. Wagner, in *Liquid Metals, Chemistry and Physics*, S. Z. Beer, ed., Marcel Dekker, New York, N. Y., 1972, p. 633.
10. A. K. Sinha and P. Duwez, J. Phys. Chem. Sol., Vol. 32, 1971, p. 267.
11. G. S. Cargill III, in *Solid State Physics*, Vol. 30, F. Seitz, D. Turnbull, and H. Ehrenreich, eds., Academic Press, New York, N. Y., 1975, p. 227.
12. D. E. Polk and D. S. Boudreaux, J. de Physique, Vol. 36, 1975, p. C2–55.
13. J. F. Sadoc and J. Dixmier, in *Proc. Second Internat. Conf. on Rapidly Quenched Metals*, Section II, N. J. Grant and B. C. Giessen, eds., Mater. Sci. Eng., Vol. 23, 1976, p. 187.
14. M. R. Hoare, in *The Glass Transition and the Nature of the Glassy State*, M. Goldstein and R. Sinha, eds., Vol. 279, New York Academy of Sciences, New York, N. Y., 1976, p. 186.
15. M. R. Hoare and P. Pal, Adv. Phys., Vol. 20, 1971, p. 161.
16. B. C. Giessen, M. Madhava, D. E. Polk, and J. Vander Sande, in *Proc. Second Internat. Conf. on Rapidly Quenched Metals*, Section II, N. J. Grant and B. C. Giessen, eds., Mater. Sci. Eng., Vol. 23, 1976, p. 145.
17. H. S. Chen and D. Turnbull, Appl. Phys. Letters, Vol. 10, 1967, p. 284.
18. H. S. Chen and D. Turnbull, J. Chem. Phys., Vol. 40, 1968, p. 2560.
19. H. S. Chen and E. Coleman, Appl. Phys. Letters, Vol. 28, 1976, p. 245.
20. F. E. Luborsky, R. O. McCary, and J. J. Becker, in *Proc. Second Internat. Conf. on Rapidly Quenched Metals*, Section I, N. J. Grant and B. C. Giessen, eds., M.I.T. Press, Cambridge, Mass., 1976, p. 37.
21. T. Masumoto and R. Maddin, Mater. Sci. Eng., Vol. 19, 1975, p. 1.
22. B. G. Bagley and D. Turnbull, Acta Met., Vol. 18, 1970, p. 857.
23. P. H. Shingu, K. Shimomura, K. Kobayaski, and R. Ozaki, in *Proc. Second Internat. Conf. on Rapidly Quenched Metals*, Section II, N. J. Grant and B. C. Giessen, eds., Mater. Sci. Eng., Vol. 23, 1976, p. 183.
24. M. Ohring and A. Haldipur, Rev. Sci. Instrum., Vol. 42, 1971, p. 530.
25. H. S. Chen and C. E. Miller, Rev. Sci. Instrum., Vol. 41, 1970, p. 1237.
26. R. Pond and R. Maddin, Trans. Met. Soc. AIME, Vol. 245, 1969, p. 2475.
27. R. E. Maringer and C. E. Mobley, J. Vac. Sci. Tech., Vol. 11, 1974, p. 1067.
28. R. B. Pond, Sr. and J. M. Winter, Jr., in *Proc. Second Internat. Conf. on Rapidly Quenched Metals*, Section II, N. J. Grant and B. C. Giessen, eds., Mater. Sci. Eng., Vol. 23, 1976, p. 87.
29. R. E. Maringer, C. E. Mobley, and E. W. Collings, in *Proc. Second Internat. Conf. on Rapidly Quenched Metals*, Section I, N. J. Grant and B. C. Giessen, eds., M.I.T. Press, Cambridge, Mass., 1976, p. 29.

30. E. M. Breinan, B. H. Kear, C. M. Baras, and L. E. Greenwald, in *Proc. Third Internat. Symposium on Superalloys* (Seven Springs, 1976), Claitor's Publishing Div., Baton Rouge (in press).
31. R. P. Allen, S. D. Dahlgren, and M. D. Merz, in *Proc. Second Internat. Conf. on Rapidly Quenched Metals,* Section I, N. J. Grant and B. C. Giessen, eds., M.I.T. Press, Cambridge, Mass., 1976, p. 37.
32. P. K. Keung and J. G. Wright, Phil. Mag., Vol. 30, 1974, p. 995.
33. A. Brenner, D. E. Couch, and C. K. Williams, J. Res. Natl. Bur. Std., Vol. 44, 1950, p. 109.
34. A. W. Goldenstein, W. Rostoker, and F. Schossberger, J. Electrochem. Soc., Vol. 104, 1957, p. 104.
35. R. Brill, Z. Krist., Vol. 75, 1930, p. 217.
36. R. C. Ruhl, Mat. Sci. Eng., Vol. 1, 1967, p. 313.
37. S. Takayama, J. Mater. Sci., Vol. 11, 1976, p. 164.
38. H. A. Davies, Phys. and Chem. of Glasses, Vol. 17, 1976.
39. M. H. Cohen and D. Turnbull, J. Chem. Phys., Vol. 31, 1959, p. 1164.
40. D. Turnbull, in *Solidification,* American Society for Metals, Metals Park, Ohio, 1971, p. 1.
41. F. Spaepen and D. Turnbull, in *Proc. Second Internat. Conf. on Rapidly Quenched Metals*, Section I, N. J. Grant and B. C. Giessen, eds., M.I.T. Press, Cambridge, Mass., 1976, p. 205.
42. M. Marcus and D. Turnbull, in *Proc. Second Internat. Conf. on Rapidly Quenched Metals*, Section II, N. J. Grant and B. C. Giessen, eds., Mater. Sci. Eng., Vol. 23, 1976, p. 211.
43. P. Duwez and S. C. H. Lin, J. Appl. Phys., Vol. 38, 1964, p. 4096.
44. L. A. Davis, R. Ray, C. P. Chou, and R. C. O'Handley, Scripta Met., Vol. 10, 1976, p. 541.
45. P. Duwez, R. H. Willens, and R. C. Crewdson, J. Appl. Phys., Vol. 36, 1965, p. 2267.
46. H. S. Chen and D. Turnbull, Acta Met., Vol. 17, 1969, p. 1021.
47. (a) M. Hansen, *Constitution of Binary Alloys*, McGraw-Hill, New York, N.Y., 1958.
(b) R. P. Elliott, *Constitution of Binary Alloys, First Supplement*, McGraw-Hill, New York, N.Y., 1965.
(c) F. A. Shunk, *Constitution of Binary Alloys, Second Supplement*, McGraw-Hill, New York, N.Y., 1969.
(d) W. G. Moffatt, *Binary Phase Diagrams Handbook*, General Electric, Schenectady, N.Y., 1976.
48. C. H. Bennett, D. E. Polk, and D. Turnbull, Acta Met., Vol. 19, 1971, p. 1295.
49. S. R. Nagel and J. Tauc, Phys. Rev. Lett., Vol. 35, 1975, p. 380.
50. R. Ray, B. C. Giessen, and N. J. Grant, Scripta Met., Vol. 2, 1968, p. 357.
51. D. E. Polk, A. Calka, and B. C. Giessen (to be published).
52. R. Ruhl, B. C. Giessen, M. Cohen, and N. J. Grant, Acta Met., Vol. 15, 1967, p. 1693.
53. R. Ray, Sc. D. Thesis, Massachusetts Institute of Technology, Cambridge, Mass., 1969.
54. K. Togano and K. Tachikawa, in *Proc. Second Internat. Conf. on Rapidly Quenched Metals*, Section I, N. J. Grant and B. C. Giessen, eds., M.I.T.

Press, Cambridge, Mass., 1976, p. 519.
55. B. C. Giessen, R. Raman, R. Ray, and M. Segnini (to be published).
56. A. Calka, M. Madhava, D. E. Polk, and B. C. Giessen, Scripta Met., Vol. 11, 1977, p. 65.
57. L. E. Tanner and R. Ray (to be published).
58. R. Ray and E. Musso, U.S. Patent No. 3981722, 1976.
59. P. Chaudhari, J. J. Cuomo, R. J. Gambino, S. Kirkpatrick, and L. J. Tao, in *Magnetism and Magnetic Materials 1974*, American Institute of Physics Conf. Proc. No. 24, AIP, New York, 1975, p. 562.
60. D. R. Uhlmann, J. Noncryst. Solids, Vol. 7, 1972, p. 337.
61. H. A. Davies, J. Aucote, and J. B. Hull, Scripta Met., Vol. 8, 1974, p. 1179.
62. A. Kerns, R. Ray, and B. C. Giessen (to be published).
63. H. S. Chen, Acta Met., Vol. 22, 1974, p. 897.
64. S. Z. Beer, ed., *Physics and Chemistry of Liquid Metals*, M. Dekker, New York, N.Y., 1972.
65. M. Marcus and D. Turnbull, in *Proc. Second Internat. Conf. on Rapidly Quenched Metals*, Section II, N. J. Grant and B. C. Giessen, eds., Mater. Sci. Eng., Vol. 23, 1976, p. 211.
66. A. Nylund, Acta Chem. Scand., Vol. 20, 1966, p. 2381.
67. B. C. Giessen, in *Developments in the Structural Chemistry of Alloy Phases*, B. C. Giessen, ed., Plenum Press, New York, N.Y., 1969, p. 227.
68. W. B. Pearson, *The Crystal Chemistry and Physics of Metals and Alloys*, Wiley-Interscience, New York, N.Y., 1972.
69. D. E. Polk, Acta Met., Vol. 20, 1972, p. 485.
70. K. Suzuki, T. Fukunaga, M. Misawa, and T. Masumoto, in *Proc. Second Internat. Conf. on Rapidly Quenched Metals*, Section II, N. J. Grant and B. C. Giessen, eds., Mater. Sci. Eng., Vol. 23, 1976, p. 215.
71. B. Aronsson and A. Nylund, Acta Chem. Scand., Vol. 14, 1969, p. 1011.
72. R. St. Amand and B. C. Giessen (to be published).
73. J. L. Finney, J. de Physique, Vol. 36, 1975, p. C2-1.
74. A. K. Sinha, J. Appl. Phys., Vol. 42, 1971, p. 338.
75. W. Hume-Rothery and E. Anderson, Phil. Mag., Vol. 5, 1960, p. 383.
76. H. S. Chen and B. K. Park, Acta Met., Vol. 21, 1973, p. 395.
77. D. Turnbull, J. de Physique, Vol. 35, 1974, p. C4-1.
78. S. R. Nagel and J. Tauc, Solid State Comm., in print.
79. S. R. Nagel and J. Tauc, in *Proc. Second Internat. Conf. on Rapidly Quenched Metals*, Section I, N. J. Grant and B. C. Giessen, eds., M.I.T. Press, Cambridge, Mass., 1976, p. 337.
80. S. R. Nagel, G. B. Fisher, J. Tauc, and B. G. Bagley, Phys. Rev., Vol. B13, 1976, p. 3284.
81. S. R. Nagel, J. Tauc, and B. C. Giessen, Solid State Comm. (to be published).
82. C. B. Shoemaker and D. P. Shoemaker, in *Developments in the Structural Chemistry of Alloy Phases*, B. C. Giessen, ed., Plenum Press, New York, N.Y., 1969, p. 107.
83. J. Barrick and B. C. Giessen (to be published).
84. H. S. Chen and T. T. Wang, J. Appl. Phys., Vol. 41, 1970, p. 5338.
85. C. A. Pampillo and H. S. Chen, Mater. Sci. Eng., Vol. 13, 1974, p. 181.
86. C. A. Pampillo and D. E. Polk, Acta Met., Vol. 22, 1974, p. 741.
87. L. A. Davis, C. P. Chou, L. E. Tanner, and R. Ray, Scripta Met., Vol.

10, 1976, p. 937.

88. M. Naka, S. Tomizawa, T. Masumoto, and T. Watanabe, in *Proc. Second Internat. Conf. on Rapidly Quenched Metals*, Section I, N. J. Grant and B. C. Giessen, eds., M.I.T. Press, Cambridge, Mass., 1976, p. 273.
89. F. E. Luborsky, in *Proc. Second Internat. Conf. on Amorphous Magnetism*, R. A. Levy and R. Hasegawa, eds., Plenum Press, New York, N.Y. (in press).
90. K. Hashimoto and T. Masumoto, in *Proc. Second Internat. Conf. on Rapidly Quenched Metals*, Section II, N. J. Grant and B. C. Giessen, eds., Mater. Sci. Eng., Vol. 23, 1976, p. 285.
91. T. B. Flanagan and B. C. Giessen (to be published).

LIQUID-QUENCHED METAL-METAL GLASSES

C. Suryanarayana*

The Research Institute for Iron, Steel and Other Metals, Tohoku University, Sendai 980.

ABSTRACT

A wide variety of amorphous alloys (metallic glasses) containing only metals has been produced during the last few years. These metastable materials exhibit interesting thermal, electrical and magnetic properties and can be prepared in a very wide composition range. The composition ranges for the formation, thermal properties and crystallization behavior of the metal-metal glasses are presented in this review.

1. INTRODUCTION

The advent of splat quenching [1] in 1960 by Professor Pol Duwez of California Institute of Technology, Pasadena, USA, ushered in a new area of investigation in physical metallurgy. This technique, capable of producing cooling rates in excess of a million degrees per second in alloys cooled from the liquid state, has been found to produce several metastable effects. These include extension of solid solubility limits, formation of metastable crystalline intermediate phases and even amorphous phases at high enough cooling rates. The last category has attracted increasing interest in recent times since it was noticed that they have superior chemical, electrical, magnetic and mechanical properties in comparison to their crystalline counterparts. Several reviews are now available on this exotic category

* Permanent address: Department of Metallurgical Engineering, Bnaras Hindu University, Varanasi-221005, India.

of metallic materials (see for example refs. 2-6). Although amorphous alloys can be produced by other techniques such as vapor deposition, sputtering, electrodeposition, etc., liquid quenching has by far been the most popular one and so the present paper will concentrate on results obtained only by this technique.

Broadly speaking, three categories of amorphous solids produced by liquid quenching are recognized:

(i) Pure metals e.g., Ge, Ni, Fe

(ii) Metal -Metalloid type e.g., $Pd_{80}Si_{20}$, $Fe_{80}P_{13}C_7$, $Fe_{81}B_{13}C_2Si_4$, $W_{35}Mo_{20}Cr_{15}$-$Fe_5Ni_5P_6B_6C_5Si_3$

(iii) Metal-Metal type e.g., Ni-Nb, Cu-Zr, Mg-Zn, Ca-Mg

Amongst these, pure metals have very limited technological potential, while the metal-metalloid type (containing approximately 20 at.% of the metalloid, irrespective of the number of components involved have very useful properties to warrant commercial exploitation. The properties of these amorphous alloys can also be suitably modified by alloy additions and a large proportion of research in the field of amorphous alloys is concerned with this category. The third category of metal-metal type is relatively less explored and the present review aims at presenting some of the salient features of this category of amorphous alloys.

Although discovered [7] simultaneously with Fe-base metal-metalloid type (FePC) glasses in 1967 [8], some doubt existed about their real 'amorphousness'. In fact, initially, these were designated as 'microcrystalline'. Later x-ray diffraction studies indicated that these are also amorphous in the same sense as the metal-metalloid type glasses. Consequently, intensive research on different aspects of metal-metal glasses began only recently.

2. FORMATION

Table I lists the various metal-metal amorphous alloys produced till now. It may be noticed that the number of alloy systems is quite large and that the composition range in which the amorphous phase forms does not show any preferred compositions similar to the conventional metal-metalloid glasses. In addition to the above, amorphous phases were also detected in localised areas in some Al-base alloys (Al-Cr[36], Al-Cu[37], Al-Ni[38], and Al-Pd[39]), Au-Pb[40], Au-Sn[41] and Pb-Au[40] systems. Generally speaking, the following criteria have been recognized to favor easy glass formation: (a) presence of deep eutectics, (b) widely differing sizes of component atoms and (c) a high value of the reduced glass transition temperature, T_{gr} (=T_g/T_m, where T_g is the glass transition temperature and T_m is the melting point of the alloy).

As mentioned previously, the metal-metalloid type amorphous alloys form when the metalloid content is about 20 at.%. Based

Table I Composition range for formation of amorphous phases in metal-metal systems

System	Solute conc. at.%	Ref.	System	Solute conc. at.%	Ref.
Ca-Ag	12.5-42.5	9	Mg-Ga	19	24
Ca-Al	12.5-47.5	9,10	Mg-Zn	25-32	25
Ca-Cu	12.5-62.5	9			
Ca-Ga	16	10	Nb-Ni	40-66	7,26
Ca-Mg	22.5-42.5	9,10	Nb-Rh	42-45	27
Ca-Zn	17.5-62.5	9,10			
			$Nb_{48}Ni_{39}Al_{13}$	–	7
Co-Ti	21-23	11			
Co-Zr	9-16	12	Ni-Zr	10-11	12
Cu-Cd	33-45	13	Sr-Al	18,30	10
Cu-Hf	40	14	Sr-Ga	30	10
Cu-Mg	?	13	Sr-Mg	30	10
Cu-Ti	30-35	15	Sr-Zn	25	10
Cu-Zr	25-60	15			
			Ta-Ni	40-70	7,26
$Cu_{40}Hf_{30}Nb_{30}$	–	14			
$Cu_{40}Ti_{30}Nb_{30}$	–	14	Ti-Be	37-41	28
$Cu_{40}Zr_{30}Nb_{30}$	–	14	Ti-Ni	30-40	29
Er-Fe	32	16	$Ti_{50}Be_{40}Zr_{10}$	–	30
Fe-Th	28-80	17	U-Co	20-40	31
Fe-Zr	9-11	12	U-Cr	27	31
			U-Fe	20-40	31
Gd-Co	40-50	18	U-Mn	20-35	31
Gd-Fe	32-50	16	U-Ni	20-40	31
			U-V	27	31
Hf-Co	22-91	19			
Hf-Ni	20-89	19	Zr-Be	30-50	28
			Zr-Co	22-90	32
La-Ag	25	20	Zr-Fe	33	33
La-Al	18-34	21	Zr-Ni	22-90	32
La-Au	18-26	22	Zr-Pd	20-35	15
La-Cu	30	22	Zr-Rh	18-26	34
La-Ga	16-28	23			
La-Ge	17-22	21	$ZrBe_xNb_y$	x=20-35	35
La-Ni	22	22		y=2.5-20	

on this observation, it was suggested that the Bernal model of dense random packing of hard spheres (DRPHS) is applicable to amorphous alloys as modified by Polk [42]. It was proposed that the larger size metal atoms form tetrahedra and the smaller size metalloid atoms occupy the interstices. Since all the vacant space can be filled up by about 19 at.% of the metalloid atoms, it was considered that this model suitably describes the amorphous structure. However, in the case of metal-metal glasses, no such strict compositional requirements are observed. The amorphous phase can be produced in some binary systems from 10 to 90 at.% of solute content. Thus, it is doubtful whether the DRPHS model is applicable to metal-metal glasses.

Waseda et al. [43] performed a detailed analysis of x-ray scattering intensities on amorphous Cu-Zr alloys and came to the conclusion that the DRPHS model is applicable in this system. It must be stressed, however, that only very few detailed x-ray investigations have been carried out on metal-metal glasses; the scantily available data indicates that these amorphous alloys are similar to the metal-metalloid type.

The glass-forming compositions in the metal-metal category encompass eutectic compositions and in some cases intermediate phases with large unit cells and complex crystal structures. Based on these observations and (a) the presence of a sharp first peak in the radial distribution function (indicative of a high degree of short range order), (b) high thermal stability, (c) high hardness and (d) the close correspondence between the densities of amorphous alloys and the corresponding crystalline intermediate compounds, Wang [44] proposed another model involving random stacking of interpenetrating Kasper polyhedra. This model successfully explained the observations made on sputtered W-Fe, Ni-Nb, etc. amorphous alloys produced at equiatomic compositions. Finney [45] raised doubts about its applicability in three dimensions.

Zielinski et al. [33] on the other hand suggested that the glass-forming tendency is related to low values of ΔT (the difference between ideal liquidus and actual liquidus temperatures) and also to low values of $\Delta H^M/R$, where ΔH^M is heat of mixing and R is the gas constant. This was shown to explain the formation of amorphous phases in Zr-base alloys quite satisfactorily.

Thus, even though different models are proposed almost all of them seem to suffer from some disadvantage or other and a universally applicable model is yet to emerge.

3. THERMAL PROPERTIES

Table II summarises the thermal properties of some metal-metal glasses. The glass transition temperature is detected only in some alloys. Another observation that can be made is that the crystallization temperature (T_x) is high. Giessen <u>et al</u>. [47] found that T_g varies linearly with the solidus temperature and also with the melting point of the CaM_2 compounds in Ca-base metal-metal amorphous alloys. Chen and Krause [48] observed that T_g increases with increasing Cu content in Zr-Cu glasses and that Young's modulus also follows a similar trend.

Table II Thermal Properties of Metal-Metal Amorphous Alloys

Alloy	T_g(K)	T_{x1}(K)	T_{x2}(K)	T_{x3}(K)	Heating Rate K/min	Ref.
$Co_{78}Ti_{22}$		777	823		5	11
$Cu_{60}Zr_{40}$	729	753	771		10	46
$Cu_{50}Zr_{50}$	677	724	731		10	46
$Cu_{46}Zr_{54}$	663	704	747		10	46
$Hf_{78}Co_{22}$		758			50	19
$Hf_{60}Co_{40}$		823			50	19
$Hf_{9}Co_{91}$		828			50	19
$Hf_{80}Ni_{20}$		738			50	19
$Hf_{62}Ni_{38}$		808			50	19
$Hf_{36}Ni_{64}$		923			50	19
$Mg_{70}Zn_{30}$		380	401	501	30	25
$Ti_{63}Be_{37}$		673	773		20	28
$U_{60}Fe_{40}$		560	584		40	31
$U_{73}Co_{27}$	527	554	563	578	40	31
$U_{73}Cr_{27}$		684			40	31
$U_{67}Ni_{33}$		599	641	677	40	31
$U_{70}Mn_{30}$			605	677	40	31
$Zr_{65}Be_{35}$	623	648	663	683	20	28
$Zr_{78}Co_{22}$		643			50	32
$Zr_{64}Co_{36}$		740			50	32
$Zr_{10}Co_{90}$		833			50	32
$Zr_{78}Ni_{22}$		611			50	32
$Zr_{64}Ni_{36}$		730			50	32
$Zr_{32}Ni_{63}$		839			50	32

4. CRYSTALLIZATION BEHAVIOR

The transformation behavior from the amorphous to the crystalline equilibrium phases was investigated only for a few systems in the metal-metal category. Even in those cases, the extent and details of investigations do not match those of the more conventional metal-metalloid type like Pd-Si, Fe-P-C, Fe-B-Si. This is rather surprising in view of the fact that metal-metal glasses may be obtained in a much wider composition range than in metal-metalloid type. One reason for this may be that in a large number of cases, under equilibrium conditions, the alloy systems consist of several complex phases, the crystal structures of some of which are not even determined. This seems to be the case particularly for Mg-Zn and Cu-Zr, two systems quite extensively investigated.

The available data on the crystallization of metal-metal glasses is summarised in the following paragraphs.

Fig. 1 shows a bright-field micrograph and the corresponding diffraction pattern of liquid-quenched Co-22 at.% Ti alloy. The lack of contrast in the micrograph and the presence of diffuse haloes in the diffraction pattern clearly indicate that the as-quenched product is amorphous. It may so happen that sometimes even though the x-ray patterns indicate an amorphous phase, high-magnification electron microscopy may reveal the presence of fine particles/microcrystals. Hence, it may be advisable always to supplement the x-ray work with electron microscopy.

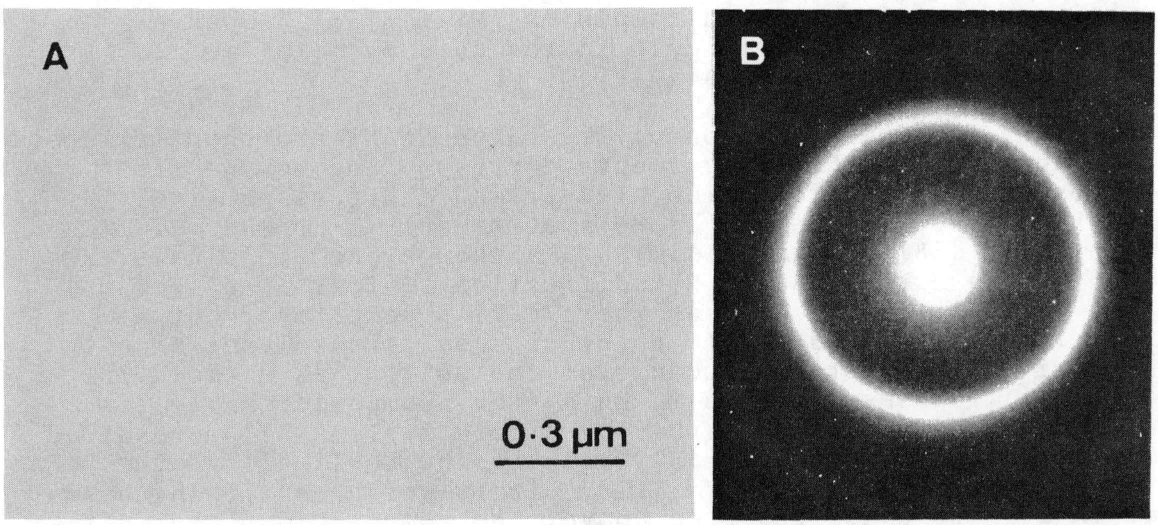

Fig. 1 (a) The as-quenched amorphous structure of a Co-22at.% Ti alloy (b) Electron diffraction pattern.

Structural relaxation at temperatures lower than the crystallization temperature, T_x, has been observed in some metal-metal glasses, e.g., Cu-Zr. However, contrary to the metal-metalloid glasses, Cu-Zr amorphous alloys remain completely ductile till the onset of crystallization [49]. Since several theories were proposed to explain temper embrittlement in metal-metalloid amorphous alloys - fast diffusion of smaller size metalloid atoms [50] or fine scale phase separation in the amorphous alloy [51] - it will be interesting to evaluate their applicability to metal-metal systems as well. Since this category of amorphous alloys does not contain any metalloid atoms, the latter theory may perhaps apply. However, the amount of data avilable is rather limited at present and so this problem may have to wait for some more time to be solved.

Vitek et al. [52] showed that $Cu_{10}Zr_7$ and a 'transformed non-crystalline structure' form when Cu-Zr amorphous alloys are annealed below T_g. Such a transformation was related to the compositional and structural reorientation. They had also observed that particles of oblong shape and about 0.1 μm in diameter are formed initially. High resolution electron microscopy investigations suggested that the crystallites had a substructure. Lattice imaging results showed the subgrains, free of dislocation arrays, to have a size of about 30-50 nm. Tilt boundaries were also detected in support of the substructure hypothesis.

All the metal-metal glasses do not show a glass transition temperature. Calorimetric investigations indicate the presence of clear exothermic peaks. The number and height of the peaks varies from system to system. In metal-metalloid systems, the first peak is generally associated with the formation of a solvent-rich phase (MSI), while the second peak is associated with the formation of metal-metalloid complexes (MSII) [53]. Such a generalization appears difficult in the case of metal-metal glasses as the following observations show.

During the decomposition of Zr-Ni and Hf-Ni amorphous alloys, it was noticed that heat treatments corresponding to the first peak resulted in the formation of β-Zr and α-Hf, respectively [32, 19]. In both Zr-Be and Ti-Be systems only α-Zr and α-Ti precipitated [28]. This is similar (although one would have expected the high temperature β modification to form in all cases) to the metal-metalloid glasses where β-Ti precipitated out at T_{x1} in Ti-Si alloys [54]. In no other case, the solvent-rich phase, with the crystal structure of the solvent is noticed in metal-metal glasses. In the Mg-Zn system, compositionally disordered and defective Mg_7Zn_3 having a relatively small unresolved grain structure formed [25, 55] while in the Co-Ti alloys the Co_2Ti phase was found to form [11]. It is worth mentioning here that in all the cases reported above, the compositions are near the eutectic composition and in some cases they are slightly hypereutectic.

Heat treatments at temperatures corresponding to the second peak in the calorimetric curves indicated that mostly the equilibrium phases are formed, although in some cases metastable phases are detected. Generally speaking, it is noticed that the alloy compositions of interest correspond to a mixture of several intermetallic phases with complex crystal structures. In some cases, the crystal structures are unknown. Thus, a complete identification of the phases formed during decomposition is tedious and often tentative.

Fig. 2 shows a bright-field micrograph of a Co-22 at.% Ti amorphous alloy crystallized at 823 K for 600 s. One can notice two types of precipitates. The finely dispersed particles belong to the Co_2Ti phase, while the larger, almost spherical, particles are of the Co_3Ti phase. Most often it is noticed that the first crystalline phase to precipitate from the amorphous matrix is in the form of very fine particles and is uniformly distributed, while the second phase particles have a much larger size.

Fig. 3 shows a typical electron diffraction pattern from a

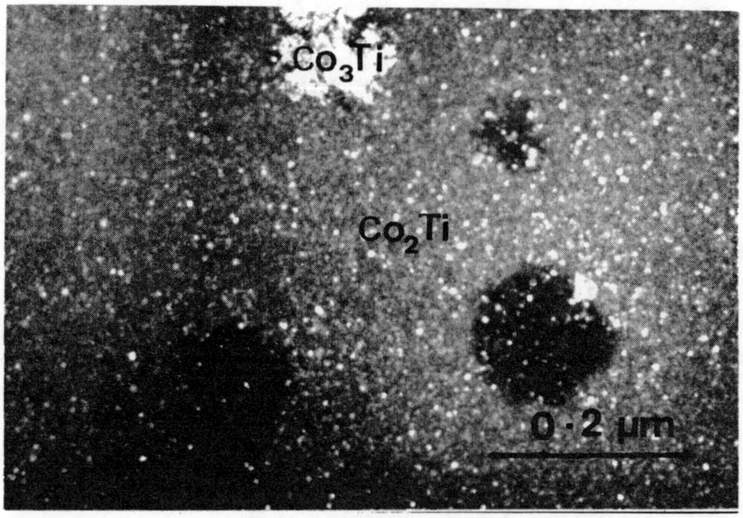

Fig. 2 Co-22 at.% Ti alloy annealed for 600 s at 823 K showing the presence of Co_2Ti and Co_3Ti phases.

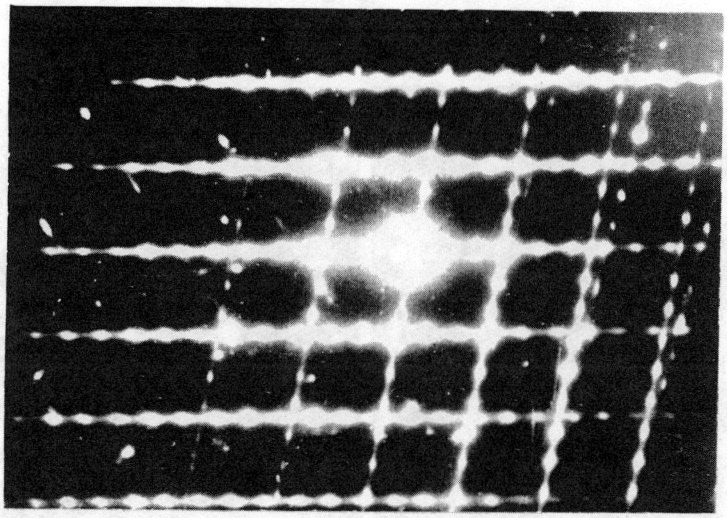

Fig. 3 Electron Diffraction pattern of a $Nb_{40}Ni_{60}$ amorphous alloy annealed for 3600 s at 873 K.

$Nb_{40}Ni_{60}$ amorphous alloy annealed for 3600 s at 873 K. Even though the diffraction pattern looks complicated, in conjunction with careful x-ray analysis, it has been shown that one of the phases present has a tetragonal structure with the lattice constants a=1.106 and c=2.664 nm [56]. This appears to be a metastable phase preceding the formation of equilibrium μ-NiNb and Ni_3Nb phases.

$Cu_{40}X_{30}Nb_{30}$ (X=Ti, Zr or Hf) were quenched into the amorphous state by melt spinning. In spite of the fact that alloys in the as-quenched state are not superconducting, it was noticed that annealing at high temperatures induces a superconducting transition in Ti- and Hf-containing alloys but not in the Zr-containing alloy. Careful electron microscopy was carried out on samples crystallized for different times at various temperatures to investigate the origin of this discrepancy. $Cu_{40}Zr_{30}Nb_{30}$ alloys, after crystallization gave rise to very complex diffraction patterns (consisting of $CuZr_2$ and at least another phase), whereas $Cu_{40}Ti_{30}Nb_{30}$ and $Cu_{40}Hf_{30}Nb_{30}$ alloys yielded apparently simpler patterns. The phase to form in both these alloys appears to be ordered on the basis of a b.c.c. unit cell with a lattice parameter of about 0.32 nm [57] [Fig. 4]. Although, it is tempting to assume that this b.c.c. phase may be the β-Ti or Hf containing niobium in solid solution, the size of the unit cell and the presence of ordering precludes this possibility. The mechanism of ordering and the crystal structure of the ordered phase are being determined at present. The detection of superconductivity in these two alloy systems appears to be related to the presence of this ordered phase.

Relatively little work has been carried out on the kinetics of crystallization in metal-metal glasses (see for example ref. 52).

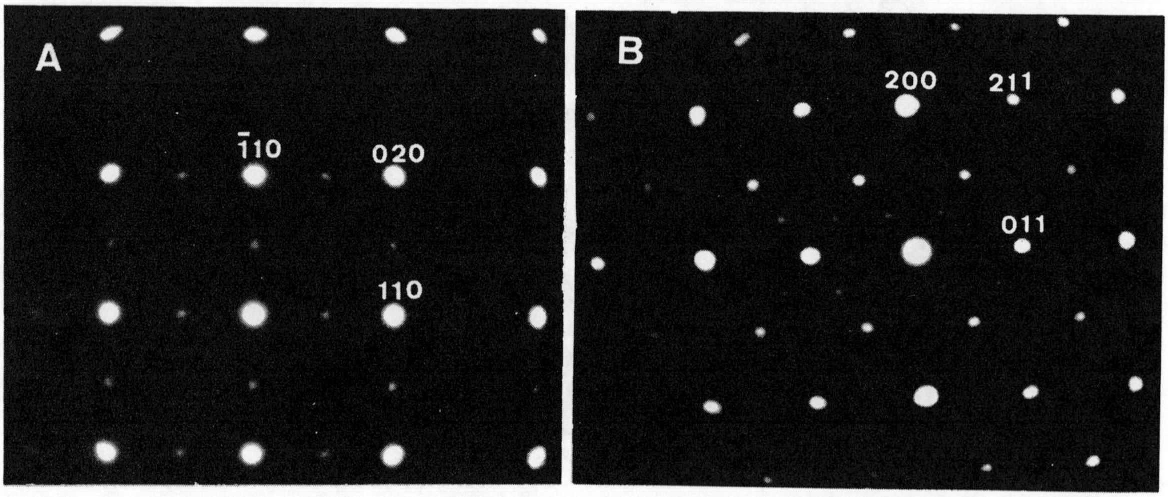

Fig. 4 Electron diffraction patterns showing the presence of ordered structures in (a) $Cu_{40}Ti_{30}Nb_{30}$ and (b) $Cu_{40}Hf_{30}Nb_{30}$ alloys.

5. SUPERCONDUCTIVITY

Although first discovered only in 1975, many metal-metal glasses have been found to exhibit superconductivity either in the as-quenched state or after crystallization. (Since a comprehensive report appears on the superconductivity of liquid-quenched amorphous alloys separately in this proceedings, only the general outlines will be given here).

Amongst the metal-metal glasses, superconductivity is detected in alloy systems based on La, Zr and Nb and the highest T_c reported so far is only 6.85 K for a $Zr_{65}Nb_{15}Be_{20}$ alloy [35] (and 4.7 K for $Nb_{58}Rh_{42}$). The highest achieved T_c for liquid-quenched amorphous alloys is 9.0 K for $Mo_{80}P_{10}B_{10}$. The presence of metalloid elements thus seems to raise the T_c value. The critical magnetic field, H_{c2}, values are reasonably high, but the critical current density is very low. Such a combination of properties seems to preclude immediate applications for these alloys even though these can be readily produced in the form of long thin ribbons of uniform cross section and possessing good bend ductility.

6. CONCLUDING REMARKS

Metal-metal glasses do not differ essentially from the more extensively investigated metal-metalloid type glasses. However, it appears that in general they can be produced in the amorphous state in a much wider composition range and seem to have higher crystallization temperatures. Further, characterization of the phases formed during crystallization may be easier since most of these involve only two or in some cases three components, while useful metal-metalloid amorphous alloys contain at least four elements. They also seem to have a high resistance to radiation damage [58]. Although, it is premature to think of industrial applications for this category of amorphous alloys, their interesting electrical [59], magnetic [16, 17] and superconducting [60] properties can give a better insight into these metastable alloys.

ACKNOWLEDGEMENTS

The author is grateful to Professor T. Masumoto for valuable advice and encouragement. He also wishes to thank the Japan Society for Promotion of Science for financial assistance.

REFERENCES

1) P. Duwez, R. H. Willens and W. Klement, J. Appl. Phys., 31 (1960), 1136.
2) N. J. Grant and B. C. Giessen (eds.), Rapidly Quenched Metals (Proceedings of the Second International Conference) Section I: M. I. T. Press, Cambridge, Mass, 1976, Section II: Mater. Sci. & Engg., 23 (1976).
3) H. Jones and C. Suryanarayana, J. Mater. Sci., 8 (1973), 705.
4) T. Masumoto and K. Suzuki (eds.) Structure and Properties of Amorphous Metals, Supplement to Sci. Rep. RITU, A27 (1978).
5) B. Cantor (ed.), Rapidly Quenched Metals III, The Metals Society, London, 1978.

6) J. J. Gilman and H. J. Leamy (eds.), Metallic Glasses, Amer. Soc. for Metals, Metals Park, Ohio, 1978.
7) R. C. Ruhl, B. C. Giessen, M. Cohen and N. J. Grant, Acta Met., 15 (1967), 1693.
8) P. Duwez and S. C. H. Lin, J. Appl. Phys., 38 (1967), 4096.
9) R. St. Amand and B. C. Giessen, Scripta Met., 12 (1978), 1021.
10) F. Sommer, G. Duddek and B. Predel, Z. Metallkde., 69 (1978), 587.
11) A. Inoue, K. Kobayashi, C. Suryanarayana and T. Masumoto, Scripta Met., 14 (1980), 119.
12) M. Nose and T. Masumoto, Sci. Rep. RITU, A28 (1980).
13) J. Latuszkiewicz, M. Strzeszewska, A. Calka, P. G. Zielinski and H. Matyja, Metals Tech., 5 (1978), 329.
14) C. Suryanarayana, A. Inoue and T. Masumoto, unpublished research.
15) R. Ray, B. C. Giessen and N. J. Grant, Scripta Met., 2 (1968), 357.
16) K. H. J. Buschow, J. Less - Common Metals, 66 (1979), 89.
17) K. H. J. Buschow and A. M. van der Kraan, Phys. Stat. Sol., (a) 53 (1979), 665.
18) K. Fukamichi, M. Kikuchi, T. Masumoto and M. Matsuura, Phys. Letters, A73 (1979), 436.
19) K. H. J. Buschow and N. M. Beekmans, J. Appl. Phys., 50 (1979), 6348.
20) N. I. Varich, A. A. Yakunin and A. B. Lysenko, Sov. Phys.-Solid State, 17 (1975), 1194.
21) K. Agyeman, R. Muller and C. C. Tsuei, Phys. Rev., B19 (1979), 193.
22) W. L. Johnson, S. J. Poon and P. Duwez, Phys. Rev., B11 (1975), 150.
23) W. H. Shull, D. G. Naugle, S. J. Poon and W. L. Johnson, Phys. Rev., B18 (1978), 3263.
24) B. Predel and K. Hulse, J. Less-Common Metals, 63 (1979), 245.
25) A. Calka, M. Madhava, D. E. Polk, B. C. Giessen, H. Matyja and J. B. Vander Sande, Scripta Met., 11 (1977) 65.
26) B. C. Giessen, M. Madhava, D. E. Polk and J. B. Vander Sande, Mater. Sci. & Engg., 23 (1976), 145.
27) W. L. Johnson and S. J. Poon, J. Appl. Phys., 46 (1975), 1787.
28) L. E. Tanner and R. Ray, Acta Met., 27 (1979), 1727.
29) D. E. Polk, A. Calka and B. C. Giessen, Acta. Met., 26 (1978), 1097.
30) L. E. Tanner and R. Ray, Scripta Met., 11 (1977), 783.
31) B. C. Giessen and R. O. Elliott, Ref. 5, Vol. 1, p. 406.
32) K. H. J. Buschow and N. M. Beekmans, Phys. Rev., B19 (1979), 3843.
33) P. G. Zielinski, J. Ostatek, M. Kijek and H. Matyja, Ref. 5, Vol. 1, p. 337.
34) K. Togano and K. Tachikawa, J. Appl. Phys., 46 (1975), 3609.
35) R. Hasegawa and L. E. Tanner, J. Appl. Phys., 49 (1978), 1196.
36) P. Furrer and H. Warlimont, Mater. Sci. & Engg., 28 (1977), 127.
37) H. A. Davies and J. B. Hull, Scripta Met., 6 (1972), 241.
38) K. Chattopaclhyay, P. Ramachandrarao, S. Lele and T. R. Anantharaman, Ref. 2, Section I, p. 157.

39) G. V. Sastry, C. Suryanarayana, O. N. Srivastava and H. A. Davies, Trans. Indian Inst. Metals, 31 (1978), 292.
40) P. Predecki, B. C. Giessen and N. J. Grant, Trans. TMS-AIME, 233 (1965), 1438.
41) B. C. Giessen, Z. Metallkde., 59 (1968), 805.
42) D. E. Polk, Scripta Met., 4 (1970), 117; Acta Met., 20 (1972), 485.
43) Y. Waseda, T. Masumoto and S. Tomizawa, Canad. Met. Quart., 16 (1977), 142.
44) R. Wang, Nature, 278 (1979), 700.
45) J. L. Finney, Nature, 280 (1979), 847.
46) R. L. Freed and J. B. Vander Sande, J. Non-Cryst. Solids, 27 (1978), 9.
47) B. C. Giessen, J. Hong, L. Kabacoff, D. E. Polk, R. Raman and R. St. Amand, Ref. 5, Vol. 1, p. 249.
48) H. S. Chen and J. T. Krause, Scripta Met., 11 (1977) 761.
49) N. A. Pratten and M. G. Scott, Ref. 5, Vol. 1, p. 387.
50) J. L. Walter, F. Bacon and F. E. Luborsky, Mater. Sci. & Engg., 24 (1976), 239.
51) G. C. Chi, H. S. Chen and C. E. Miller, J. Appl. Phys, 49 (1978), 1715.
52) J. M. Vitek, J. B. Vander Sande and N. J. Grant, Acta Met., 23 (1975), 165.
53) T. Masumoto and R. Maddin, Mater. Sci. & Engg., 19 (1975), 1.
54) C. Suryanarayana, A. Inoue and T. Masumoto, J. Mater. Sci. (in press).
55) P. G. Boswell, Mater. Sci. & Engg., 34 (1978), 1.
56) B. Padhy, O. N. Srivasta and S. Ranganathan, Banaras Hindu University, unpublished research.
57) C. Suryanarayana, A. Inoue and T. Masumoto, Paper to be presented at the Spring Meeting of the Japan Institute of Metals, April 1980.
58) M. D. Rechtin, J. B. Vander Sande and P. M. Baldo, Scripta Met., 12 (1978), 639.
59) S. R. Nagel, J. Vassiliou, P. M. Horn and B. C. Giessen, Phys. Rev., B17 (1978), 462.
60) A. J. Drehman and W. L. Johnson, Phys. Stat. Sol., (a) 52 (1979), 499.

ELECTRICAL AND MAGNETIC PROPERTIES OF RAPIDLY QUENCHED METALS

C.C. Tsuei

IBM Thomas J. Watson Research Center
Yorktown Heights, New York

1.0.0 INTRODUCTION

A variety of metastable crystalline and amorphous metallic phases can be formed with the tehniques of rapid-quenching such as splat-cooling or vapor deposition, etc. The purpose of this article is to review some of the electrical and magnetic properties of materials thus obtained. The emphasis will be on those phenomena associated with the metastable nature of these metals and alloys.

The types of alloys resulting from rapid quenching can be summarized as follows:[1,2]

1] Supersaturated solid solutions:

If the rate of quenching is sufficiently high to retard the nucleation and growth of a second solid phase, solute atoms may be retained in a parent lattice in excess of the equilibrium concentration. The effect of rapid quenching is thus to extend the terminal solid solubility limit or the range of homogeneity of an intermediate solid phase. For example, the solubility of magnesium in aluminum is increased from 18.9 to 37at.% by liquid quenching, and that of aluminum in magnesium from 11.6 to 23at.%.[2] Of particular interest, it has been demonstrated that by liquid quenching, complete solubility can be achieved in certain alloy systems such as Cu-Ag, GaSb-Ge, etc.[2] Another interesting example is the extension of the homogeneity range of the A-15 compound Nb_3Ge by sputtering.[3] More examples of alloys with extended solute solubility can be found in several review articles.[1,2]

2] Non-equilibrium crystalline phases:

In this case, the effect of rapid quenching is to form a crystalline phase which does not exist under equilibrium conditions. The simple cubic alloys (such as Au-Te,[4] Pd-Sb[5] alloys) obtained by liquid quenching are examples of this type of alloy.

3] Amorphous phases:

The crystallization process can be by-passed by rapid quenching from a melt, vapor, or solution, providing that certain structural and kinetic conditions[6] are satisfied. The resulting metastable phases are characterized by the absence of any long-range structural order in the atomic arrangement beyond nearest neighbors. An amorphous state has been achieved in pure elements as well as well as in alloys by using an appropriate technique of rapid quenching.[2,7] The occurrence and structure of amorphous phases have been well documented in the literature.

2.0.0 METASTABLE SIMPLE CUBIC ALLOYS

One of the most novel non-equilibrium crystalline phases obtained by rapid quenching is probably the metastable simple cubic phase formed in splat-cooled alloys. The simple cubic crystal structure (space group Pm3m, one atom/unit cell), although used frequently to demonstrate calculations involved in various solid-state physics problems, is rare in equilibrium. With the exception of one allotropic form of the element Po, this structure does not occur in any known equilibrium elements or alloys. Metastable simple cubic phases have been obtained in a number of alloy systems by liquid quenching.[2] Attempts to prepare simple cubic alloys by the techniques of vapor deposition or sputtering have not been successful.

Metastable simple cubic phases were first found in liquid-quenched Te-Au and Te-Ag binary alloys by Luo and Klement.[4] It was also found that several Sb-base alloy systems tended to form this phase. The simple cubic alloys have been studied with the Mössbauer effect and by measuring the lattice parameter, superconducting transition temperature, thermoelectric power and magneto-resistance. The results of these studies can be summarized as the following:

1] All simple cubic alloys exhibit metallic conduction. In fact, they are also <u>all</u> superconducting with transition temperatures ranging from 1.3° to ∿7°K. [See Table, following page.]

2] Anomalies in the variation of lattice parameter, thermoelectric power, and superconducting transition temperature with concentration in the simple cubic Te_xAu_{1-x} alloys (x = 60 to 85at.%) can be qualitatively explained in terms of a Fermi-surface-Brillouin-zone interaction.[12,17]

Those who are interested in the result [2] are referred to the original references. The significance of superconductivity in simple cubic alloys will be discussed in some detail.

Source: Rapidly Quenched Metals, November 1975, 441-458

TABLE: Critical Temperature and Lattice Parameter of Simple Cubic Alloy Phases

COMPOSITION (at. %)	T_c (°K)	Lattice Parameter (Å)
$Au_{25}Sb_{75}$	6.7[8]	2.93[9]
$Pd_{16.5}Sb_{83.5}$	4.9[8]	3.00[9]
$In_{25}Sb_{75}$	4.1[8]	3.06[10]
$Au_{16.7}Pd_{16.6}Te_{66.7}$	4.6[8]	2.93[8]
Au_xTe_{100-x} (x=15 to 45)	1.6 to 3.0[11,12]	3.10-2.93[4]
$Ag_{25}Te_{75}$	2.6[11]	3.06[4]
$In_{50}Te_{50}$	2.19[13]	3.07[14]
P	4.7[15] (110 Kb)	----
$Sn_{25}Te_{75}$	1.7[16]	3.15[16]

First of all, the fact that metallic conduction prevails in simple cubic alloys containing as much as ∿90at.% Te requires some explanation. This can be qualitatively interpreted as a result of the dilution of the covalent bonding that characterizes the semiconducting hexagonal tellurium. The tellurium atom has six outer electrons (s^2p^4) beyond the closed shells. Two of these electrons are paired in the s-orbital, two in one of the three p-orbitals, while the remaining two are available for covalent bonding in half-filled p-orbitals. In the hexagonal tellurium, the atoms form spiral chains with three atoms per turn and the axes of these chains are located at the corner of a hexagon. Each tellurium atom forms two covalent bonds with its two nearest neighbors. The chain structure, however, does not exist in the simple cubic lattice. The coordination number is six instead of two, giving approximately 0.67 electrons per bond. Therefore, it is believed that an important part of the binding energy of the simple cubic alloy results from the mutual sharing of the unpaired electrons. As a consequence of this unsaturated covalent bonding, electrons can move with ease from one atom to another and hence give rise to the metallic conduction.

To provide more insight to this problem, augmented plane-wave energy-band calculations[18] have been performed for Te having a hypothetical

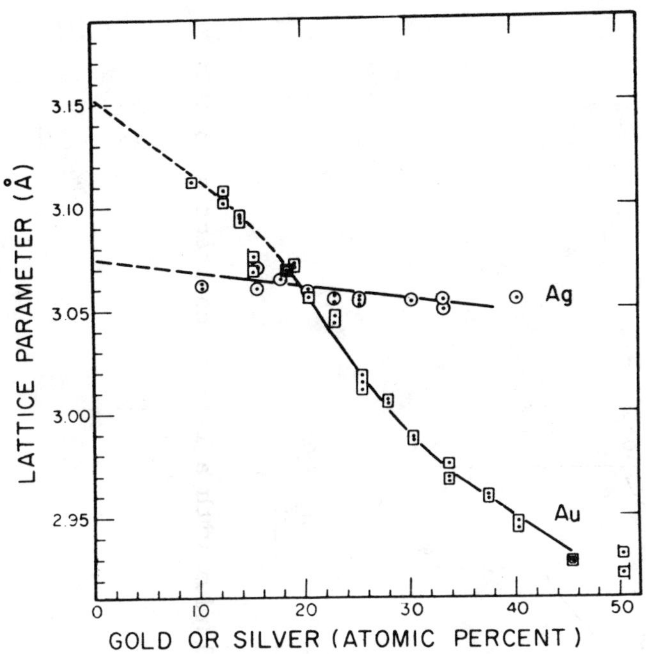

Figure 1. Lattice parameter as a function of composition for simple cubic Au-Te and Ag-Te alloys.[4]

simple cubic crystal structure with a lattice parameter based on extrapolation of the Te-Au experimental data as shown in Figure 1. The energy bands along the five segments joining the symmetry points of the first Brillouin zone are shown in Figure 2 for a simple cubic tellurium with a lattice parameter of 3.18Å. From these energy-band calculations, one concludes that the Fermi energy (E_F) of the simple cubic Te probably falls in the region $0.40<E_F<0.60$ Ry and it should be a metallic conductor. Furthermore, it is estimated that the density of states at the Fermi energy $N(E_F)$ is of the order of 12-15 electron states/atom/Ry. This value is about three or four times larger than the $N(E_F)$ for a metal like Cu, and about 50 to 60% as large as that for most of the A-15 type superconducting compounds. It should be pointed out that it has been proposed that the three-orthogonal-chain structure in the A-15 type is partially responsible for the attractive electron pairing[19] and the relatively high $N(E_F)$[20] found in this type of material. The simple cubic structure also has three orthogonal chains of atoms, although the interatomic distance in the chain is not as short as those found in the A-15 compounds. This analogy does suggest, however, that the simple cubic structure favors the occurrence of superconductivity in a similar

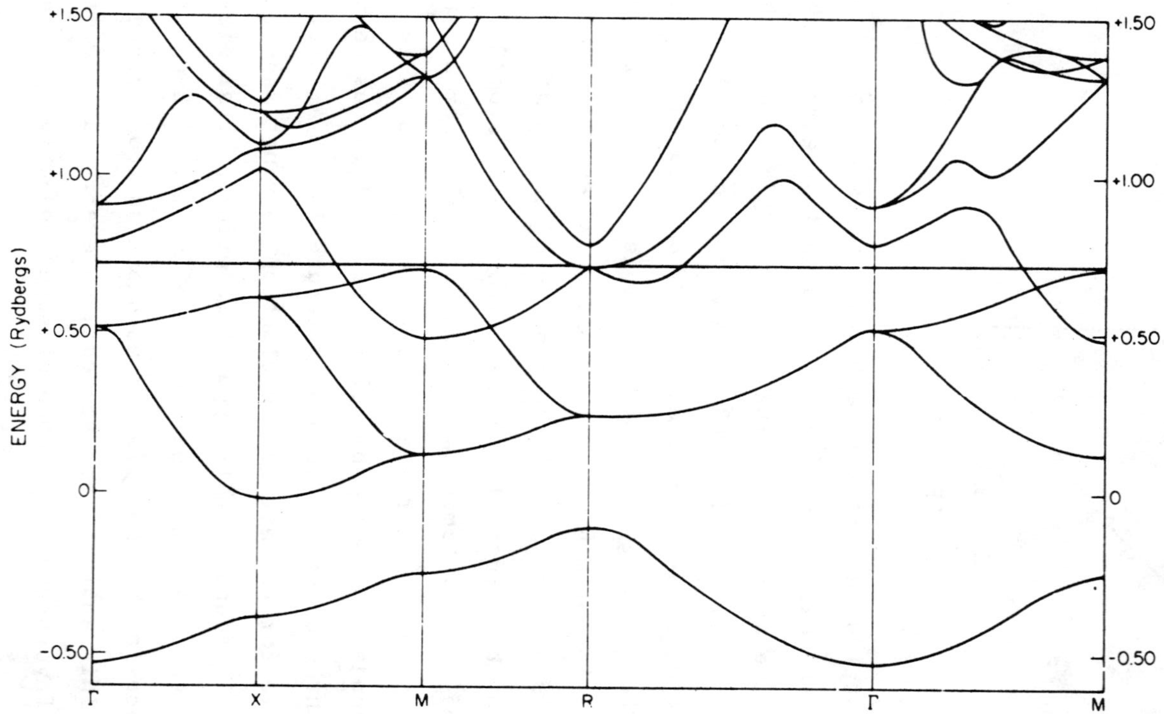

Figure 2. APW energy bands for hypothetical simple cubic tellurium with a lattice constant of 3.18Å.[18]

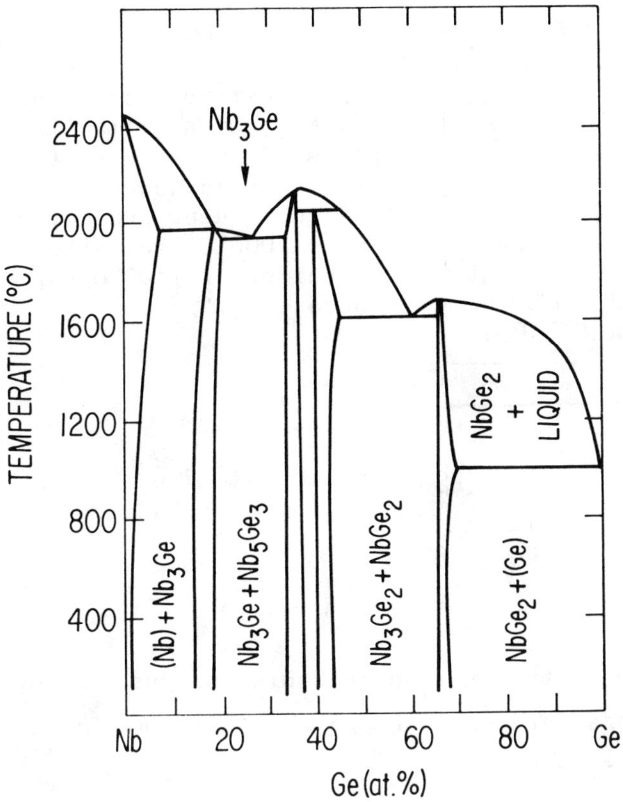

Figure 3. Nb-Ge phase diagram.[23]

manner as the A-15 structure, and is consistent with the fact that all the simple cubic materials are superconductors.

3.0.0 THE SUPERCONDUCTING TRANSITION TEMPERATURE OF Nb_3Ge

As mentioned in the Introduction, one of the effects of rapid-quenching is to produce supersaturated solid solutions and intermetallic compounds with extended homogeneity ranges. There has been considerable effort devoted to the study of the electrical and magnetic properties of these metastable alloys. Although the crystal structure of these alloys is not altered by rapid quenching, an extension of the composition range and/or an alteration in morphology can sometimes lead to some interesting properties which do not exist in the equilibrium alloys. For instance, superconductivity has been observed as a bulk property of some eutectic alloys[21] (e.g., Al-Si and Al-Ge). Transition temperatures (T_c) well in excess of those for the pure metals were found. Quenching the alloys from the liquid state was found to increase the superconducting transition temperature and critical field (H_c), and decrease the characteristic size of the metal and semiconductor domains. A correlation between the Fermi energy of the metal and the enhancement of T_c was observed. On the other hand, in metal-metal eutectic systems (e.g., Al-Al_2Cu) with comparable microstructure no enhancement of T_c was found. This suggests that a metal-semiconductor interface is necessary for the T_c enhancement. These experimental find-

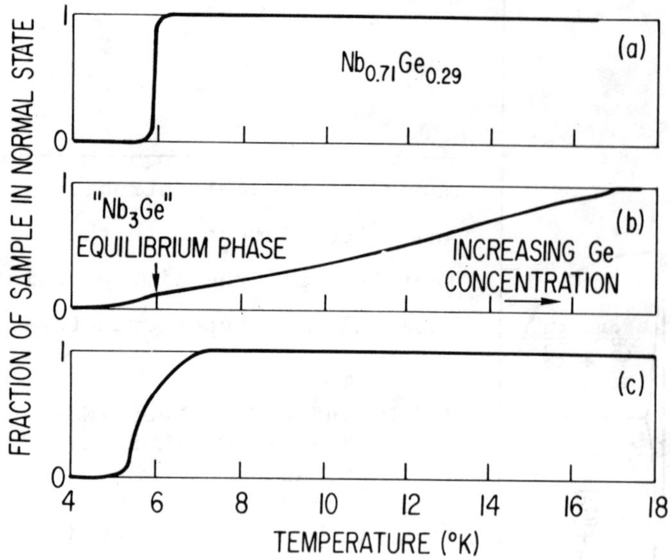

Figure 4. Superconducting transitions of an alloy $Nb_{0.79}Ge_{0.21}$ (a) as prepared in the arc, (b) after liquid quenching and subsequent annealing up to 1000°C, (c) after annealing at 1100°C for three days.[25]

ings were tentatively attributed to the exciton mechanism discussed by Ginsburg and by Allender, Bray and Bardeen although there is some uncertainty and controversy about this mechanism.

Another example is the high T_c phase found in Nb-Ge alloys. Recently, Gevaler[3] and Testardi[22] showed that T_c values as high as 23°K can be achieved by low energy sputtering onto hot substrates (∿700-900°C). It should be pointed out that liquid quenching has played a significant role in the development of achieving Nb_3Ge with high T_c. As indicated in the Nb-Ge equilibrium phase diagram[23] (Figure 3), the A-15 phase normally exists over a non-stoichiometric composition range (∿14-18at.% Ge). The non-stoichiometry of the equilibrium Nb_3Ge phase was, in fact, pointed out by Geller[24] based on the lattice parameter data. The T_c of the equilibrium A-15 compound is ∿6°K (Figure 4a). Realizing that rapid quenching can extend the homogeneity range of an intermetallic compound, Matthias, et al.,[25] liquid-quenched Nb-Ge alloys with nearly stoichiometric composition and obtained a broad superconducting transition with an onset at 17°K (Figure 4b). The equilibrium value of T_c of this metastable alloy was restored by annealing at 1100°C for three days (Figure 4c). Experimental evidence was also presented to show that this unusual increase in T_c can be attributed to an optimization of the Nb:Ge stoichi-

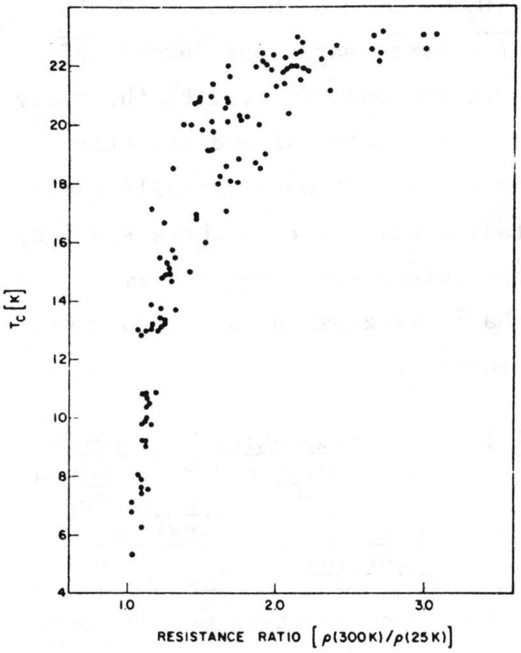

Figure 5. Correlation between T_c and resistance ratio [ρ(300°K)/ρ(25°K)] for Nb-Ge films prepared under various conditions.[22]

ometry achieved by liquid quenching. It was also recognized that there is a detrimental effect on T_c and its transition width due to the disorder in the atomic arrangement, an unavoidable by-product of rapid-quenching from the melt. It was concluded[25] that "it is likely that stoichiometric-ordered Nb_3Ge would have a transition of at least 18°K," a prediction that was borne out by Gavaler's discovery. The importance of the exact 3:1 Nb:Ge ratio on the highest possible T_c is, however, controversial[3,22,26] at the present time. On the other hand, the disorder along a chain in the A-15 structure is probably quite crucial in controlling T_c. Testardi, et al.[22], showed there is a simple correlation of T_c and electrical resistance ratio (Figure 5, supra). This correlation is essentially independent of all sputtering conditions and composition. The estimated electronic mean free path for films with resistance ratios between 2 to 3 (T_c > 21°K) is about 50Å which is about the same order of magnitude as a typical coherence length of most A-15 compounds. This observation suggests that order in the atomic arrangement (especially the chain integrity) that extends over more than one coherence length is rather important in achieving high T_c, regardless of how this condition is satisfied. A mean free path much longer than the coherence length probably will lead to a T_c not much higher than 23°K, as implied by Figure 5.

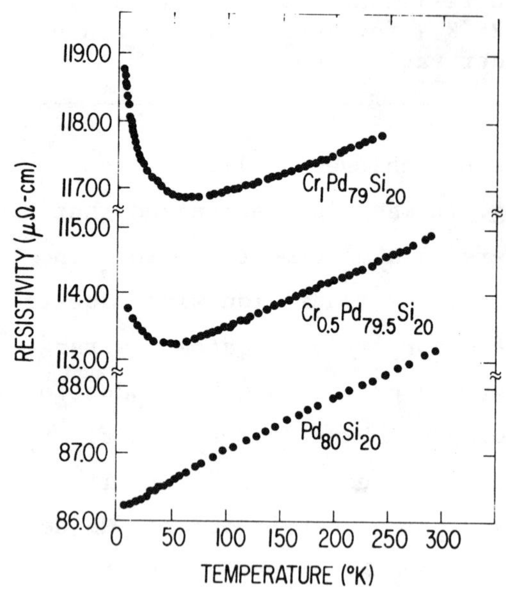

Figure 6. Resistivity as a function of temperature for amorphous Pd-Si alloys containing a small amount of Cr (0 to 1%).[28]

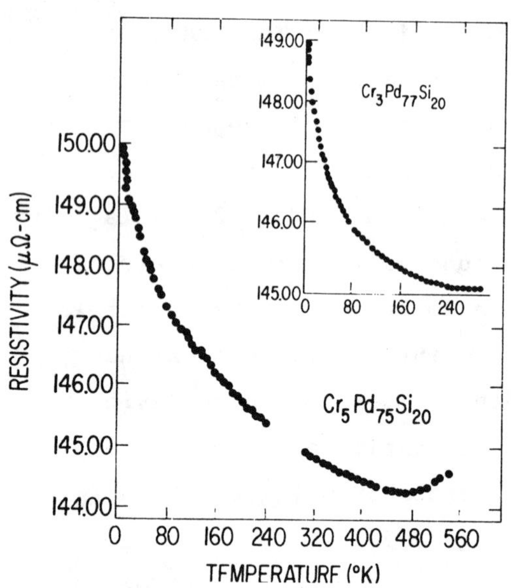

Figure 7. Resistivity as a function of temperature for amorphous Pd-Si alloys containing a small amount of Cr (3 to 5%).[28]

4.0.0 AMORPHOUS METALLIC PHASES

In recent years considerable effort has been devoted to the study of the electrical and magnetic properties of non-crystalline metallic alloys. From these studies, two interesting features emerge characterizing the amorphous metallic state.

4.1.0 The Atomic-Like Characteristics Play a More Prominent Role in the Amorphous Phase than in Its Crystalline Counterpart

There is no first-principle theory as yet that predicts that the above statement should be true. The fact that the electron mean free path is of the same order of magnitude as the interatomic distance suggests that the electrical and magnetic properties of amorphous metallic alloys preferentially probe the details of the atomic environment over other long-range properties. This effect also leads to a relatively weak correlation between spins or electrons in the amorphous solids, some examples of which follow:

4.1.1 Kondo-type resistance minimum: A well-defined minimum in the electrical resistivity versus temperature curve is found for a number of amorphous alloys (e.g., Pd-Si alloys) containing small amounts of magnetic elements such as Fe, Cr, etc. (Figures 6 and 7).[28] Below the resistivity-minimum temperature, the resistivity difference $\Delta\rho$ varies as $-\ell nT$ as predicted by the Kondo theory[29] for exchange scattering between conduction electrons and localized moments. At lower temperature, $\Delta\rho$ tends to level off, suggesting the formation of the spin-compensated state (Figure 8). The resistivity-minimum anomaly is always accompanied by a negative magneto-resistivity approximately proportional to the square of the magnetization and a susceptibility obeying the Curie-Weiss law for a wide temperature range. It should be mentioned that this Kondo-type effect has been observed in amorphous alloys containing as much as 7at.% Cr. This is quite a contrast to the case of most crystalline Kondo systems[30] in which the resistivity-minimum is completely suppressed due to the spin-spin correlation, when the magnetic impurity content exceeds ∿0.1at.% or less. From the point of view of the single-impurity based Kondo theory, the most unexpected result is probably the observation that the Kondo-type resistance minimum can co-exist with ferromagnetism in amorphous alloys. The existence of a Kondo-type effect

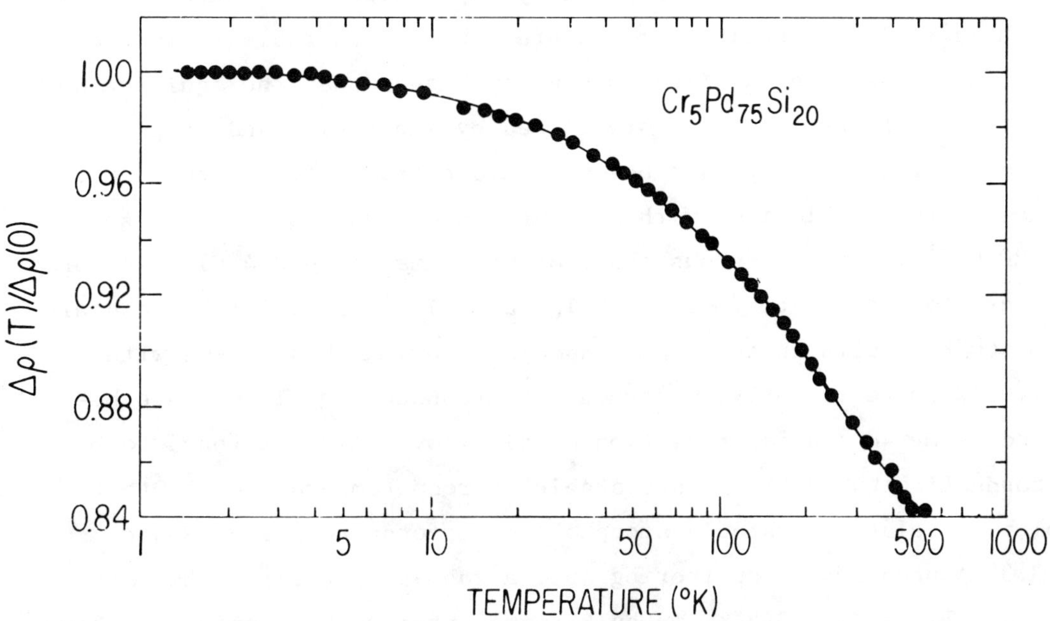

Figure 8. $\Delta\rho(T)/\Delta\rho(o)$ versus ℓn T plot for amorphous $Cr_5Pd_{75}Si_{20}$.[28]

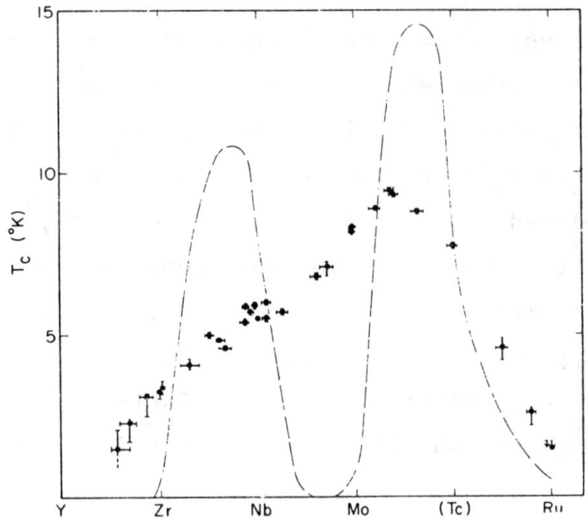

Figure 9. The superconducting transition temperature for amorphous 4d metals and their alloys.[37]

in a ferromagnet was first observed in the Co-Pd-Si alloy system.[31] Similar results were also reported in several amorphous ferromagnetic alloys with relatively high Curie temperatures.[32,33] Based on the experimental findings just described, one can conclude that the experimental results obtained so far suggest strongly that the effect of the d-d spin correlation in reducing the s-d exchange effect is relatively weak in amorphous alloys.

4.1.2 <u>The systematics of T_c for amorphous transition-metal superconductors</u>: Over two decades ago, it was demonstrated that a number of amorphous superconductors can be prepared by vapor deposition onto a substrate held at liquid helium temperature. These metastable films are superconducting with a T_c which is usually higher than the equilibrium value. Most of these films are stabilized by impurities and transform to one or more crystalline phases below liquid nitrogen temperature. For example, amorphous Bi films with a T_c ∿6°K are found to crystallize above 20°K,[35] and the amorphous phase of Ga films (T_c = 8.4°K) transforms into a crystalline state (β-Ga with T_c ∿6.4°K) at 15°K. Further annealing at ∿60°K results in the room-temperature stable phase, α-Ga with a T_c = 1.08°K. More recently, amorphous superconducting alloys have been reported; some 4d and 5d transition metal alloy films are found to be superconducting to ∿8 °K and are stable at room temperature.[37] Other examples of relatively stable amorphous superconductors are: Nb_3Ge (T_c ∿3.3°K) prepared by sputtering onto a room-temperature substrate[38] and La-Au alloys (T_c ∿3.8°K, in bulk form) obtained by liquid quenching.[39]

As discussed in Anderson's theory of dirty superconductors, the mechanism for electron-electron attraction in an amorphous superconductor is based on pairing each one-electron state in the presence of structural and compositional disorder scattering with its exact time reverse, a generalization of the $\vec{k}$ up, $-\vec{k}$ down pairing of the BCS theory which is independent of such scattering. In general, then, amorphous superconductors are not expected to behave in a fundamentally different way from their crystalline counterparts. There are, however, some general characteristics which can be associated with the amorphous nature of the material. One of such features is the trend of T_c in amorphous transition-metal (4d and 5d series) superconducting metals and alloys: Recent work by Collver and Hammond[37] shows that the variation of T_c as a function of the valence electron per atom ratio (z) has a sharp triangular peak in the middle of the transition-metal series (Figure 9). For the 4d transition metals and their alloys, the peak is located at $\sim z = 6.4$ which corresponds to a valence electron per atom ratio at which the d-shell of the atom is half-filled with five electrons with parallel spins. Therefore, the systematics of T_c for the amorphous superconductors is believed to be a manifestation of the dominance of atomic-like parameters in determining the superconducting properties of these amorphous materials. There are two attempts to explain the T_c dependence of z in the amorphous state.[41,42] Both conclude that the observed trend in T_c reflects the dependence of a local atomic parameter $\eta \approx N(o)\langle I^2\rangle$ on z, where N(o) is the electronic density of states at the Fermi level, and $\langle I^2\rangle$ the average squared electron-phonon interaction strength. This type of atomic parameter was first used by Hopfield for transition-metal superconductors.[43] To carry this concept one step further, one can assume that N(o) varies smoothly or even is constant for the most part of the transition-metal series as a result of simple smearing or averaging of the electronic density of states or other effects as discussed by Collver.[44] The T_c of the amorphous superconductors is then characterized essentially by the strength of electron-phonon coupling, an atomic-like parameter.

Source: Rapidly Quenched Metals, November 1975, 441-458

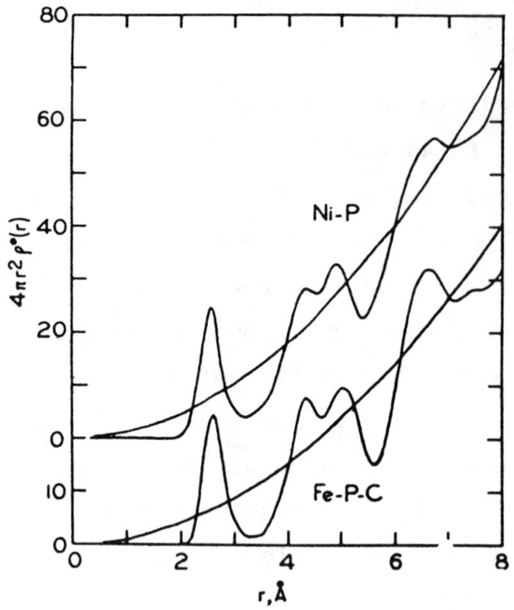

Figure 10. Radial distribution functions for $Ni_{77}P_{23}$ and $Fe_{80}P_{13}C_7$ (taken from Reference 46).

4.2.0 <u>Many Electrical and Magnetic Properties of Amorphous Metallic Solids can be Described in Terms of a Distribution Function for Certain Physical Parameters.</u>

In amorphous metallic alloys, there are at least two kinds of disorder which are responsible for the broadening effect on many characteristic properties: (1) Compositional disorder; (2) structural disorder. The former exists in both the crystalline and amorphous alloys while the latter is unique for the non-crystalline state. The structure of amorphous metals and alloys has been studied extensively by x-ray and electron diffraction.[45,46] The radial distribution functions (RDF) of the amorphous metallic alloys are found to be quite similar if the differences in the sizes are taken into consideration. The RDF's for two amorphous metallic alloys are shown in Figure 10. In general, it is found that there are about 12-13 nearest neighbors for a given metal atom. The interatomic distances are on the average of $\sim$2.6Å (for the amorphous Fe-P-C alloys) and with a spread of $\sim$0.5Å. Undoubtedly, such a distribution in the atomic arrangement will have a significant effect on many physical properties. Results of magnetization and Mössbauer measurements on a typical amorphous ferromagnet $Fe_{75}P_{15}C_{10}$ prepared by splat-cooling illustrate this point:

4.2.1 <u>Magnetization and fluctuations in the exchange interaction</u>: The spontaneous magnetization as a function of temperature up to the Curie temperature T_c (=596.5°K) has been measured for the amorphous alloy $Fe_{75}P_{15}C_{10}$.[47] The spontaneous magnetization at 0°K for this alloy is found to be 151 emu/g (or 1.8μ_B per Fe atom). The reduced magnetization $[M_s(T) / M_s(o)]$ versus reduced temperature (T/T_c) is shown in Figure 11.

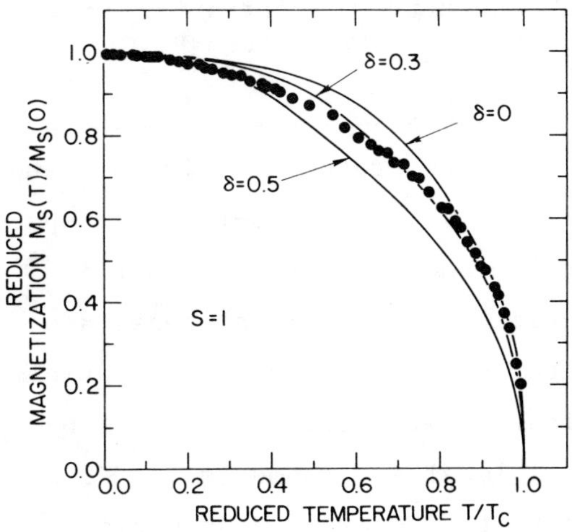

Figure 11. Magnetization as a function of temperature for amorphous $Fe_{75}P_{15}C_{10}$ alloy as compared with the Handrich prediction for the cases of S=1; δ=0, 0.3 and 0.5.[47]

It is found that the corresponding states curve is considerably lower than that of Fe_3P (the closest crystalline counterpart of the amorphous alloy) or crystalline Fe. This is apparently a general property common to many amorphous ferromagnets.[48,49] To gain a better understanding of the mechanism that is responsible for the temperature dependence of magnetization, the experimental data are compared with Handrich's theory for non-crystalline ferromagnets. This theory is based on a molecular-field approach and assumes that the compositional and structural disorder can lead to fluctuations in the exchange-interaction strength (J_{ij}) between the neighboring spins. The effect of these fluctuations on the reduced magnetization can be expressed by the following formula:

$$\sigma(T) \equiv M_s(T) \;/\; M_s(0) = \frac{1}{2}\{B_S[(1+\delta)\chi] + B_S[(1-\delta)\chi]\}$$

where $\chi = 3S\sigma T_c \;/\; (S+1)T$, and δ is a measure of the degree of disorder and is defined as the root mean square of deviation from an average exchange integral between two nearest-neighbor spins: $\delta^2 = \langle\Delta J^2\rangle \;/\; \langle J\rangle^2$.

As shown in Figure 11, the theoretical curve with $S = 1$, $\delta = 0.3$ gives the best overall fit to the experimental data. In terms of the Handrich theory,[50] the experimental results ($\delta \simeq 0.3$) for amorphous $Fe_{75}P_{15}C_{10}$ suggest that the average fluctuation in J can be the same order of magnitude as J itself.

4.2.2 Mössbauer spectrum and hyperfine field distribution: The room temperature Mössbauer spectrum for the amorphous alloy $Fe_{75}P_{15}C_{10}$ is shown in Figure 12. For comparison, the Mössbauer spectrum for crystalline

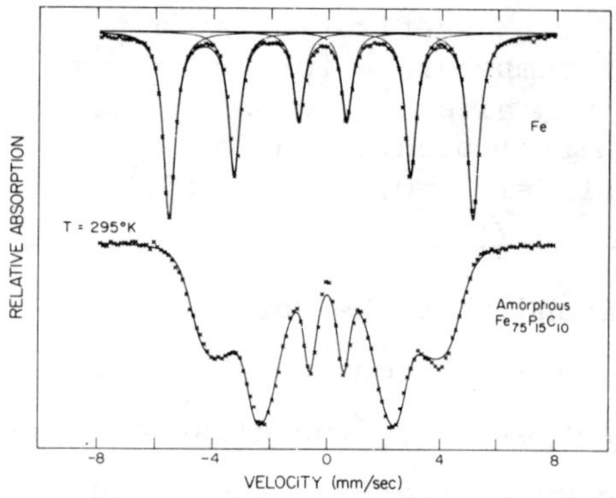

Figure 12. Mössbauer spectra for crystalline Fe and amorphous $Fe_{75}P_{15}C_{10}$ alloys.

iron at room temperature is also presented in Figure 12. The well-resolved spectrum for the amorphous ferromagnet is characterized by sharp inner peaks and broad outer ones. These characteristics are indicative of a distribution of hyperfine fields which reflects all possible local environments for Fe atoms. As discussed previously[49] the apparent absence of quadrupole interactions below T_c is probably due to a directional range of electric field gradients, with respect to the hyperfine field, which gives a line broadening rather than a line shift. The Mössbauer spectrum can therefore be fitted in terms of a distribution of hyperfine fields P(h), an average isomer shift and a line width which allows for the effect of quadrupole broadening. A reasonably good fit to the experimental data can be obtained by using the following model[49] for P(h):

$$P(h) = \begin{cases} A[(h-h_o)^2 + (\Delta_o/2)^2]^{-1} & \text{for } 0 \leq h \leq h_o, \\ \frac{4A}{\Delta_o^2} \exp[-(h-h_o)^2/2\Delta_1^2] & \text{for } h > h_o \end{cases}$$

where A is a normalization constant determined by the condition:

$$\int_0^\infty P(h)\,dh = 1$$

The calculated spectrum based on this model is shown as the solid curve in Figure 12. The fitting parameters are:

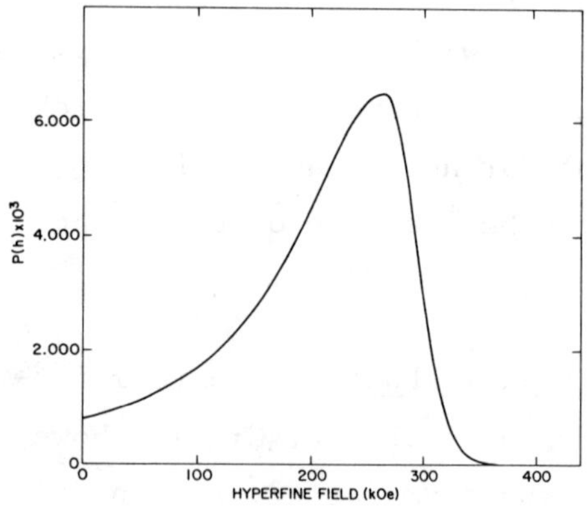

Figure 13. The hyperfine field distribution function P(h).[47]

$h_o = 264.8$, $\Delta_o = 195.6$, and $\Delta_1 = 27.6$ (all in units of kOe). The P(h) curve obtained from the fitting is shown in Figure 13. Using these numbers, one can calculate the following quantities:

$$\bar{h} = \int^{\infty} P(h)hdh \qquad \text{and} \qquad \Delta^2 = \int_0^{\infty} (h-\bar{h})^2 P(h)dh$$

It is found that the ratio $\Delta/\bar{h}$ for the amorphous $Fe_{75}P_{15}C_{10}$ is equal to 0.36. If the magnetic hyperfine field seen by a given Fe^{57} atom is proportional to its local magnetization, the quantity $\Delta/\bar{h}$ is then a measure of the ratio of root mean square deviation to the average of the magnetization. The value for $\Delta/\bar{h} = 0.36$ is consistent with the estimate of $\delta \sim 0.3$ from the magnetization data.

5.0.0 SUMMARY

In summary, three types of metastable metals and alloys have been obtained with various techniques of rapid quenching; supersaturated solid solutions, non-equilibrium crystalline phases, and amorphous metallic phases. Some of the electrical and magnetic properties of several metastable phases selected from the above categories were discussed. Simple cubic phases were found to be favorable for the occurrence of superconductivity and the origin of this phenomenon was explored. The development of achieving high T_c in Nb_3Ge was used as an example to demonstrate how rapid quenching can significantly change the physical properties of an intermetallic compound or a solid solution without altering the crystal structure. From the study of amorphous phases, one concludes that the atomic-like characteristics play a more prominent role in the amorphous phase than in its crystalline counterpart, and that many electrical and magnetic properties of amorphous metallic solids can be described in terms of a distribution function for certain physical parameters.

REFERENCES

1. Duwez, Pol, Trans. ASM 60, 607 (1967).
2. Jones, H., Rep. Prog. Phys. 36, 1425 (1973); Jones, H. and C. Suryanarayana, J. Mater. Sci. 8, 705 (1973); and the references in these two review articles.

3. Gavaler, J.R., J. Appl. Phys. Lett. 23, 480 (1973).
4. Luo, H.L. and W. Klement, Jr., J. Chem. Phys. 36, 1870 (1962).
5. Giessen, B.C. in: "Developments in the Structural Chemistry of Alloy Phases", B.C. Giessen, Ed., Plenum Press, New York (1969), 227.
6. Turnbull, D., Contemp. Phys. 10, 473 (1969).
7. Chopra, K.L., "Thin Film Phenomena," McGraw-Hill, New York (1969).
8. Tsuei, C.C., Huan-Chun Yen, and P. Duwez, Phys. Lett. 34A, 80 (1971).
9. Giessen, B.C., U. Wolff, and N.J. Grant, Trans. AIME 242, 597 (1968).
10. Jordan, C.B., J. Chem. Phys. 39, 1613 (1963).
11. Luo, H.L., M.F. Merriam and D.C. Hamilton, Science 145, 581 (1964).
12. Tsuei, C.C. and L.R. Newkirk, Phys. Rev. 183, 619 (1969).
13. Bommel, H.E., A.J. Darnell, W.F. Libby, B.R. Tittmann, and A.J. Yencha, Science 141, 714 (1963).
14. Darnell, A.J., A.J. Yencha, and W.F. Libby, Science 141, 713 (1963).
15. Wittig, J. and B.T. Matthias, Science 160, 994 (1968).
16. Johnson, W.L. and S.J. Poon, J. Appl. Phys. 45, 3683 (1974).
17. Chen, W.Y.K. and C.C. Tsuei, Phys. Rev. B5, 901 (1972).
18. Newkirk, L.R. and C.C. Tsuei, Phys. Rev. B4, 2321 (1971).
19. Westbrook, J.H., "Intermetallic Compounds," Wiley, New York (1967), 587.
20. Weger, M., J. Phys. Chem. Solids 31, 1621 (1970).
21. Tsuei, C.C. and W.L. Johnson, Phys. Rev. B9, 4742 (1974).
22. Testardi, L.R., J.H. Wernick, and W.A. Royer, Solid State Comm. 15 (1974); Testardi, L.R., R.L. Meek, J.M. Poate, W.A. Royer, A.R. Storm and J.H. Wernick, Phys. Rev. B11, 4304 (1975).
23. Pan, V.M., V.I. Latysheva, and E.A. Shishkin, "Physics and Metallurgy of Superconductors," ed. by E.M. Savitskii and V.V. Baron (Consultants Bureau, New York) [1970], 179.
24. Geller, S., Acta Crystallogr. 9, 885 (1956).
25. Matthias, B.T., T.H. Geballe, R.H. Willens, E. Corenzwit, and G.W. Hull, Jr., Phys. Rev. 139, A1505 (1965).
26. Gavaler, J.R., M.A. Janocko and C.K. Jones, J. Appl. Phys. 45, 3009 (1974); Santhanam, A.T. and J.R. Gavaler, J. Appl. Phys. 46, 3633 (1975).
27. The quantity $\Delta\rho$ is defined as the change in resistivity due to the addition of magnetic elements to the host alloy.
28. Hasegawa, R. and C.C. Tsuei, Phys. Rev. B2, 1631 (1970).
29. Kondo, J., Solid State Physics, ed. by F. Seitz, D. Turnbull and H. Ehrenreich, Vol. 23, p. 183, Academic Press, New York (1975).
30. Heeger, A.J., ibid, p. 283.
31. Tsuei, C.C. and R. Hasegawa, Solid State Comm. 7, 1581 (1959).
32. Maitrepeirre, P.L., Ph.D. Thesis, CA Inst. Tech. (1969).
33. Hasegawa, R. and J. Dermon, Phys. Lett. 42A, 407 (1973).
34. Buckel, W. and R. Hilsch, Z. Physik 138, 109 (1954).
35. Barth, N., Z. Physik 142, 58 (1955); Buckel, W. and R. Hilsch, Z. Physik 146, 27 (1956).
36. Wühl, N., J.E. Jackson, and C.V. Briscoe, Phys. Rev. Lett. 20, 1496 (1968).
37. Collver, M.M. and R.H. Hammond, Phys. Rev. Lett. 30, 92 (1973).
38. Chencinski, N. and F.J. Cadieu, J. Low Temp. Phys. 16, 507 (1974).
39. Johnson, W.L., S.J. Poon, and P. Duwez, Phys. Rev. B11, 150 (1975).

40. Anderson, P.W., J. Phys. Chem. Solids 11, 26 (1959).
41. Hammond, R.H. and M.M. Collver, in: Low Temperature Physics—LT13, ed. by W.J. O'Sullivan and E.F. Hammel; Plenum Press, New York (1974), 3, 532.
42. Kerker, G. and K.H. Bennemann, Z. Physik 264, 15 (1973).
43. Hopfield, J.J., Phys. Rev. 186, 443 (1969).
44. Collver, M.M., Lawrence Berkeley Laboratory Rpt. No. LBL-167 (1971).
45. Cargill, G.S. III, Solid State Physics, ed. by F. Seitz, D. Turnbull and H. Ehrenreich; Academic Press, New York (1975), 30, 225.
46. Giessen, B.C. and C.N.J. Wagner in Physics and Chemistry of Liquid Metals, ed. by S.Z. Beer; Marcel Dekker, New York (1972), 633.
47. Tsuei, C.C. and H. Lilienthal [to be published in Phys. Rev. B].
48. Rhyne, J.J., J.H. Schelleng, and N.C. Koon, Phys. Rev. B10, 4672 (1974).
49. Sharon, T.E. and C.C. Tsuei, Phys. Rev. B5, 1047 (1972).
50. Handrich, K., Phys. Stat. Sol (b) 53, K17 (1972).

State-of-the-Art and Prospects for Magnetic, Electronic and Mechanical Applications of Amorphous Metals

A Synopsis of the ONR Materials Workshop at Northeastern University, Boston, Mass., November 20 - 21, 1975

D. E. POLK and B. C. GIESSEN

Institute of Chemical Analysis, Applications and Forensic Science, Northeastern University, Boston, Mass. (U.S.A.)

F. S. GARDNER

Office of Naval Research, Boston, Mass. (U.S.A.)

SUMMARY

A workshop on the State-of-the-Art and Prospects for Magnetic, Electronic and Mechanical Applications of Amorphous Metals was held on November 20 - 21, 1975, to provide a forum for discussions of the potential applications of amorphous metals and their related physical properties. The 150 attendees of the ONR Materials Workshop at Northeastern University, Boston, included representatives from academia, industry and government, many of whom had also participated in the immediately preceding Second International Conference on Rapidly Quenched Metals held at M.I.T. Half-day sessions were devoted to: (1) preparation and characterization, (2) mechanical applications and (3) magnetic and electronic applications. While statements at the workshop were not for attribution, this report seeks to present a synopsis of the views which were expressed. Although amorphous metals are clearly highly promising for technology, it is indicated that a large amount of essential research and development lies ahead.

INTRODUCTION

Amorphous metals are attracting ever-increasing attention from the world's technological community, perhaps because they offer something for everyone. At this point in time, many studies have concentrated on characterizing and understanding the magnetic, electronic and elastic/plastic behavior of these metallic solids which are distinguished from other metals by their lack of long-range structural periodicity. Other studies have considered the kinetics and thermodynamics which are related to the formation and transformations of these metastable phases. Other interest has been attracted by the likelihood that the desirable magnetic, mechanical and corrosion-resisting properties of selected amorphous alloys will find commercial applications.

It was to address the potential and associated problems of amorphous metals that the ONR Materials Workshop at Northeastern University was convened on November 20 - 21, 1975. The Workshop had been arranged by the authors and was co-chaired by F. S. Gardner, ONR Boston, and B. C. Giessen, Northeastern University. The minimal program structure was designed to maximize free discussion. Three half-day sessions were conducted as follows:

I. Preparation and Characterization of Amorphous Metals.
   Co-chairmen: R. E. Maringer, Battelle Memorial Institute and G. S. Cargill III, IBM.
   Keynote Speaker: D. E. Polk, Northeastern University.

II. Mechanical Applications of Amorphous Metals.
   Co-chairmen: B. H. Kear, Pratt & Whitney Aircraft and T. Masumoto, Tohoku University.

Reprinted with permission from Materials Science and Engineering, June 1976, 309-316, © 1976 Elsevier Sequoia S.A.

Keynote Speaker: J. C. M. Li, University of Rochester.

III. Magnetic and Electronic Applications of Amorphous Metals.

Co-Chairmen: J. J. Rhyne, N. B. S. and C. D. Graham, University of Pennsylvania.

Keynote Speakers: G. S. Cargill III, IBM and C. D. Graham, University of Pennsylvania.

PREPARATION TECHNIQUES

Amorphous metals are generally produced by techniques divided into two classes: those in which the corresponding liquid is rapidly cooled and those involving an atomic deposition process. The liquid-quench techniques typically use a metal substrate to absorb the heat, while sputtering is now the most commonly utilized deposition technique.

While simple in principle, the liquid-quench techniques often suffer from problems in producing a product which is uniform enough to meet commercial requirements. For example, the cooling rate may vary along the sample due to uneven surface contact which can then lead to internal stresses or localized crystalline regions. Interface effects, due to substrate surface roughness or to surface films, thus become of major importance in attempts to optimize the process. In addition, the macroscopic dimensions of a fiber often must be carefully controlled if additional processing is to be avoided.

It was suggested that the Taylor process could be used to achieve quench rates of about 10^5 °C/sec and hence result in an amorphous metal. In this process, a liquid metal is held in a glass tube selected to be viscous at the operating temperature; the combination is then stretched out to give a fine wire filament of high quality which can be as small as one micron in diameter. The method was believed to be practised in Russia and Italy, but American experience with the technique was not known.

Direct melt-spinning of wire, where the molten metal is ejected from an orifice into free-flight, is also used to form wires of amorphous metals. Typically, wires of 3 to 5 mils are produced, while very small sizes are apparently prevented by operational problems. The most severe problem is in keeping the nozzle open since the orifice tends to clog with impurity particles. Thus, the cleanliness of the melt becomes the overriding consideration.

It should be possible to prepare the final amorphous metal product directly in many more shapes than the range from round wire to narrow ribbons which is presently produced. Shaped substrates can be used to form a product with angles. Wide sheets may be possible by directing a stream of metal at a solid surface such that it broadens out into a uniform sheet of liquid which then impinges on a rapidly moving substrate. A wide screen can be made at 10 or 20 ft./sec by using multiple orifices whose jets are independently aimed and cross over each other at the substrate while the metal is still molten.

High rate sputtering, by which films are made at the rate of about a micron/minute so that samples several mm thick are readily produced, has been developed. It is thought that several hundred square inches of material can be made at one time. Even for difficult alloys, the technique should be capable of achieving production costs of a few dollars/in^2 for material ½ mm thick, a reasonable cost for electromagnetic components. It was noted that this process had an electrode configuration designed to minimize substrate heating. However, a good technique to measure substrate temperature is still needed.

Thick amorphous materials have also been produced using plasma gun techniques. In one case, the substrate rotates and the gun oscillates to cover a wider substrate area and thus reduce substrate heating. In the other case, a gas jet at right angles to the plasma sweeps away the hot plasma gas. Samples which are several mm thick can be made this way, though they have about 10 - 15% porosity. However, the thickness appears to be limited by the internal strains which increase with thickness until the deposit cracks or separates from the substrate.

For any of these preparation techniques, the quench rates that are actually achieved depend upon the process being used and the actual process variables being applied. For melt quenching, the rate depends strongly on the thickness and on the heat transfer coefficient, which itself depends on surface oxide films, trapped gas, surface roughness, etc.

Source: Materials Science and Engineering, June 1976, 309-316

The surface heat transfer coefficient becomes the dominant factor determining cooling rate as the thickness gets very small. It was suggested that approximate upper limits are 10^{10} K/sec for the gun technique, 10^8 for the piston and anvil method, 10^6 for the melt extraction, and 10^4 for atomization. Atomic deposition techniques, by consideration of their energy and likely equilibration times, can be thought of as producing "cooling rates" of $\sim 10^{15}$ K/sec.

Sputtered films can have a wide range of entrapped gas content, ranging from undetectable levels up to about 10%. The amount incorporated is affected by the biasing conditions, the substrate temperature and the sputtering rate.

In any application of amorphous metals, the uniformity of engineering properties, as dependent on processing, is of importance. Any potential commercial process must give predictable properties. An examination of melt-spun ribbons produced in the U.S. showed compositional uniformity along the ribbon, but magnetic domain studies showed large scale periodic fluctuations which were thought to be due to fluctuations in internal stress. This periodicity, which is on a scale of somewhat less than a mm, appears to be unrelated to surface features due to the quench surfaces. It also appears on ribbons made in Japan. The periodicity might be related to an instability in the stream of molten metal; this origin can be tested since it would also lead to a dimensional fluctuation. Variations in saturation magnetization, coercive force and other magnetic properties from spool to spool have also been seen, possibly due partly to compositional variation. It may be that those variations due to process variables can be eliminated by annealing. In addition, the edge and the center portions of such a ribbon were reported to have different behavior in the differential scanning calorimeter (DSC), presumably due to different cooling rates. Such unexplained periodicities also appear on ribbons made by melt extraction; these may be due to variations in the heat transfer along the surface.

Alternatively, these periodicities may be related to the vortex rings left behind when two metals collide during explosive welding. The propagation rates in explosive welding are not unlike the speeds with which the liquid metal contacts the substrate in these liquid-quench processes. The turbulence at the quench interface could be periodic. The ripples in explosive welding of Cu and Al are also of the order of 1 mm.

It is of interest to determine the effect of changing the process variables, *e.g.* changing the substrate speed by a factor of 2, on the periodicity.

An important point concerning the processing techniques used to prepare amorphous metals is that they may also be valuable when microcrystalline material is desired. For example, a rapidly quenched fine powder, preferably of a narrow size spectrum, is of great value for producing superplastic material through powder metallurgical techniques such as pressing and sintering. Similarly, wide sheets which were superplastic at high temperatures could be used to fabricate very strong materials into intricate shapes of many kinds by blow forming.

Electro and chemical deposition were suggested to deserve much more attention than they have so far received. It was noted that such deposition of amorphous Ni–P was already in commercial use for protective coatings at the time liquid quenching to produce amorphous metals was developed in the 60's.

Interpretation of the contrast observed in transmission electron microscopy (TEM) remains uncertain. However, it was suggested that what some workers have seen in the past using tilt-field illumination, *e.g.* the reported "lattice fringes" in amorphous metals, are essentially artifacts. The origin of contrast seen in high resolution TEM using axial illumination is still not clear.

Auger spectroscopy has been used to examine the fracture surface of a Ni–Fe–P–B alloy, which had been annealed before fracture at low temperature to cause "embrittling". An enrichment of the fracture surface in phosphorus was reported.

A more complete characterization of amorphous metal samples used in property studies is seen as a general need. Thus, it has been observed that samples which appear identical using the standard X-ray examination can behave quite differently in other experiments, *e.g.* they can have variations in ductility and magnetic properties. Techniques which may be useful in detecting differences between samples include DSC, TEM and small angle X-ray scattering (SAXRS). Any measurement providing information about the local atomic configuration would be of special usefulness in this regard.

FORMATION/ANNEALING EFFECTS/KINETICS

We still lack the ability to predict with the required accuracy which compositions can be made amorphous with a given process or what quenching rate is required to form an amorphous metal for a given composition. More thermodynamic and kinetic data are needed since glass formation is apparently governed both by the free energy difference between the glass and the competing crystalline phases and by the rates of nucleation and of crystal growth at different temperatures. Of special interest would be the free energy difference between the homogeneous supersaturated solid solution of simple structure and the amorphous phase.

For melt spinning with good surface contact, the cooling rate for material near the substrate can be higher than that for material at the free surface, leading in some cases to structural differences. In one study, a ribbon for which electron microscopy could detect no differences between the substrate and the free side was annealed; annealing caused some crystallization on the free side but not on the substrate side, indicating quenched-in differences.

It is still not clear if this is due to oxygen contamination, as suggested by studies of the vapor deposition of "pure" metals which show that small amounts of oxygen can have a large effect on the crystallization rate. The effects of the oxygen contained within the amorphous metals is an area which has not received enough attention. Alloys rich in rare-earth elements are of special difficulty in this regard, and such materials, as produced in the amorphous state without further precautions, may contain up to 2% oxygen due both to original material impurity and to alloy preparation and quenching.

For amorphous Fe–C, it appears that the growth of the crystal interface into the amorphous material is not simply controlled by diffusion. The growth is very rapid initially and then decreases to a slow rate. It was also noted that the crystallization behavior of amorphous $Fe_{80}C_{20}$ is changed drastically by the addition of only 1% Si.

It is not known whether there are conditions of practical interest for which the bulk of an amorphous metal would be unstable with respect to a local fluctuation, *i.e.* where an externally induced localized crystallization would advance through the entire sample.

It has been observed that Fe and Ni–Cr based alloys containing only B as the metalloid species appear to have better stability at temperatures well below T_g than similar alloys containing P. In particular, the low temperature embrittlement reported to precede observable crystallization in Fe–Ni–P–B alloys is not seen in the corresponding P-free alloys. However, it is not yet clear whether P segregation is the cause of embrittlement in Fe–Ni–P–B alloys upon annealing. Structural studies have suggested a strong short range order, with P-atoms supposedly avoiding each other; the concept of P-rich regions would seem to be inconsistent with this view. However, this problem does not arise in one suggested hypothesis according to which P migrates to sites where it can form a lower energy precursor.

MECHANICAL PROPERTIES/APPLICATIONS

The relatively high strength and hardness of amorphous metals are of interest for commercial applications. Their yield strength is generally $E/50$ to $E/80$ (where E is Young's modulus), while E is 60 to 75% of that of the corresponding equilibrium crystalline phases. For an Fe–B alloy, strengths exceeding 500 k.s.i. and a hardness of 1100 kg/mm^2 have been obtained for as-quenched material. Further, these high strength amorphous metals generally exhibit ductility, *e.g.* they can be bent sharply or rolled down.

When measuring the fracture strength of an amorphous metal, one should be aware that the tensile strength measured on an as-produced liquid quenched fiber is typically 10 to 15% less than the true yield strength measured in tension on a specimen with polished edges and appropriate cross section. However, the fracture strength can be quite uniform along an amorphous metal fiber; for 3 kg of ribbon, the fracture strengths of a random distribution of samples were within ±3% of the average value.

It was suggested that terminology should be standardized as follows: *shear band* — the planar regions of localized deformation; *shear fracture* or *shear band fracture* — the mode of fracture occurring along a shear band after partial offset; *veins* — the features due to intersecting cracks seen on shear band fracture surfaces.

The relative merits of the two models for the plastic deformation of amorphous metals, the free volume model and the dislocation model, were discussed at length, though no concensus was reached.

A question deserving attention is whether one gets unique properties, *e.g.* enhanced ductility, if one heats the amorphous metal to temperatures near the crystallization temperature while at the same time subjecting it to deformation.

Amorphous metal fibers which are stable up to 400 to 500 °C would be useful for the straightforward reinforcement of aluminum alloys. Further, amorphous metals may also be useful as a hard coating on aluminum alloys; adherent amorphous metal layers have been put on such alloys by spraying, though their wear resistance and the spraying method were not divulged.

In composites, it may be possible to get even better properties by putting an as-quenched, ductile amorphous metal into a sufficiently stable matrix and then heat treating the composite to partially crystallize the amorphous metal so as to increase its modulus and strength.

It should be possible to produce amorphous fibers with strengths of at least 600,000 p.s.i.; these could be encased in plastic to make a composite especially suitable, because of the fiber toughness, for applications like automobile fenders. Forming would be easy upon gentle heating, which may possibly be produced by resistance heating the fibers. The advantage of the amorphous metal fiber compared to a glass fiber is its much greater toughness.

Amorphous metal composites made with an epoxy matrix were reported to have imperfect bonding between the fibers and the matrix. More work is needed to solve this problem, probably by an appropriate surface coating.

Amorphous metals may be useful as a precursor material; a suitable amorphous alloy would be heated to crystallize it to a desired microstructure by controlling the time/temperature characteristic of the annealing. This route can be attractive when a compositionally uniform alloy of very small grain size is desired. This may also be a preferred way of obtaining a given metastable crystalline phase or a phase which will not form directly from the melt.

CORROSION BEHAVIOR/APPLICATIONS

It was reported that amorphous metals containing chromium have been made which are "extremely corrosion resistant", both under static and dynamic conditions, as compared to ordinary stainless steels. For such alloys, tensile specimens exposed to corrosion-active liquids, such as sea water, did not show evidence of stress corrosion cracking.

It is, however, true that amorphous metal alloys can be susceptible to hydrogen embrittlement. Here the hydrogen may be produced in the alloy by corrosion; also it was found that some as-prepared specimens contained about 18 ppm of hydrogen. The degree of hydrogen embrittlement occurring in the presence of Cl^- ions in the solution was discussed.

MAGNETIC AND ELECTRONIC PROPERTIES/ APPLICATIONS

Materials for magnetic bubble applications

Thin films of amorphous metals formed by sputtering have been shown to be suitable for application in magnetic bubble memories in a manner similar to crystalline garnets. These films are anisotropic, with the magnetic easy axis perpendicular to the film. The striped magnetic domains which occur naturally are converted to cylindrical domains ("bubbles") by the application of a magnetic field. The alloys typically contain a rare earth and a transition metal; the composition and preparation parameters must be chosen so as to produce the proper magnetization, anisotropy energy and exchange energy. Suitable films have been made with compositions near $Gd_{60}Co_{40}$, with the addition of Mo.

Molybdenum is included to produce a film with a relatively small temperature variation of magnetization and bubble size (corresponding to a similar constancy of anisotropy and exchange energy) near the operating temperature of ~320 K.

At this time, the origin of the magnetic anisotropy is not fully understood on an atomic scale, nor is its temperature dependence. Attempts are being made to determine its origin using X-ray measurements. It is noted, however, that the magnetic anisotropy is complex and that its degree is governed by the operating parameters used in the sputtering process.

Generally, anisotropic Gd–Co films are produced by imposing a bias during sputtering; however, some rf sputtered films have perpendicular anisotropy without having been biased during deposition. One can get perpendicular anisotropy in vapor deposited Gd–Fe, while vapor deposited Gd–Co has inplane anisotropy.

For any bubble application, the thermal stability of the film becomes of paramount importance. One must know what changes will occur in the film during the further processing needed to turn the film into an operating device, which involves some relatively high temperature steps, and during its lifetime in use.

Permanent magnet materials

Another area of major interest also involves rare earth (RE)–transition metal amorphous alloys, with compositions such as $(RE)Fe_2$. These materials can be made over a wide composition range by sputtering, *e.g.* in the Tb–Fe system at least over the range from 11% Tb to 88% Tb (though an amorphous alloy with 2% Tb was also reported in a different study). Such amorphous alloys are presently of interest partly as an intermediate material which can be crystallized to a desirable, very fine grained structure of potential use as a permanent magnet with high energy product, as discussed below.

Amorphous $(RE)Fe_2$ alloys are also reported to have interesting magnetic properties at low temperature. At 4 K, amorphous $TbFe_2$ has a coercive field of 30,000 oersteds and saturation magnetization of over 100 emu/g, leading to one of the highest values of the energy product, $B \cdot H_{max}$, ever measured. However, the large coercive field drops off rapidly with increasing temperature so that the observed high energy product is not of practical use. The decrease appears to be due to an activation process related to a weak coupling between magnetic clusters in amorphous metals.

Unlike the bubble materials, samples of amorphous $TbFe_2$ do not show a macroscopic anisotropy. It is thought that each rare earth atom experiences a separate "crystal field" due to its environment, such that each rare earth atom possesses a randomly oriented easy magnetization direction and hence will have a higher energy when it reorients under an applied field. Thus, each direction can be a hard direction, with a large coercive field, although the sample is macroscopically isotropic. A model for this effect assumes superparamagnetic regions of about 100 to 300 $Å^3$. The existence of such magnetic clusters has been demonstrated by low temperature small angle neutron scattering. Measurements above the Curie temperature demonstrate that such clusters are not due to compositional fluctuations; their origin is thus unknown.

To produce useful permanent magnet materials, amorphous metals such as $TbFe_2$ with room temperature coercive fields of about 100 to 150 oersteds are annealed at >400 °C to give a crystalline sample of about 100 Å grain size with a coercive field of 3,500 to 4,000 oersteds. One can select the heat treating to give the most desirable microstructure and resultant magnetic properties, including the energy product.

Additionally, if one crystallizes in a magnetic field at a temperature somewhat below the temperature of rapid crystallization, (*e.g.* at 350 to 400 °C), one can induce magnetic anisotropy in the crystallized sample even though $TbFe_2$ is cubic. This apparently occurs because the domains have anisotropic shapes, although this has not yet been demonstrated.

For the composition Sm_2Co_{17}, the beginning stages of crystallization of the amorphous metal appear to be accompanied by compositional fluctuations which then disappear upon further annealing. However, these films seem to be relatively stable since long-time anneals at temperatures up to

200 °C cause no change in the magnetic behavior.

Since liquid quenching is much less expensive than sputtering, it would be desirable to determine whether or not alloys such as $TbFe_2$ can be quenched to an amorphous metal from the liquid.

Soft magnet materials

The type of magnetic behavior found in amorphous metals which now attracts most attention is the "soft" ferromagnetism, *i.e.* the high magnetic permeability, shown by liquid-quenched alloys containing about 80 at.% Fe–Co–Ni and 20 at.% B–C–Si–P. The saturation magnetization ranges up to about 20,000 gauss while the coercive fields generally range downward from 0.1 oersted to values below the limit of measurability. Such materials may find use as replacements for the crystalline soft magnets now in use; their mechanical hardness and low cost compared to permalloy are attractive features. The ribbon shape generally produced is especially appropriate for permalloy-type applications.

Generally, as-produced ribbons of these materials exhibit a very rapid magnetization up to about half of the saturation value, followed by a slow approach to saturation at 20 to 30 oersted. This seems to be due to internal stresses interacting with the magnetostriction. One can remove this undesirable characteristic by: (1) superimposing a uniaxial stress on the ribbon; (2) annealing the material to get stress relaxation (though in some cases this can cause embrittlement); and (3) selecting a material with zero magnetostriction. It was noted that the recently announced Fe–B alloys can be annealed to relieve the internal stresses, and thus improve their as-formed magnetic properties, without causing embrittlement.

For iron–nickel alloys, stress relaxation generally occurs at about 300 °C without detectable crystallization. An annealed sample of such an amorphous metal was said to be as good a soft d.c. magnetic material as is known to man. While its a.c. properties are not completely established, it appears that they are similar to those of oriented permalloy. Further, there is no evidence that the soft magnetic properties of amorphous metals have yet been optimized, both in terms of composition and processing variables.

It was noted that one can get an induced anisotropy by annealing in a magnetic field, though this appears to be unrelated to the stress relaxation or embrittlement effects of annealing. This magnetic annealing may be due to a change in the pair ordering. Another type of anisotropy may be that due to the flow patterns within the metal during quenching. This effect is reported to appear even in materials of zero magnetostriction and thus to be apparently not due to internal stresses.

While the ability to produce a desirable magnetic anisotropy and domain pattern by annealing in a magnetic field is an asset, the same effect may cause difficulties at in-service temperatures. The appearance of magnetic anisotropy upon annealing an Fe–Ni alloy in a field is observed to be a first-order process with an activation energy of 1.4 eV and a pre-exponential of $10^{11} sec^{-1}\ cm^{-1}$. For such a rate, one would see appreciable changes in the d.c. characteristics after only 1000 h at 100 °C in the worst possible situation, *i.e.* starting in the low remanence state. However, if used in their high remanence state even presently available materials would have a sufficiently long useful life. Further, other alloys may be more stable in this regard; for example, for an Fe–P–C alloy an activation energy of 2.2 eV was reported so that this alloy would be expected to be more stable in service at 100 °C.

The list of potential uses of soft ferromagnetic materials is quite large. One can consider them as relating to one of the magnetic characteristics of the material:

(a) minimum energy loss: transformer type applications;

(b) very high permeability: sensors, such as flux meters, recorder heads, ground pole detectors; shielding;

(c) square loop: switch cores, low frequency inverters; magnetic amplifier cores; and

(d) constant permeability: filter chokes, transducers, inductors.

Additionally, as-formed amorphous metals containing internal stresses may be useful in a stress-measuring device. Because of the good mechanical properties, the materials may find use in magnetostrictive delay devices. Amorphous metals may also be inexpensive enough to be used as a replacement for silicon–iron in motors and generators, where they could reduce the power loss. Losses as low as 40 mW/lb. have been observed for magnetically annealed material, though this sample had a saturation field of only 8,000 gauss.

The amorphous metals may not be attractive for use in large power transformers because the low packing fraction to be expected with several mil thick material and the corresponding relatively low flux density would unacceptably increase the size and weight of the transformer. However, pole-type distribution transformers are a more promising potential application.

In terms of research, more information on the thermal stability of such materials and a better understanding of the changes that occur upon low temperature annealing is needed. Creep and internal friction studies may also provide further insight into magnetic behavior.

Resistors

The resistivity behavior of amorphous metals may also be of use. It may be possible to make use of the near-constant resistivity associated with the shallow Kondo-like minimum measured on various amorphous metals or of the relatively linear temperature dependence of the resistivity over a wide temperature range. There was discussion concerning the origin of the resistivity minimum; it was suggested that this effect is intrinsic to amorphous metals rather than being a Kondo effect since Pd–Si alloys with two different trace Fe contents had identical resistivity minima.

REFERENCES

The following publications (1 - 6) and conference proceedings (7, 8) are recommended as sources of additional information on amorphous metals.

1 B. C. Giessen and C. N. J. Wagner, in S. Z. Beer (Ed.), Liquid Metals, Chemistry and Physics, Marcel Dekker, New York, 1972, p. 633.
2 H. Jones, Rept. Prog. Phys., 36 (1973) 1425.
3 J. J. Gilman, Physics Today, 28 (5) (1975) 46.
4 T. Masumoto and R. Maddin, Mater. Sci. Eng., 19 (1975) 1.
5 A. K. Sinha, B. C. Giessen and D. E. Polk, in N. B. Hannay (Ed.), Treatise on Solid State Chemistry, Vol. III, Plenum Press, New York, 1976, p. 1.
6 G. S. Cargill III, in C. D. Graham Jr., G. H. Lander and J. J. Rhyne (Eds.), Magnetism and Magnetic Materials — 1974, AIP Conf. Proc. No. 24, Amer. Inst. of Physics, New York, 1975, p. 138.
7 In N. J. Grant and B. C. Giessen (eds.), Proc. Second Intern. Conf. on Rapidly Quenched Metals, MIT, 1975, Section I, MIT Press, Cambridge, Mass., 1976.
8 In N. J. Grant and B. C. Giessen (eds.), Proc. Second Intern. Conf. on Rapidly Quenched Metals, MIT, 1975, Mater. Sci. Eng., 23 (2/3) (1976).

LIQUID-QUENCHED AMORPHOUS SUPERCONDUCTORS*

POL DUWEZ and W. L. JOHNSON

W. M. Keck Laboratory of Engineering Materials, California Institute of Technology, Pasadena, Calif. 91125 (U.S.A.)

(Received June 8, 1978)

Summary

Since the first amorphous superconducting alloy obtained by liquid quenching was reported in 1975, a systematic search for new alloy compositions has been carried out. Thirteen alloys are now known to be superconductors; their transition temperatures range from 1.5 to 9.0 K.

The first amorphous superconductor was reported by Buckel and Hilsch in 1954 [1]. Amorphous bismuth with a transition temperature of 6 K was obtained as a thin film by evaporation on a helium-cooled substrate. Since then many investigators have used this method to obtain amorphous alloys as well as metals. The superconductors produced by evaporation on a cold substrate are unstable and generally crystallize below room temperature. In 1975, an amorphous superconducting alloy was obtained by rapid quenching from the liquid state [2]. The composition of this alloy, $La_{80}Au_{20}$, was chosen firstly because La is a superconducting element and secondly because there is a relatively low melting point eutectic at 20 at.% Au in the La–Au phase diagram. This low eutectic condition was pointed out as early as 1961 [3] as a favorable condition for obtaining an amorphous alloy by liquid quenching. However, it is not a necessary condition and some liquid quenched amorphous alloys do not correspond to eutectic compositions. The foils obtained by liquid quenching are about 50 μm thick and can be considered as bulk films in comparison with evaporated films. In addition, superconducting alloys prepared by liquid quenching are very stable at room temperature and above. A few binary alloys prepared by sputtering have been reported to be amorphous and superconducting. These alloys are typically 1 - 5 μm thick and are stable at room temperature. Because of the very limited information published on sputtered amorphous superconductors, they will be mentioned only briefly in this paper. The discussion will center mainly on liquid-quenched materials.

Table 1 is a list of amorphous superconductors reported in the literature. Only qualitative remarks can be made about a correlation between the chemical composition of these alloys, the fact that they are amorphous after liquid

---

*Dedicated to Professor B. T. Matthias in celebration of his 60th birthday.

Reprinted with permission from Journal of the Less-Common Metals, Vol. 62, November-December 1978, 215-220, © 1978 Elsevier Sequoia S.A.

TABLE 1

Superconducting properties of amorphous alloys

Alloy composition	T_c (K)	Width (K)	dH_{c1}/dT (Oe K^{-1})	dH_{c2}/dT (kOe K^{-1})	Ref.
Liquid-quenched					
$La_{80}Au_{20}$	3.5	–	–	–	2, 4
$La_{76}Au_{24}$	3.3	0.03	–	24.0	2, 4
$La_{70}Cu_{30}$	3.5	0.03	–	–	2
$La_{80}Ga_{20}$	3.8	0.1	–	22.5	5
	3.84[a]	0.13[a]	–	–	5
$La_{78}Ni_{22}$	3.0	~0.1	–	–	2
$Zr_{70}Pd_{30}$	2.4	<0.4	–	26.5	4, 6
$Zr_{75}Rh_{25}$	4.55	0.005[b]	6.5 - 8.0	26.3	7, 8, 12
$Zr_{70}Be_{30}$	2.80	~0.01	–	23.8	9
$Nb_{58}Rh_{42}$	4.7	~0.2	–	31.0	4
$Nb_{60}Ni_{40}$	1.5	<0.4	–	–	10, 4
$Nb_{58}Rh_{26}Ni_{16}$	3.4	–	–	31.5	4
$Mo_{80}P_{10}B_{10}$	9.0	~0.1	4.2	17.6	11, 12
$(Mo_{40}Ru_{60})_{80}P_{20}$	4.68	~0.1	17.5	26.8	11, 12
$(Mo_{60}Ru_{40})_{80}P_{20}$	6.18	~0.1	6.5	25.5	11, 12
$(Mo_{80}Ru_{20})_{80}P_{20}$	7.31	~0.1	–	24.5	11
$(Mo_{60}Ru_{40})_{80}B_{20}$	5.78	~0.1	–	28.0	13
$(Mo_{80}Re_{20})_{80}P_{10}B_{10}$	8.71	~0.1	–	24.2	11
Sputtered					
$Re_{70}Mo_{30}$	7.62	0.021	–	–	14
Nb_3Ge	3.34	0.014	–	–	14

[a]Calorimetric measurements. [b]Critical phenomena study.

quenching and the existence of superconductivity. The first seven alloys listed in Table 1 are of the La–Au type. They all contain about 70 - 80 at.% of a relatively large atom; the balance is an atom of smaller radius. This is a condition which has been recognized as favorable to the amorphous state on the basis of the random packing of hard spheres model [15]. In addition, these alloys satisfy the low temperature eutectic condition mentioned for $La_{80}Au_{20}$. The two alloys $Nb_{58}Rh_{42}$ and $Nb_{60}Ni_{40}$ are quite different from the previous seven because they are close to the equiatomic composition. However, they are still located near a low melting point eutectic between two compounds in the phase diagram. The ternary alloy $Nb_{58}Rh_{26}Ni_{16}$ is just a mixture of the two superconducting alloys Nb–Rh and Nb–Ni and it is not surprising that it is also a superconductor.

The last six alloys containing three or more elements belong to a class of amorphous alloys consisting of transition metals and metalloids. As has been pointed out [16], a general formula for this class of amorphous alloys is $T_{75-80}G_{25-20}$ in which T represents the sum of the transition metals and G the sum of the metalloids or glass-formers. A typical example is the rather complex ferromagnetic amorphous alloy $Ni_{49}Fe_{29}P_{14}B_6Si_2$. It is probable that more amorphous superconducting alloys of the type T_{75-80}-

G_{25-20} will be obtained by liquid quenching. The metal–metalloid approach to the problem appears very promising because, as shown in Table 1, these alloys have relatively higher transition temperatures than those containing only metals, *i.e.* in the range of about 4.7 - 9.0 compared with 1.5 - 4.7 K. The highest value of T_c reported so far for these materials (9 K) is still very low compared with the transition temperatures achieved for some of the crystalline alloys (as high as 23 K). From theoretical studies of superconductivity, it cannot be concluded that lack of periodicity in a solid will always result in a low superconducting transition temperature. In fact, in the case of thin films of pure f.c.c. or h.c.p. metals deposited at liquid helium temperature the T_c of the amorphous film is higher than that of the crystalline metal (see Table 1 of ref. 17). The possibility of obtaining a high temperature amorphous alloy by liquid quenching cannot be ruled out. A comparison between the transition temperatures of an alloy in the amorphous and crystalline states is meaningful only if the crystalline alloy is single phase. Unfortunately this is not generally the case and among the sixteen alloys listed in Table 1 only one, $Nb_{60}Ni_{40}$, crystallizes in a stable μ phase. The transition temperature has not been measured for the crystalline alloy but is known to be below 1.4 K compared with 1.5 K [4]. The only other cases found in the literature refer to two sputtered amorphous alloys listed in Table 1. For $Re_{70}Mo_{30}$ the amorphous T_c is within the range found for the crystalline σ phase [18], but T_c for amorphous Nb_3Ge is much lower than that of the A15 crystalline phase [14, 19].

The intrinsic properties of amorphous superconductors have been reported for only a limited number of alloys but the results clearly show the effect which the absence of lattice periodicity has on these properties. There is a difference, for example, in the effect of a magnetic field on the transition temperature. A typical example is shown in Fig. 1 for an amorphous La_{76}-Au_{24} alloy compared with the same alloy heat treated into a highly disordered microcrystalline state which is probably a mixture of La and La_2Au. The important feature of the H_{c2} *versus* T curve for the amorphous alloy is its linearity down to a relatively low temperature compared with the concave downward curvature of the crystalline alloy. While the curve for the crystalline alloy lies below the relationship based on Maki's theory, the linear portion of the curve for the amorphous alloy clearly extends above the Maki curve. This is not unique to the La–Au alloy but has been found in more than ten amorphous alloys studied so far. Because of temperature and field limitations, it is not known how far down the temperature scale the H_{c2} *versus* T relation remains linear; it is hoped that experiments will be carried out soon below 1.5 K and at higher fields. From such measurements it will be possible to obtain reliable values of H_{c2} at 0 K.

Because of their highly disordered structure, amorphous alloys have a high residual resistivity ρ_0 and a short electronic mean free path. For La_{78}-Au_{32}, for example, ρ_0 is about 200 $\mu\Omega$ cm [2] and the mean free path is of the order of the interatomic distances. The small mean free path in an amorphous superconductor leads to a small coherence length ξ, a large

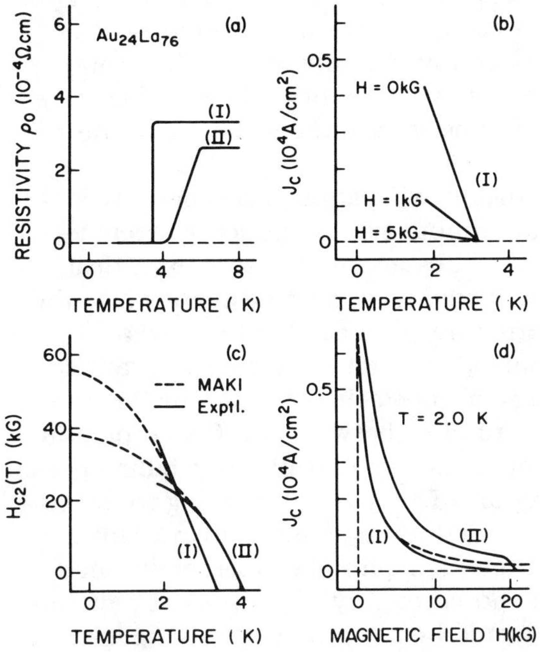

Fig. 1. Measurements of the electrical resistivity, critical magnetic field and transport properties of amorphous (I) and highly crystallized (II) $La_{76}Au_{24}$ alloys.

penetration depth λ and a large Ginzburg–Landau parameter κ. These parameters can be estimated from experimental measurements of the temperature dependence of the upper and lower critical fields $H_{c2}(T)$ and $H_{c1}(T)$. Values of the gradients $(dH_{c2}(T)/dT)_{Tc}$ and $(dH_{c1}(T)/dT)_{Tc}$ are given in Table 1. The determination of ξ, λ and κ on the basis of these data is discussed in ref. 12. Typical values of κ range from 40 to 100. Such values of κ are among the largest observed for any Type II superconducting materials. Characteristic values of ξ are in the range 30 - 100 Å while λ is typically 2000 - 10 000 Å. Amorphous superconductors are extreme Type II materials.

Because the lattice is disordered on a scale much smaller than the coherence length, favorable centers for flux pinning do not exist. As a consequence the magnetic properties are rather reversible, the critical d.c. current density J_c decreases very rapidly with applied magnetic field as illustrated in Fig. 1 for $La_{76}Au_{24}$ and the J_c values are much too low for practical applications.

From this short review of the limited number of presently known amorphous superconductors it appears that the amorphous state leads to relatively low transition temperatures, high values of H_{c2} and extremely low critical current densities. Excluding practical applications, the amorphous superconductors constitute a class of materials of great scientific interest. There is no theoretical reason why the absence of a lattice should automatically lead to a low transition temperature; this problem should be of interest to the theoretical physicists. Another problem which should be of interest to theoreticians is the dependence of the critical field on temperature. The relation between H_{c2} and T remains linear down to low temperatures and hence higher fields than predicted by the existing theories. It would be most interesting to find out if this behavior can be related theoretically to the absence of a lattice.

The very sharp drop in critical current with applied magnetic field which is a characteristic property of amorphous superconductors excludes their use in many practical applications. Substantial increases in critical currents have already been obtained by introducing pinning centers into the otherwise homogeneous amorphous structure. This can be done by a suitable heat treatment allowing a controlled number of microcrystals to nucleate in the amorphous matrix. It can also be achieved by adjusting the composition of an alloy to be on the borderline between the field separating the single-phase (amorphous) region and an amorphous plus crystalline phase region in the non-equilibrium phase diagram. The reason for trying to increase the critical current of amorphous superconductors is more than academic and may someday answer the need for ductile high field superconducting alloys which do not deteriorate when exposed to very high doses of radiation. A high field magnet for fusion power would be a typical engineering application. The properties of crystalline superconductors are affected by radiation, a typical example being Nb_3Sn magnet winding tape whose transition temperature progressively decreases with increasing doses of radiation from 18 to about 3 K [20, 21]. No direct evidence for the insensitivity of superconducting amorphous alloys to high radiation doses has been reported. However, Lesueur [22] has shown that the electrical resistivity of an amorphous $Pd_{80}Si_{20}$ alloy does not change when irradiated by fission fragments (f.f.) up to a dose of at least 5.6×10^{14} f.f. cm^{-2}. During that time the X-ray diffraction pattern, which was typical of the amorphous state, did not change. When a specimen of the same alloy in the fully crystallized state was exposed to the same dose of radiation, its resistivity increased progressively and approached that of the amorphous alloy. Also, the X-ray diffraction pattern, which consisted initially of a series of very sharp Bragg peaks, changed progressively to a smoothed-out pattern typical of the amorphous state.

In 1970 [23] Professor Matthias pointed out that "most high transition temperature superconductors are not very stable, they are metastable at best". This statement is certainly applicable to the amorphous superconductors which are probably an extreme case of metastability. They are not, however, high temperature superconductors. Will it be possible to raise their transition temperatures and if so what are the most promising avenues of research? There is no well-established theoretical reason why the transition temperatures of amorphous superconductors should always be lower than those of crystalline alloys. As for the possibility of formulating new amorphous alloy compositions with transition temperatures higher than those reported in this paper, the approach will probably be the same as that which led to the discovery of higher T_c values in crystalline alloys. Here again we can quote

Professor Matthias [23] discussing the progressive increase in T_c during the last twenty-five years: "this was done through purely experimental techniques relying exclusively on an enlightened empirical metallurgy". Exactly the same statement can be made about the amorphous superconductors reported in this paper.

Acknowledgments

This work was supported by the Department of Energy, Contract No. EY-76-C-03-0822.

References

1 W. Buckel and R. Hilsch, Z. Phys., 138 (1954) 109.
2 W. L. Johnson, S. J. Poon and P. Duwez, Phys. Rev. B, 11 (1975) 150.
3 M. H. Cohen and D. Turnbull, Nature (London), 189 (1961) 131.
4 W. L. Johnson and S. J. Poon, J. Appl. Phys., 46 (1975) 1787.
5 W. H. Shull, D. G. Naugle, S. J. Poon and W. L. Johnson, Phys. Rev. B, 18 (1978).
6 J. E. Graebner, B. Golding, R. J. Schultz, F. S. L. Hsu and H. S. Chen, Phys. Rev. Lett., 39 (1977) 1480.
7 K. Togano and K. Tachikawa, J. Appl. Phys., 46 (1975) 3609.
8 W. L. Johnson, C. C. Tsuei and P. Chaudhari, Phys. Rev. B, 17 (1978) 2884.
9 R. Hasegawa and L. E. Tanner, Phys. Rev. B, 16 (1977) 3925.
10 R. Ray, B. C. Giessen and J. J. Grant, Scr. Metall., 2 (1968) 357.
11 W. L. Johnson, S. J. Poon and P. Duwez, Phys. Rev. B, 18 (1978) 206.
12 Ellen R. Domb and W. L. Johnson, J. Low Temp. Phys., Aug. (1978).
13 W. L. Johnson and J. Bennett, unpublished data, 1978.
14 C. C. Tsuei, W. L. Johnson, R. B. Laibowitz and J. M. Viggiano, Solid State Commun., 24 (1977) 615.
15 G. S. Cargill III, Solid State Phys., 30 (1975) 227.
16 Pol Duwez, Annu. Rev. Mater. Sci., 6 (1976) 83.
17 P. Chaudhari and D. Turnbull, Science, 199 (1978) 11.
18 J. K. Hulm and R. D. Blaugher, Phys. Rev., 123 (1961) 1569.
19 J. R. Gavaler, Appl. Phys. Lett., 23 (1973) 480.
L. R. Testardi, J. H. Wernick and W. A. Royer, Solid State Commun., 15 (1974) 1.
20 A. R. Sweedler, D. Cox, D. G. Schweitzer and G. W. Webb, IEEE Trans. Magn., 11 (2) (1975) 163.
21 D. M. Parkin and A. R. Sweedler, IEEE Trans. Magn., 11 (2) (1975) 166.
22 M. D. Lesueur, C. R. Acad. Sci., 266 (1968) 1038.
23 B. T. Matthias, Comments Solid State Phys., 3 (1970) 93.

MELT-QUENCHED AMORPHOUS SUPERCONDUCTING ALLOYS

Akihisa Inoue and Tsuyoshi Masumoto

The Research Institute for Iron, Steel and Other Metals, Tohoku University, Sendai 980.

ABSTRACT

This paper reviews our current works on melt-quenched amorphous superconducting alloys. Amorphous alloys of Nb-Si, Nb-Mo-Si, Nb-Si-M(M=B, C and Ge), Mo-Si-B, W-Si-B and Ti-Nb-Si system have been found to be superconductors with transition temperatures ranging from 4.2 to 6.8 K. Superconducting properties(T_c, H_c and J_c) are remarkably improved after crystallization upon annealing. The formation of continuous amorphous tapes with highly ductile nature suggests the prospect of alternative methods for the production of superconducting materials.

1. INTRODUCTION

Superconducting materials possessing good mechanical properties as well as a high transition temperature(T_c) are desirable for their applications. Good superconducting materials such as A-15 compounds[1], NaCl-type carbides[2] and C-15 type Laves compounds[3] are extremely brittle at ambient temperature. Such poor mechanical properties prevent these materials from being used under high electrical/magnetic fields and the production of ductile superconductors can alleviate this difficulty. Since it was found [4] that amorphous alloys possess both high strengths and good ductility compared to the crystalline alloys, the expectation that a superconducting material possessing a high T_c as well as good mechanical properties may be obtained in the amorphous state has aroused increasing interest in the study on amorphous superconducting alloys.

The first amorphous superconductor was reported by Buckel and Hilsch[5] in 1954. Amorphous bismuth with a T_c of 6 K was obtained as a thin film by evaporation on a helium-cooled substrate. Since then many investigators have used this method to obtain amorphous metals and alloys[6]. The superconducting metals produced by evaporation on a cold substrate are unstable and generally crystallize below room temperature. In 1975, an amorphous superconducting alloy was obtained by rapid quenching from the liquid state[7]. The samples obtained by melt-quenching can be considered as bulk films in comparison with evaporated films and are very stable at room temperature and above, but they are in the form of circular foils of approximately 1-3 cm diameter and 20-60 μm thickness. Subsequently, in 1978, we have succeeded[8] in producting several refractory metal-based amorphous alloys in the form of a continuous ribbon by using a modified single roller-type quenching apparatus. The continuous ribbon formation of the superconducting materials is very important for practical uses. In the course of further investigations[9-15], we have observed that melt-quenched Nb-, Ti-Nb-based alloys are amorphous and highly ductile. These amorphous alloys have also been found to have a T_c above 4.2 K. This paper reviews our recent results[9-13] on melt-quenched amorphous superconductors, especially the amorphous phase formation ranges of Nb-Si, Nb-M-Si(M=V, Zr, Mo, Ta, W, B, C and Ge), Mo-Si-B, W-Si-B, Ti-Nb-Si and Ti-V-Si systems and the superconducting and mechanical properties of these amorphous alloys.

2. ALLOY COMPOSITIONS AND QUENCHING APPARATUS [8-15]

The specimens reported in the present paper are $Nb_{75-85}Si_{15-25}$ binary alloys and $Nb_{80-x}M_xSi_{20}$(M=V, Zr, Mo, Ta and W), $Nb_{80}Si_{20-x}M_x$(M=B, C and Ge), $Mo_{70}Si_{20}B_{10}$, $W_{70}Si_{20}B_{10}$, $Ti_{85-x}Nb_xSi_{15}$, $Ti_{70-x}Nb_{30}Si_x$ and $Ti_{85-x}V_xSi_{15}$ ternary alloys. All compositions are expressed in atomic percent as subscripts. Mixtures of each pure metal(Ti, V, Zr, Nb, Mo, Ta and W), silicon, boron, graphite and germanium were melted in an arc furnace on a water-cooled copper hearth with a tungsten electrode. The melting was accomplished in a purified and gettered argon atmosphere at pressures of about 8×10^4 Pa. Each alloy was melted 4 or more times to ensure complete mixing of the two or three elements. Weight loss after melting was less than 15 mg in a 30 g ingot, then the composition of alloys reported is expressed by the nominal value.

Continuous ribbons of about 1-2 mm width and 0.02-0.04 mm thickness were prepared from these mother alloys under a protective argon atmosphere using a modified single roller quenching apparatus designed for high melting alloys by the present authors. A schematic illustration of the apparatus is shown in Fig. 1. The alloy is levitation melted at an argon pressure of about 3 MPa, and by opening the shutter and, at the same time, by cutting the coil current off, the molten alloy is ejected by the difference in pressure between the two chambers, through the nozzle at the end of a quartz tube onto the surface of a copper roll which is rotating at a high speed. The ejected melt is solidified on the roll

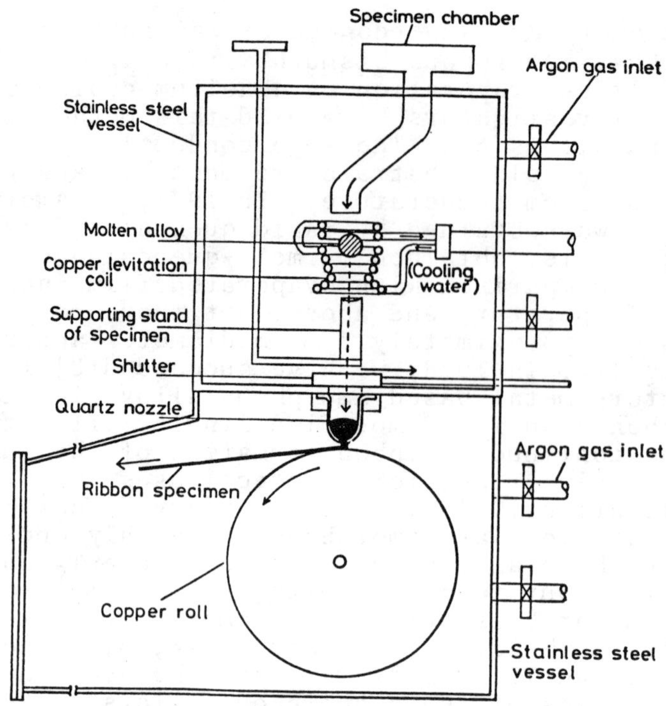

Fig. 1 Schematic illustration of the rapid quenching apparatus used in the present work.

surface in the form of a continuous ribbon. Typically, the amount of alloys melted in one run is about 5 g and the rotation speed of the roll(20 cm in diameter) is about 4000 rpm. This modified melt-quenching apparatus is useful for refractory metals with high melting temperatures and for active metals which easily react with a quartz tube and/or oxygen in air.

3. FORMATION RANGE OF AMORPHOUS PHASE [9-12]

Figure 2 shows the composition range in which the amorphous phase formed without any trace of crystallinity for the Nb-Si binary system. Identification of the as-quenched phases was made by conventional method using nickel-filtered Cu Kα radiation. The ribbons were classified amorphous when the X-ray intensity as a function of the diffraction angle showed a typical liquid like structure. The amorphous phase forms in a narrow range of 17-21 at% Si around a eutectic point. Also, the amorphous single phase of the Nb-Si alloy is kept by the replacement of silicon with boron up to 12 at%, carbon up to 8 at% and germanium up to 4 at%. Additionally, the formation ranges for $Nb_{80-x}M_xSi_{20}$(M=V, Zr, Mo, Ta and W), Ti-Nb-Si and Ti-V-Si ternary systems are shown in Figs. 3 to 5. In the figures, the formation of amorphous phase is seen in the alloy composition ranges of 0-20 at% V, 0-25 at% Zr, 0-30 at% Mo, 0-10 at% Ta and 0-10 at% W for Nb-M-Si system, of 0-43

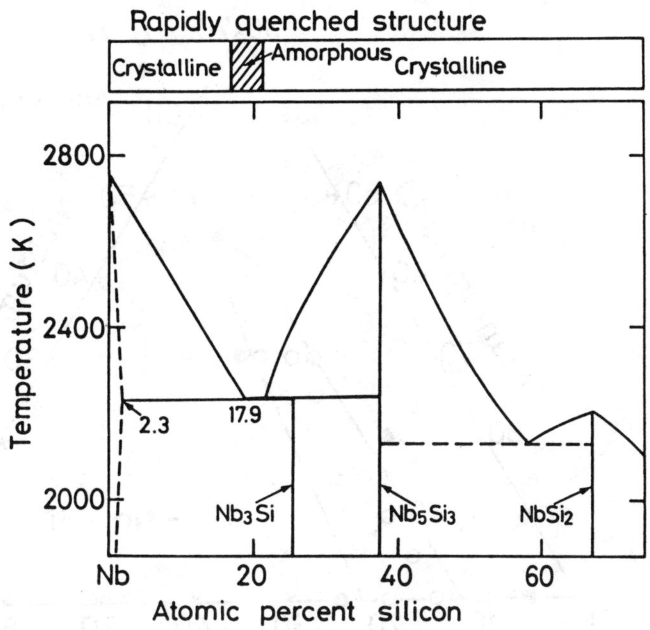

Fig. 2 Composition range for formation of amorphous phase in Nb-Si binary system. Binary phase diagram was obtained from Ref. [17].

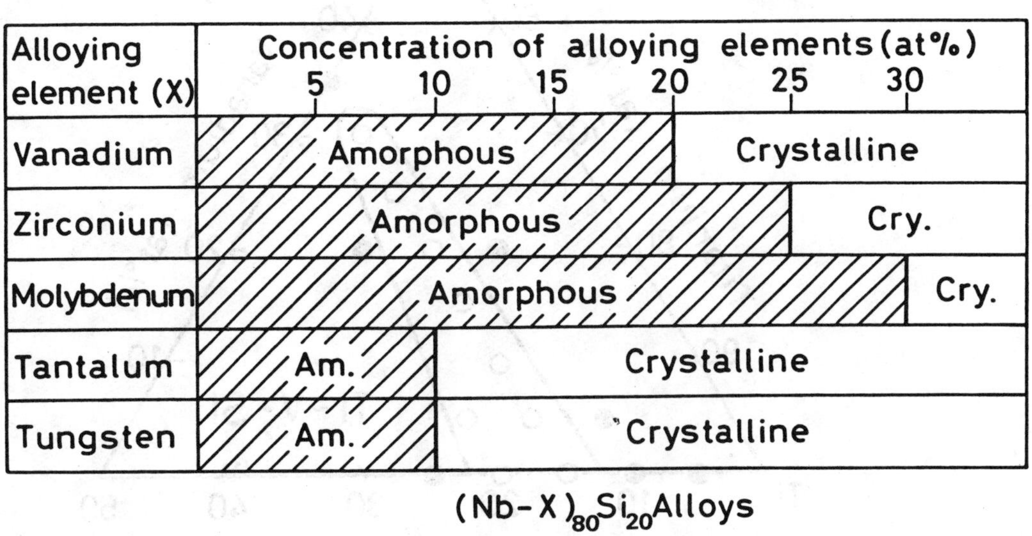

Fig. 3 Composition ranges of amorphous phase formation for the $(Nb\text{-}M)_{80}Si_{20}$ (M=V, Zr, Mo, Ta and W) alloy systems.

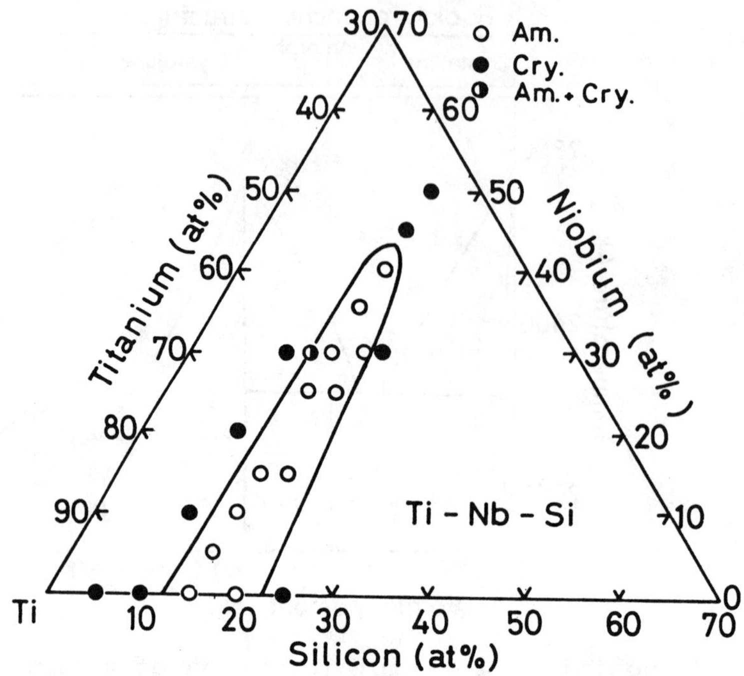

Fig. 4 Composition range for formation of amorphous phase in Ti-Nb-Si ternary system.

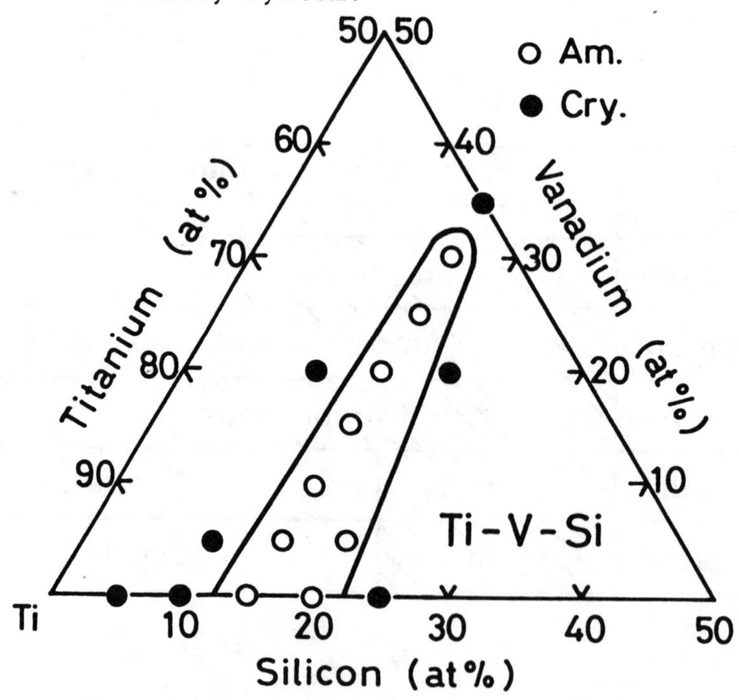

Fig. 5 Composition range for formation of amorphous phase in Ti-V-Si ternary system.

at% Nb and 13-23 at% Si for Ti-Nb-Si system and of 0-32 at% V and 13-23 at% Si for Ti-V-Si system.

In general, the composition range of amorphous alloys obtained by melt-quenching technique is located around a deep eutectic in the alloy phase diagram. Turnbull[16] has noted that a deep eutectic can be interpreted in terms of a comparatively large negative heat of formation of the liquid alloy. Hence, it is predicted that alloys having compositions near a deep eutectic are most prone to the formation of an amorphous phase. Each phase diagram [17] of the Nb-Si, Ti-Si and V-Si binary systems features a eutectic reaction at the composition of $Nb_{82}Si_{18}$, $Ti_{86}Si_{14}$ and $V_{87}Si_{13}$ with a eutectic temperature of 2216 K, 1606 K and 2053 K, respectively. As seen in Figs. 2, 4 and 5, the present amorphous-forming composition ranges also fall near the trough of the eutectics in the Nb-Si binary system and the Ti-Nb-Si and Ti-V-Si ternary systems, as is common in other melt-quenched amorphous alloys, and thus the ability to form an amorphous phase seems to be closely related to the large negative heat of formation of the liquid alloy.

4. MECHANICAL PROPERTIES AND CRYSTALLIZATION TEMPERATURE [9-13]

Vickers hardness(Hv), tensile fracture strength(σ_f), crystallization temperature(Tx) and critical fracture temperature(T_f) of $Nb_{80}Si_{20}$, $Nb_{80-x}M_xSi_{20}$(M=V, Zr, Mo, Ta and W), $Ti_{85-x}Nb_xSi_{15}$ and $Ti_{85-x}V_xSi_{15}$ amorphous alloys are presented in Table 1. Here, Hv and σ_f were measured by a Vickers microhardness tester with a 100 g load and an Instron-type tensile testing machine at a strain rate of 1.7×10^{-4} /s, respectively. The crystallization temperature of the alloys was examined at a heating rate of 8.33×10^{-2} K/s by a differential thermal analyzer(DTA). T_f was defined as the temperature of aging for 6000 s which lead to the fracture of alloys in a simple bend test[18]. The hardness values are in the range of 640-900 DPN for the Nb-based alloys, 550-580 DPN for Ti-Nb-Si alloys and 570-630 DPN for Ti-V-Si alloys and the tensile strengths are in the range 2010-2160 MPa for the Ti-based alloys. Also, the crystallization temperatures are in the ranges of 953-1027 K for the Nb-based alloys, 720-762 K for Ti-Nb-Si alloys and 688-770 K for Ti-V-Si alloys. As seen in the table, there is a tendency that Hv and Tx increase by the replacement of Nb with V, Zr, Mo or W and of Ti with Nb or V. Further, one can notice that the strength and Tx of the Nb-based alloys are superior to those for the Ti-based alloys. This tendency agrees well with that of the bonding forces between Nb-Si and Ti-Si which are presumed from the melting points[19] of Nb_5Si_3 and Ti_5Si_3 compounds.

Tensile fracture occurs on the shear plane at 45-55 deg to the tensile axis in the direction of thickness, and the fracture surface consists of a smooth part produced by shear slip and a vein-like part produced by plastic instability, similar to the characteristics[20] of general fracture morphology for other transition metal-based amorphous alloys.

Table 1 Vickers hardness(Hv), tensile fracture strength(σ_f), crystallization temperature(Tx) and critical fracture temperature(T_f) for several refractory metal-based amorphous alloys.

Alloy system (at%)	Hv(DPN)	σ_f(MPa)	Tx(K, 5 K/min)	T_f(K,×6000s)
$Nb_{80}Si_{20}$	640	2090	953	780
$Nb_{70}V_{10}Si_{20}$	840	-	980	-
$Nb_{70}Zr_{10}Si_{20}$	715	2180	987	750
$Nb_{70}Mo_{10}Si_{20}$	870	-	994	-
$Nb_{60}Mo_{20}Si_{20}$	895	-	1008	-
$Nb_{70}Ta_{10}Si_{20}$	805	-	966	-
$Nb_{70}W_{10}Si_{20}$	895	-	1027	-
$Mo_{70}Si_{20}B_{10}$	1000	-	928	-
$W_{70}Si_{20}B_{10}$	1300	-	1033	-
$Ti_{70}Nb_{15}Si_{15}$	560	2060	718	670
$Ti_{55}Nb_{30}Si_{15}$	580	2060	798	670
$Ti_{50}Nb_{35}Si_{15}$	625	2010	870	-
$Ti_{75}V_{10}Si_{15}$	570	2080	738	-
$Ti_{65}V_{20}Si_{15}$	610	2160	-	600
$Ti_{60}V_{25}Si_{15}$	645	2210	762	600

Furthermore, these amorphous alloys possess a good bend ductility. As an example, deformation structure of $Ti_{55}Nb_{30}Si_{15}$ amorphous alloy bent completely by pressing against the edge of a razor blade is shown in Fig. 6. Numerous deformation markings can be seen

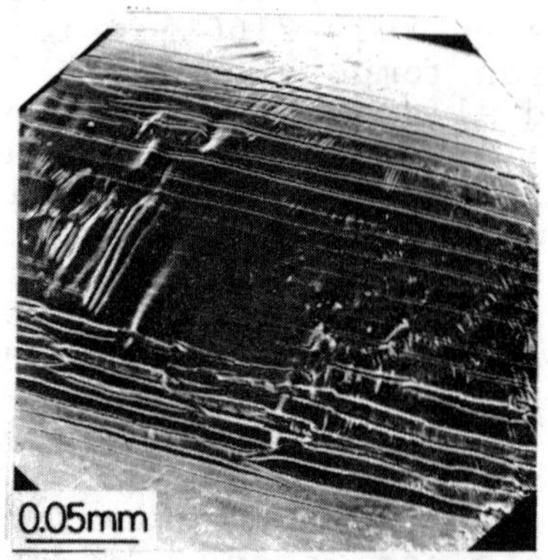

Fig. 6 Scanning electron micrograph showing deformation markings at the tip of $Ti_{55}Nb_{30}Si_{15}$ amorphous alloy bent through 180°.

near the bent edge, and no cracks are observed even after such a severe deformation.

Ductile-brittle transition behavior for $Nb_{80}Si_{20}$, $Nb_{70}Zr_{10}Si_{20}$, $Ti_{70}Nb_{15}Si_{15}$, $Ti_{55}Nb_{30}Si_{15}$ and $Ti_{65}V_{20}Si_{15}$ amorphous alloys was examined as functions of aging time and temperature. As an example, their embrittlement behavior during isochronal aging for 1 h is shown in Fig. 7. The strain on the outer surface required for fracture, ε_f, is estimated from the relation $\varepsilon_f = t/(2r-t)$, where r is the radius of curvature of bent samples at fracture and t is the thickness of ribbon specimens. The temperature for beginning of embrittlement is about 800 K for $Nb_{80}Si_{20}$, 770 K for $Nb_{70}Zr_{10}Si_{20}$ and 600 K for $Ti_{65}V_{20}Si_{15}$ and is lower by about 150-200 K than the crystallization temperatures, while that for $Ti_{55-70}Nb_{15-30}Si_{15}$ is about 700 K and is nearly equal to the Tx, indicating that Ti-Nb-Si amorphous alloys remain ductile until the start of crystallization. This result suggests that the decrease in T_f by mixing depends largely on the kind of constituent elements.

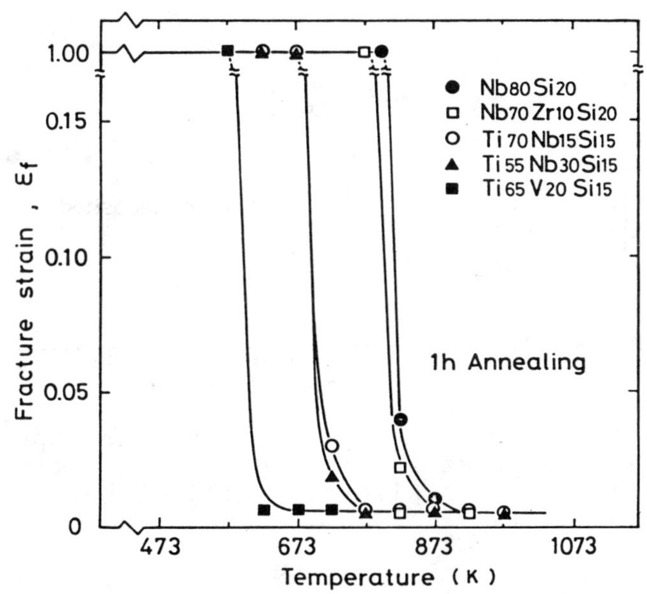

Fig. 7 Change in fracture strain by annealing for 1 h at various temperatures for $Nb_{80}Si_{20}$, $Nb_{70}Zr_{10}Si_{20}$, $Ti_{85-x}Nb_xSi_{15}$ and $Ti_{65}V_{20}Si_{15}$ amorphous alloys.

5. SUPERCONDUCTING PROPERTIES OF THE AMORPHOUS ALLOYS [9-15]

(1) Transition temperature(T_c) in the as-quenched state

The superconducting transition was monitored by measuring the resistivity of the material by a conventional four-probe method. The specimen current used is 1 mA. The temperature was measured using a calibrated Au-Fe + Chromel thermocouple with an accuracy of ±0.05 K. In Fig. 8, the T_c and ΔT_c of the amorphous alloys for Nb-Si binary system are plotted as a function of silicon content. The value of T_c is taken as the temperature at which R/Rn=0.5, where Rn is the resistance in the normal state. The transition width, ΔT_c, corresponds to the temperature difference between 0.1 and 0.9 of R/Rn. The T_c is 4.4 K for $Nb_{82}Si_{18}$ and has a tendency to decrease slightly with increasing silicon content. If a linear relation is assumed to exist between T_c and silicon content, T_c of $Nb_{75}Si_{25}$ amorphous alloy with Nb_3Si stoichiometric composition is estimated to be about 3.9 K. This extrapolated value agrees well with that($T_c \simeq 3.9$ K)[21] of $Nb_{75}Si_{25}$ amorphous alloy prepared in thin film by covaporation or sputtering. Also, ΔT_c is of the order of 0.3 K and is much smaller compared with that of the crystalline superconducting alloys indicative of a single phase.

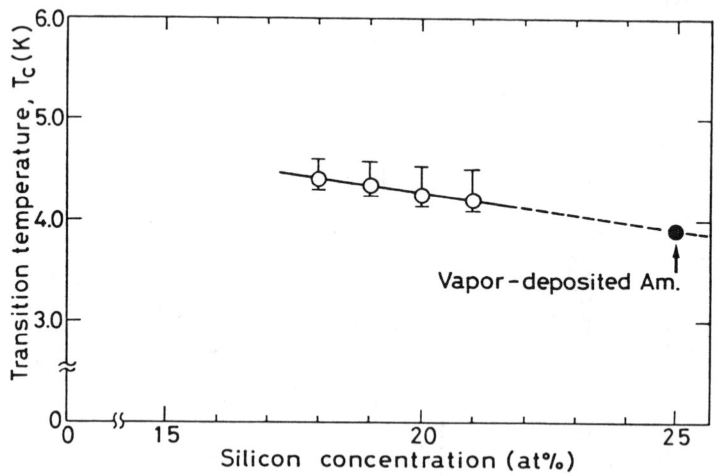

Fig. 8 Silicon concentration dependence of the superconducting transition temperature(T_c) and the transition width(ΔT_c) for Nb-Si amorphous alloys. Vertical bars represent the transition width.

Figures 9 and 10 show the reduced electrical resistance curves measured in no applied magnetic field in the vicinity of T_c for $Nb_{80}Si_{20}$, $(Nb\text{-}M)_{80}Si_{20}$(M=V, Zr, Mo, Ta and W), $Mo_{70}Si_{20}B_{10}$ and $W_{70}Si_{20}B_{10}$ amorphous alloys. The T_c of $(Nb\text{-}Mo)_{80}Si_{20}$ system increases with increasing amount of Mo and attains 5.5 K for the $Nb_{50}Mo_{30}Si_{20}$ amorphous alloy. The addition of V, Zr, Ta or W decreases T_c below the liquid helium temperature(4.2 K), and the onset of the superconducting transition is about 4.1 K for $Nb_{70}V_{10}Si_{20}$ and about 4.2 K for $Nb_{70}W_{10}Si_{20}$, while it can not be detected for $Nb_{60}V_{20}Si_{20}$, $Nb_{70}Zr_{10}Si_{20}$ and $Nb_{70}Ta_{10}Si_{20}$. Thus, the addition of Mo which is at the right hand side of Nb in the periodic table results in an increase in T_c. On the other hand, the addition of V, Zr, Ta and W being at the left side of or in dif-

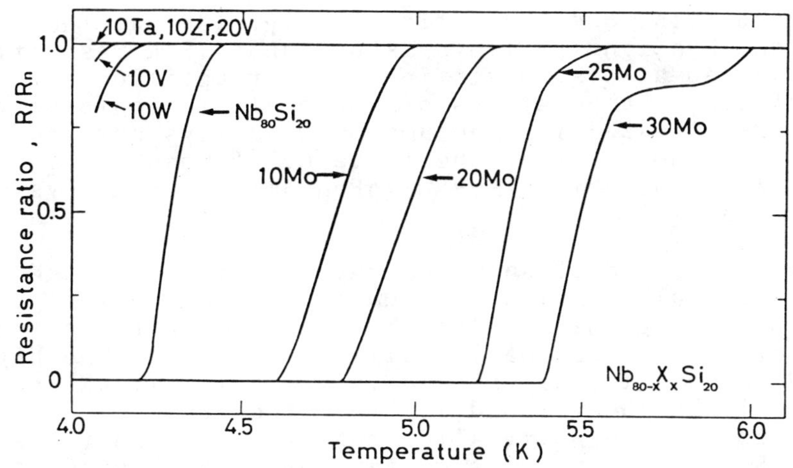

Fig. 9 Resistance ratio R/Rn as a function of temperature for $Nb_{80}Si_{20}$ and $Nb_{80-x}M_xSi_{20}$ (M=V, Zr, Mo, Ta and W) amorphous alloys.

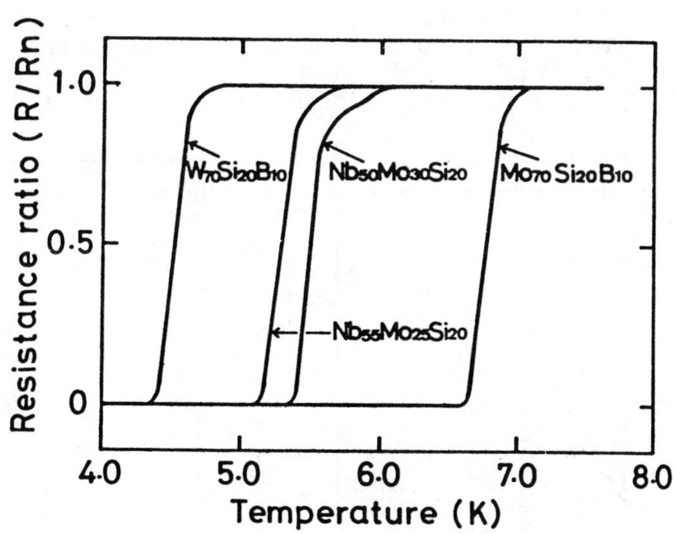

Fig. 10 Resistance ratio R/Rn as a function of temperature for $Mo_{70}Si_{20}B_{10}$ and $W_{70}Si_{20}B_{10}$ amorphous alloys. Also, the data for $Nb_{55}Mo_{25}Si_{20}$ and $Nb_{50}Mo_{30}Si_{20}$ amorphous alloys are shown for comparison.

ferent periods from Nb results in a decrease in T_c.

Further, the effect of metalloid elements (B, C and Ge) on the T_c of $Nb_{80}Si_{20}$ amorphous alloy was examined. The replacement of Si by B, C or Ge up to 4 at% resulted in a slight increase in T_c. The T_c attained was 4.5 K for $Nb_{80}Si_{16}C_4$ and 4.7 K for both of $Nb_{80}Si_{16}B_4$ and $Nb_{80}Si_{16}Ge_4$.

In addition, one can notice in Fig. 10 that the T_c of $Mo_{70}Si_{20}B_{10}$ and $W_{70}Si_{20}B_{10}$ alloys is 6.8 and 4.5 K respectively. The T_c of Mo-based alloys consisting of 4d transition metal is higher than that for W-based alloys of 5d metal, in conformity with the previous results[22] for the amorphous alloys obtained by vapor-deposition method. Also, the T_c value of $Mo_{70}Si_{20}B_{10}$ is at the same level as those for $(Mo\text{-}Ru)_{80}P_{20}$ amorphous alloys obtained by Johnson et al.[23].

Figures 11 and 12 show the changes of the T_c and ΔT_c of Ti-Nb-Si amorphous alloys with niobium or silicon content in the case of no applied magnetic field. T_c increases with increasing niobium content or with decreasing silicon content. The highest T_c attained in this ternary system is 5.1 K for $Ti_{45}Nb_{40}Si_{15}$. More recently, it has been found[15] that the replacement of silicon by boron for $Ti_{45}Nb_{40}Si_{15}$ alloy results in an increase of T_c from 5.1 to 5.5 K. In specimens containing less than 10 at% Nb, normal state resistance was observed to remain down to the temperature of liquid helium. The Ti-V-Si amorphous alloys also have T_c down to the temperature of liquid helium even for the specimens containing more than 30 at% V. The T_c value for Ti-Nb-Si-B alloys is much higher than the highest value(4.4 K) reported for ductile binary amorphous Nb-Si alloys and is, within the authors knowledge, the highest among ductile amorphous alloys having completely bent nature. However, the value is about half of those[24,25] reported

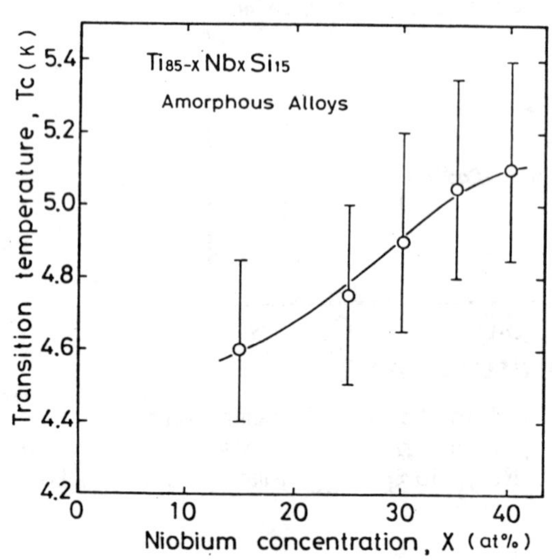

Fig. 11 Niobium concentration dependence of the superconducting transition temperature(T_c) and the transition width(ΔT_c) for $Ti_{85-x}Nb_xSi_{15}$ amorphous alloys. Vertical bars represent the transition width.

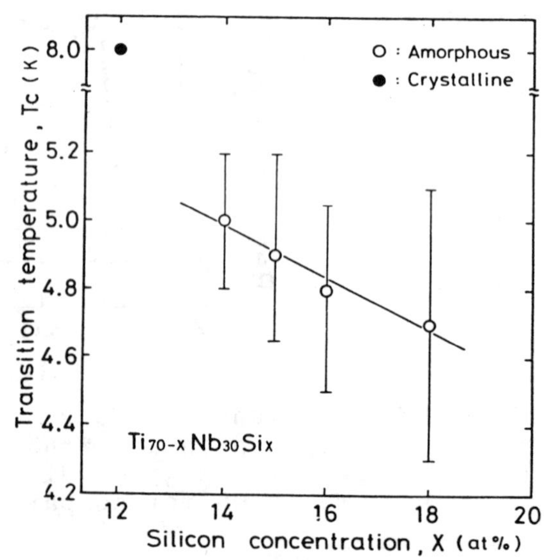

Fig. 12 Silicon concentration dependence of the superconducting transition temperature(T_c) and the transition width(ΔT_c) for $Ti_{70-x}Nb_{30}Si_x$ amorphous alloys. Vertical bars represent the transition width.

for conventional Ti-Nb superconducting materials, indicating that the amorphous state leads to a reduction in T_c, similar to many other amorphous superconducting alloys[26] except Mo-based alloys. The effect of additional elements on the T_c of the Ti-Nb based amorphous alloys has not been extensively studied yet and a further enhancement of T_c may be expected by a better design of alloy composition. Also, as seen in Figs. 11 and 12, the transition occurs within a temperature width of about 0.5 K, which is slightly larger compared with the data of the amorphous superconducting alloys[26] reported previously as well as the Nb-Si alloys shown in Fig. 8. Although the reason for such a large transition width in the present alloys is not certain, it may be responsible for the existence of oxygen in the alloys because there is a close correlation[27] between oxygen content and broad transition width.

According to the study on vapor-deposited amorphous metals and alloys by Collver and Hammond[22], the T_c in the amorphous state is related closely to the average outer electron concentration(e/a) showing a broad maximum around [e/a]=6.2-6.4 for the 4d and 5d transition-metal series. This result indicates that the T_c value of the present amorphous alloys increases in the order of Ti, Nb and Mo systems. However, the present result disagrees with the expected order because Ti-based alloys possess higher T_c than Nb-based alloys. In order to examine the reason for such a disagreement, the T_c values of the Nb-Si, $(Nb\text{-}Mo)_{80}Si_{20}$, $Nb_{80}(Si\text{-}M)_{20}$ (M=B, C and Ge), $Mo_{70}Si_{20}B_{10}$ and Ti-Nb-Si amorphous alloys are plotted as a function of the average [e/a] of the transition metal components in Fig. 13 and compared with the variation of T_c for amorphous transition metals obtained by Collver and Hammond.

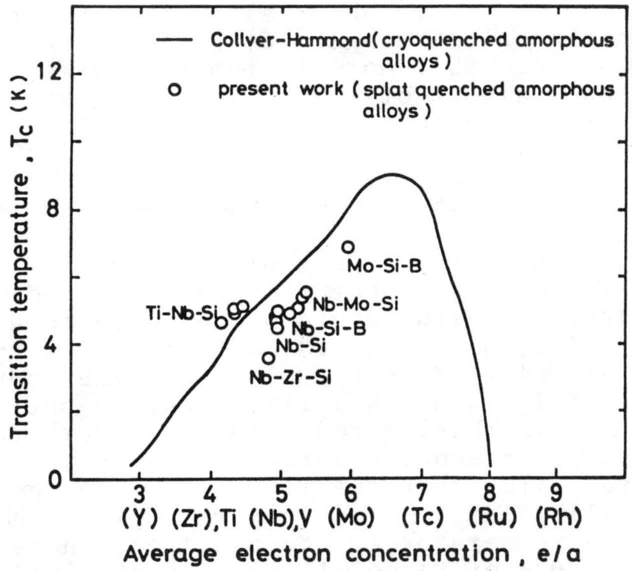

Fig. 13 Superconducting transition temperature(T_c) versus electron per atom ratio for the 4d transition-metals series. The data for the amorphous vapor-quenched films were taken from Ref. [26].

From this figure, it is seen that the variation of T_c with alloy composition for each alloy system is related apparently to the average e/a. Johnson[23] have shown for $(Mo\text{-}Ru)_{80}P_{20}$ amorphous alloys that such a change of T_c with e/a corresponds well to the change of the electron-phonon bare density of states at the Fermi level(E_f=0) N(0) with e/a. It has been shown by McMillan[28] that N(0) is one of the most important factors in determining T_c. Additionally, one can notice that the T_c values of Nb-Si, Nb-Mo-Si and Mo-Si-B alloys lie below the Collver-Hammond curve. The tendency that T_c values for melt-quenched alloys lie below the Collver -Hammond curve is similar to the results for the other amorphous alloys such as Mo-Ru-P[23], Mo-Ru-B[26], Mo-Re-P-B[26] and Zr-Rh [28,29] systems; namely, when the transition metal components in the alloys are widely separated in e/a and/or the alloys contain metalloid elements, the T_c tends to fall below the Collver-Hammond curve for metals and alloys of neighboring metals. On the other hand, the T_c values of Ti-Nb-Si alloys are higher compared with the Collver-Hammond curve. Although the reason for such an abnormal variation of T_c with e/a is uncertain at present, its origin may be found in one or two of the following two factors: (1) the mixing of the metal elements in different transition metal series, and (2) the oxygen content. Previous reports[26] have shown that the mixing between metal elements in different series results in the decrease of T_c of amorphous alloys. On the other hand, it has been reported[27] that the addition of a small amount of oxygen apparently raises T_c of crystalline Ti-Ta alloys significantly. This fact suggests that the T_c higher than the value expected from Collver-Hammond curve may be due to a small amount of oxygen contained in the Ti-Nb-Si amorphous alloys. The occurrence of such a high T_c has been also recognized[15] for Ti-Nb-Si-B amorphous alloys. It is hoped that more systematic investigations on the effect of a small amount of impure elements such as oxygen or nitrogen on T_c will be made in near future.

(2) Effect of cold-working on the transition temperature(T_c)

Investigation on the effect of cold-working on T_c of amorphous alloys is very important for practical uses, because of the possibility that the alloy would be deformed to a final shape. The T_c values of cold-rolled $Ti_{70}Nb_{15}Si_{15}$ and $Ti_{55}Nb_{30}Si_{15}$ amorphous alloys are plotted as a function of the reduction in area in Fig. 14. The T_c value is a little change until 15% reduction, but with further increasing reduction it lowers rapidly and fall to about 4.2 K after about 40% reduction. However, it can be noticed in Fig. 14(b) that the T_c recovers to the nearly previous level by annealing for 1 h at 573 K which is much lower than the crystallization temperatures. In addition, there was no change in good ductile nature of these alloys. Consequently, the lowering of T_c by cold-rolling seems not to be a serious problem for practical uses of amorphous superconductors. We believe that such a lowering of T_c is due to the internal strain introduced in amorphous structure by cold-working.

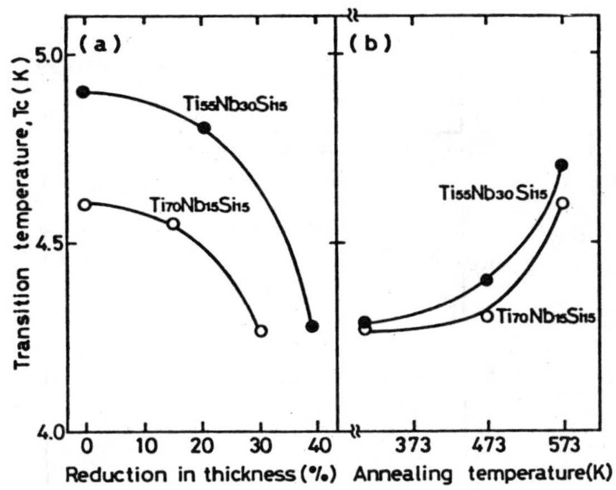

Fig. 14 (a) Change in superconducting transition temperature(T_c) of $Ti_{70}Nb_{15}Si_{15}$ and $Ti_{55}Nb_{30}Si_{15}$ amorphous alloys by cold rolling.
(b) Change in superconducting transition temperature(T_c) of cold-rolled $Ti_{70}Nb_{15}Si_{15}$ and $Ti_{55}Nb_{30}Si_{15}$ amorphous alloys by annealing.

(3) Effect of low-temperature annealing on the transition temperature(T_c)

It has been pointed out[30-32] that Fe- and Co-based amorphous alloys obtained by melt-quenching exhibit improved magnetic properties when annealed at temperatures sufficiently low as to produce no crystallization. Stress relaxation[32], embrittlement [16,32] and a change in the atomic pair correlation function[33] have also been observed as a result of such low-temperature annealing. Considering such an effect by low-temperature annealing we expected a similar effect on the superconducting properties which are strongly affected by the microscopic structure of the material.

A change in the T_c of $Ti_{55}Nb_{30}Si_{15}$ amorphous alloy with annealing temperatures is shown in Fig. 15, wherein the structure and the ductility after annealing are also represented for comparison. "Ductile" implies that complete bending through 180° is possible without breaking and "brittle" implies that the alloy fractures during bending. The T_c remains practically unchanged up to about 500 K. However, the value begins to decrease around about 550 K and shows a minimum(about 4.1 K) at about 670 K followed by a rapid increase. As seen in the figure, the remarkable increase in the T_c at higher temperatures above about 700 K is due to the precipitation of a crystalline phase(Ti(Nb) solid solution with bcc structure)[10]. Figure 16 shows bright field images and selected area diffraction patterns of the as-quenched (a and b) and annealed(c and d) $Ti_{55}Nb_{30}Si_{15}$ amorphous alloys. As seen in the diffraction pattern(d), the amorphous structure remains unchanged

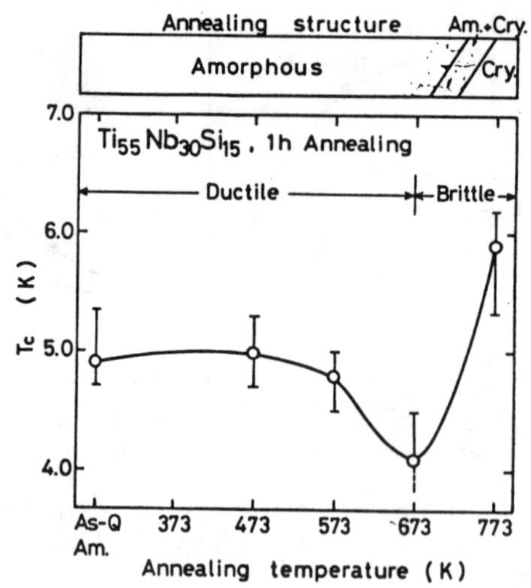

Fig. 15 Change in superconducting transition temperature(T_c) of $Ti_{55}Nb_{30}Si_{15}$ amorphous alloy upon annealing for 1 h at various temperatures.

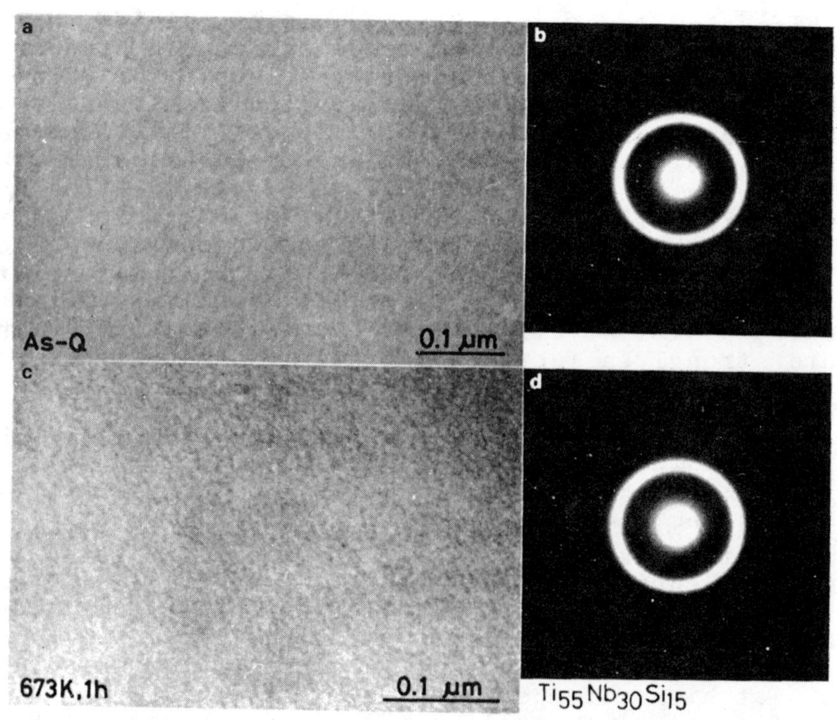

Fig. 16 Transmission electron micrographs and selected area diffraction patterns showing the as-quenched structure(a and b) and the structure after annealing for 1 h at 673 K (c and d) of $Ti_{55}Nb_{30}Si_{15}$ amorphous alloy.

even after annealing for 1 h at 673 K. But in the highly magnified bright field image(c), a fine weak contrast can be seen innumerably all over the photograph, indicating a microscopic change in the amorphous structure. Hence, it is suggested that the decrease in the T_c upon low-temperature annealing is due to the structure relaxation or short-range ordering of the amorphous structure. Thus, an improvement of superconducting properties can not be achieved by annealing treatments below the crystallization temperature. The decrease in T_c upon low-temperature annealing has also been recognized for amorphous Zr-Rh[34] and Nb-Si[9] binary alloys.

In general, it has been known that the T_c is a function of several microscopic parameters which characterize the d-band structure. However, an explanation of the low-temperature annealing effect on T_c seems to be very difficult because one would need further knowledges of the microscopic atomic configurations and their theoretical connection with the microscopic parameters which play an important role in determining the T_c.

(4) Critical magnetic field(H_c) and critical current density(Jc)

The lower- and upper-critical magnetic fields, H_{c1} and H_{c2} were measured at liquid helium temperature by a standard four-probe method using a superconducting solenoid for magnetic fields up to about 7.2×10^6 A/m(90 kOe) applied transversely to the specimen in a liquid helium bath. Typical resistance-magnetic field curves for the Nb-, Mo-, W- and Ti-Nb-based amorphous alloys with T_c above 4.5 K are shown in Figs. 17 to 19. The H_{c2}(onset) value obtained is about 2.3×10^6 A/m for $Nb_{55}Mo_{25}Si_{20}$, 9.2×10^5 A/m for $Nb_{70}Mo_{10}Si_{20}$, 6.4×10^5 A/m for $Nb_{80}Si_{16}Ge_4$, 6.0×10^6 A/m for $Mo_{70}Si_{20}B_{10}$, 6.4×10^5 A/m for $W_{70}Si_{20}B_{10}$, 2.6×10^6 A/m for $Ti_{55}Nb_{30}Si_{15}$ and 3.1×10^6 A/m for $Ti_{45}Nb_{40}Si_{15}$, following a general tendency[35] that the higher the T_c, the higher is H_{c2}. In addition, one can notice in these figures that the H_{c2} values are much higher compared with the H_{c1} values.

From Ginzburg-Landau(GL) theory[36], H_{c2} and H_{c1} are expressed as

$$H_{c2} = \sqrt{2}\,\kappa\, H_c \qquad (1)$$

$$H_{c1} = (1/\sqrt{2}\,\kappa)(\log \kappa + 0.08)\, H_c, \qquad (2)$$

where H_c is the critical magnetic field at 0 K and κ is the Ginzburg-Landau parameter given by the following equation;

$$\kappa = \frac{\lambda(T)}{\xi(T)} = 0.725\frac{\lambda_L(0)}{\ell} \qquad (3)$$

λ is the penetration length of magnetic field, ξ the coherence length, λ_L the penetration length of London, and ℓ the mean free path of electrons. It has been reported[37] that the value of κ for amorphous superconductors is very large(50-100) because of a small value of the electron mean free path. As is evident from the equations (1) and (2), it is presumed that the Nb-, Mo-, W- and Ti-Nb-based amorphous alloys have a high H_{c2} value and a low

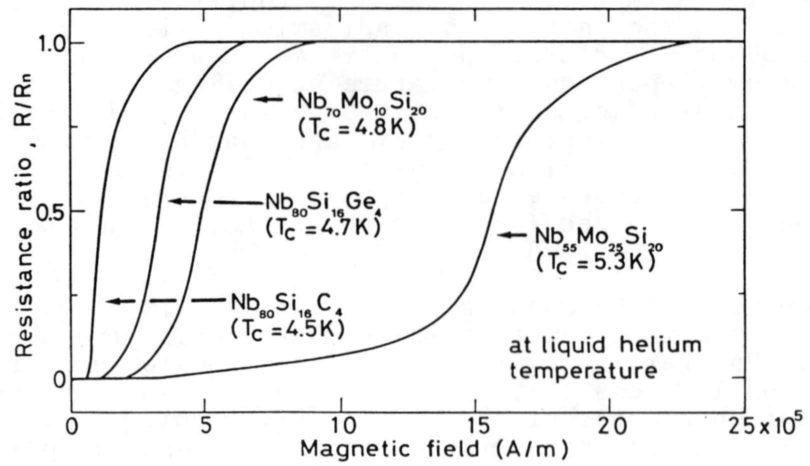

Fig. 17 Resistance ratio R/Rn as a function of magnetic field at liquid helium temperature for several Nb-based amorphous alloys.

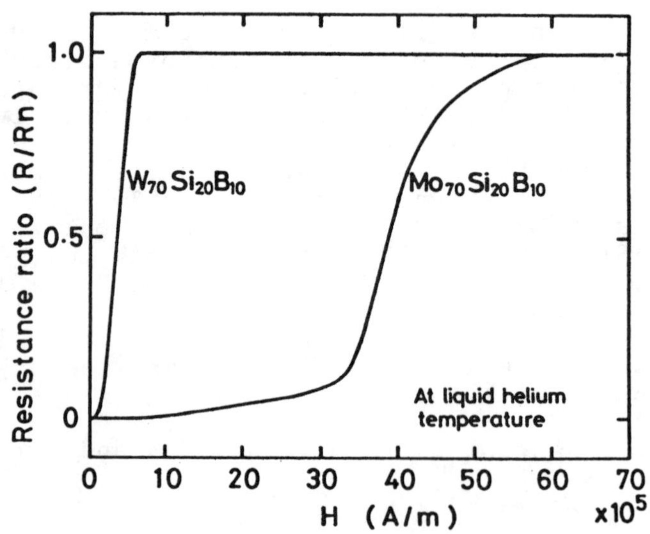

Fig. 18 Resistance ratio R/Rn as a function of magnetic field at liquid helium temperature for $Mo_{70}Si_{20}B_{10}$ and $W_{70}Si_{20}B_{10}$ amorphous alloys.

H_{c1} value due to the large κ value. These results indicate that the present amorphous alloys can be classified as type-II superconductors, similar to the melt-quenched other amorphous alloys.

In our investigations, no information has been obtained about the temperature dependence of H_{c2} for the Nb-, Mo-, W- and Ti-Nb-based amorphous alloys. However, other papers[23,29] have shown that there exists a linear relationship between H_{c2} and tempera-

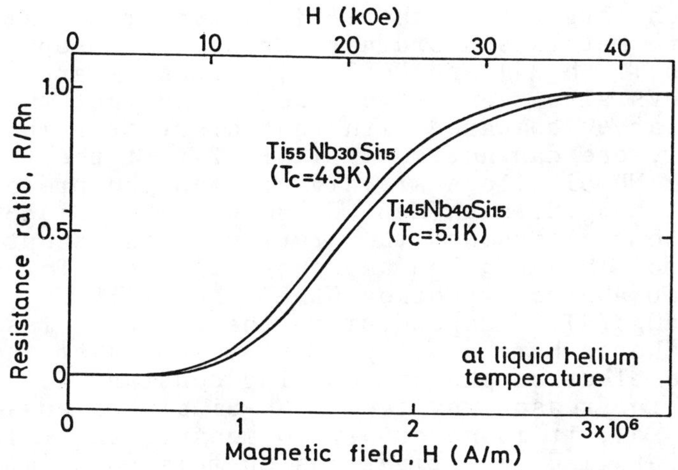

Fig. 19 Resistance ratio R/Rn as a function of magnetic field at liquid helium temperature for $Ti_{55}Nb_{30}Si_{15}$ and $Ti_{45}Nb_{40}Si_{15}$ amorphous alloys.

ture. Based on this fact, the value of dH_{c2}/dT can be estimated for the amorphous alloys of Nb-Mo-Si, Nb-Si-B, Nb-Si-C, Nb-Si-Ge, Mo-Si-B, W-Si-B and Ti-Nb-Si. The results are shown in Table 2 together with other superconducting properties. The value of

Table 2 Superconducting transition temperature(T_c), transition width(ΔT_c), critical current density(J_c), upper critical magnetic field(H_{c2}) and temperature dependence of H_{c2} (dH_{c2}/dT) for several refractory metal-based amorphous alloys.

Alloy system (at%)	T_c (K)	ΔT_c (K)	J_c (A/cm^2) ×10^3	H_{c2} (A/m) ×10^5	dH_{c2}/dT (A·m^{-1}/K) ×10^5
$Nb_{70}Mo_{10}Si_{20}$	4.8	0.3	-	9.2	23
$Nb_{60}Mo_{20}Si_{20}$	5.0	0.3	4.6	-	-
$Nb_{55}Mo_{25}Si_{20}$	5.3	0.4	11	23	23
$Nb_{80}Si_{16}B_4$	4.7	0.3	-	-	-
$Nb_{80}Si_{16}C_4$	4.5	0.3	-	4.3	22
$Nb_{80}Si_{16}Ge_4$	4.7	0.3	-	6.4	19
$Mo_{70}Si_{20}B_{10}$	6.8	0.3	5	60	23
$W_{70}Si_{20}B_{10}$	4.5	0.3	3	6.4	21
$Ti_{55}Nb_{30}Si_{15}$	4.9	0.5	0.27	26	60
$Ti_{45}Nb_{40}Si_{15}$	5.1	0.5	0.30	31	48

dH_{c2}/dT is about 2.2×10^6 $A \cdot m^{-1}/K$ for the Nb-, Mo- and W-based alloys and about 5.4×10^6 $A \cdot m^{-1}/K$ for Ti-Nb-Si alloys. These values are much larger than those[6] for vapor-deposited amorphous transition pure metals and ordinary crystalline superconducting alloys such as Ti-Nb[24] and Nb_3Sn[1]. Such a large value of the amorphous alloys also seems to be due to the fact that the value of κ is much larger compared with that of crystalline superconductors. Further, one can notice in Table 2 that the values of dH_{c2}/dT for Ti-Nb-Si alloys are higher than the present data for $(Nb-Mo)_{80}Si_{20}$, $Nb_{80}(Si-M)_{20}$ (M=B, C and Ge), $Mo_{70}Si_{20}B_{10}$ and $W_{70}Si_{20}B_{10}$ amorphous alloys and the previous data (about 2.1×10^6 $A \cdot m^{-1}/K$)[23] for $(Mo-Ru)_{80}P_{20}$ amorphous alloys. For an extended Ginzburg-Landau-Abrikosov-Gorkov(GLAG) theory[38,39], it is expressed that dH_{c2}/dT is dependent on the residual resistivity(ρ), the electron-phonon bare density of states N(0) at the Fermi level ($E_f=0$) and the electron-phonon coupling constant λ. Hence, for understanding the reason why the Ti-Nb-Si alloys possess the large values of dH_{c2}/dT, it is necessary to measure the values of ρ, N(0) and λ. Detailed investigations on this point are in progress at present.

The critical current density J_c was measured for the amorphous alloys with T_c above 4.7 K. The critical current was defined as the current at which a measurable voltage(1 μV) appeared across a 55 mm-length of the specimen in a liquid helium bath at no applied field. The critical current density J_c was obtained by dividing the critical current with the cross-sectional area of the specimen. The J_c value obtained is about 4.6×10^3 A/cm^2 for $Nb_{60}Mo_{20}Si_{20}$, about 1.1×10^4 A/cm^2 for $Nb_{55}Mo_{25}Si_{20}$, about 4.4×10^3 A/cm^2 for $Nb_{80}Si_{16}B_4$, 5×10^3 A/cm^2 for $Mo_{70}Si_{20}B_{10}$, 3×10^3 A/cm^2 for $W_{70}Si_{20}B_{10}$ and 3×10^2 A/cm^2 for $Ti_{45}Nb_{40}Si_{15}$. Further, J_c was measured in external applied magnetic fields(H) in a liquid helium bath. As an example, Fig. 20 shows the critical current density $J_c(H)$ as a function of H for $Ti_{55}Nb_{30}Si_{15}$ amorphous alloy. For H=0, J_c is found to be about 270 A/cm^2. For H>0, J_c is observed to fall very rapidly with increasing H. For example, at $H \simeq 0.5\ H_{c2}$, J_c is of the order of 150 A/cm^2. Such small values seem to result from the fact that the amorphous phase does not contain the pinning centers such as grain boundaries, dislocations, precipitates and other defects and is quite homogeneous on the scale of coherence length(ξ=3-10 nm)[37].

However, $J_c(H)$ value improves by the introduction of a small volume fraction of crystalline precipitates into the amorphous superconducting matrix. Figure 21[40] shows $J_c(H)$ as a function of H at 4.2 K for amorphous $Ti_{55}Nb_{30}Si_4B_{11}$ and $Ti_{57}Nb_{30}Si_{10}B_3$ samples containing crystalline phases with bcc structure. The critical current decreases much more slowly with increasing magnetic field compared with that shown in Fig. 20. For H=0, J_c is of order of 7×10^4 A/cm^2 for $Ti_{55}Nb_{30}Si_4B_{11}$ and 2×10^4 A/cm^2 for $Ti_{57}Nb_{30}Si_{10}B_3$, and also H_{c2} is more than 6.4×10^6 A/m(80 kOe) for $Ti_{55}Nb_{30}Si_4B_{11}$. These results indicate that the introduction of crystalline phase into the amorphous matrix results in a remarkable increase of flux pinning forces.

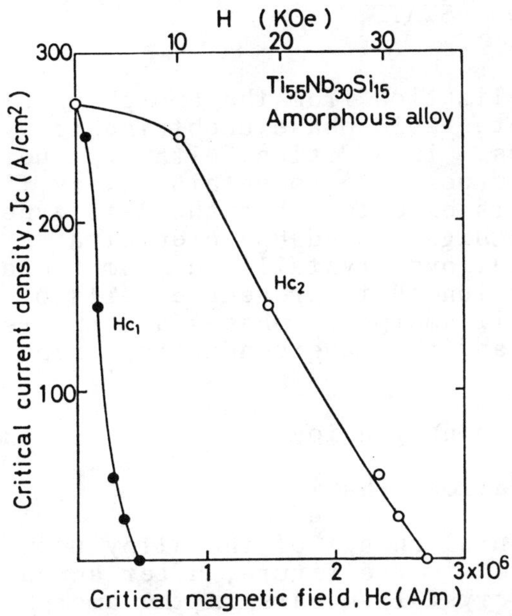

Fig. 20 Critical current density(J_c) of $Ti_{55}Nb_{30}Si_{15}$ amorphous alloy as a function of magnetic field applied normal to the direction of current flow.

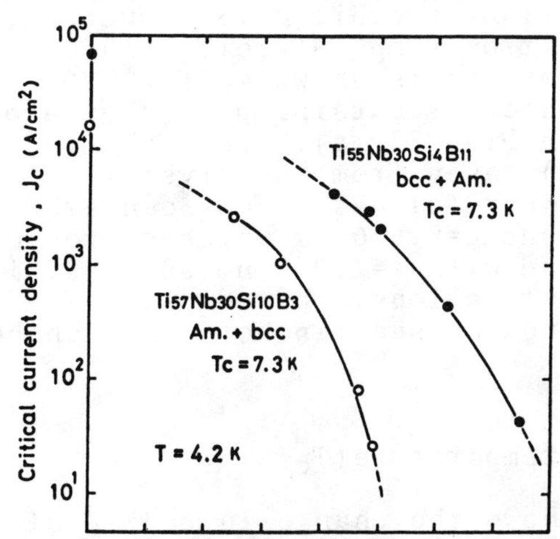

Fig. 21 Critical current density(J_c) of $Ti_{55}Nb_{30}Si_4B_{11}$ and $Ti_{57}Nb_{30}Si_{10}B_3$ amorphous alloys as a function of magnetic field applied normal to the direction of current flow.

6. SUPERCONDUCTING PROPERTIES OF THE ALLOYS CRYSTALLIZED FROM THE AMORPHOUS STATE

By crystallization from the amorphous state it is possible to produce a metastable phase unobtainable by standard metallurgical techniques. In addition, with a proper choice of heat treatments it is possible to obtain a very fine grain structure. Especially, it is expected that the A-15 type Nb_3Si[1] or V_3Si[1] compound with a high T_c and H_{c2} precipitates in the Nb-Si, Ti-Nb-Si and Ti-V-Si alloys crystallized from the amorphous state. The aim of this section is to present details of the crystallization behavior of their amorphous phases and the relation between the microstructure and the superconducting properties of these alloys.

(1) $Nb_{80}Si_{20}$ binary alloy

(a) Crystallization behavior

The structural change of the alloy upon annealing is shown in Fig. 22. As seen in the figure, after annealing at 873 K for 10 h an oval-shaped crystal appears in the amorphous matrix. With further heating the crystal grows rapidly in the remaining amorphous phase and then complete devitrification occurs at about 973 K, as shown in Fig. 22 (a)-(c). Above about 973 K, this crystal decomposes into two stable phases of bct Nb_3Si and bcc Nb, as shown in Fig. 22 (d). The detailed analysis of oval-shaped metastable crystal was made due to the possibility that the compound Nb_3Si, if it was formed in the cubic A-15 structure, would have a superconducting transition temperature(T_c) in excess of 25 K[1]. One example of the results is shown in Fig. 23. The crystals with 0.3-0.5 μm in diameter containing a large amount of internal defects are seen in Fig. 23 (a). The selected area diffraction patterns (b) and (c) taken from the crystals in Fig. 23 (a) show clearly that the crystal has a body-centered tetragonal structure with a=1.017 nm and c=0.516 nm which corresponds to equilibrium bct Nb_3Si compound with a=1.021 nm and c=0.519 nm[41]. Within the present investigations, the compound Nb_3Si with the cubic A-15 structure could not be seen in both as-quenched and annealed states.

(b) Transition temperature(T_c)

Figure 24 shows the change in the T_c of $Nb_{80}Si_{20}$ amorphous alloy upon heating together with the crystal structure identified by X-ray diffraction. An open circle represents the amorphous single phase, semiopen circle the coexistence of bct Nb_3Si compound and amorphous phase, and filled circle the equilibrium phases-bct Nb_3Si and bcc Nb. The value in bracket represents the T_c. As seen in the figure, in the aged amorphous state and in the mixed state of amorphous and Nb_3Si compound the values of T_c are

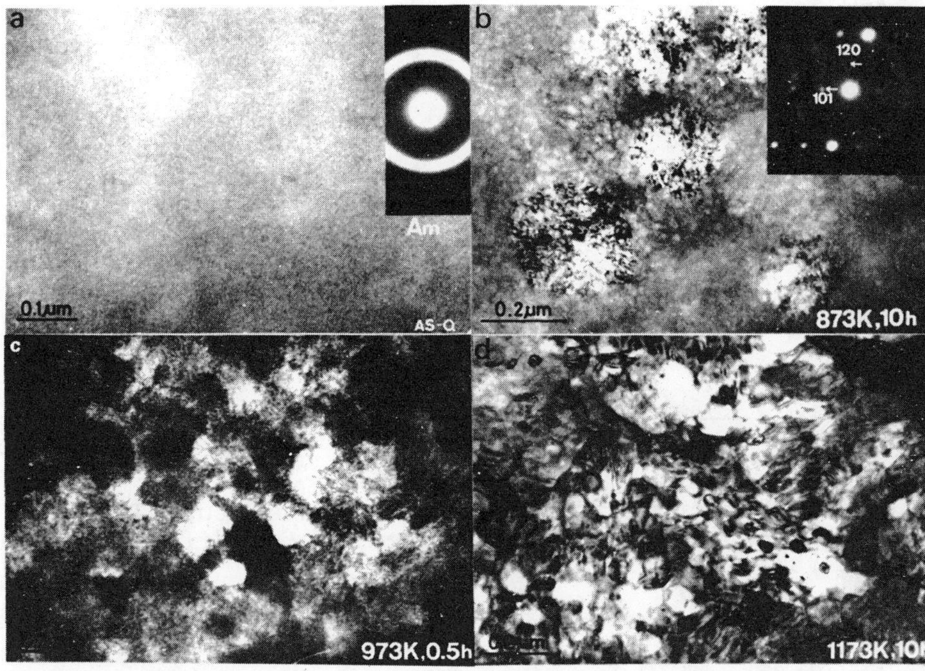

Fig. 22 Transmission electron micrographs and selected area diffraction patterns showing the microstructures of $Nb_{80}Si_{20}$ amorphous alloy heated for different periods at various temperatures.

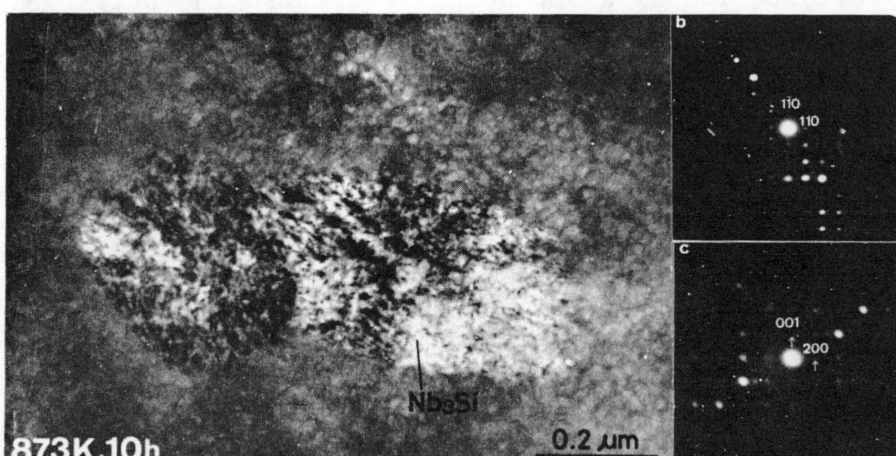

Fig. 23 The morphology and crystalline structures of Nb_3Si compound for $Nb_{80}Si_{20}$ amorphous alloy heated for 10 h at 873 K. (a) bright field image, (b) and (c) selected area diffraction patterns taken from the crystals in the photograph (a). The normals to the crystals are [001] and [010], respectively.

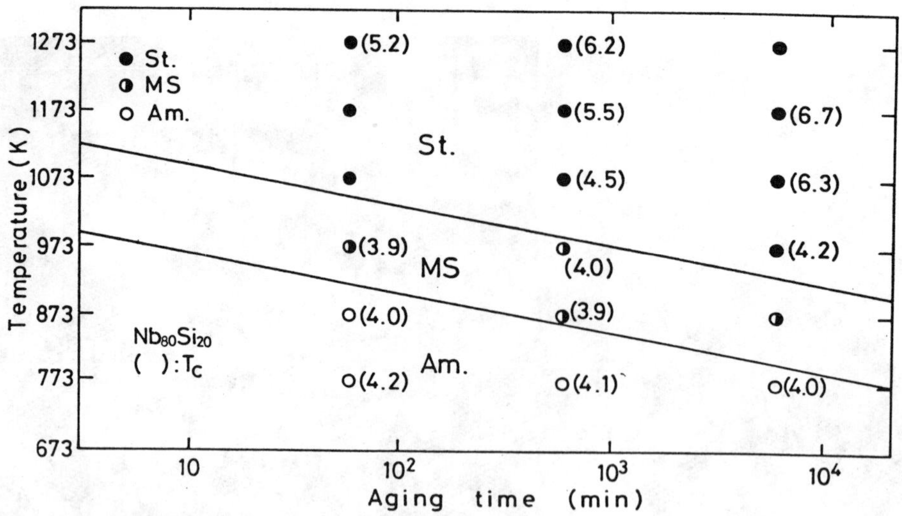

Fig. 24 The superconducting transition temperature(T_c) of $Nb_{80}Si_{20}$ amorphous alloy annealed for different periods at various temperatures.

in the range of 3.9-4.2 K and are slightly lower compared with that of the as-quenched amorphous phase. The transformation to the equilibrium phases results in an increase in T_c from 4.0 K to about 6.7 K. Thus, the T_c in the crystalline state is fairly high compared with that in the amorphous state with the same alloy composition. The increase in T_c after crystallization is attributed to the precipitation of a bcc Nb-phase because the bct Nb_3Si compound is nonsuperconducting above 4.2 K.

Hasegawa and Tanner[42] have reported that partially crystallized amorphous $Zr_{65}Nb_{15}Be_{20}$ alloy obtained by liquid quenching exhibits a T_c value of 6.85 K. This high T_c value has been attributed to a proximity effect arising from a dispersion of crystalline Zr-rich Zr(Nb) bcc particles having 100-300 nm diameter and average spacings of 200 nm or less embedded in a amorphous Be-Zr-Nb matrix. This result implies that in the Nb_3Si-plus-amorphous state the Nb-Si alloys would be a superconductor with a high -transition temperature, if Nb_3Si had a A-15 structure and dispersed in an appropriate particle size and spacing state in the amorphous matrix.

(2) Ti-Nb-Si and Ti-V-Si ternary alloys

(a) Crystallization behavior

The general features of the DTA curves for Ti-Nb-Si and Ti-V-Si alloys are exemplified in Figs. 25 and 26, wherein two exothermic peaks are seen, indicating that there exist two stages of crystallization for both the alloy systems. Figures 27 and 28

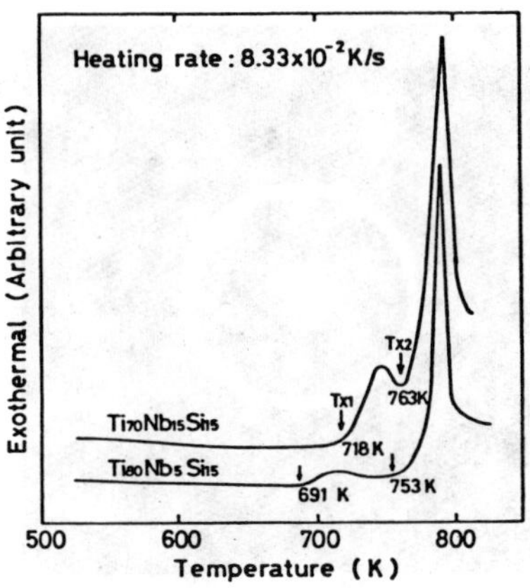

Fig. 25 Differential thermal analysis curves of $Ti_{80}Nb_5Si_{15}$ and $Ti_{70}Nb_{15}Si_{15}$ amorphous alloys.

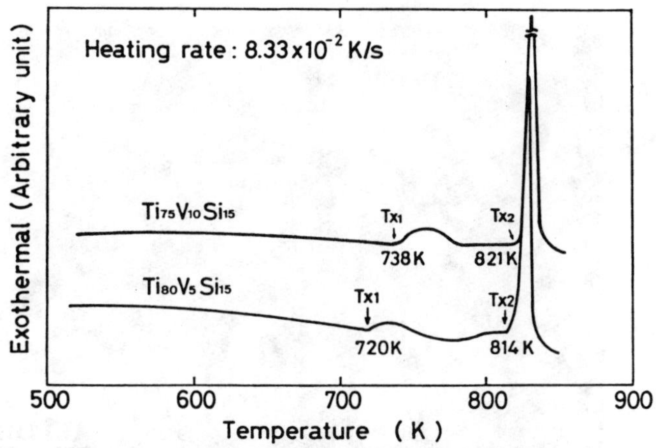

Fig. 26 Differential thermal analysis curves of $Ti_{80}V_5Si_{15}$ and $Ti_{75}V_{10}Si_{15}$ amorphous alloys.

show transmission electron micrographs of $Ti_{70}Nb_{15}Si_{15}$ and $Ti_{80}V_5Si_{15}$ alloys annealed for different periods at specified temperatures, which correspond to the beginning temperatures of the first and the second exothermic peaks on the DTA curve. Featureless structure and several haloes, which are typical feature of the amorphous material, are seen in the figures (a) and (b). There is not any observable change in the amorphous structure up to about 710 K for $Ti_{70}Nb_{15}Si_{15}$ and about 720 K for $Ti_{80}V_5Si_{15}$. In each specimen annealed for 10 min at 763 and 723 K, fine globular or elliptical precipitates are seen over the entire area of their specimens, as seen in the figures(c). The growth rate of these crystals is extremely slow and their average size is less than about 15 nm. In the diffraction patterns(d), some reflection spots from these crystals are superimposed in the rings from amorphous phase. Both the crystalline phases have a bcc structure which corresponds to a β-Ti(Nb) or β-Ti(V) solid solution. The lattice parameter is a=0.328 nm for $Ti_{70}Nb_{15}Si_{15}$ and a=0.326 nm for $Ti_{80}V_5Si_{15}$. Since any phase with a bcc structure does not exist in Ti-Nb and Ti-V equilibrium phase diagrams at room temperature, these phases can be considered a metastable phase(MS-I) representing the enhanced solid solubility of Nb or V in each β-Ti. The extremely slow growth of the MS-I phase is supposed to be probably due to an enrichment of silicon content in the amorphous phase. In both the specimens annealed for 60 min at 763 K for $Ti_{70}Nb_{15}Si_{15}$ and at 813 K for $Ti_{80}V_5Si_{15}$ oval-shaped crystals (MS-II phase) are seen over the entire area of the figures(e), in addition to the MS-I phase. From the electron diffraction patterns (f), the MS-II phase is Nb_3Si having a bct structure with a=1.021 nm and c=0.519 nm for $Ti_{70}Nb_{15}Si_{15}$ and Ti_5Si_3 having a hexagonal structure with a=0.747 nm and c=0.516 nm for $Ti_{80}V_5Si_{15}$.

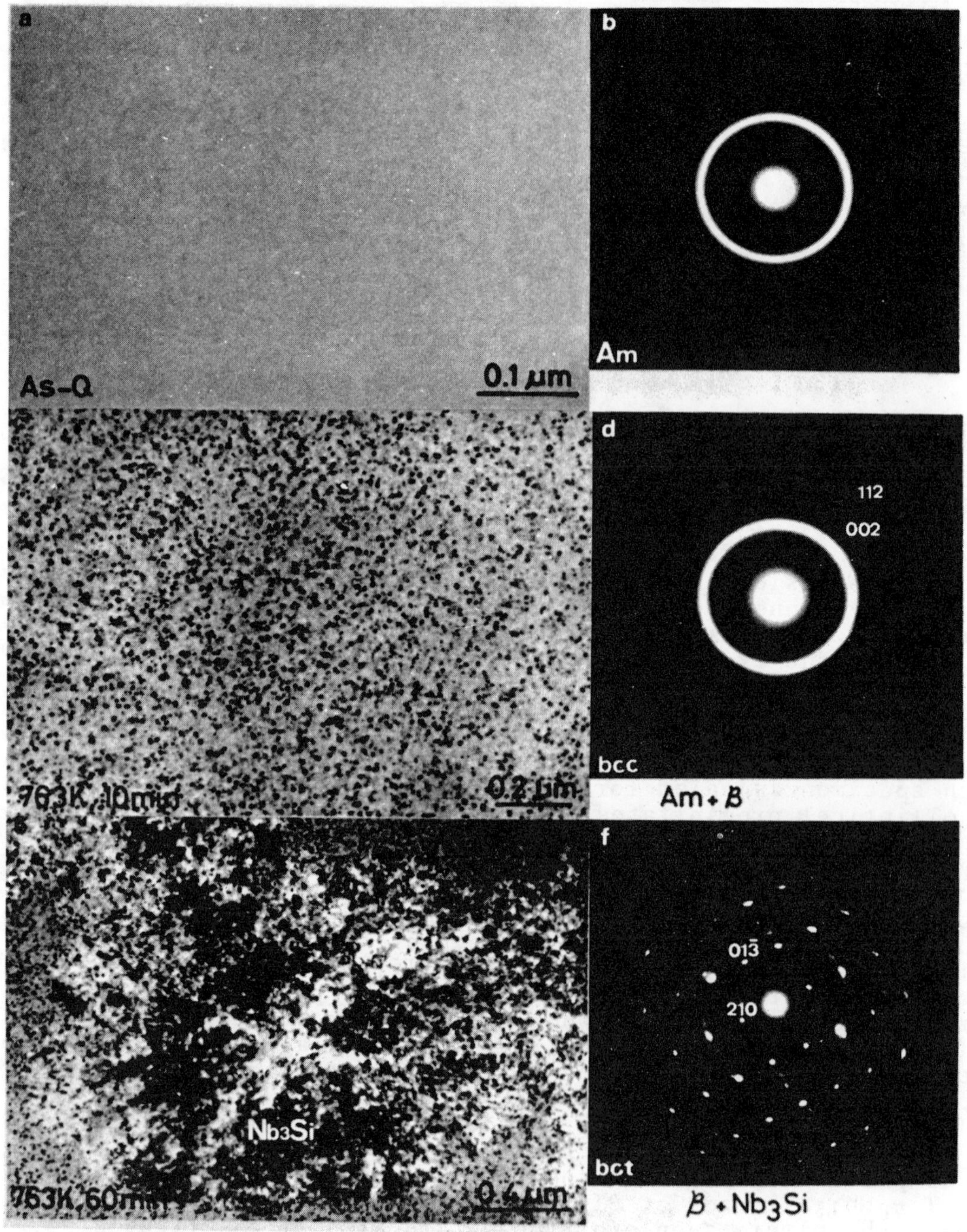

Fig. 27 Transmission electron micrographs and selected area diffraction patterns showing the as-quenched(a and b) and annealed(c, d, e and f) structure of $Ti_{70}Nb_{15}Si_{15}$ amorphous alloy.

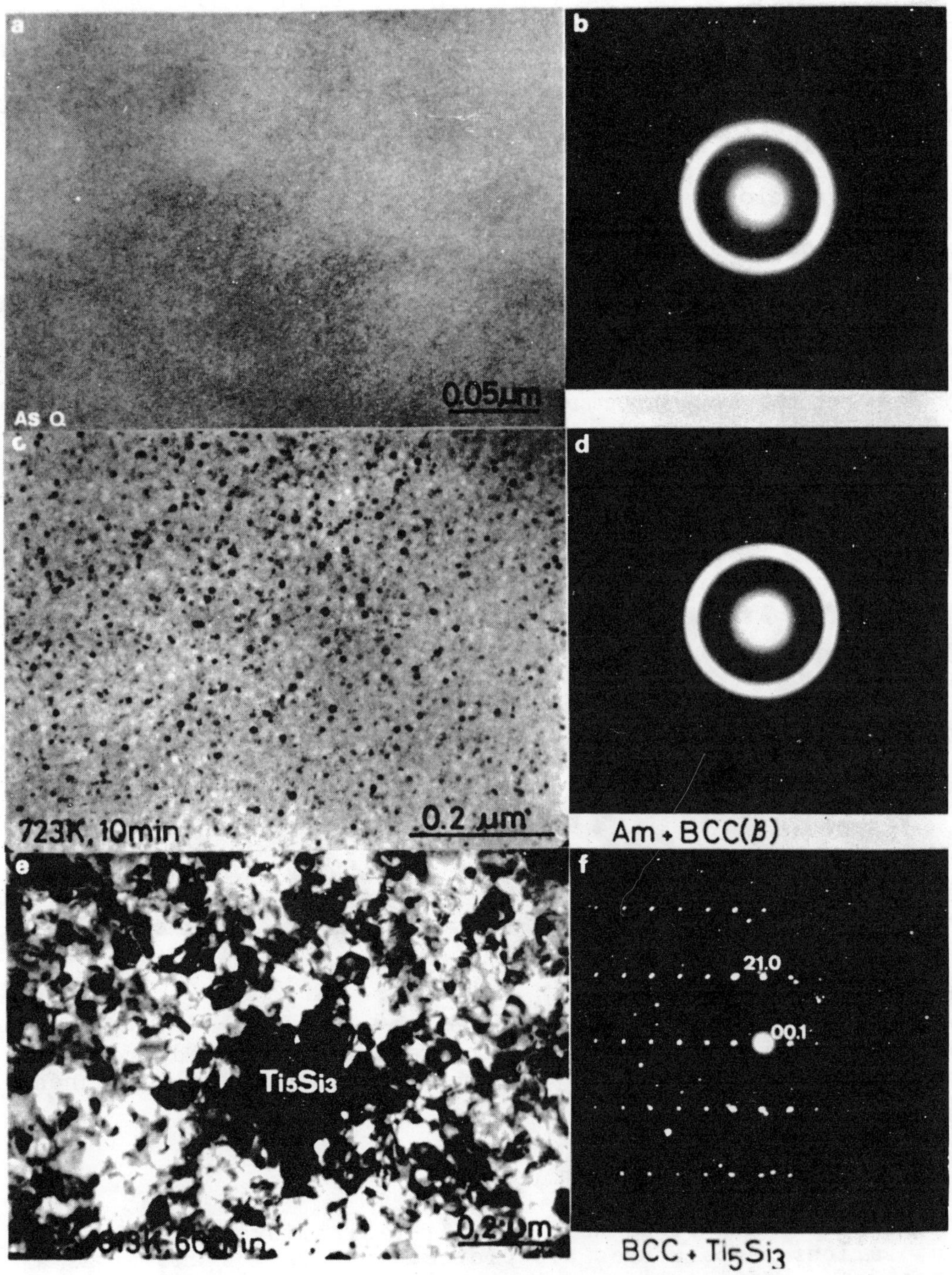

Fig. 28 Transmission electron micrographs and selected area diffraction patterns showing the as-quenched(a and b) and annealed(c, d, e and f) structures of $Ti_{80}V_5Si_{15}$ amorphous alloy.

At higher annealing temperatures, their MS-I phases transform to the hcp α-Ti stable phase. As an example, microstructure consisting of α-Ti and Nb_3Si in $Ti_{70}Nb_{15}Si_{15}$ alloy annealed for 1 h at 833 K is shown in Fig. 29, wherein the α-Ti phase contains a large number of fine planar faults. Additionally, the hexagonal omega (ω) phase with a=0.295 nm and c=0.467 nm was sometimes observed in the temperature range of 700 to 900 K, which corresponds to the transformation range from the bcc β-phase to the hcp α-phase. However, we could not obtain detailed information about the transformation process among β, α and ω phases. Within this investigation, the compound Nb_3Si or V_3Si with the cubic A-15 structure could not be seen in the annealed samples. The above observations indicate that a broad, low intensity peak at the low temperature side in the DTA curves corresponds to the precipitation of the bcc β-phase(MS-I) from the amorphous phase and a narrow, high intensity peak at the higher temperature side is due to the transition of the remaining amorphous phase to the MS-II phase, i.e., bct-Nb_3Si for Ti-Nb-Si alloy and hexagonal Ti_5Si_3 compound for Ti-V-Si alloy. Such a crystallization process, i.e., Am. ⟶ Am. + MS-I ⟶ MS-I + MS-II ⟶ Stable, agrees well with those of metal-metalloid system [4,43] as well as metal-metal system of Co-Ti [44].

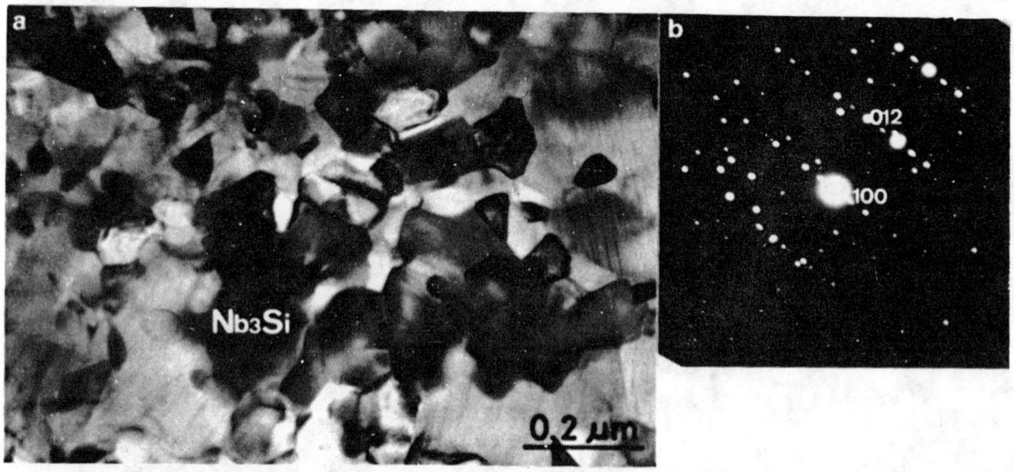

Fig. 29 Transmission electron micrograph and selected area diffraction pattern of $Ti_{70}Nb_{15}Si_{15}$ amorphous alloy annealed for 1 h at 833 K.

(b) Transition temperature(T_c)

The T_c and ΔT_c of Ti-Nb-Si and Ti-V-Si alloys annealed for 1 h at various temperatures are plotted in Figs. 30 and 31 as a function of annealing temperature. As seen in these figures, T_c increases rapidly with increasing annealing temperature after crystallization, reaches a maximum value at appropriate temperatures and with continued annealing decreases. The maximum T_c for $Ti_{85-x}Nb_xSi_{15}$ system is about 4.8 K for x=5 at%, 7.3 K for x=15

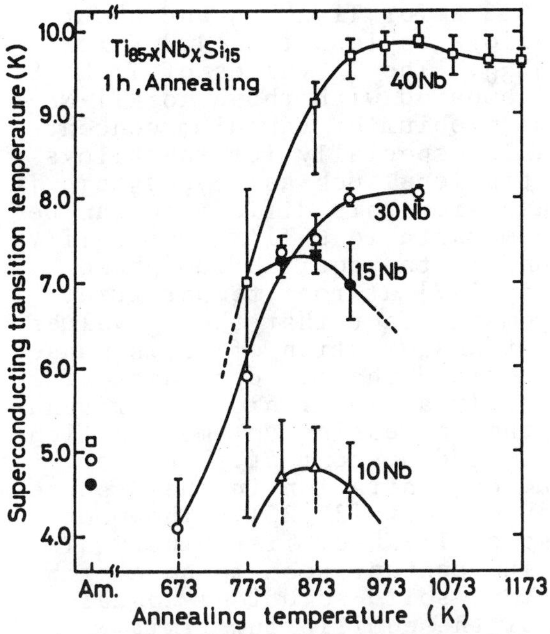

Fig. 30 Changes in superconducting transition temperature(T_c) and transition width(ΔT_c) for $Ti_{85-x}Nb_xSi_{15}$ amorphous alloys with annealing temperature. Vertical bars represent the transition width.

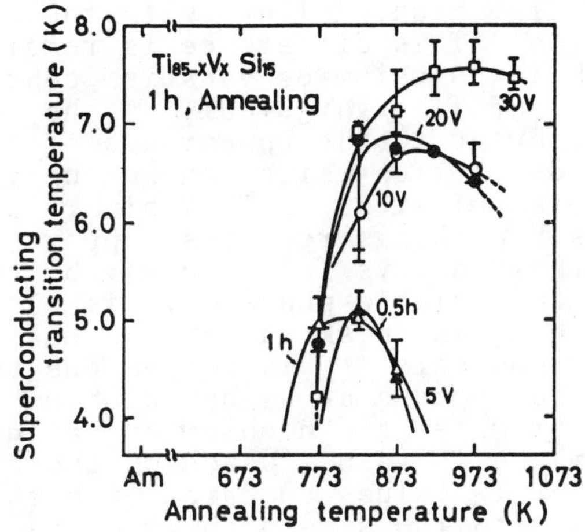

Fig. 31 Changes in superconducting transition temperature (T_c) and transition width(ΔT_c) for $Ti_{85-x}V_xSi_{15}$ amorphous alloys with annealing temperature. Vertical bars represent the transition width.

at%, 8.1 K for x=30 at% and 9.9 K for x=40 at% and it for $Ti_{85-x}V_xSi_{15}$ system is 5.1 K for x=5 at%, 6.7 K for x=10 at%, 6.9 K for x=20 at% and 7.6 K for x=30 at%. Thus, there is a clear tendency that the T_c increases with increasing niobium or vanadium content. On the other hand, the ΔT_c is considerably large in the mixed state of amorphous and crystalline phases, but decreases rapidly after crystallization of the amorphous phase and is less than about 0.5 K at appropriate temperatures where T_c shows a maximum value.

The above-described transmission electron microscopic observations indicate that the crystalline superconductors consist of bcc β-Ti(Nb) and bct Nb_3Si for the Ti-Nb-Si alloys and bcc β-Ti(V) and hexagonal Ti_5Si_3 for the Ti-V-Si alloys. Since the bct Nb_3Si[41] and hexagonal Ti_5Si_3 [45] phases are nonsuperconducting above 4.2 K, it is concluded that the high T_c values of the crystalline Ti-Nb-Si and Ti-V-Si alloys are attributed to the β-Ti(Nb) and β-Ti(V) solid solution, respectively, similar to the mechanism for the conventional superconducting Ti-Nb[25] and Ti-V[46] binary alloys. It has been shown[24,25] that T_c and H_c depend largely on the Nb or V content in the β phase. According to their results,

T_c of crystalline Ti-Nb binary alloys is less than 4.2 K for $Ti_{90}Nb_{10}$ and $Ti_{85}Nb_{15}$ and is about 6.8 K for $Ti_{70}Nb_{30}$ and about 8.0 K for $Ti_{60}Nb_{40}$ and it of Ti-V alloys is less than 4.2 K for $Ti_{80}V_{20}$ and is about 6.2 K for $Ti_{70}V_{30}$. The T_c values of Ti-Nb-Si and Ti-V-Si alloys are much higher compared with those for Ti-Nb and Ti-V binary alloys with the same niobium or vanadium concentration. This difference is remarkable especially for the alloys with low niobium or vanadium concentrations such as $Ti_{70-75}Nb_{10-15}Si_{15}$ and $Ti_{75-80}V_{5-10}Si_{15}$. The reason for this difference can be attributed to the appearance of the metastable β-Ti(Nb) or β-Ti(V) solid solution which can not be found in the equilibrium phase diagram of Ti-Nb or Ti-V binary alloys[47] at room temperature. This hypothesis receives support from the fact that the T_c values of their alloys fall rapidly by the transformation of bcc β-phase to hcp α-stable phase. It is also noticed that T_c decreases after reaching the peak value more drastically at lower niobium or vanadium contents. This may be due to the decreasing volume fraction of the α-phase at higher niobium or vanadium content. Further, we estimated the niobium or vanadium concentration in the β-phase from the T_c values based on the previous data[22,23]. The much higher T_c values compared with those of Ti-Nb or Ti-V binary alloys with the same niobium or vanadium content seem to be attributable to the fact that the metastable β solid solutions contain niobium or vanadium much in excess of the equilibrium limit.

The maximum T_c values of Nb-Si, Ti-Nb-Si and Ti-V-Si crystal-

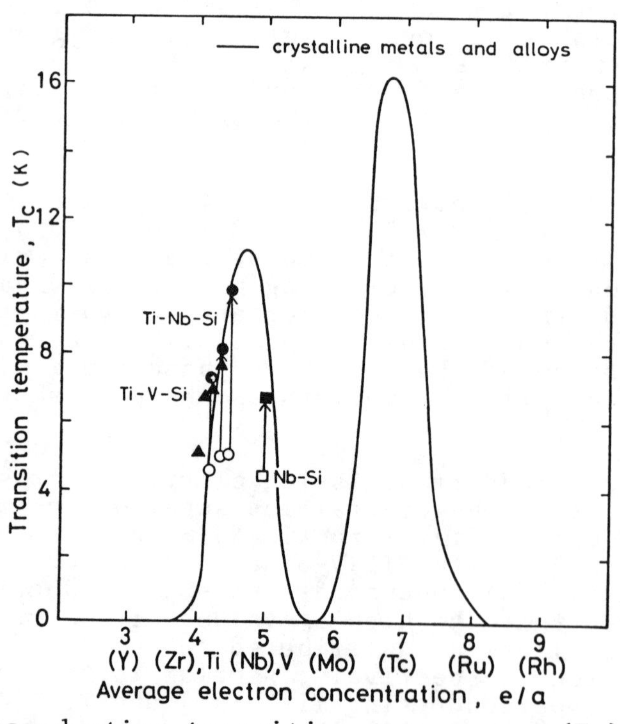

Fig. 32 Superconducting transition temperature(T_c) vs. electron per atom ratio(e/a) for the 4d transition-metal series.

line alloys are plotted as a function of the average outer electron concentration[e/a] of the transition metal components in Fig. 32, wherein the values of T_c for amorphous Nb-Si and Ti-Nb-Si alloys are also included. Further, these data are compared with the variation of T_c for crystalline transition metals and alloys which is well known as Matthias rule. From this figure, it is seen that the T_c of the amorphous alloys increases remarkably upon crystallization and their values agree well with those expected by the Matthias rule.

(c) Critical magnetic field(H_c) and critical current density(J_c)

The critical current density J_c was measured at external applied magnetic field(H) in a liquid helium bath. Figures 33 to 37 show the critical current density J_c(H) as a function of H for $Ti_{70}Nb_{15}Si_{15}$, $Ti_{55}Nb_{30}Si_{15}$, $Ti_{50}Nb_{35}Si_{15}$, $Ti_{65}V_{20}Si_{15}$ and $Ti_{55}V_{30}Si_{15}$ amorphous alloys annealed for different periods at various temperatures. As seen in these figures, Ti-Nb-Si alloys possess much higher J_c(H) values compared with Ti-V-Si alloys. For example, for $H=5.6 \times 10^6$ A/m(70 kOe), the J_c is about 5×10^3 A/cm^2 for $Ti_{70}Nb_{15}Si_{15}$, about 1×10^4 A/cm^2 for $Ti_{55}Nb_{30}Si_{15}$ and about 4.7×10^4 A/cm^2 for $Ti_{50}Nb_{35}Si_{15}$ and about 2×10^3 A/cm^2 for $Ti_{55}V_{30}Si_{15}$. Also, one can notice the tendency that J_c(H) values

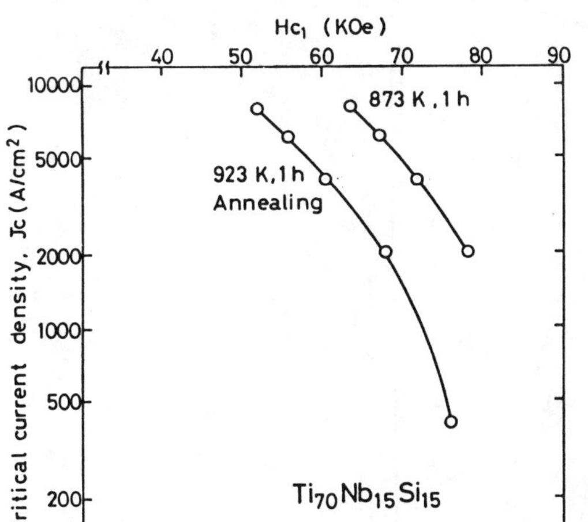

Fig. 33 Critical current density(J_c) of $Ti_{70}Nb_{15}Si_{15}$ amorphous alloy annealed for 1 h at 873 and 923 K as a function of magnetic field applied normal to the direction of current flow.

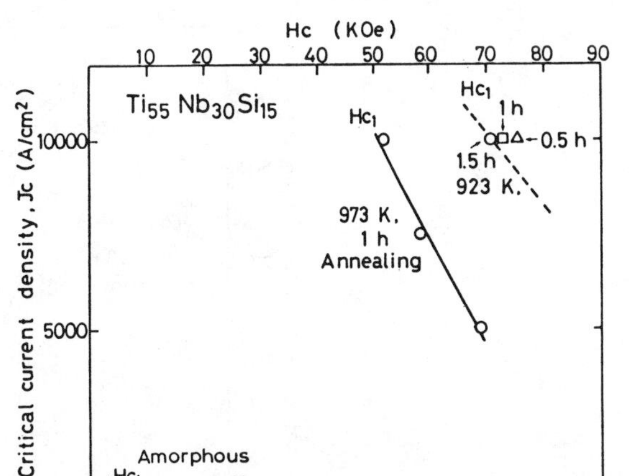

Fig. 34 Critical current density(J_c) of $Ti_{55}Nb_{30}Si_{15}$ amorphous alloy annealed for 0.5 and 1 h at 923 and 973 K as a function of magnetic field applied normal to the direction of current flow.

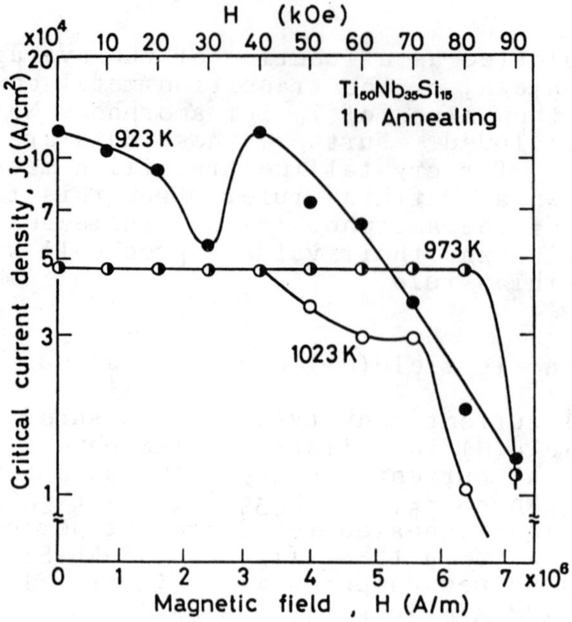

Fig. 35 Critical current density(J_c) of $Ti_{50}Nb_{35}Si_{15}$ amorphous alloy annealed for 1 h at various temperatures as a function of magnetic field applied normal to the direction of current flow.

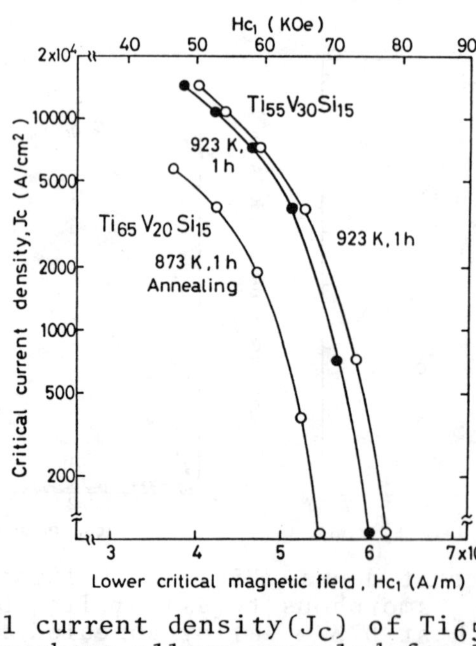

Fig. 36 Critical current density(J_c) of $Ti_{65}V_{20}Si_{15}$ and $Ti_{55}V_{30}Si_{15}$ amorphous alloys annealed for 1 h at 873 and 923 K as a function of magnetic field applied normal to the direction of current flow.

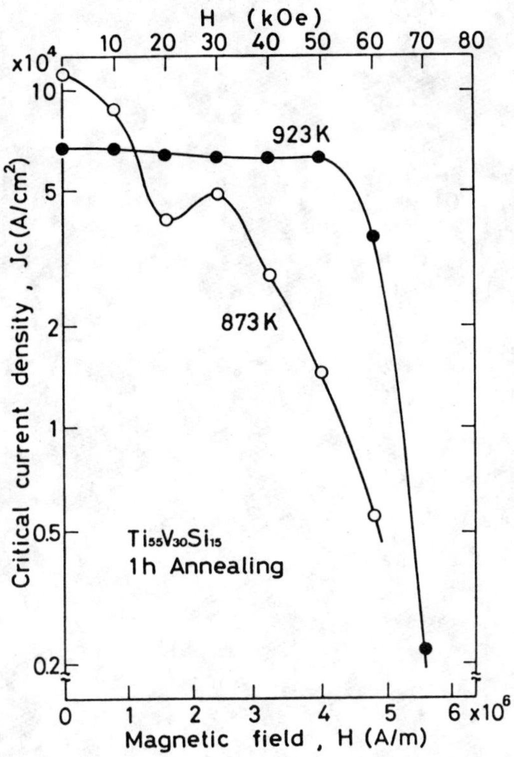

Fig. 37 Critical current density(J_c) of $Ti_{55}V_{30}Si_{15}$ amorphous alloy annealed for 1 h at 873 and 923 K as a function of magnetic field applied normal to the direction of current flow.

increase with increasing niobium or vanadium content. In addition, as seen in Fig. 35, the pinning force which is given by the area under J_c-H curve shows a maximum value at an appropriate temperature (about 973 K for $Ti_{50}Nb_{35}Si_{15}$ alloy). The underaging at 923 K results in high J_c of about 1×10^5 A/cm^2 for H less than about 3.2×10^6 A/m(40 kOe) and the overaging at 1023 K results in J_c values less than 5×10^4 A/cm^2 over all the H range. In general, it is well known that $J_c(H)$ is closely related to the microstructure of the alloy. Figure 38 shows the structure of $Ti_{50}Nb_{35}Si_{15}$ amorphous alloys annealed for 1 h at 923, 973 and 1023 K. All structures consist of bcc β-Ti(Nb) and bct Nb_3Si phase and the average particle diameter of the Nb_3Si phase is about 50 nm at 923 K, 90 nm at 973 K and 200 nm at 1023 K. These structures indicate that the high $J_c(H)$ value for $Ti_{50}Nb_{35}Si_{15}$ alloy annealed for 1 h at 973 K is attributed to a dispersion effect of the Nb_3Si particles, having about 90 nm diameter and an average spacing of 100 nm or less, in bcc β-Ti(Nb) phase.

Source: Structures and Properties of Amorphous Metals II, 165-200

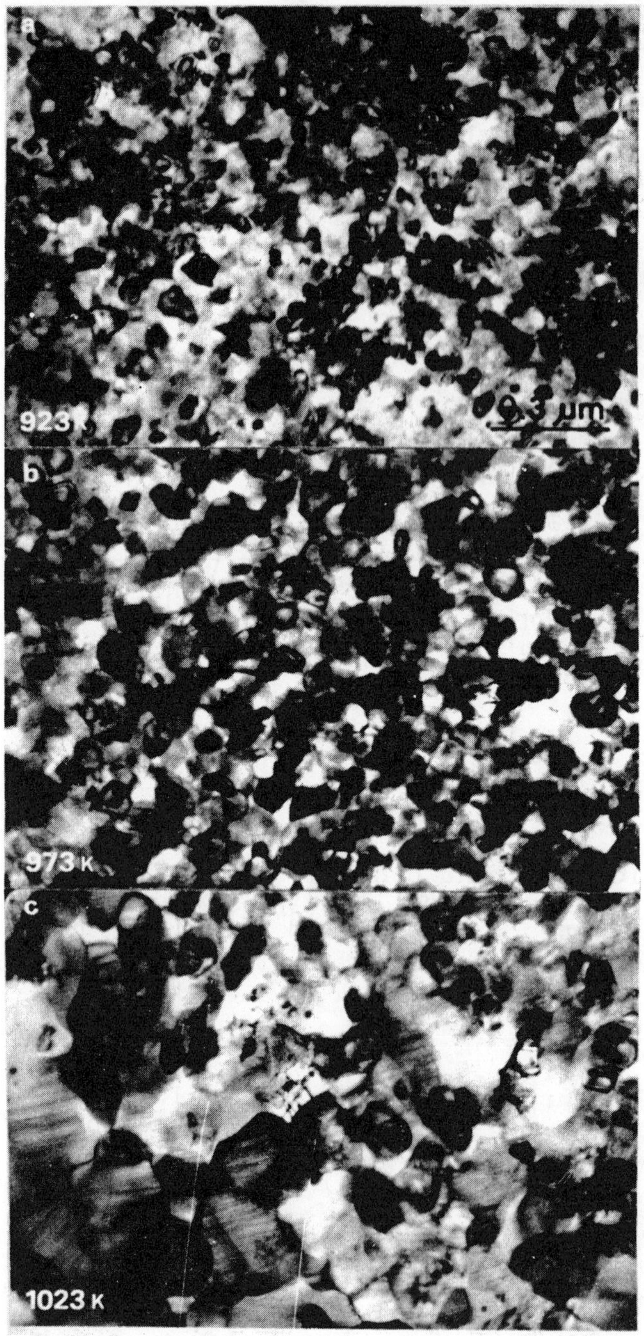

Fig. 38 Transmission electron micrographs of $Ti_{50}Nb_{35}Si_{15}$ amorphous alloy annealed for 1 h at 923 K(a), 973 K(b) and 1023 K(c).

7. CONCLUSION

The superconducting properties of the amorphous alloys characterized in the present paper are still insufficient for practical use. However, there is no well-established theoretical reason why the T_c of amorphous superconductors should always be lower than those of crystalline alloys. We believe that it would be possible to find new amorphous alloys with transition temperatures higher than those reported in this review as well as good ductility through more extensive investigations. In addition, from the technological point of view, it is to be noted that melt quenching makes the processing of superconducting materials much simpler. Especially, the existence of a readily obtainable amorphous phase in the Ti-Nb-Si system seems to offer alternative means of fabricating superconducting tapes and wires.

ACKNOWLEDGEMENTS

For the preparation of this review, the authors rely mainly upon results obtained by the authors and their coworkers, i.e., Messrs. S. Sakai, H. M. Kimura and A. Hoshi and Dr. Suryanarayana. The authors would like to express their grateful acknowledgement to them.

REFERENCES

1) D. Dew-Hughes: Cryogenics, 15(1975), 435.
2) R. H. Willens, E. Buehler and B. T. Matthias: Phys. Rev., 159(1967), 327.
3) B. T. Matthias, V. B. Compton and E. Corenzwit: J. Phys. Chem. Solids, 19(1961), 130.
4) T. Masumoto and R. Maddin: Acta Met., 19(1971), 725
5) W. Buckel and R. Hilsch: Z. Phys., 138(1954), 109.
6) G. Bergmann: Phys. Reports, 27C(1976), 161.
7) W. L. Johnson, S. J. Poon and P. Duwez: Phys. Rev. B, 11(1975) 150.
8) S. Sakai, H. M. Kimura, A. Inoue and T. Masumoto: Papers represented at the Meetings of the Japan Institute of Metals, October 1978, p. 236; April 1979, p. 190.
9) T. Masumoto, A. Inoue, S. Sakai, H. M. Kimura and A. Hoshi: Trans. JIM, 21(1980), 115.
10) A. Inoue, H. M. Kimura, T. Masumoto, C. Suryanarayana and A. Hoshi: J. Appl. Phys., to be published.
11) A. Inoue, C. Suryanarayana, T. Masumoto and A. Hoshi: Sci. Rep. Res. Inst. Tohoku University, A28(1980), No. 2, in press.
12) A. Inoue, H. M. Kimura, C. Suryanarayana, T. Masumoto and A. Hoshi: J. Mater. Sci., to be published.
13) A. Inoue, S. Sakai, H. M. Kimura, T. Masumoto and A. Hoshi: Scripta Met., 14(1980), No. 2, in press.
14) S. Sakai, A. Inoue, H. M. Kimura and T. Masumoto: Scripta Met., 14(1980), No. 3, in press.
15) A. Inoue, T. Masumoto, C. Suryanarayana and A. Hoshi: to be presented at the 4th Intern. Conf. on Liquid and Amorphous Metals, Grenoble, France, July 1980.
16) D. Turnbull: J. Phys. (Paris) Colloq., 35(1974), 1.
17) C. J. Smithells: Metals Reference Book, 5th Edition, p. 725, 780, 782, Butterworths, (1976).
18) A. Inoue, T. Masumoto and H. M. Kimura: J. Japan Inst. Metals, 42(1978), 303; Sci. Rep. Res. Inst. Tohoku University, A27 (1979), 159.
19) W. D. Forgeng and W. D. Forgeng, Jr.: Metals Handbook eighth edition, Metallography, Structures and Phase Diagram, p.283, 334, ASM, (1973).
20) For example, T. Masumoto: Sci. Rep. Res. Inst. Tohoku University, A26(1977), 246.
21) C. C. Tsuei, W. L. Johnson, R. L. Laibowitz and J. M. Viggiano: Solid State Commun., 24(1977), 615.
22) M. M. Collver and R. H. Hammond: Phys. Rev. Lett., 30(1973), 92.
23) W. L. Johnson, S. J. Poon, J. Durand and P. Duwez: Phys. Rev. B, 18(1978), 206.
24) T. G. Berlincourt and R. R. Hake: Phys. Rev., 131(1963), 140.
25) J. V. Vetrano and R. W. Boom: J. Appl. Phys., 36(1965), 1179.
26) For example, W. L. Johnson: J. Appl. Phys., 50(1979), 1557.
27) For example, W. DeSorbo: Phys. Rev., 130(1963), 2177.
28) K. Togano and K. Tachikawa: Phys. Lett., 54A(1975), 205.
29) E. R. Domb and W. L. Johnson: J. Low Temp. Phys., 33(1978), 29.

30) H. Fujimori, T. Masumoto, Y. Obi and M. Kikuchi: Japan. J. Appl. Phys., 13(1974), 1889.
31) T. Egami, P. J. Flanders and C. D. Graham, Jr.: Appl. Phys. Letters, 26(1975), 128.
32) H. S. Chen, C. D. Ferris, E. M. Gyorgy, H. J. Leamy and R. C. Sherwood: Appl. Phys. Letters, 26(1975), 405.
33) T. Egami and P. J. Flanders: A.I.P. Conf. Proc., 29(1976), 220.
34) A. J. Drehman and W. L. Johnson: Phys. Stat. Sol., (a)52(1979), 499.
35) K. Maki and T. Tsuzuki: Phys. Rev. 139(1965), 868.
36) N. R. Werthamer, E. Helfand and P. C. Hohenberg: Phys. Rev., 147(1966), 295.
37) W. L. Johnson: Rapidly Quenched Metals III, Ed. B. Cantor, Vol. 2, The Metals Society, London (1978), p. 1.
38) G. Eilenberger and V. Ambegaokar: Phys. Rev., 168(1967), 332.
39) D. Rainer, G. Bergmann and V. Eckhardt: Phys. Rev. B, 8(1973), 5324.
40) A. Inoue, A. Hoshi, C. Suryanarayana and T. Masumoto: Scripta Met., to be published.
41) D. K. Deardorff, R. E. Siemens, P. A. Romans and R. A. McCune: J. Less-Common Metals, 18(1969), 11.
42) R. Hasegawa and L. E. Tanner: J. Appl. Phys., 49(1978), 1196.
43) For example, T. Masumoto, H. M. Kimura, A. Inoue and Y. Waseda: Mater. Sci. Eng., 23(1976).
44) A. Inoue, K. Kobayashi, C. Suryanarayana and T. Masumoto: Scripta Met., 14(1980), No. 1, in press.
45) A. Inoue, C. Suryanarayana and T. Masumoto: unpublished research.
46) C. H. Chang, K. P. Gupta, E. C. Van Reuth and P. A. Beck: Phys. Rev., 126(1962), 2030.
47) C. J. Smithells: Metals Reference Book, 5th Edition, p. 726, 793, Butterworths, (1976).

SILICON RIBBONS MADE BY ROLLER QUENCHING METHODS

N. Tsuya, K. I. Arai, T. Takeuchi, K. Ohmori, Research Institute of Electrical Communication, Tohoku University, Sendai 980

and

T. Ojima A. Kuroiwa, Research and Developement Laboratory, TDK Electronics Co. Ltd., Ichikawa 272, Japan

Polycrystalline silicon ribbons were prepared by Roller Quenching Methods. Ribbons were obtained flexible and of virtually continuous length, 20 μm to 200 μm in thickness, and 1 mm to 50 mm in width. Microscopic observations of etched ribbon surfaces showed that the average grain sizes were 20 μm to 30 μm. Cross-sectional views showed the ribbons to have a columnar grain structure. Unwanted impurities were undetectable by Auger analysis, except very thin oxygenated layer was present on the surface of the ribbons. By means of DC conductivity and Hall measurements of intentionally doped silicon ribbons, it was found that the active carrier concentration of the ribbons was nearly the same as the doped carrier concentration, from $10^{14}/cm^3$ to $10^{20}/cm^3$, the resistivity of the ribbons were about one order of magnitude higher than that of silicon single crystals, and the Hall mobility parallel to the ribbon axis and perpendicular to the ribbon plane was 1 to 150 $cm^2 \cdot V^{-1} \cdot sec^{-1}$ and 50 to 700 $cm^2 \cdot V^{-1} \cdot sec^{-1}$, respectively. A solar cell using the present ribbon was made by means of a CVD silicon deposition method. Satisfactory diode characteristics and a conversion efficiency of about 5 % were obtained.

Key words : solar cell, silicon, rapid quenching, silicon ribbon.

Introduction

Among various kinds of solar cells, silicon cells attract many investigations because of advantages such as reliability, unlimited supply of the raw materials and so on. For practical use of solar cells, it is necessary to achieve a great reduction in cost and high speed production of cells with long life and high efficiency. The conventional technology, based on cutting large ingots into wafers, is accompanied by severe waste of materials and fabrication time. Efforts have been made to grow silicon ribbons, and the EFG process[1], the web-Dendric process[2], the Stepanov process[3], the Ribbon to Ribbon process[4] and the Horizontal Pulling process[5] were proposed. The main attraction of these processes lies in the fact that the direct production of ribbon-shaped crystals from the melt permit significant cost reductions over the usual methods of manufacturing silicon solar cells. When quick roller quenching methods can be used to form molten silicon into thin crystalline ribbons, a large reduction in cost and fabrication time is expected.

We have applied the above-mentioned roller quenching methods to prepare at high speed long flexible ribbon-shaped magnetic materials such as Sendust[6,7] and high silicon-iron alloys with good magnetic and mechanical characteristics even though these materials are mechanically brittle in their bulk states.

We tried to extend these methods to prepare at high speed silicon polycrystalline ribbons with good electronic and mechanical properties, and very brief results of these experiments were reported[8]. Flexible polycrystalline

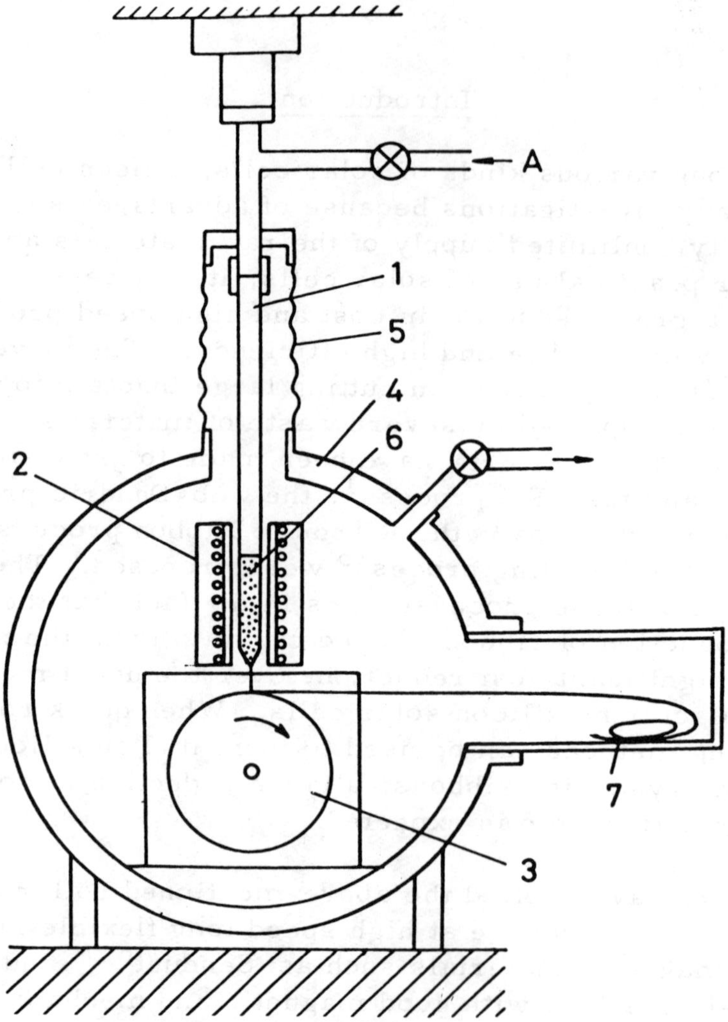

Fig. 1 Schematic arrangements of single roller quenching method.

1. heat resisting tube 2. furnace
3. roller 4. chamber 5. bellows
6. silicon charge 7. silicon ribbon
A. inert gas.

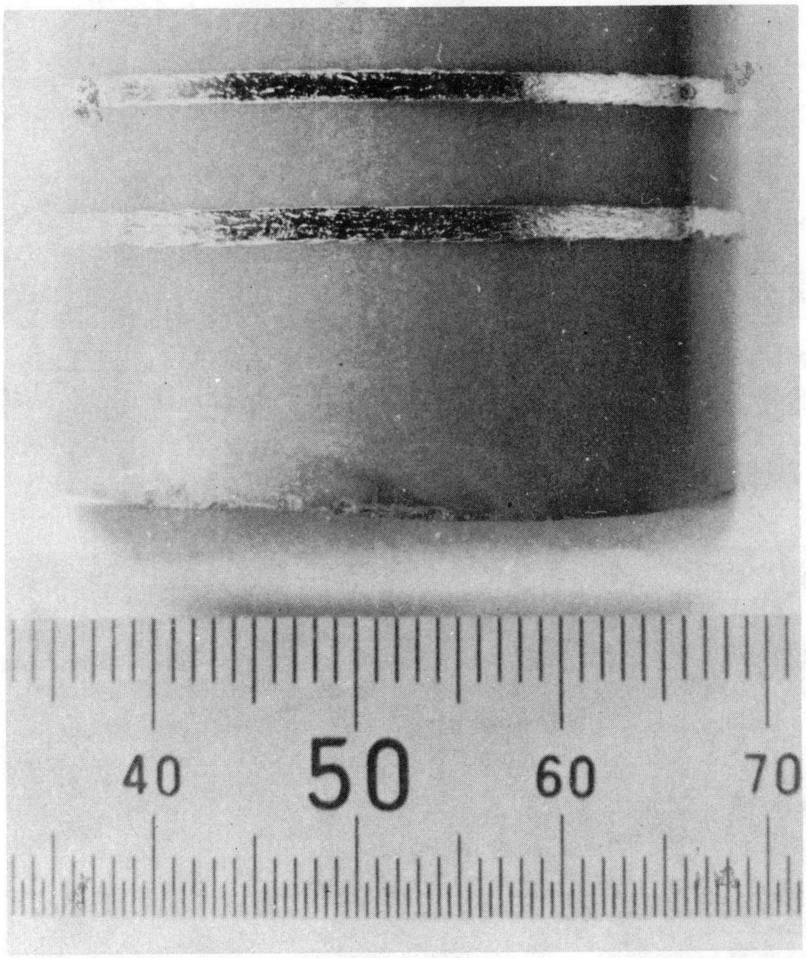

Fig. 2 Roller quenched silicon ribbon wound on a 34 mm rod, whose thickness is 30 μm and width is 2 mm.

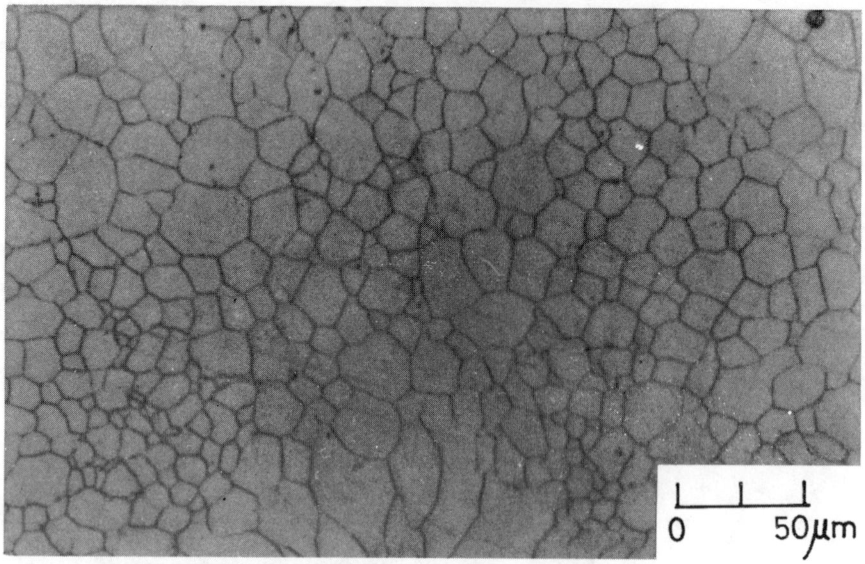

(A) Surface of the ribbon.

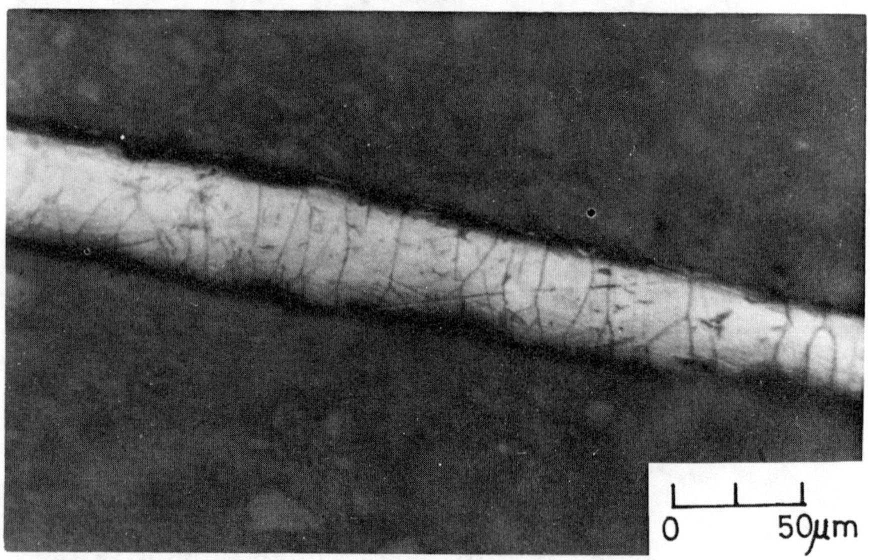

(B) Cross section of the ribbon.

Fig. 3 Micrograph of the etched silicon ribbon surface.

silicon ribbons with thicknesses from 20 μm to 200 μm were produced at a very high speed of around 30 m/sec.

In this paper we report more fully on the roller quenching method and the resulting ribbon characteristics such as morphology, impurity analysis, electrical properties and preliminary results on the fabrication of the diodes and solar cells.

Experiments

In these experiments, n- and p-type silicon ribbons were made by direct roller quenching methods. In these methods the molten silicon material was injected between a pair of rapidly rotating rolls or onto the surface of a single roll, and cooled immediately into a long crystalline ribbon.

One form of the experimental apparatus is shown in Fig. 1. The starting silicon charge 6 was put in a heat-resisting tube with a small nozzle 1, and melted with a small furnace 2. A stream of the molten silicon charge was ejected onto the surface of the roll 3 through the nozzle by a rapid increase of Ar pressure 4 in the tube. At this moment the molten charge was quenched very rapidly by the roll into a long crystalline silicon ribbon. Some ribbons were prepared in an Ar atmosphere. The thickness and width of the ribbon could be controlled by changing the roll material, the roll size, the speed of rotation, the ejection pressure, and the temperature of the melt. One representative condition which proved satisfactory was as follows : the roll material was copper, roll diameter about 40 cm, rotational speed about 1000 rpm, ejection pressure about 0.5 atmosphere, temperature of the melt about 1500°C, tube material quartz, and nozzle diameter 0.5 mm.

Auger spectra of silicon ribbons were observed by means of a PHI 545 Auger Electron spectrometer. DC

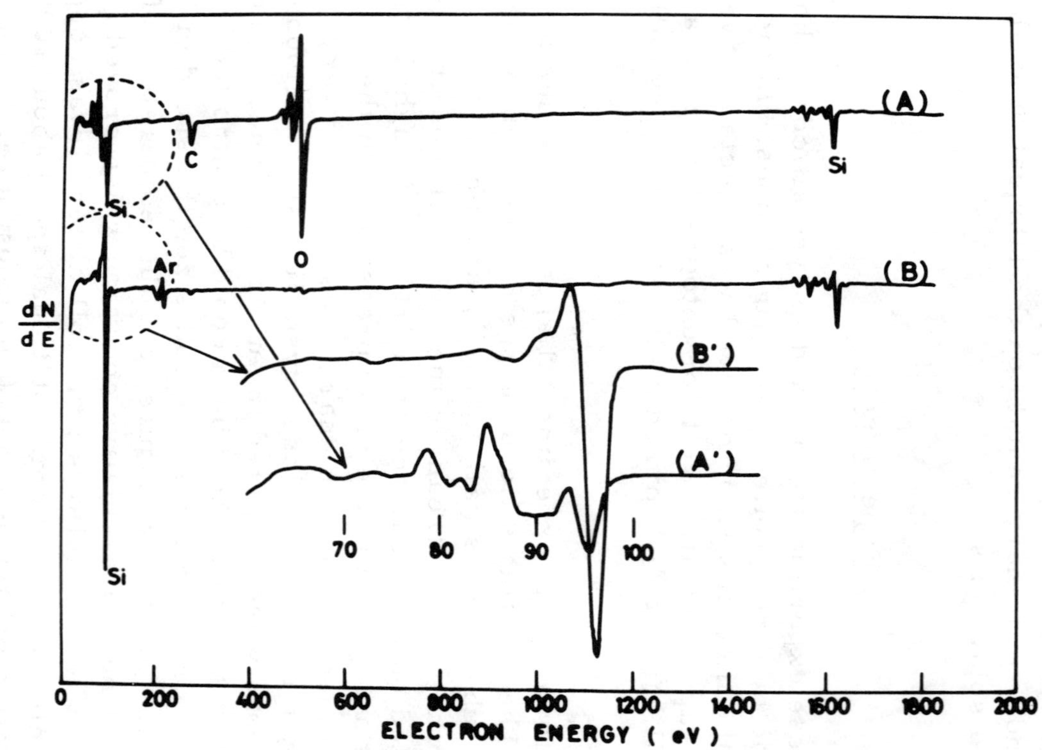

Fig 4 Auger spectra of the silicon ribbon roller quenched in Argon 3.3×10^4 Pa. atmosphere.
(A) as rolled ribbon
(B) etched by argon in bombardment.

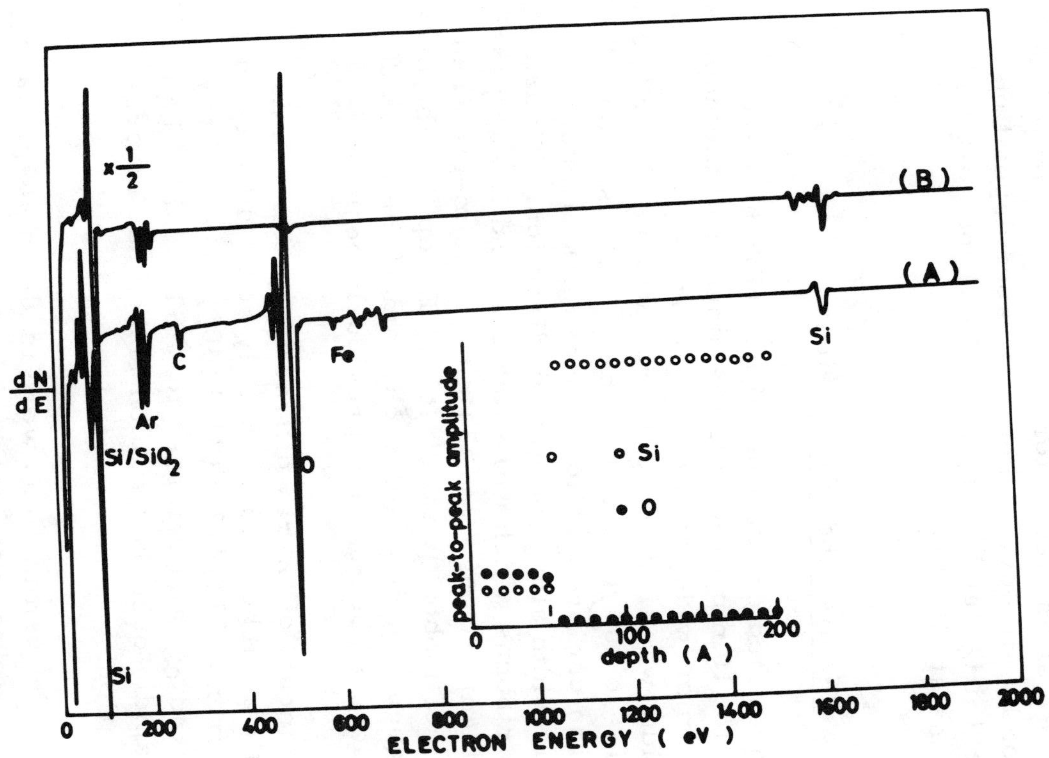

Fig. 5 Auger spectra of the ribbon rolled in air and depth profile.
(A) as rolled ribbon
(B) etched by argon in bombardment.

conductivity and Hall measurements were made along the ribbon axis and across the thickness. In these measurements ohmic contacts were made using Al on p-type ribbons and AuSb on n-type ribbons by vacuum deposition and annealing at 550°C and 350°C for 10 minutes respectively. We attempted to form a p-n junction on the p-type ribbon by employing chemical vapor deposition using silane (SiH_4) and phosphine (PH_3) on p-type silicon ribbons at 1000°C.

Results and Discussions

The roller quenched crystalline silicon ribbons were typically 1 to 10 mm wide for the single roll method, and up to 50 mm wide for the pair-of-rolls method in which the width was limited by the roll width. The ribbons were 20 μm to 200 μm in thickness. The silicon ribbon obtained in one preparation process has constant thickness and width along the growth direction. Figure 2 shows these ribbons which were flexible, and it was possible to wind them around a rod 34 mm in diameter when the thickness of ribbon was 30 μm. Except for the present roller quenched ribbons, such a small thickness of reported silicon ribbon has not previously been.

Measurements of X-ray diffraction pattern on as prepared silicon ribbon showed nearly equal diffraction contribution from (111) (220) (311) (400) ----. X-ray source was CuK_α and referred to ASTM data. This means the rapidly quenched silicon ribbon has no preferred orientation.

Microscopic examinations were made on as-prepared ribbons. Before this examination, the surface was mechanically polished, and etched for ten seconds in a mixture of 95 % nitric acid 5 % hydrofluoric acid. Fig. 3 shows typical grain structure of the silicon ribbon surface and of the cross-section. Grain sizes were measured with the line-intercept-method[9] yielding the average grain size of about 20 μm on the surface. In the cross-section of the ribbon, we observed a colmnar structure perpendicular to the ribbon plane with grains extending through the entire or half of the thickness of the ribbons depending on either the single roll or a pair of rolls methods employed, respectively.

Auger Spectroscopy

In Fig. 4 (A) are shown the Auger spectra for the free surface of the silicon ribbon prepared by the single roll method in an atmosphere of 3.3×10^4 Pa argon gas. The 92 eV peak corresponds to elemental Si[10], and Carbon, Oxgen peaks are also observed. After argon ion bombardment for one minute at 1.0 kV and 10 mA ion current corresponding to 7 A etching, the 92 eV peak becomes sharp and the oxygen peak fades out as shown in (B). It is noted that the peaks correspond to SiO_2[11] and unwanted impurities were not observed except oxygen contamination in a region about 7 A from the ribbon surface. Fig. 5 shows Auger spectra from the free surface of the silicon ribbon prepared in air by single roller method. The 92 eV peak is entirely missing in (A) but appears after argon ion bombardment for 3 min at 1.5 kV and 10 mA ion current as shown in (B). Insertion in Fig. 5 shows the depth profile of Si and O from the free surface. This shows about 50 A thickness of the surface layer is oxydized to SiO_2 during the rapid quenching process when silicon ribbon is prepared in air atmosphere.

Electrical Properties

DC conductivity and Hall measurements of the intentionally boron and phosphorous doped silicon ribbons were made. Carrier concentrations of the ribbons were directly determined from these measurements at room temperature. In Fig. 6 are plotted the active carrier concentra-

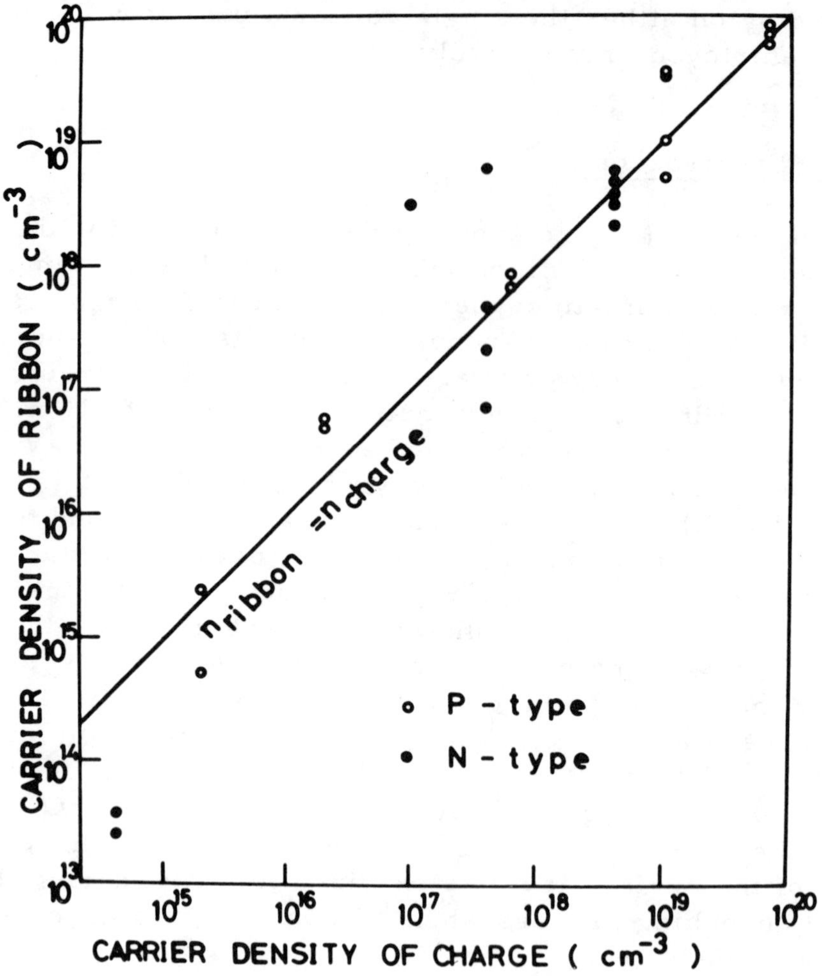

Fig. 6 Active carrier density of the silicon ribbon versus density of silicon charge.

● : P-doped ribbon
o : B-doped ribbon

Solid line shows $n_{ribbon} = n_{charge}$.

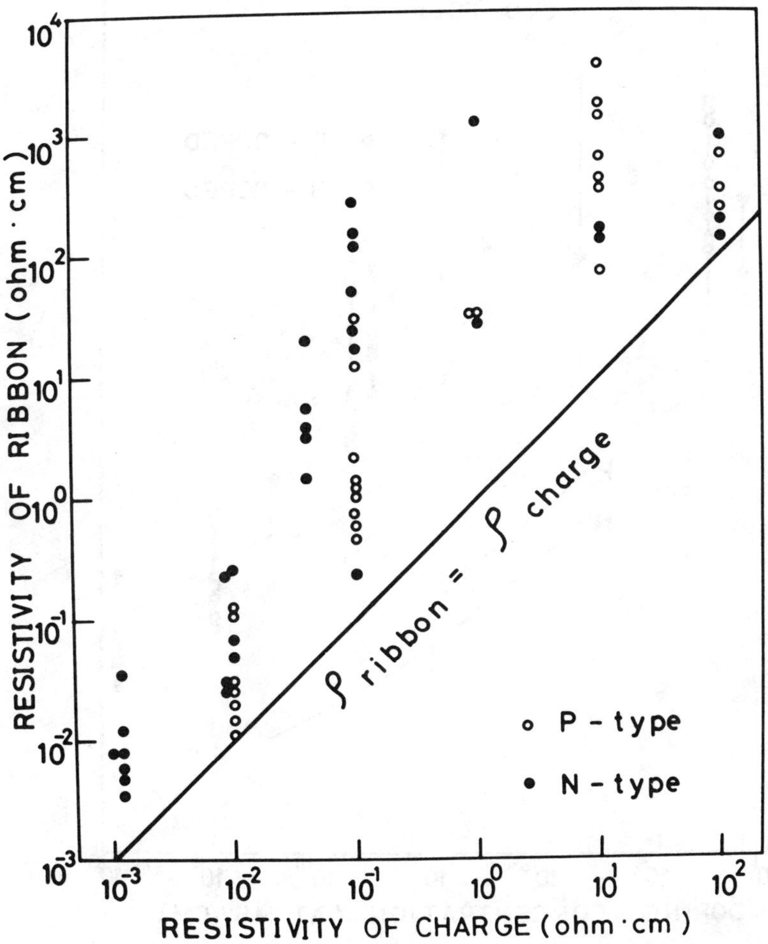

Fig. 7 Resistivity of the ribbon versus that of the silicon charge.
Solid line shows $\rho_{ribbon} = \rho_{charge}$.

o : p-type ribbon doped with boron
● : n-type ribbon doped with phosphorous

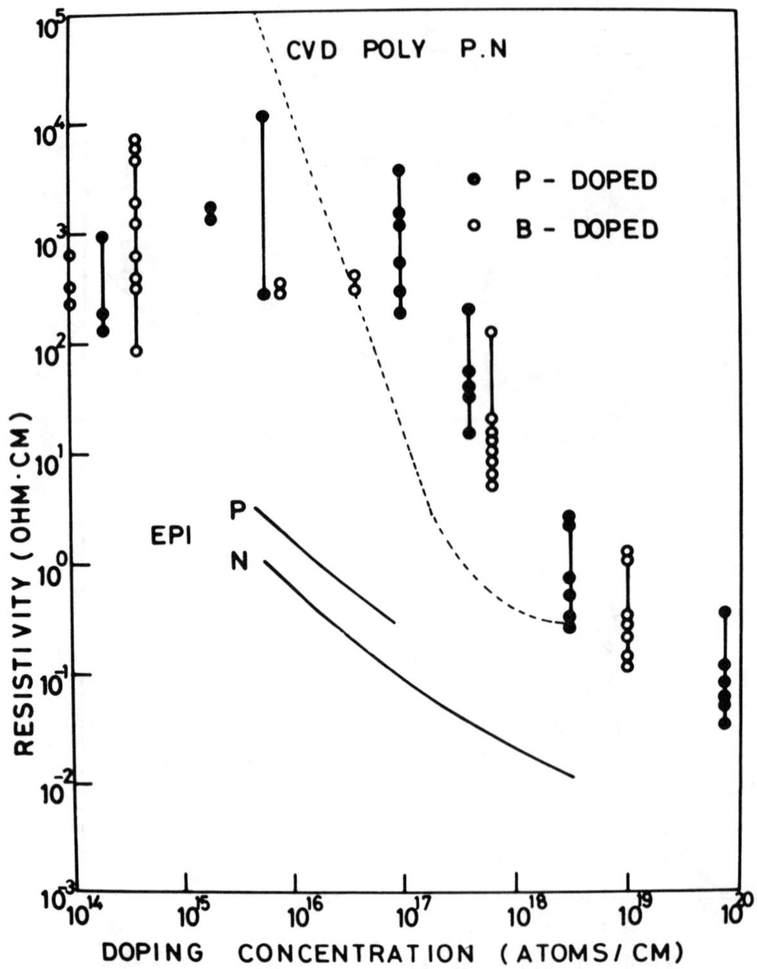

Fig. 8 Resistivity of the ribbon versus the doping concentration of the charge deduced from the Irvin curve[15]

o : B-doped ribbon ● : P-doped ribbon

-----: CVD films[13] EPI : single crystals[13]

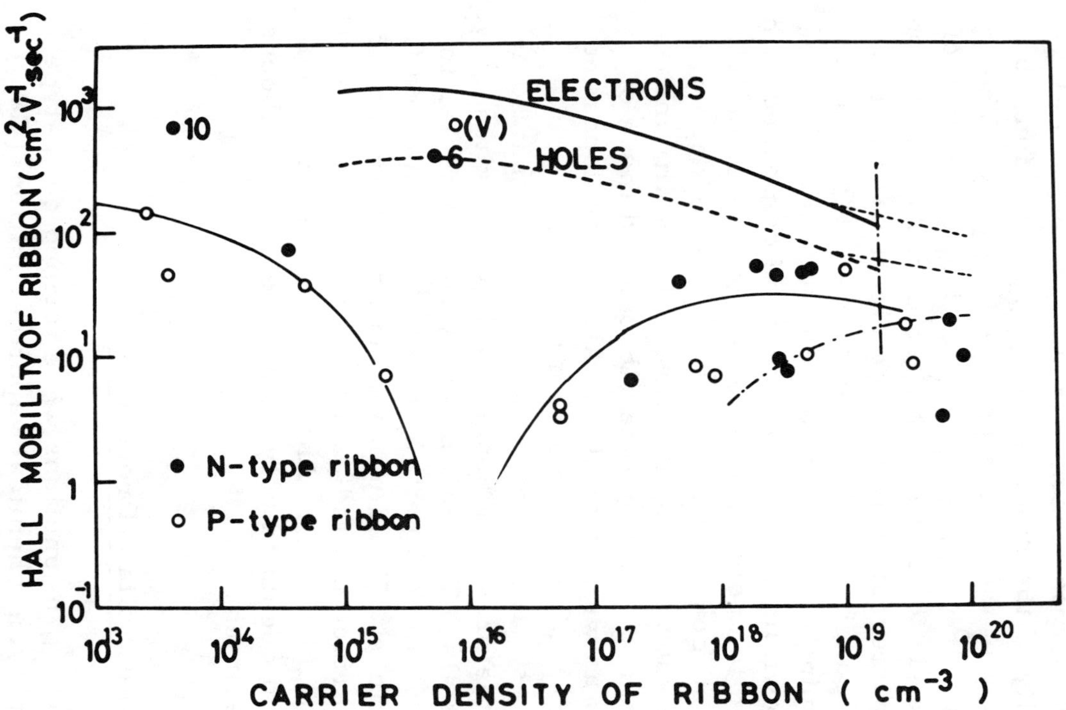

Fig. 9 Hall mobility versus active carrier density, parallel to the ribbon surface. ——— : n-type single crystal[15], ----- : p-type single crystal[15], ●: n-type silicon ribbon O : p-type silicon ribbon , ●10, ●6 : P-type silicon charges O(V) : mobility of the silicon ribbon, perpendicular to the ribbon surface, —·— : CVD poly silicon[14].

tion versus silicon charge doping concentrations. The solid line corresponds to the active carrier concentration of the ribbon when it remains unchanged by the present roller quenching process. From this figure, except for the lower doping region ($< 10^{14}/cm^3$) of n-type ribbons, no large deviation from the solid line was found for both the p- and n-type ribbons. Fig. 7 shows the electrical resistivity of the ribbons, ρ_{ribbon}, as a function of the resistivity of the silicon charge, ρ_{charge}. The solid line shows $\rho_{ribbon} = \rho_{charge}$. The resistivity of the ribbons are 1 to 10 times higher than that of silicon charges. It is interesting to plot the electrical resistivities of ribbons ranging from $10^{-2}\Omega cm$ to $10^{4}\Omega cm$ as a functions of doping density due to the Irvin curve[12]. The results are shown in Fig. 8. These were compared with curves obtained for polycrystalline silicon and epitaxial single crystal silicon grown by chemical vapor deposition[13]. In Fig. 9 are shown the data of the Hall mobility measurements of both p- and n-type ribbons and those of single crystals as well as of chemically vapor deposited polycrystalline[14] silicon films as a function of active carrier densities. It is interesting to remark that some of the data on the present ribbons are lying near the curves for single crystals. The Hall mobility perpendicular to the films was 50 to 700 $cm^2 \cdot V^{-1} \cdot sec^{-1}$, as expected from the grain structure. The dip of the Hall mobility parallel to the ribbon surface plotted as a function of active carrier density is found in Fig. 9. Analytical transport properties on the rapid quenched polycrystalline silicon ribbons are under experiments.

Diodes and Photo-Voltaic Properties

In this work, p-n junctions were formed on p-type ribbons by employing chemical vapor deposition (CVD). The n-type silicon was deposited on p-type ribbons ($1\Omega cm$) by the pyrolysis of silane (SiH_4) with phosphine (PH_3) as a dopant. The horizontal type quartz reaction tube ($7 \times 10 \times 50\ cm^3$) was used. The ribbons were heated upto 980

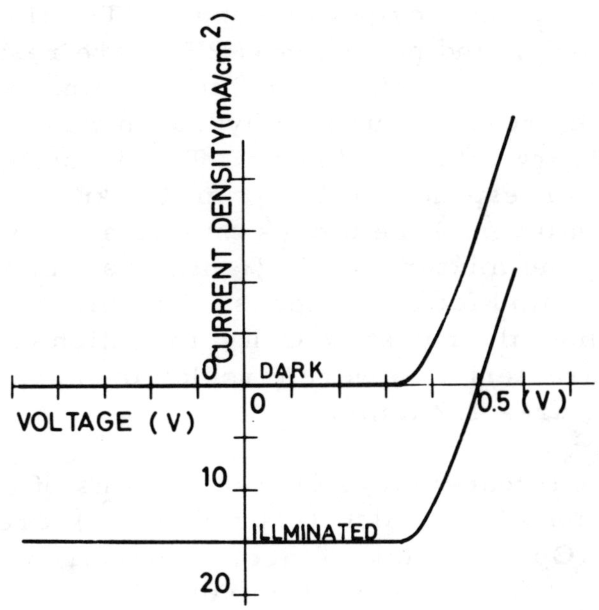

Fig. 10 Current-voltage characteristics for rapidly quenched polycrystalline silicon cell illuminated in AM 1 sunlight (94 mW/cm^2).

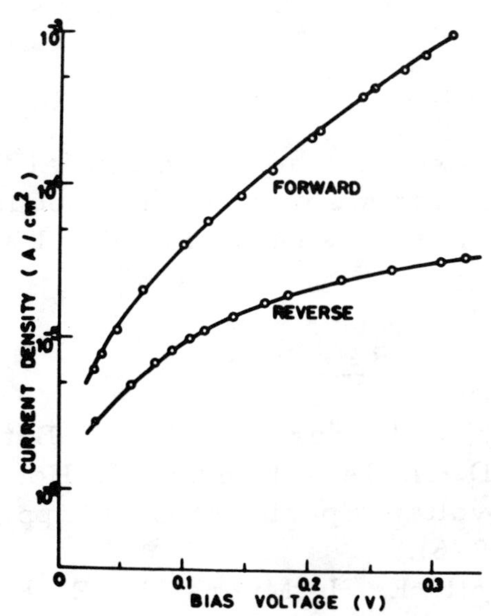

Fig. 11 Current-voltage characteristics of CVD cell on rapidly quenched polycrystalline silicon.

°C by high frequency power supply. The flow rates of the silane (SiH_4) and phosphine (PH_3) are respectively 1 x 10^{-3} mol/min. and 0.5 x 10^{-3} mol. /min. and unchanged during the growth. Purified hydrogen was used as a carrier gas (flow rate 20 l/min). The deposition time of 1.5 minutes corresponds to 0.3 μm thickness of n-type deposition layer. The diodes thus obtained were mesa-etched by the mixture of 75 % nitric acid and 25 % hydrofluoric acid for 30 seconds. The electrodes were made using silver paste. Cells dimention exhibiting the diode characteristics were circular of 2.3 mm in diameter (area 4.2 mm^2).

The current-voltage characteristics of the cell under illumination of an AM 1 (94 mW/cm^2) are shown in Fig. 10. Open-circuit voltage, short-circuit current density and fill factor were 0.50 volt, 15 mA/cm^2 and 0.63, respectively, which corresponds to a conversion efficiency of 5 %. Fig. 11 shows the current-voltage characteristics of the larger cell (area 1 cm^2) made with the same conditions.

Acknowledgements

The authors express their reverential thanks to Prof. Dr. T. Takahashi for his most warm encouragements and to Prof. Dr. J. Nishizawa for his helpful advices.

References

1. K. V. Ravi, H. B. Serrege, H. E. Bates, A. D. Morrison, D. N. Jewett and J. C. Ho, Proc. 12 th IEEE Photovoltaic Specialist Conf. pp. 299-302, May 6-8 (1976).
2. D. L. Barrett. E. H. Myers D. R. Hamilton and A. I. Bennett, J. Electrochem. Soc., 118 952 (1971)
3. S. V Tsvinskii and A. V. Stepanov, Soviet Phisics-Solid State, 7 148 (1965)

4. I. A. Lesk, A. Baghdadi, R. W. Gurtler, R. J. Ellis, J. W. Wise and M. G. Coleman, Proc. 12 th IEEE Photovoltaic Specialist Conf. pp. 173, May 6-8 (1976)
5. B. Kudo Kinzoku Jihyo, No. 770. 3 (407), (1975)
6. N. Tsuya and K. I. Arai, Solid State Physics, 13 237 (1978) in Japanese.
7. N. Tsuya and K. I. Arai, 3M Conf., Cleveland, 6A-1 (1978)
8. N. Tsuya and K. I. Arai, Jpn. J. Appl. Phys., 18 207 (1979)
9. R. L. Fullman, Trans. AIME 197 447 (1953)
10. C. C. Chang, Surf. Sci., 23 283 (1970)
11. H. Ibach and J. E. Rowe, Phys. Rev. B10 710 (1974)
12. J. C. Irvin, Bell Syst. Tech. J. 41 387 (1962)
13. A. L. Fripp and L. H. Slack, J. Electrochem. Soc. 120 145 (1973)
14. Zvy Lieblich and Adir Bar-Lev, IEEE Trans. ED-24 1025 (1977)
15. H. Wolf, Silicon Semiconductor Data., (Pergamon Press) (1967)

Ferrous Metallic Glasses

By John J. Gilman

A NEW CHAPTER IN FERROUS METALLURGY began in 1972 when Ho-Sou Chen and Donald Polk spun the first ductile glassy ribbons of ferrous alloys.[1] Since then, such ribbons have been found to combine outstanding strength and toughness, biaxial strength, high corrosion resistance, very low acoustic attenuation, high electrical resistivity, great ease of magnetization, and inexpensive processing into wires and strips as well as ribbons.

Before 1972, metallic glasses were known to exist thanks to the pioneering work of Duwez and his coworkers,[2] but they had not been produced in useful shapes nor as ferrous alloys having significant ductility. These features were needed in combination to make them attractive as engineering materials.

Direct quench casting on a continuous basis is now used to make ferrous glass ribbons by the mile at speeds ranging from 1500 to 6000 ft/min (450 to 1800 m/min). The ribbons are 1 to 5 mils (25 to 130 μm) thick and from 3 mils to 2 in. (75 μm to 50 mm) wide. Their compositions are distinctly different from those of conventional steels, being much higher in metalloid content (see Table I). And their molecular structures differ markedly from the structures of polycrystalline metals.

Novel compositions and structures supply remarkable combinations of chemical and physical properties to metallic glasses, but interesting properties alone would not cause these materials to have significant engineering interest. The fact that they can be economically produced is critical.

Ferrous Glasses Are Made Via a 'One-Step' Process

A glass is usually defined as a supercooled liquid in which the atomic mobility is so low that the liquid strongly resists shear. In principle, any normal liquid can be converted into a glass provided it can be cooled fast enough to bypass crystallization. In the case of polymeric liquids (silicates and methacrylates, for example), this is easy to do because their liquid phases are relatively viscous and they are slow to crystallize. Metals, on the other hand, form low viscosity liquids that very rapidly crystallize. Quenching rates in excess of 10^{10} C/s (or F/s) are needed to convert pure metals into glasses. The metallic glasses of current interest require much lower rates: 10^5 to 10^6 C/s (or F/s).

The alloys that are good glass formers are usually eutectics; and the deeper the eutectic trough the better. The most simple ferrous ones are binary combinations such as Fe-B, Fe-C, and Fe-P. However, the composition is often made more complex to enhance either processing or property characteristics. Thus, nickel may be added to improve ductility, chromium for corrosion resistance, molybdenum for strength, cobalt for magnetization, and other elements for processability.

Process Advantages: To cast a continuous ribbon, a small jet of liquid alloy may be squirted onto a cold rotating wheel. If conditions are right, the jet spreads as it strikes the wheel and solidifies into a ribbon.

Immediately after this quench casting, metallic glasses have good mechanical, magnetic, and chemical properties. Little or no secondary processing is required. Labor and plant size needs are thus less than those required for conventional metallurgical processing of strip material.

An important cost element in any manufacturing process is the life expectancy of key parts. The prime example in metallic glass production is the spinning nozzle. Long lifetimes are now being obtained, with much of the credit going to the selection of eutectic glass forming alloys which melt (and can be cast) at relatively low temperatures.

The shaping of filaments by direct quench casting is resisted only by the viscosity of the liquid metal, not by the

Source: Metal Progress, July 1979, 42-47, © 1979 American Society for Metals

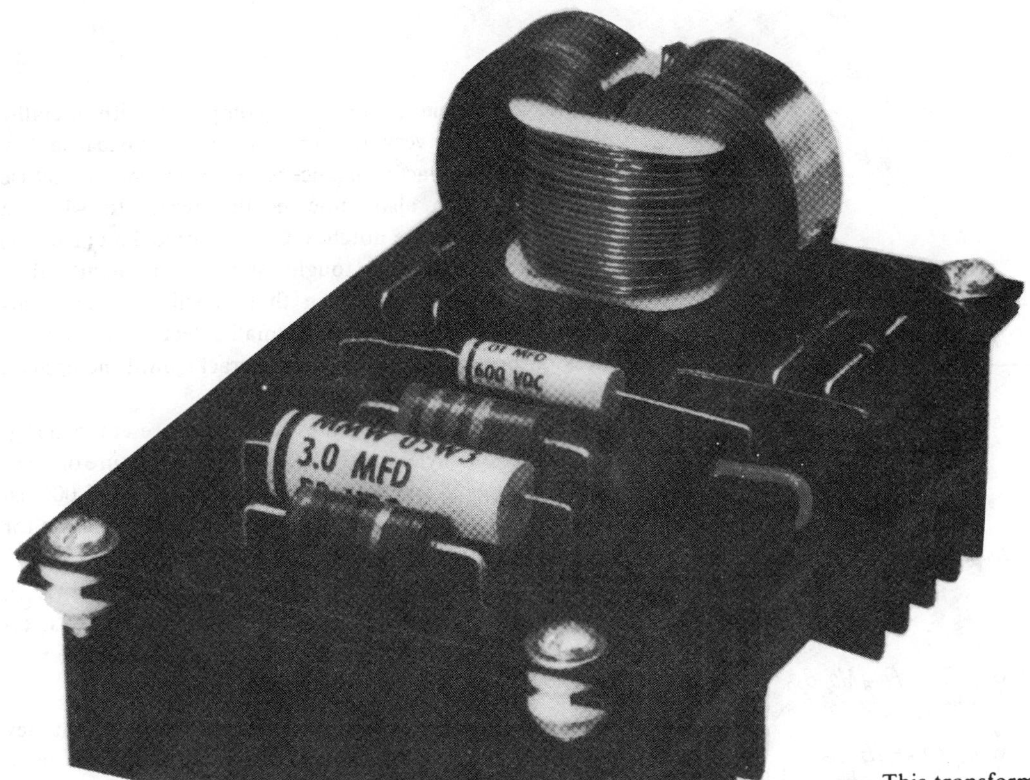

This transformer with a metallic glass core is part of a dc-to-dc (20 to 400 V) power supply. The transformer's operating frequency is 10 kHz at 100 W.

yield (or flow) stress of the solid. Thus, the forces required for shaping are small, and shaping machinery can be relatively light in weight. This reduces the amount of capital investment needed, and can reduce energy consumption by a factor of four to five.

A process disadvantage is that the need for very fast heat extraction forces at least one dimension of the cast material to be small — usually 0.010 in. (0.25 mm) or less. Thus, powders, wires, ribbons, sheets, and thin shells can be cast, but bulk shapes can't.

Once formed, ferrous metallic glasses are stable at room temperature. On heating, however, a temperature is reached — the glass transition temperature — at which atomic mobility becomes high. At a slightly higher temperature, the materials devitrify to form two or more crystalline phases. For example, the devitrification temperature of $Fe_{40}Ni_{40}P_{14}B_6$ is about 375 C (710 F). Atomic motion is very slow, however, at temperatures below about 200 C (390 F). Other compositions, such as $Fe_{80}B_{20}$ have higher devitrification temperatures.

The molecular structures of metallic glasses will not be discussed here, suffice it to say that no evidence of long-rang atomic order has been found, but there is appreciable evidence that they contain short-range order.[3]

Ferrous Glasses Combine Strength, Ductility, Toughness

No other ferrous materials have yield stresses as high as those of metallic glasses, although some steels have comparable breaking strengths (Table I). For example, the yield stress of $Fe_{80}B_{20}$ glass is in excess of 525 000 psi (3620 MPa). This is reflected by high hardness (about dph 1100) and good abrasion resistance. Some glasses are even stronger: $Fe_{75}B_{25}$ has a hardness of dph 1300, which corresponds to a yield stress of 575 000 psi (3960 MPa).[4] The strain hardening that occurs after yielding is quite small, making glasses plastically unstable in tension, although they can sustain large plastic strains in bending, shear, and compression.

Yield stress is essentially independent of temperature below about 0.7 of the glass transition temperature. Strain rate also has little effect on yield stress. This provides impact resistance. Furthermore, the high yield stresses are intrinsic because dislocation mobilities are very low as a result of the disordered structures.[5] More extensive information on mechanical properties is contained in reviews by Masumoto[6] and Davis.[7]

Fracture Toughness: Silicate glasses have high strengths and relatively low densities, and they are derived from low-

Source: Metal Progress, July 1979, 42-47

cost raw materials. However, compared with metallic glasses they have very little impact resistance because they lack ductility. Impact resistance is best measured as fracture toughness which also indicates the degree to which a material can tolerate notches. Comparative data (Table I) indicate that the fracture toughness of ferrous metallic glass is substantially greater than that of silicate glass. This means that scratches and other small defects in the metallic glass do not become propagating cracks until the applied stress reaches a high value.

Measured fracture toughness values are presented in Fig. 1 as a function of yield stress.[8] A remarkable feature of these data is that even when the yield stress reaches 525 000 psi (3620 MPa), the fracture toughness of metallic glasses for plane strain crack propagation (K_{Ic}) is about 11 ksi$\sqrt{in.}$ (12 MPa$\sqrt{m}$); for tearing or anti-plane strain cracking (K_{45}), it's about 42 ksi$\sqrt{in.}$ (46 MPa$\sqrt{m}$). For high strength steels at the same yield stress level, fracture toughness appears to nearly vanish; as it certainly does for silicate glasses.

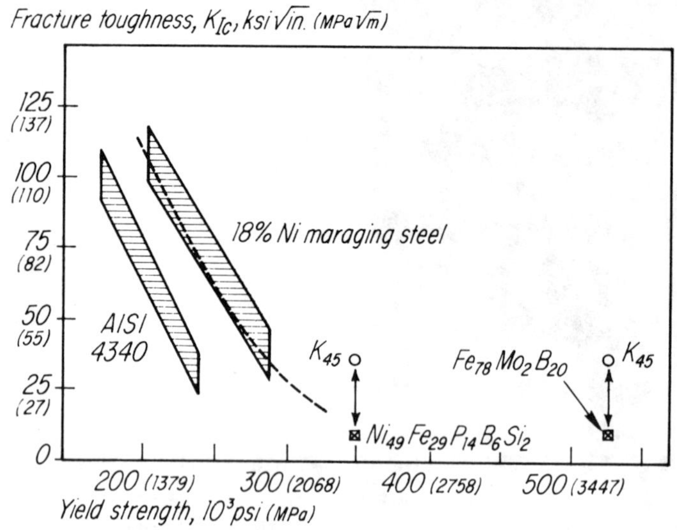

Fig. 1 — Fracture toughness of metallic glass and high strength steel. Squares are for plane strain crack propagation (K_{Ic}); circles are for anti-plane strain cracking, or tearing (K_{45}).

Table I — Mechanical Properties of Selected Metallic Glasses, Steel, and Silicate Glass

Material	Yield Strength, 10^3 psi (MPa)	Fracture Toughness, ksi $\sqrt{in.}$ K_{Ic}	(MPa $\sqrt{m}$) K_{45}
Metallic glasses			
$Fe_{40}Ni_{40}P_{14}B_6$	350 (2410)	8.5 (9.3)	37 (41)
$Fe_{80}B_{10}Si_{10}$	350 (2410)	—	—
$Fe_{80}B_{20}$	525 (3620)	11 (12)	42 (46)
$Fe_{60}Cr_6Mo_6B_{28}$	650 (4480)	—	—
$Ni_{69}Cr_6Fe_3B_{14}Si_8$	410 (2830)	21 (23)	48 (53)
Conventional steels			
Type 32 (cold worked)	200 (1380)	—	—
AISI 4340	240 (1650)	30 (33)	—
Ausformed 18% Ni	290 (2000)	35 (38)	—
Tire cord (0.8% C)	400 (2760)	—	—
Silicate glass			
E-glass	500 (3450)[(1)]	0.3 (0.3)	—

1. Fracture strength.

Biaxiality: A unique feature of metallic glasses is that they can be produced as strip materials with biaxial yield strengths in excess of 500 000 psi (3500 MPa). Strong fibers can be laid parallel to each other to form strips with high uniaxial strengths, and polycrystalline metals can be rolled to obtain high strengths in the rolling direction, but until now it has not been possible to make a strip material with high biaxial yield strength.

Biaxially strong strips have many design advantages in composite structure applications.[9] They include:

1. Wrapping ease — strip material can be wrapped more quickly than filaments.
2. Improved packing fraction — the maximum possible packing fraction for circular filaments is 91%; for rectangular strips, it's 100%.
3. Greater relative adhesion (pull out strength) — if a round filament and a rectangular strip have equal strengths and cross sectional areas, the adhesive force of the strip can be much larger than that of the filament. This is because the strip's perimeter is substantially longer than the filament's.
4. Minimized stress concentrations — for a given amount of reinforcing material, thin strips can be wrapped with overlaps to give relatively few stress concentrations at their edges as compared with filament reinforcement. Also unlike filaments, cracks that form at stress concentrations will be locally confined by adjacent strips.
5. Shear stiffness — a metallic glass ribbon will resist shearing forces that act parallel to its edges with the full stiffness and strength of the metal, whereas a unidirectional fiber reinforced tape has little more shear resistance than that of its polymeric matrix. This shear resistance (combined with the high perimeter-to-cross sectional area ratio) allows matching matrix shear strength to ribbon strength, optimizing composite performance.
6. Penetration resistance — filaments can be pushed

aside by a penetrator, strips cannot (for penetrators of a size equal to or smaller than the strip width).

7. Impact and scratch resistance — the ductilities of metallic glasses allow them to absorb impacts through plastic work. This is expressed quantitatively by their fracture toughness parameters.

Applications: Metallic glass structures can be made by weaving ribbons into fabrics, braiding them into tubes, and stacking them like bricks in parallel or orthogonal patterns, or in helical patterns to form cylinders. Shapes like these can be used for cables and cable sheaths, torque transmission shafts, pressure vessels, flywheels, transmission belts, rocket casings, and tire belts. A metallic glass tube is shown in Fig. 2.

Bear in mind that service temperature must be well below the devitrification temperature.

Ferrous Glasses Resist Chlorides and Sulfates

Protective oxide films form very readily on some chromium-containing metallic glasses, making them highly resistant to corrosion, particularly in chloride environments.[10] The film consists of hydrated chromium oxyhydroxide, and is more effective than the ones that form on polycrystalline stainless steels. Three factors may be responsible:

1. The absence of grain boundaries which tend to cause discontinuities and to contain segregated inpurity phases.
2. The irregular structure of the glass which facilitates oxide nucleation.
3. The presence of high concentrations of metalloid atoms which promote glass formation in the oxides, allowing thicker high density films to form.

Regardless of the reason, metallic glasses are indeed often exceptionally corrosion resistant. For example, after 20 h in 6% $FeCl_3$ at 60 C (140 F), stainless steels are severely attacked while metallic glass emerges unscathed:

Alloy	Composition	Weight Loss, %
Type 430	Fe 18Cr	87
Type 304	Fe 20Cr 8Ni	6
Type 316	Fe 18Cr 10Ni 2Mo	1.5
Metallic glass	Fe 14Cr 35Ni 12P 6B	nil

A more general indication of the superior corrosion resistance of metallic glasses is provided by polarization curves (Fig. 3) which plot the currents flowing through specimens at various constant applied potentials. The current is proportional to the rate of corrosive attack. Note that the metallic glass specimen is passivated at much lower potentials than any of the stainless steels. Also, its maximum corrosion rate is one hundred times smaller than that of type 430 and ten times smaller than that of type 316 stainless steel.

Applications: Because they resist chloride (and sulfate) solutions, metallic glasses may be attractive for marine cables, naval aircraft control cables, chemical filters, reactors, electrodes, and other chemical engineering components.

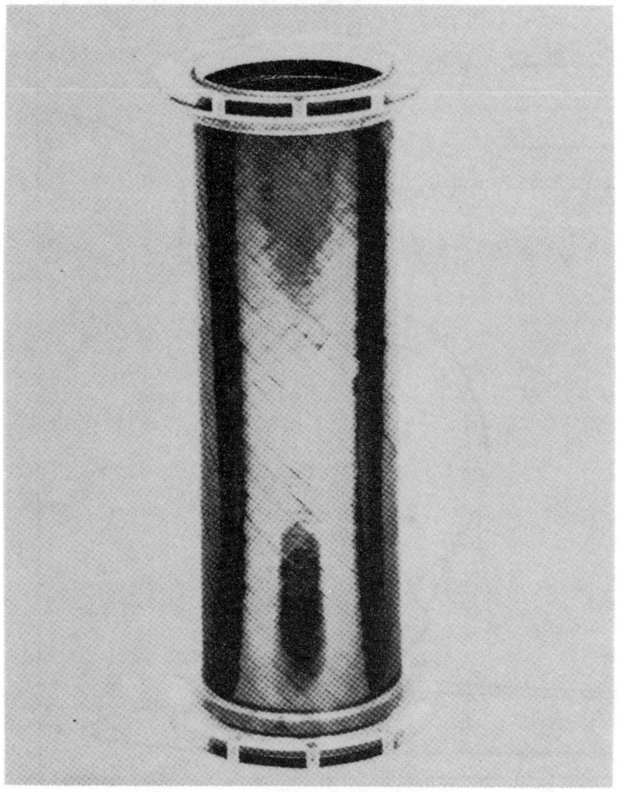

Fig. 2 — A tube made from metallic glass strips. Fabrics, plates and cylinders can also be fabricated.

Ferrous Glasses Boast Outstanding Magnetic Properties

Ferrous glasses are among the most easily magnetized of all ferromagnetic materials. Some can be magnetized by the application of fields in the millioersted (~100 mA/m) range — fields one hundred times smaller than the Earth's. Previously, only materials like Permalloy (39 to 81% Ni, 61 to 19% Fe) could be so easily magnetized. These alloys, however, are mechanically soft and more difficult to manufacture.

Despite their irregular structures, ferrous glasses exhibit high saturation magnetizations.[11] This is illustrated in Fig. 4 which shows how saturation magnetization varies with composition in the series: Fe, Co, Ni. In the glasses, the magnetic moment per transition metal atom is decreased by the irregular structure, and by the presence of metalloid atoms (boron, for example, causes less of a shift in the curve than phosphorus).

Ease of magnetization implies that magnetic domain walls readily move through metallic glasses. This has been

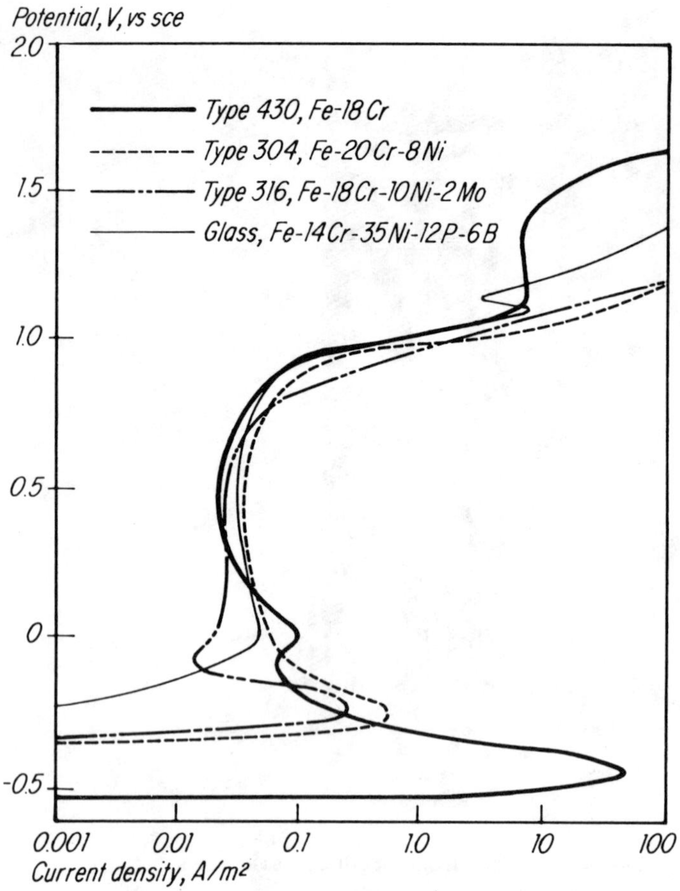

Fig. 3 — Corrosion rates (current densities) at various applied potentials for specimens immersed in 1N H_2SO_4 at 22 C (72 F). Compared with the stainless steels, the metallic glass corrodes much more slowly at low potentials (with a much smaller peak rate). All four materials behave similarly in the passive region (0 to 1 V).

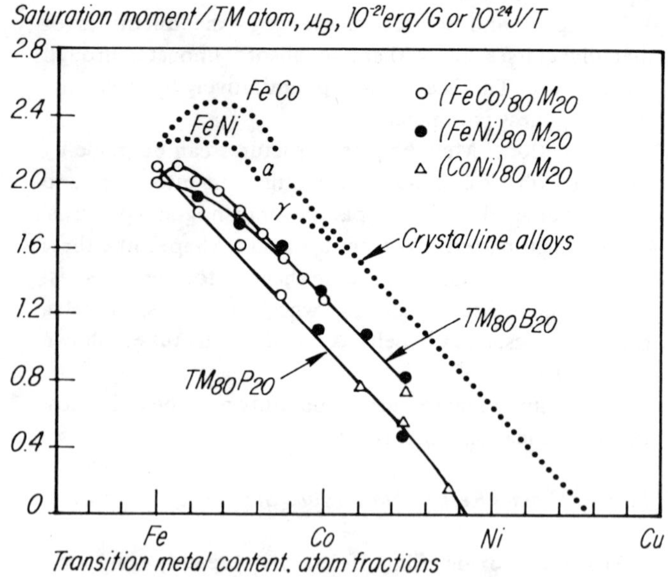

Fig. 4 — Effect of transition metal (TM) content on saturation magnetization; metallic glasses versus crystalline alloys. The curves are shifted toward smaller d-electron concentrations in the transition metals. A possible explanation: the metalloid (M) atoms, boron and phosphorus, contribute electrons to the d-bands.

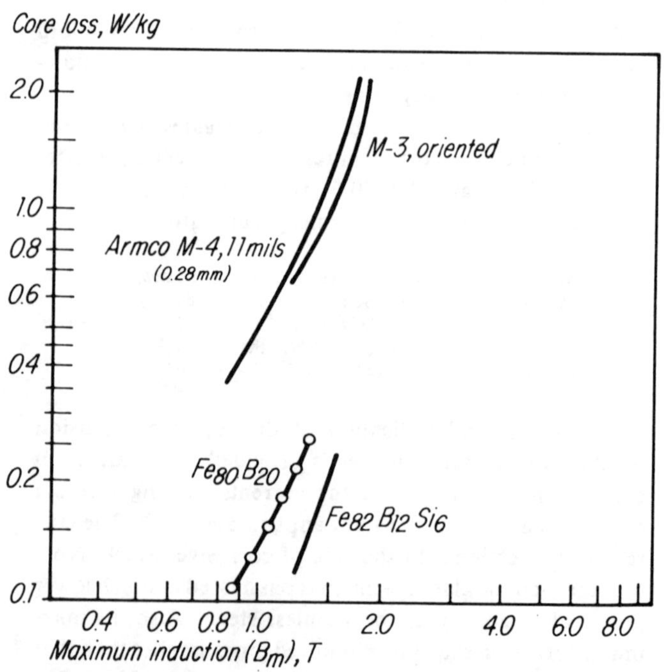

Fig. 5 — Core losses of 2 mil (0.05 mm) thick transformer materials at power line frequency (60 Hz-sin B) and various induction levels.

confirmed by measurements of domain wall velocities as a function of applied driving field,[12] which yield damping constants (viscous drag) for domain wall motion. Comparative data for a few materials show that domain walls do indeed move through metallic glasses with exceptional ease:

Material	Domain Wall Damping Constant, dyne·s/cm³ (N·s/m³)
Ni-Fe glass	0.5 (5.0)
Fe whisker monocrystal	26.5 (265)
60Ni-40Fe wire	160.0 (1600)
Si-Fe frame crystal	270.0 (2700)

Some possible reasons:

1. The absence of obstacles such as grain boundaries.
2. The high electrical resistivity which damps eddy currents.
3. The absence of crystal anisotropy (although some short-range anisotropy is present).

Applications: Domain wall mobility translates into excellent macroscopic magnetic properties (Table II). These properties combined with ease of manufacture should make ferrous glass ribbons suitable for motors, generators, transformers, amplifiers, switches, memories, recording heads, delay lines, transducers, and shielding. Some of these applications take advantage of the material's mechanical hardness to minimize wear or to maximize acoustic wave propagation.

Transformer core applications (see lead photo) are particularly attractive. A manufacturing cost reduction could be coupled with an energy conserving reduction in hysteresis loss. Core losses for conventional materials and several ferromagnetic glasses are compared in Fig. 5.

For More Information: You are invited to contact the author directly by letter or telephone. Dr. Gilman is director, Corporate Development Center, Allied Chemical Corp., P. O. Box 1021R, Morristown, N.J. 07960; tel.: 201/538-2170.

References

1. U.S. Patent 3 845 805, Reported in *Chem. and Eng. News*, 19 November 1973, p. 24.
2. "Non-Crystalline Structure in Solidified Gold-Silicon Alloys," by W. Klement, R. H. Willens, and P. Duwez, *Nature*, Vol. 187, 1969, p. 869.
3. "Structures of Ferrous Eutectic Liquids," by J. J. Gilman, *Phil. Mag.*, Vol. 37, No. 5, 1978, p. 577.
4. "Correlation Between Elastic Constants and Flow Behavior in Metallic Glasses," by H. S. Chen, *J. Appl. Phys.*, Vol. 49, 1978, p. 462.
5. "Micromechanisms of Deformation and Fracture," by J. C. M. Li, *Metallic Glasses*, Ed. by Gilman and Leamy, ASM, Metals Park, Ohio, 1977, p. 224.
6. "Mechanical Characteristics of Amorphous Metals," by T. Masumoto, Report 1673 of the Research Institute for Iron, Steel and Other Metals, Tohoku University, Sendai, Japan, 1976.
7. "Strength, Ductility and Toughness," by L. A. Davis, *Metallic Glasses*, Ed. by Gilman and Leamy, ASM, Metals Park, Ohio, 1977, p. 190.
8. "Fracture of Nickel-Iron Base Metallic Glasses," by L. A. Davis, *J. Mat. Sci.*, Vol. 10, 1975, p. 1557.
9. "Deformation and Fracture Mechanisms in Aluminum Reinforced by High Strength Steel Ribbons," by E. A. Anderson and R. Bode, *Failure Modes in Composites I*, Ed. by I. Toth, AIME, New York, 1973, p. 289.
10. "High Corrosion Resistance of Chromium-Bearing Amorphous Alloys in Neutral and Acidic Solutions Containing Chloride," by M. Naka, K. Hashimoto, and T. Masumoto, *Corrosion*, Vol. 32, 1976, p. 146.
11. "Magnetic Properties of $TM_{80}P_{20}$ Glasses," by R. C. O'Handley, R. Hasegawa, R. Ray, and C.-P. Chou, *J. Appl. Phys.*, Vol. 48, 1977, p. 2095.
12. "Domain Wall Kinetics in Soft Ferromagnetic Metallic Glasses," by R. C. O'Handley, *J. Appl. Phys.*, Vol. 46, 1975, p. 4996.

Table II — Magnetic Properties of Selected Metallic Glasses and Magnetic Alloys

Material	Saturation Induction-B, kG (T)	DC Permeability at 20 Oe (1.6 kA/m)	Core Loss at 60 Hz-Sin B, W/kg 13 kG (1.3 T)	16 kG (1.6 T)
Metallic Glasses				
$Fe_{40}NI_{40}P_{14}B_6$	7.6 (0.76)	14 000	—	—
$Fe_{40}Ni_{38}Mo_4B_{18}$	8.8 (0.88)	20 000	—	—
$Fe_{80}B_{20}$	15.6 (1.56)	4000	0.22	—
$Fe_{82}B_{10}Si_8$, 2 mil (0.05 mm)	16.3 (1.63)	9000	0.13	0.23
Magnetic Alloys				
Mo-Permalloy (79% Ni, 17% Fe)	7.8 (0.78)	35 000	—	—
Deltamax (50Ni-50Fe)	16 (1.6)	500	0.77	—
M-4 (Fe-3% Si), 11 mil (0.28 mm)	19.7 (1.97)	1000	0.78	1.30
H-2 (Fe-3% Si), 12 mil (0.30 mm)	19.7 (1.97)	1000	0.80	1.25

Bulk Microcrystalline Alloys From Metallic Glasses

By Ranjan Ray

A new family of rapid solidification processed (RSP) metallic glass alloys, based on iron, nickel, and/or cobalt, has been recently synthesized. When these materials are heat treated sufficiently above the crystallization temperature (T_C), they become devitrified into multiphase, microcrystalline alloys having: a tensile strength generally above 180 000 psi (1240 MPa), good ductility, and good thermal stability up to at least 700 C (1290 F). Additionally, the alloys can be fabricated into bulk shapes suitable for engineering and structural applications.

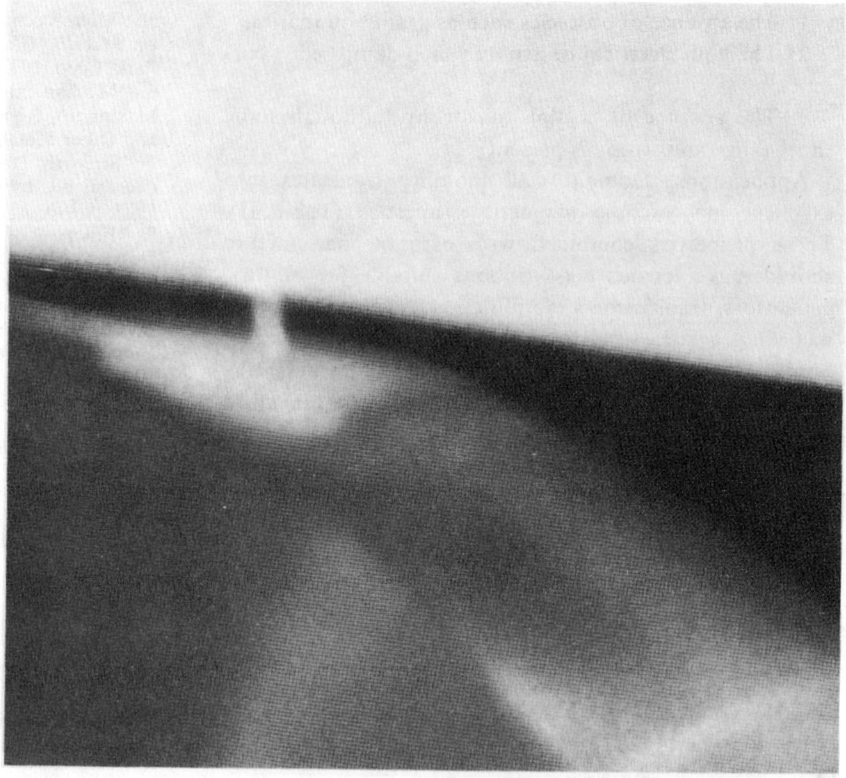

Melt-spin-chill casting of metallic glass. A molten stream is impinged onto a copper drum moving at a surface speed of about 6000 ft/min. (30 m/s). The liquid instantly solidifies when it comes in contact with the moving chill substrate, forming a thin glassy ribbon. Cooling rate is about 10^6 C/s (1.8×10^6 F/s).

Drawbacks of Metallic Glasses Are Remedied

Metallic glasses belong to a novel class of RSP alloys well known for their unique compositions, amorphous structures, and excellent mechanical, magnetic, and corrosion resistance properties.[1,2] Rapid solidification techniques such as melt spinning are capable of fabricating metallic glasses in large quantities as continuous ribbons and tapes.[3]

The materials have, however, one important drawback. They undergo transformation by crystallization (devitrification) into very brittle phases at relatively low temperatures — 350 to 500 C (660 to 930 F). This rules out their use at high temperatures and, to date, no noteworthy applications have been developed for the brittle, crystalline alloys devitrified from glassy phases.

Also, because of the need for extremely fast heat extraction during processing, metallic glasses can only be fabricated as thin bodies. Their application in bulk shape products is not yet feasible.

Unique Compositions — The new metallic glasses are formulated with certain critical combinations of transition metals and metalloids, which allow the devitrified material to attain its ductility upon heat treatment at high temperatures. Most metallic glasses have eutectic/hypereutectic compositions of approximately 20 at. % of boron, phosphorus, carbon, and/or silicon. The new versions contain these metalloids in ranges of 5 to 13%, with boron being the dominating metalloid (see Table I for possible compositions).

Ultimate tensile strength values in excess of 180 000 psi (1240 MPa) have been obtained with the crystalline alloys in ribbon form prepared by devitrification of melt spun glassy ribbons. The required heating time depends upon the temperature used, and may range from about 0.01 to 100 h, and preferably between 1 and 20 h at a temperature between 0.7 and 0.8 of the solidus temperature. Optimization of compositions and heat treatment conditions leads to higher tensile strength values of the devitrified alloys in ribbon form, typically in the range of 250 000 to 350 000 psi (1725 to 2415 MPa).

Microstructure — The superior mechanical properties of the new devitrified alloys have their origin in their microstructures. The isotropic glassy phase, which has a critically balanced composition upon heat treatment at temperatures sufficiently above T_C, and preferably between 0.6 and 0.95 of the solidus temperature, gives rise to a unique microstructure consisting of an aggregate of microcrystalline phases. TEM studies reveal that the devitrified alloys exhibiting high tensile strength values consist of extremely fine scale, homogeneous microstructures (Fig. 1). The matrix based on iron, nickel, and/or cobalt is constituted typically of ultrafine grains, 0.2 to 0.3 μm (8 to 12 micro-in.), that are uniformly dispersed with 0.1 to 0.2 μm (4 to 8 micro-in.) intermetallic phases (e.g. borides).

The present microcrystalline alloys can be classified as new generic alloys based on composition/microstructure/processing principles. They have evolved from a series of hypoeutectic metallic glass alloys with significantly low contents of metalloid elements. Low boron is critical in achieving the desirable microstructures of the devitrified alloys consisting of an optimum content (e.g. approximately 20 to 30 vol %) of finely dispersed intermetallic phases.

Fabrication of Bulk Microcrystalline Alloys

The important finding that strong, thermally stable microcrystalline alloys could be produced from metallic glasses of specific compositions led to the next step: hot consolidation of metallic glasses into bulk microcrystalline alloys on a best effort basis. The objective was to examine the properties of the bulk microcrystalline alloys with respect to potential engineering-structural applications.

One method of fabrication of bulk microcrystalline alloys involves hot isostatic pressing and/or hot extruding tape wound rolls of metallic glass ribbons, with rolls packed in a sealed can under vacuum. Alternatively, the novel metallic glasses can be processed as powders, followed by subsequent powder metallurgical consolidation to bulk microcrystalline alloys.

Several approaches to make RSP powders have been developed.[4,5] Generally, atomization of liquid is utilized, followed by various cooling procedures. Such processes have a limitation: a range of liquid droplet sizes is produced. Droplets of different sizes experience a range of cooling rates, leading to a nonuniform product. Further, larger droplets may experience cooling rates lower than the critical rate (i.e. 10^6 C/s [1.8 x 10^6 F/s]) necessary to form metallic glasses. Screening to separate only the finest particles leads to a lower yield and a less economical process.

Pulverizing — The metalloid-lean metallic glasses of the present category were found, unexpectedly, to form brittle ribbons upon melt spinning. This behavior is in striking contrast to the previously known metalloid-rich metallic glass ribbons having high tensile strength and excellent bend ductility.

The brittle ribbons of the present glassy alloys can be, therefore, readily pulverized into powders by known standard equipment such as hammer mills. Efficient high volume production of such metallic glass powders, necessary for commercialization, would involve inline pulverization of a large number of ribbons spun simultaneously at 6000 ft/min (30 m/s) off the chilled rotating substrate. The quench rate experienced by the liquid is much more uniform in the melt spinning process.

In powders made from pulverized ribbons, particles of all sizes have experienced essentially the same quench history and the consolidated product will be highly uniform. The melt spinning-pulverization procedure can be

Table I — Selected Groups of the New Metallic Glasses

Alloy System at. %	Examples
$R_{30-75}R'_{10-30}Cr_{0-30}M_{0-15}B_{5-12}(P,C,Si)_{0-2.5}$ R is one of the group consisting of Fe, Ni, and Co; R' is one or two elements of the group consisting of Fe, Ni, and Co other than R; and M is an element of the group consisting of Mo, W, Cb, and Ta. The sum of Cr, R', and M must be at least 12%, and the boron content is 80% or more of the combined metalloid content (B,P,C,Si).	$Fe_{50}Cr_{25}Ni_{10}Mo_5B_{10}$ $Fe_{45}Cr_{20}Ni_{15}Mo_{12}B_8$ $Ni_{40}Cr_{25}Co_{10}Fe_{10}Mo_5B_{10}$ $Co_{40}Cr_{30}Ni_{10}Fe_{10}B_{10}$ $Co_{55}Ni_{10}Fe_{15}W_6Mo_6B_8$
$Fe_{30-80}Cr_{0-40}(Co,Ni)_{0-2}(Mo,W)_{0-20}B_{5-12}(P,C,Si)_{0-2.5}$ The sum of Cr, Co, Ni, Mo, and/or W cannot be less than 10%, and when the content of Mo and/or W is less than 10%, Cr content must be equal to or greater than 8%. The maximum metalloid content should not exceed about 12%.	$Fe_{71}Cr_{15}Mo_4B_{10}$ $Fe_{63}Cr_{22}Ni_3Mo_2B_8C_2$ $Fe_{75}Cr_8Mo_5W_2B_{10}$ $Fe_{70}Cr_{10}Ni_3Mo_7B_8C_1Si_1$
$Ni_{48-75}Cr_{0-20}Mo_{10-30}B_{5-12}$ Mo is larger than 20%, and Cr must be equal to or less than 15%.	$Ni_{55}Cr_{15}Mo_{20}B_{10}$ $Ni_{65}Mo_{25}B_{10}$ $Ni_{60}Mo_{30}B_{10}$ $Ni_{57}Cr_{10}Mo_{25}B_8$

Table II — Hot Hardness Data After 30 Min. at Temperature

Alloy	Hot Hardness, HRC			
	Room Temperature	600 F (315 C)	1000 F (540 C)	1200 F (650 C)
$Fe_{70}Cr_{18}Mo_2B_9Si_1$	44	43	43	42.5
AISI H21 steel (0.3 C, 0.4 V, 3.25 Cr, 9.5 W, bal Fe)	49.5	45	34.5	19
AISI H26 steel (0.55 C, 1.0 V, 4.0 Cr, 18 W, bal Fe)	46.5	42	32.5	20

Fig. 1 — Devitrification of melt spun glassy ribbons can produce microcrystalline alloys having tensile strengths higher than 180 000 psi (1240 MPa). A typical structure is shown in this bright-field TEM photomicrograph. Dark regions represent the intermetallic phase; white regions represent matrix grains consisting of microtwins.

Fig. 2 — SEM photomicrograph of metallic glass powder prepared by pulverization of ribbons. The powder particles' platelet shape morphology contributes to excellent flowability.

practiced so as to have a high yield (e.g. >95%) of a relatively fine powder (e.g. 100 U.S. mesh [150 μm]) suitable for subsequent powder metallurgical handling and processing.

Metallic glass powder particles have platelet shape morphology with relatively smooth edges and surfaces resulting from fracture of hard and brittle ribbons (see Fig. 2). The powder particles, because of their shape characteristics, show excellent flowability, which make them highly desirable for subsequent powder handling and packing.

New Bulk Alloys Boast Excellent Properties

A number of metallic glasses within the scope of the present research were converted into fully dense microcrystalline bars by hot extrusion and hot isostatic pressing of glassy powders (100 U.S. mesh [150 μm] at temperatures between 900 and 1050 C (1650 and 1920 F). Initial test data indicated good prospects for potential engineering and structural applications of microcrystalline alloys both at room and elevated temperatures.

Microcrystalline alloys in the form of hot extruded bars, and having compositions like $Fe_{70}Cr_{13}Ni_6Mo_1B_9Si_1$ and $Fe_{62.5}Cr_{16}Mo_{11.5}B_{10}$, exhibited tensile strengths of 200 000 to 230 000 psi (1380 to 1585 MPa). The consolidated microcrystalline alloys based on iron and chromium and/or molybdenum and tungsten had high hot strength/hardness at 1000 to 1200 F (540 to 650 C), excellent oxidation resistance up to 1300 F (705 C), and good Charpy V-notch impact energy at elevated temperatures up to 1000 F (540 C).

Carbon Additions — The high strength of the microcrystalline alloys is essentially derived from the ultrafine grain structure of the matrix, stabilized by intermetallic phase particles which are predominantly dispersed uniformly at grain boundaries. The matrix grains can be additionally strengthened by fine carbides in a way similar to martensitic strengthening. These alloys have the composition (at. %): $Fe_{bal}Cr_{5\text{-}10}V_{1\text{-}3}W_{3\text{-}5}Mo_{3\text{-}7}B_{7\text{-}8}C_{2\text{-}2.75}Si_{0.5\text{-}1}$.

Devitrified alloys containing small additions of carbon can be heat treated to change their hardness and ductility in the same manner that the hardness and ductility of steel may be changed with heat treatment. Hardness values in excess of 68 HRC can be obtained. Such high hardness is believed to be due to unique microstructures consisting of ultrafine boride particles in an ultrafine grained martensitic matrix.

These alloys can also be devitrified or hot consolidated in bulk forms from the corresponding glassy state (powders). After suitable heat treatment (like annealing), they can be easily machined into any desired form, such as cutting tools or wear resistant parts. Machined parts can be heat treated for specific properties (e.g. high hardness, toughness). The extremely homogeneous, fine scale microstructures of these alloys enhances the grindability of tools made from them.

Corrosion Resistance—A general indication of the superior corrosion resistance of microcrystalline alloys was obtained by studies of polarization curves. Current densities at various applied potentials were determined for specimens immersed in 1N H_2SO_4 at 22 C (72 F). The current is proportional to the rate of corrosive attack. A bulk microcrystalline alloy, $Fe_{63}Cr_{22}Ni_3Mo_2B_8C_2$, was passivated at much lower potentials than any of the commercial stainless steels. Also, its maximum corrosion rate was ten times smaller than that of AISI type 316 stainless steel.

Applications — Potential applications of microcrystalline alloys include hot work die tools. Table II shows superior hot hardness properties of an iron based microcrystalline alloy as compared with commercial tungsten bearing hot work tool steels.

Nickel base microcrystalline alloys containing large amounts of molybdenum and/or chromium have potential in applications involving ultrahigh strength and/or high hardness alloys at room and elevated temperatures. Typically, these microcrystalline alloys in ductile ribbon form have a microhardness value of 600 HV.

For More Information: You are invited to contact the author directly by letter or telephone. Dr. Ray is president, Marko Materials Inc., 144 Rangeway Rd., North Billerica, Mass. 01862; tel: 617/663-2210.

References

1. *Metallic Glasses*, edited by J.J. Gilman and H.J. Leamy: American Society for Metals, Metals Park, Ohio 44073, 1978.
2. *Rapidly Quenched Metals III, 3rd International Conference, Vol 1 & 2*, edited by B. Cantor: The Metals Society, London, 1978.
3. "Ferrous Metallic Glasses," by J.J. Gillman: *Metal Progress*, July 1979, p 42.
4. "Rapid Solidification Effects of Micron-Size Droplets," by M.R. Glickstein, R.J. Patterson II, and N.E. Shockey: proceedings of the Second International Conference on Rapid Solidification Processing, Reston, Va., March 1980, p 46.
5. "A Review of Various Atomization Processes," by N.J. Grant: proceedings of the Second International Conference on Rapid Solidification Processing, Reston, Va., March 1980, p 230-240.

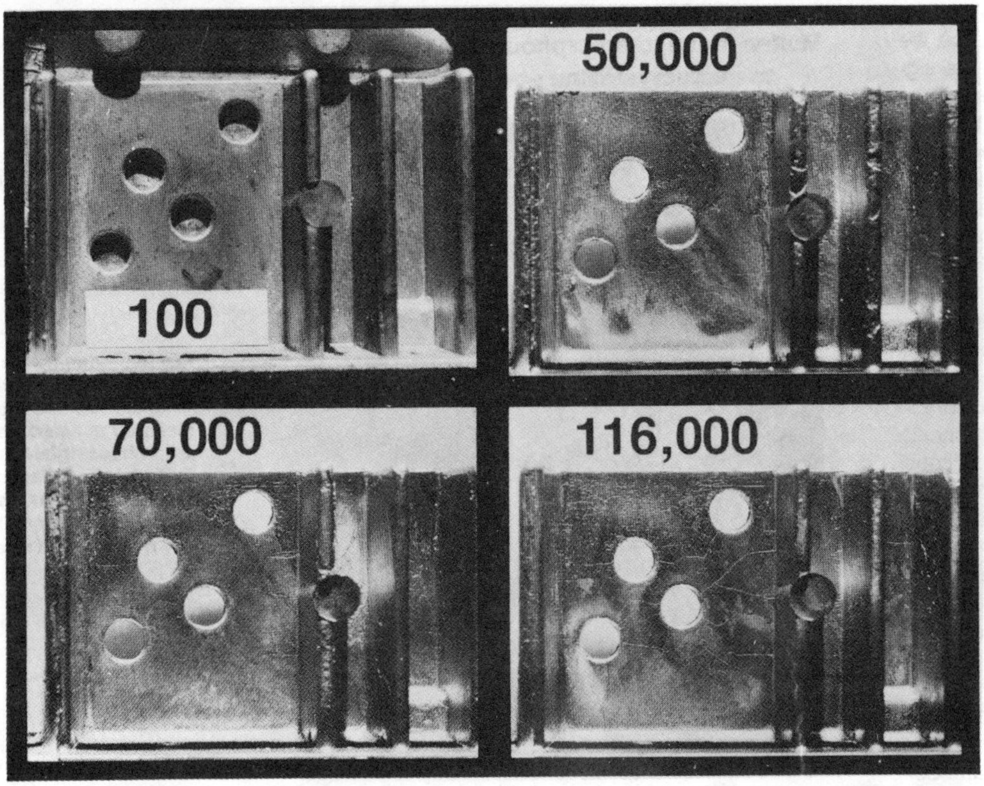

Minimum heat checking and freedom from erosion are illustrated in this series of photos of a die insert of Devitrium used by Woodstock Die Casting to produce die castings for the telecommunications industry. More than 125,000 shots have been made using this die insert in the gate area.

New P/M material extends die casting die life

After more than 125,000 shots, aluminum die casting die insert is still going strong. Next question: Is it cost effecitve?

Among the technical advances which the die casting industry is exploring to improve its productivity is an innovation in die materials made possible by a totally new class of products of powder metallurgy.

The new die material which has already been field tested for long-running aluminum die casting die cavity inserts, is based on rapidly solidified metal powders that are consolidated into dense die blocks with superior hot hardness.

The rapid solidification technology (RST) alloy powders are a development of the Consolidated Metal Products (CMP) Department of Allied Corporation. The RST metal powders derive their unique performance characteristics from their structure which is a homogeneous solution of metals in solid form rather than the crystalline aggregates typical of most metals.

Still in its developmental stages, with the economics yet to be determined, the new hot work die material may yield enough savings in machining costs, downtime and productivity to offset the higher materials cost.

Called Devitrium[1] alloy 7025, the P/M material obtains its attractive properties without the use of such strategic metals as cobalt, tungsten or chromium.

Woodstock Die Casting, an Eltra Company, a subsidiary of Allied Corp., has used the new material to replace an H13 die insert in running aluminum die castings for the telecommunications industry. At latest report, the die insert of Devitrium 7025 had passed 125,000 shots showing minimal heat checking and no sign of erosion, compared to previous normal life experience of about 50,000 shots with H13 die inserts. The castings were originally produced in a solid H13 die, but as portions of the die eroded, they were replaced with successive H13 inserts. Then the 7025 alloy insert was introduced and is still in use.

The die material, which is expected to have about a 10-to-1 cost premium over conventional hot work die steel, may have its greatest potential use for long run aluminum die cavities and inserts where die sinking costs, die life and die maintenance problems are significant. No work has yet been done with brass die casting dies, but they are believed to offer similar extended die life to that demonstrated for aluminum.

CMP's die material has also been production tested for dies used in extruding copper and brass alloys for screw machine stock. Preliminary results appear promising but there is need for further study of process variables.

The RST metal powders, which originate as an amorphous ribbon produced by the Planar Flow casting process invented at Allied, are transformed to microcrystalline ferrite grains and micron sized boride particles. This results in the exceptional mechanical and physical properties of the material.

The 7025 alloy powders are hard and lack inherent compressibility by normal cold compacting techniques. Instead, shapes are formed by hot isostatic pressing, hot extrusion or hot compacting, depending upon the shape desired. The resulting die material not only has superior hot hardness, but also excellent wear resistance, modulus of elasticity and strength.

The hardness of the Devitrium alloy makes it necessary to machine cavities by EDM using carbon or wire electrodes, unlike conventional hot work steels that are machined in the annealed condition and then hardened. However, speed of EDM and resultant surface quality are equal to that using the regular die steels.

The material can be machined using carbide or, as more recently attempted, Borazon cutters. Cutting or grinding can be done with an alumina wheel, 1K hardness. CMP has built a substantial body of information about machining practice and the tools for machining the hard die material.

Molten metal to amorphous ribbon to microcrystalline powder

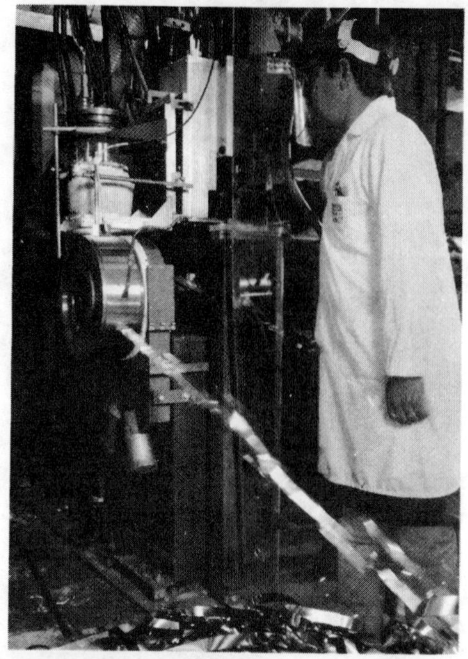

Key to producing die material is this small scale casting line. "Planar Flow Casting" process invented at Allied yields 2 in. amorphous ribbon at rates up to 5,000 ft/min. Induction heated melt is "extruded" through a slot onto a copper alloy wheel and quenched at rates up to 10^6C°/sec.

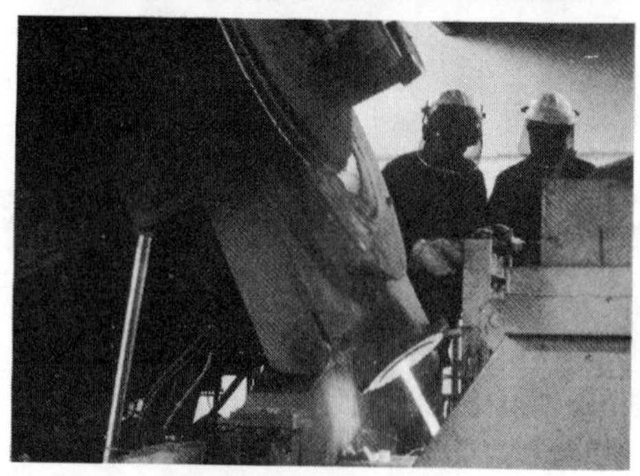

Magnetic alloy for transformer core materials is produced in this melting furnace, shown charging the casting head on a pilot line. A metal ribbon 17 cm wide is wound at more than 3,000 ft/min.

Steps in the sequence of producing the billet for making dies: (clockwise from the left) amorphous ribbon, comminuted metal powder, canned powder for HIPping, HIPped and machined billet.

Experimental use of Devitrium 7025 is encouraged by CMP who make the material available for development work. Among other suggested applications for this material in the metalworking industry are forging and heading dies, cutting tools, drill bits, sleeves, bushings and bearings, automotive valves, industrial knives, and swaging, forming, reaming, bending and upsetting tools.

Could this die material be used for complete die blocks? Probably not for most applications because of the high cost, but a die with a complex configuration which is expected to produce many hundreds of thousands of parts could conceivably be suited for it because of the high cost of machining additional dies. It is a decision which would be evaluated on a cost effectiveness basis.

Size of original die block is limited only by the size of HIPping equipment available. Maximum size at CMP presently is a block 6 in. in diameter × 12 in. long, but outside HIP facilities could shape a die block up to 40 in.

No special considerations are needed

Devitrium powder, shown in macrophoto prior to consolidation, is amorphous. Each particle, irrespective of size, has identical composition and is completely homogeneous. The ribbon from which this powder is produced can be considered a single, large particle 25-75μ thick, 2-5 mm. wide, and thousands of meters long.

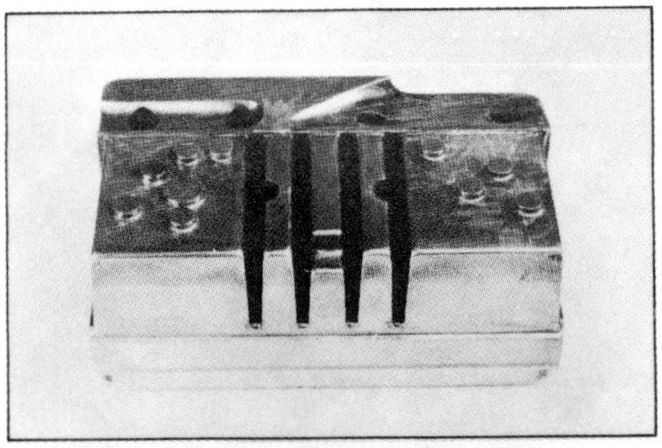

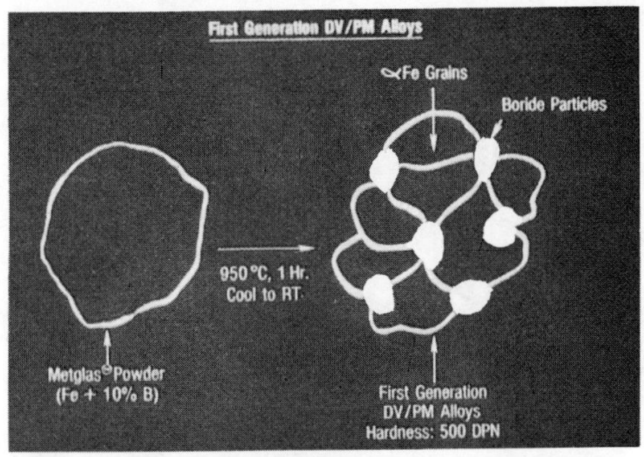

Transformation of powder from amorphous state to microcrystalline ferrite grains and micron-sized boride particles at the triple points, is illustrated by this schematic of Devitrium grain structure. Conventional alloys of the same composition would consist of varying large size grains of ferrite surrounded by brittle eutectic boride network, making the material too brittle for most engineering uses.

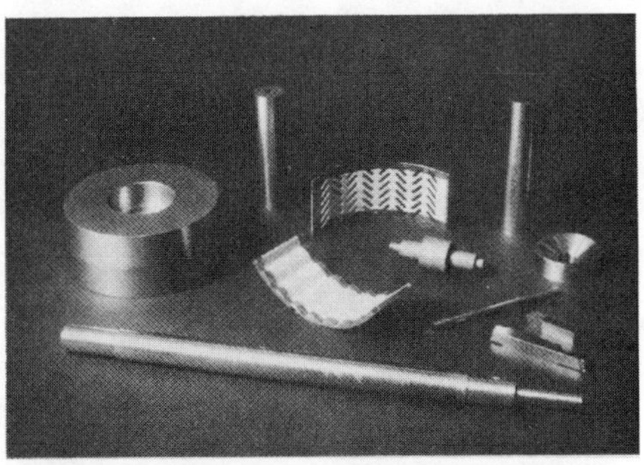

Some of the applications for CMP Devitrium are displayed here, including hot extrusion die for copper alloys, wear/corrosion resistant coatings applied by powder spraying techniques, machine tool cutters with performance equal to tungsten carbide and double that of M2/M42 steels, punches and rollers for various tooling uses.

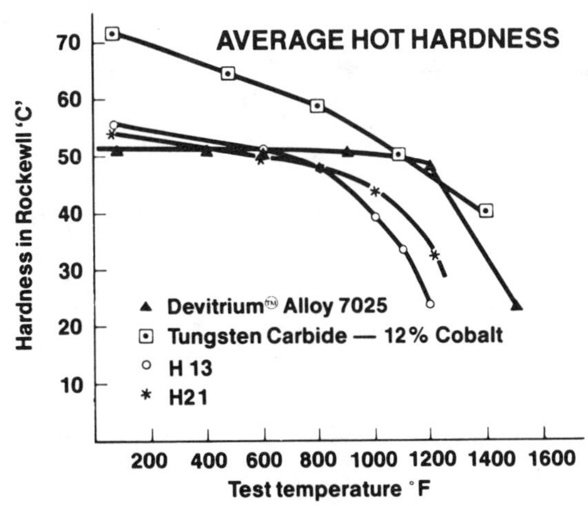

for operation of the die, with standard practice for lubrication, cleaning and maintenance.

The raw materials from which the dies are produced are ultra fine grain intermetallic compounds resulting from more than ten years of work by the Materials Research Center of Allied Corporation, a leading company in rapid solidification technology. The rapid cooling, at rates of a million degrees centigrade per second, results in an amorphous (noncrystalline) ductile ribbon or filament. Quenched at these rates, the molten metal does not tend to segregate or crystallize.

The amorphous metallic ribbon is extruded through a slot orifice onto copper alloy wheel and the continuous filament of supercooled homogeneous alloy is wound in-line at a rate of more than 3000 ft/min. The ribbon is subsequently comminuted into an ultra fine powder, transforming it from its amorphous state to microcrystalline grains with exceptional properties.

(1)Registered trademark of Allied Corporation **PM**

Source: Precision Metal, March 1982, 19-21

SECTION V

Microcrystalline and Crystalline Materials

LIQUID-METAL ATOMIZATION FOR HOT WORKING PREFORMS

N. J. GRANT and R. M. PELLOUX
Massachusetts Institute of Technology
Cambridge, Massachusetts

Rapid quenching of a liquid metal by atomization or splat cooling overcomes the major limitation of most solidification processes, namely the segregation of alloying elements, impurities, and constituent phases. The cooling rates of different atomizing processes are related to the dendrite arm spacings and to the microstructure of the atomized powders. The increased solubility limits and the formation of metastable compounds in splat-cooled alloys are discussed. Consolidation of the powders by hot isostatic compaction, hot extrusion, or hot forging and rolling processes yields billets with properties equivalent to or better than those of the wrought alloys. The application of this powder processing technology to high-performance alloys is reviewed.

INTRODUCTION

Today powder metallurgy is taking on a new meaning: the demand is for fully dense materials in low-, intermediate- or high-alloy classes, fully heat treated with hot- or cold-worked structures. The requirements for higher tonnages of powders has encouraged a multitude of atomization processes, some of which offer real promise for production of high-quality powders at costs that can be competitive with the similar alloys produced by ingot technology.

A list of the different atomization processes includes:

1. High-pressure air and water atomization
2. Low-pressure steam atomization for coarse powders
3. Inert-gas atomization by argon, helium, or nitrogen
4. Rotating-electrode atomization of high-alloy materials in an inert-gas atmosphere
5. Soluble-gas (usually hydrogen) atomization in a vacuum
6. Ultrasonic atomization
7. Magnetic atomization

8. Splat cooling or rapid quenching of small liquid-metal droplets against a high-conductivity metallic substrate.

Emerging from the list of atomization techniques is the awareness of an important feature of these processes, namely the relatively rapid to extremely rapid cooling rates of the liquid metal. These high cooling rates diminish or overcome the major faults of most solidification processes—the segregation of alloying elements, impurities, and all excess phases—and they prevent the formation of a very coarse dendrite and grain structure which is difficult to hot work, or to work at all.

Cooling Rates and Solidification Structures

Figure 1 shows the effect of cooling rate on dendrite arm spacing (DAS) for aluminum[1], and the degree of structure control which is attainable through various atomization methods. The similar curve for copper is steeper but less well defined.[2] For highly alloyed materials, the curve can be expected to flatten because of the lower thermal conductivity of the liquid metal.

At the one extreme, ingot cooling rates lead to coarse grain size and large dendrite arm spacing, coarse inclusions, and second-phase particles; furthermore, large ingots always show significant porosity and a severe three-

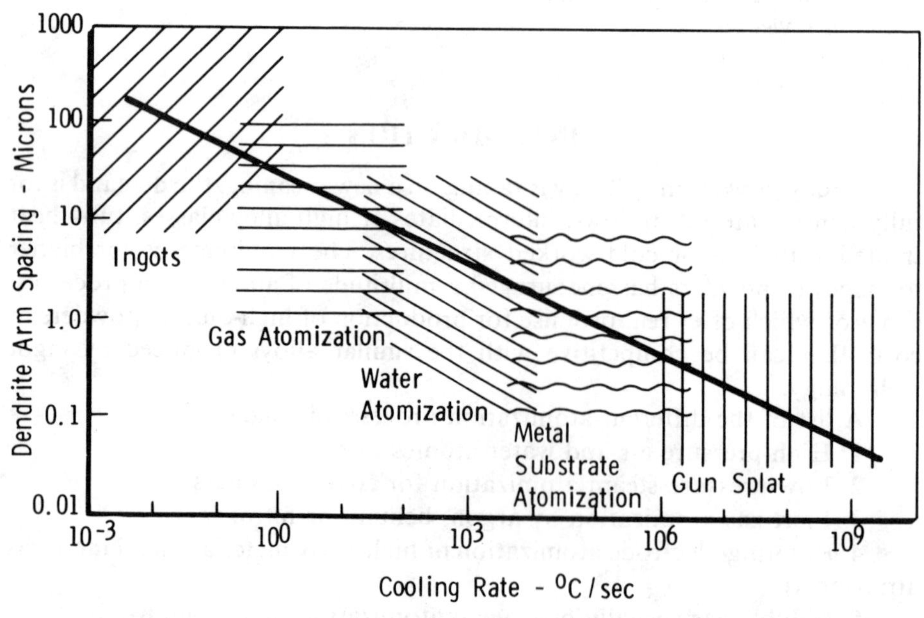

Figure 1. Dendrite arm spacing as a function of cooling rate for aluminum and aluminum alloys.

dimensional segregation pattern. With inexpensive, relatively plastic alloys these faults can be tolerated and largely corrected by working and homogenization of the end product.

Before looking at some of the improvements in the control of the structure and properties of actual alloy systems, it is worthwhile to consider rapid quenching of liquid metals somewhat more broadly.

It was only about 10 years ago that Duwez[3-6] and co-workers reexamined and significantly advanced the state of the art and potential of splat cooling. Shortly thereafter, Predecki, Mullendore, and Grant[7] found experimentally that the cooling rates attainable in splat cooling were 10^7–10^8 C/sec. These measurements were supported by calculations and by a calibration of cooling-rate data suggested by Dean[8]. Matyja *et al.*[1] confirmed the extrapolated cooling-rate data shown in Figure 1. A similar curve has now been measured for copper-base alloys[2] and further calculations have confirmed the approximate maximum cooling rates attainable[9-12].

At cooling rates greater than about 10^6 C/sec it is possible to change the solute solubility limit by an order of magnitude[13,14] (see Giessen[15]) and to form metastable compounds[15]. The size of nonequilibrium phases and of inclusions particles is also refined by an order of magnitude. Cooling rates as high as 10^7-10^9 C/sec can result in grain sizes of less than 0.01 micron[16,17,18] or in amorphous structures[19-23].

In the ultimate case of amorphous structures retained from the liquid phase, all segregation is essentially prevented. Unfortunately, to achieve these cooling rates, the resultant splats are only 0.1-10 microns thick; the foils are irregular, are difficult to process and to consolidate, and will react with the ambient atmosphere. The finished product would be expensive, to say the least.

Consider, however, the central portion of Figure 1, where practical, feasible cooling rates are attainable for commercial atomized powders. The cooling rate and DAS areas for each class of atomization depend on the powder diameter, conductivity of the atomized metal, and quench capacity of the coolant. For finely atomized powders, gas atomization will result in DAS values of about 10 microns in the case of aluminum. Water (or steam) atomization offers further structure refinements. By quenching against a metallic substrate, coarse flakes 50 microns thick by several millimeters in diameter can be produced with DAS values as small as 1 micron. This atomization against a cold metallic substrate is relatively new, but successful trials using rotating disks or copper rolls[2,11,24] as the substrate have been reported.

Source: Solidification Technology, 1972, 317-336

Powder Cleanliness

The contamination of the atomized powders will be critical, since large surface areas are generated in most of the atomization processes. The inert-gas methods, assuming suitable control of gas purity, result in only small increases in oxygen content. In the case of argon atomization of superalloys, the oxygen content increases from that of the vacuum master melt, which is typically 30-75 ppm, to a final content of 50-125 ppm, depending upon the leak rate of the atomizing system.

The rotating-electrode atomization method[25] results in an increase of less than 25 ppm oxygen over that of the electrode material because the alloy is in the molten state for a very short time.

In splat cooling coarse powders in air, the oxygen content increases by several hundred parts per million in spite of the relatively short exposure times in the oxidizing atmosphere. Because of the large surfaces produced by splat-cooling methods, inert-gas atmospheres will be required.

The water atomization and steam atomization of superalloys to produce coarse powders (0.4-4 mm in diameter) result in extensive oxidation with oxygen content of 3,000-4,000 ppm. Nevertheless, if the resultant particles are spherical in nature, without folds or creases, this oxygen content, which is largely on the surface, can be lowered to 200-300 ppm by chemical cleaning of the powders. For instance, cobalt-base superalloys strengthened by high carbon content and high-alloy iron-base materials (tool steels, high-speed tool steels, stainless steels) are successfully steam atomized—provided they are free of significant amounts of aluminum, titanium, and zirconium. The powders can then be mechanically and chemically cleaned to achieve oxygen contents of less than about 300 ppm.

The obvious advantage of water or steam atomization, wherever it can be used, is in the large cost saving. Finally, since all or most atomized materials are reexposed to the ambient atmosphere during storage, handling, and processing, the oxygen content will increase—to the detriment of the final product. A vacuum degassing of the powders in a temperature range 500 to 1000 F prior to consolidation appears to be a general processing requirement.

Powder Consolidation

The powders must be consolidated into fully dense bodies to achieve strength, toughness, and ductility. The higher the quench rates utilized to produce the powders, especially at rates greater than about 10^5 C/sec (see Figure 1), the greater is the tendency for structural changes to take place during consolidation and processing of the compact or bloom. In the case of splat-cooled powders, these changes may call for selection of hot-pressing temperatures and extrusion or rolling temperatures that will produce the desired structures and properties.

In general, the processes emerging for consolidation and densification are the following:

1. *Forging and rolling preforms.* Preforms may be prepared by cold pressing and sintering, hot mechanical pressing, and to a lesser extent by hot isostatic pressing; the preform is generally not fully dense. This process looks attractive for iron and steel powder parts for the large market areas of automotive parts, household appliances, and general hardware.

2. *Hot isostatic pressing (HIP).* Growing in parallel with developments in high-alloy powder atomization, this excellent densification method should be able to provide suitably large ingots to compete on a commercial basis with ingot practice for forging. The powder must be sheathed in an evacuated container for compaction by HIP. Maximum pressures are about 30,000 psi (2,000 atm) at temperatures as high as 1425 C. The working volume for these temperature and pressure levels is a chamber about 14″ in diameter by about 40″ high. Larger pressing units are technologically feasible and can be made as soon as there is a demand; cost per pound of material in such large units is attractive for high-alloy materials. Billet compacts, preforms, dies, vanes, and blades for jet engines can be manufactured by this process.

3. *Hot extrusion and other hot-working processes.* Powders may also be packed into containers without cold or hot pressing, evacuated, sealed and then hot extruded at some nominal extrusion ratio (from 6:1 to 20:1) to achieve full densification. Hot extrusion is preferred to forging or rolling of canned powders, but it is usually limited by the minimum extrusion ratio and by the capacity of the extrusion press. On average, lower temperatures can be used for hot extrusion of powders than are necessary for extrusion of solid compacts or ingot or bloom sections; thus the microstructure refinements can be more easily retained.

STRUCTURE AND PROPERTIES OF ALLOYS PROCESSED FROM QUENCHED POWDERS

In the discussion which follows, the structures and properties of several commercial, semicommercial, or experimental alloy products made of powders solidified at rates from 10^2-10^6 C/sec are examined. These cooling rates encompass rates achieved by gas atomization, steam atomization, and metal-substrate atomization.

Source: Solidification Technology, 1972, 317-336

Aluminum Alloys

Binary aluminum alloys, splat cooled to achieve rates of about 10^8 C/sec, show large increases in solid solubility. Table 1 shows the increase for seven binary alloys.[13,26] Suppression of precipitation during solidification beyond the maximum solubility limit in each of these systems opens the possibility of the development of precipitation hardening; in the case of the Al-Cu system, one gains the possibility of increasing the amount of finely dispersed aging phase by a factor of 2-7.

Table 1. Maximum Metastable Solubilities Observed at Cooling Rate of $\simeq 10^8$ C/sec

System	Max Equil. Sol., at. %	Max Extrap. Metastable Sol., at. %	Equil. Eutect. Comp.,at. %	Solubility Increase
Al-Pd	Nil	7.5	7.5	
Al-Cu	2.5	18	17.3	7X
Al-Mn	0.7	9	1.0	13X
Al-Fe	0.025	4	0.9	160X
Al-Co	0.02	0.5	0.45	25X
Al-Ni	0.023	1.2	2.7	50X
Al-Si	1.59	11	11.3	7X

1. *Al-Si alloy.* The potential of this class of alloy was demonstrated by splat quenching up to 1,000-gram melts against a rotating copper disk of an Al-7% Si alloy.[16] The DAS indicated a quench rate of 10^4-10^5 C/sec, with the silicon content precipitated as submicron particles. The foils, measuring about 50 microns in thickness by 10 mm in the other dimensions, were canned, evacuated, and hot extruded at only 300 C. Table 2 presents a comparison of the properties of this binary alloy with those of a commercial 5% Si alloy (No. 43) in three cast conditions. In addition to an increase in yield strength, the ductility of the cast alloy is ~3-6 times greater than that for the cast commercial material.

Table 2. Al-Si Alloys Tested at 20 C

Alloy	Condition	0.27% YS Ksi	0.27% YS Kg/mm²	UTS Ksi	UTS Kg/mm²	Elong., %	RA, %
Al-7% Si	Air atom., as extrud.	17.3	12.3	27.6	19.6	26	41
Alloy 43 (5% Si)	Die cast	13	9.2	27	19.1	4	-
	Sand cast	9	6.4	18	12.8	6	-
	Perm. mold	10	7.1	22	15.6	8	-

Significant coarsening of the silicon precipitate did not occur at temperatures less than 400 C, thus showing that the structure is resistant to overaging.

2. *Aluminum alloy 2024.* The properties of this commercial high-strength alloy (4.5% Cu, 1.45% Mg, 0.57% Mn, 0.28% Fe, 0.10% Si, balance Al) were compared for three different processing histories: bar formed from a full-scale ingot, air-atomized coarse powder, and as-splat-cooled (rotating copper disk) coarse flake powder.[16] The powders were canned, evacuated, and hot extruded at 300 C at an extrusion ratio of 12:1. All three products were of the same composition, and were given identical heat treatments prior to testing (495 C solution treatment plus room-temperature aging: T4 condition).

The structural characteristics of the three products in the hot-worked or hot-extruded condition are compared in Table 3. A comparison of the hot-worked structures of the ingot product and the splat powder product is presented in Figure 2.

It has been demonstrated that Alloy 2024 contains coarse nonequilibrium constituent particles which are cracked or serve as crack-initiation sites during fatigue.[27,28] Improvements in fatigue life can be expected from the structure refinement. Table 4 presents the results of tension and fatigue tests (axial, zero mean stress). The final grain sizes of all three products were approximately equal in the T4 condition.

Splat cooling completely suppressed the formation of the brittle, complex (CuMgFeAl) intermetallic phase and only the $CuMgAl^2$ phase was observed. This phase did not coarsen during the 495 C solution treatment. In no case was there evidence of cracking or debonding of the micron-size intermetallic phases during fatigue testing.

At 205 C the splat-atomized product did not show intergranular creep fracture in creep tests of at least 400 hours, whereas the ingot product showed intergranular fracture at about 80 hours; the net effect was an increase in stress for 1,000 hours rupture life from 8.6 to 12.5 ksi (6.1–8.9 Kg/mm^2).

Table 3. Structural Characteristics of 2024 Alloy

Product	Quench Rate, C/sec	Dendrite Size, microns	Constituent Particle Size, microns
Ingot bar	10^1	60	10-20
Air atomized	10^2	7	4-6
Splat atomized	10^5	1-2	0.5-1

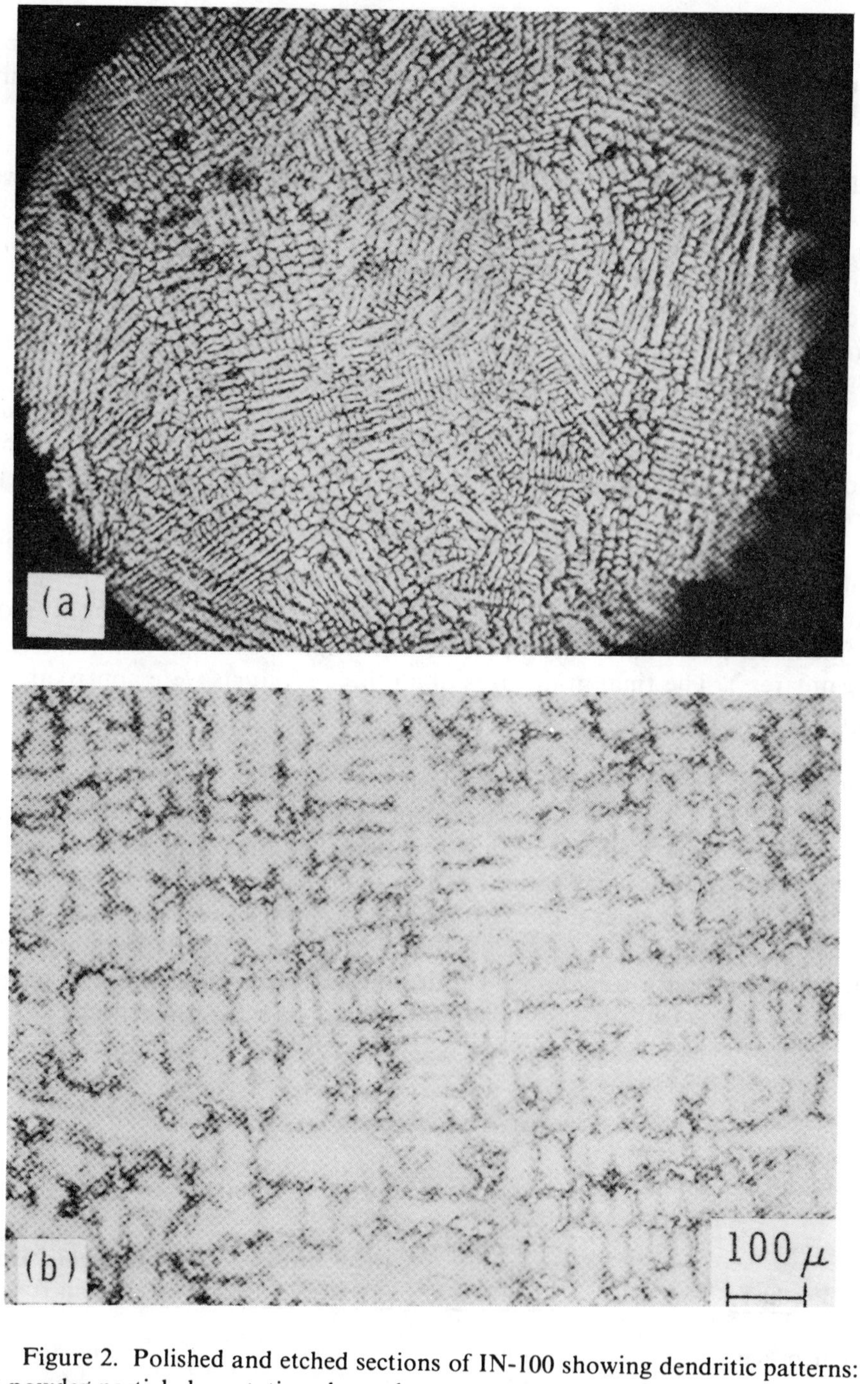

Figure 2. Polished and etched sections of IN-100 showing dendritic patterns: (a) powder particle by rotating electrode process, 200X; (b) precision cast tensile bar, 100X. (Reduced 6%)

Table 4. Room-Temperature Properties of Aluminum Alloy 2024-T4

Product	0.2% YS Ksi	0.2% YS Kg/mm²	UTS Ksi	UTS Kg/mm²	Elong., %	RA, %	Fatigue Life, cycles at 30 ksi (21.2 kg/mm²)
Ingot bar	40.2	28.5	67.2	48.7	23	37	100
Air atomized	42	29.7	70	49.5	25	33	300
Splat atomized	47	33.3	76	54	24	34	700

IN-100

IN-100 is one of the strongest nickel-base alloys available as a cast material; it is not forgeable. A typical composition, in weight percent, is: 10.5Cr, 15.4Co, 3.02Mo, 5.55Al, 4.72Ti, 0.16C, 0.015B, 0.06Zr, 1.05V, 0.94Fe, <0.10Mn, <0.1Si, 0.007S, balance Ni.

Powders from three sources showed the following oxygen content in parts per million

Powder Source	Remelt Stock	Powder
Argon atomization	108	158
Soluble-gas atomization	50	53
Rotating electrode	74	79

The cooling rate of the powders was about 10^2 C/sec, yielding a DAS of about 5 microns compared with a DAS of about 70 microns for the precision-cast structure. The latter had a grain-size range of 1,500-3,000 microns. After hot extrusion of the powders, the final grain size of the alloy ranged from 6 to 8 microns.

The dendritic patterns of a powder particle and a precision casting are compared in Figure 2. Note the relatively coarse carbide particles in the precision-cast dendritic structure.[29]

One of the problems encountered with HIP compacts of IN-100 is the presence of a carbide network (TiC) along prior powder-particle boundaries, as shown in Figure 3. This carbide distribution makes subsequent recrystallization for grain coarsening much more difficult, and also decreases the ductility of the alloy. During extrusion of the alloy, the preferential carbide and oxide patterns are broken up, but the formation of stringers can be quite detrimental.[30]

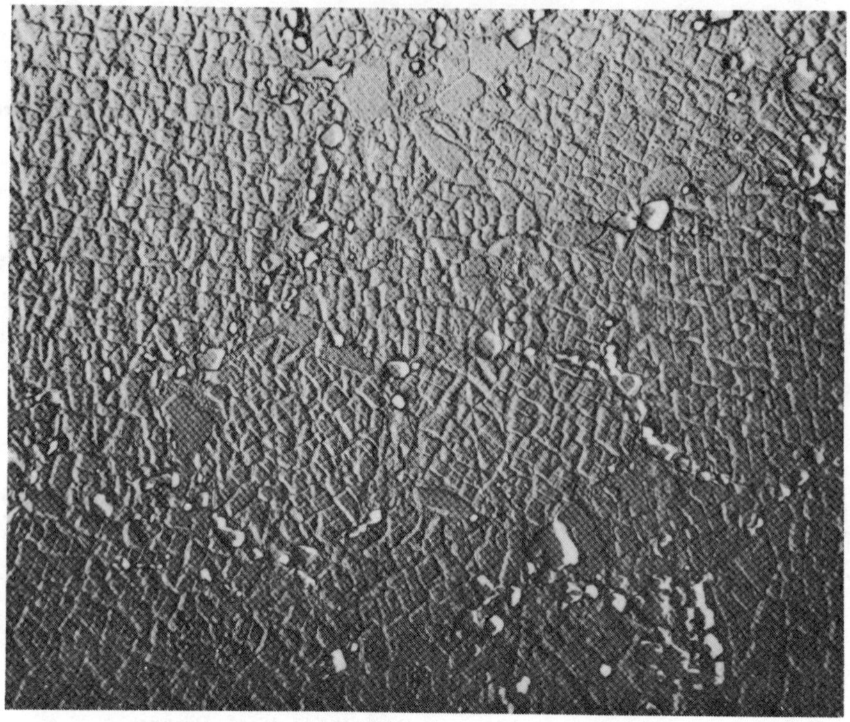

Figure 3. Replica electron micrographs of as-HIP FM powder showing extensive carbide phase surrounding prior particle boundaries. 4000X.

Room-temperature properties for the as-cast, HIP, extruded, and extruded-plus-grain-coarsened states are compared in Table 5. In the HIP condition, there is a small increase in ultimate strength associated with the increased ductility over the as-cast condition. By contrast, the extruded structures show large increases in yield strength, ultimate tensile strength, and ductility compared with those of the cast product. Even after recrystallizing to a grain size of about 100 microns (from 8 microns), the ultimate tensile

Table 5. Room-Temperature Properties of IN-100

Condition	0.2% YS		UTS		Elong.,	RA,
	Ksi	Kg/mm²	Ksi	Kg/mm²	%	%
As cast (1500 μ)	136	96	143	101	4	8
As HIP	137	97	163	116	8	10
As extrd. (8 μ)	175	124	244	172	20	16
Extrd., recryst.	171	121	238	168	21	17
100 μ grain coarsened	137	97	188	133	14	7

strength of the extruded structures remains considerably higher than that of the casting because of the higher ductility.

These highly attractive low-temperature properties of the extruded structures, which continue to be superior to those of the casting to about 750-800 C, unfortunately do not give the best creep strength at 900-1000 C where a coarse grain size stabilized by coarse carbides appears to be superior. In the HIP condition, the fracture path follows the inter-powder-particle boundaries where carbides and oxide films lead to a lower strength (Figure 4a). By comparison, in a similar test at 1150 C, IN-100 in the extruded condition undergoes normal intergranular cracking. The presence of carbides and oxide films is a major problem in using this class of alloy in the HIP condition.[29]

A comparison of the stress for a 100-hour rupture life at 982 C (1800 F) for the as-cast condition with that for various grain sizes of extruded material is presented in Table 6.

Grain coarsening beyond about 500 microns is difficult, especially when the stable coarse-carbides network is present.

Perhaps one of the most interesting applications made possible by the powder approach is that of utilizing the ultrafine-grain size to achieve superplasticity for easy forming and then to heat treat the structure to achieve the grain size desired for the particular application. Figure 5 shows the presence of the superplastic state for IN-100 in tests performed on the as-extruded condition (grain size 8 microns) at 1038 C (1900 F).

Table 6. Stress-Rupture Data at 982 C for IN-100

Condition	Stress for 100-Hour Life		
	Ksi	Kg/mm^2	Elong., %
As cast	22	15.6	10
As extrd. (8 μ g.s.)	2	2	43
Extrd., recryst. (100 μ g.s.)	12	8.5	5
Extrd., recryst. (500 μ g.s.)	17	12.1	4

MAR-M509 and a Hafnium Modification

These cobalt-base alloys are of a particular interest because of the possibility of utilizing steam or water atomization to produce coarse powders (0.4-4 mm) which can be cleaned to a level of oxygen content of about 250 ppm.[30] The water quench, being faster than the inert-gas quench, yields a DAS of about 5 microns with a much larger diameter of the powders. The

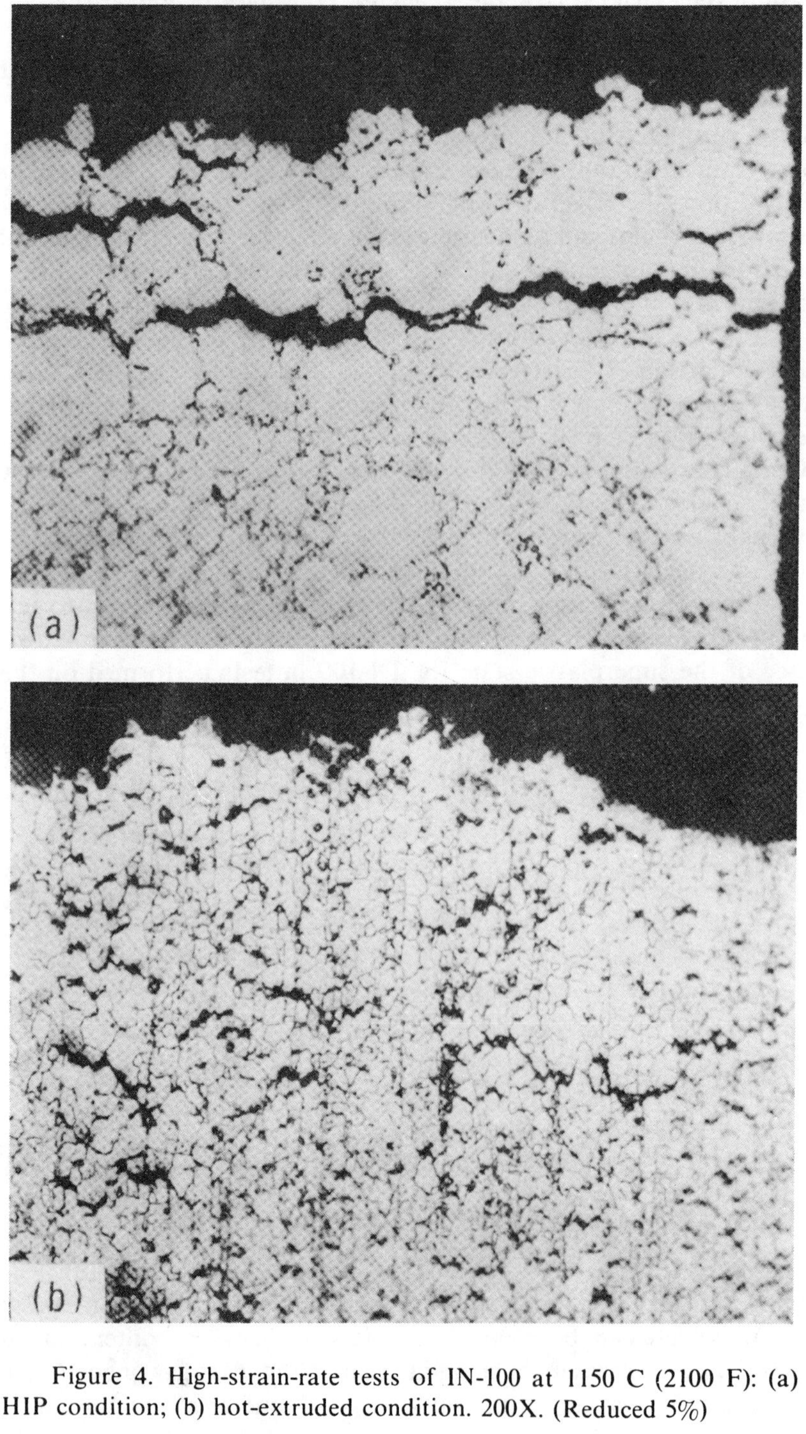

Figure 4. High-strain-rate tests of IN-100 at 1150 C (2100 F): (a) HIP condition; (b) hot-extruded condition. 200X. (Reduced 5%)

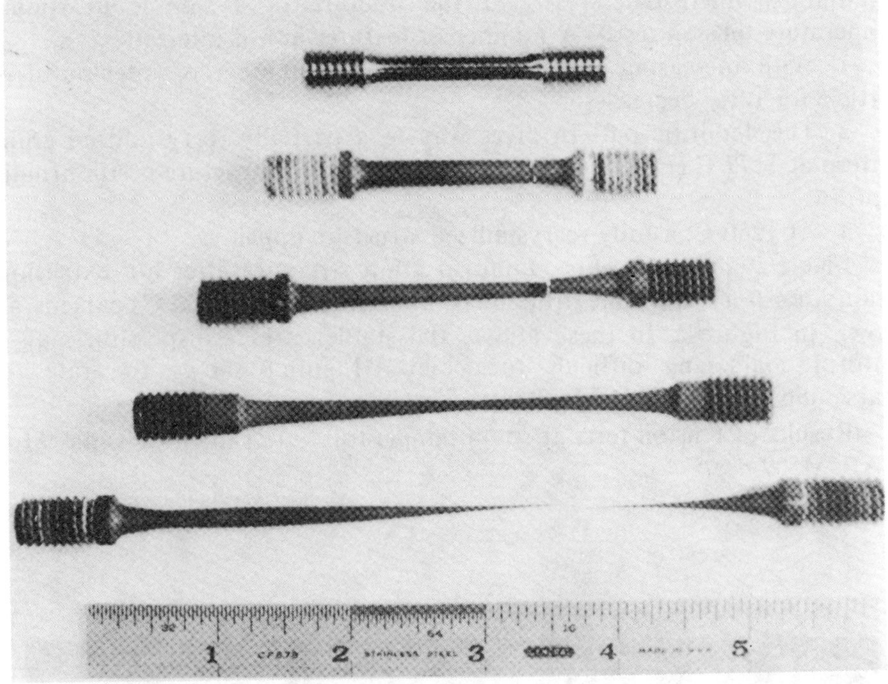

Figure 5. As-extruded test bars superplastically deformed at 1038 C (1900 F) at strain rates from 0.5 to 0.01 min^{-1}. Bar at top is size of undeformed tensile specimen.

compositions of these alloys are, in weight percent:

	C	Ni	Cr	W	Mo	Ta	Zr	Hf	Co
MAR-M509	0.6	10	22	7	—	3.5	0.7	—	Bal
Hf-Mod. Co	0.2-0.75	10	22	—	7	—	—	2-9	Bal

Once cleaned, the powders were canned, evacuated, sealed, and consolidated by HIP. In addition, powders consolidated by HIP were subsequently extruded or hot rolled.

The powder approach to making preforms or final shapes depends on one's ability to bond the powders by sintering, HIP, or hot mechanical pressing. The presence of films or hard phases at powder surfaces interferes with bonding, as shown in Figure 4 for IN-100. In the case of MAR-M509, coarse powders were processed by HIP at 28,000 psi (2,000 atm) and at temperatures of 1093, 1177, and 1260 C (2000 F, 2150 F, and 2300 F). Figure 6 shows the longitudinal microstructures near the fracture of a sample in room-temperature tension test.[31] A number of features are of interest:

1. With increasing HIP temperatures, the number of inter-powder-particle fractures decreases.

2. The dendritic pattern gives way to a partially recrystallized grain pattern at 1177 C (Figure 6); however, the dendritic structure is still prominent.

3. At 1260 C, a fully recrystallized structure appears.

Figure 7 shows the fine, uniform grain structure after hot extrusion (grain size 4-6 μ). The same structure reexposed to HIP at 1260 C coarsens, as shown in Figure 7. In these alloys, the stable carbide dispersion makes uniform coarsening difficult to achieve. Figure 8 shows the rate of coarsening for the MAR-M509 alloy.

Results of tension tests at room temperature are shown in Table 7 for MAR-M509.[31]

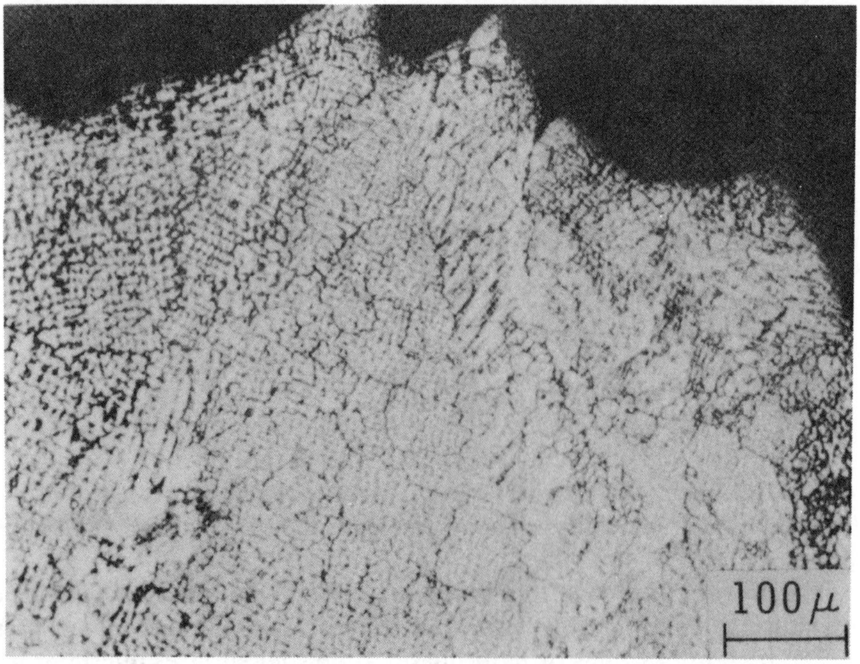

Figure 6. MAR-M509 P/M HIP at 1260 C (2300 F) and 2000 atmospheres (28 ksi) for 2 hours. Tension test at 20 C, longitudinal section at fracture.

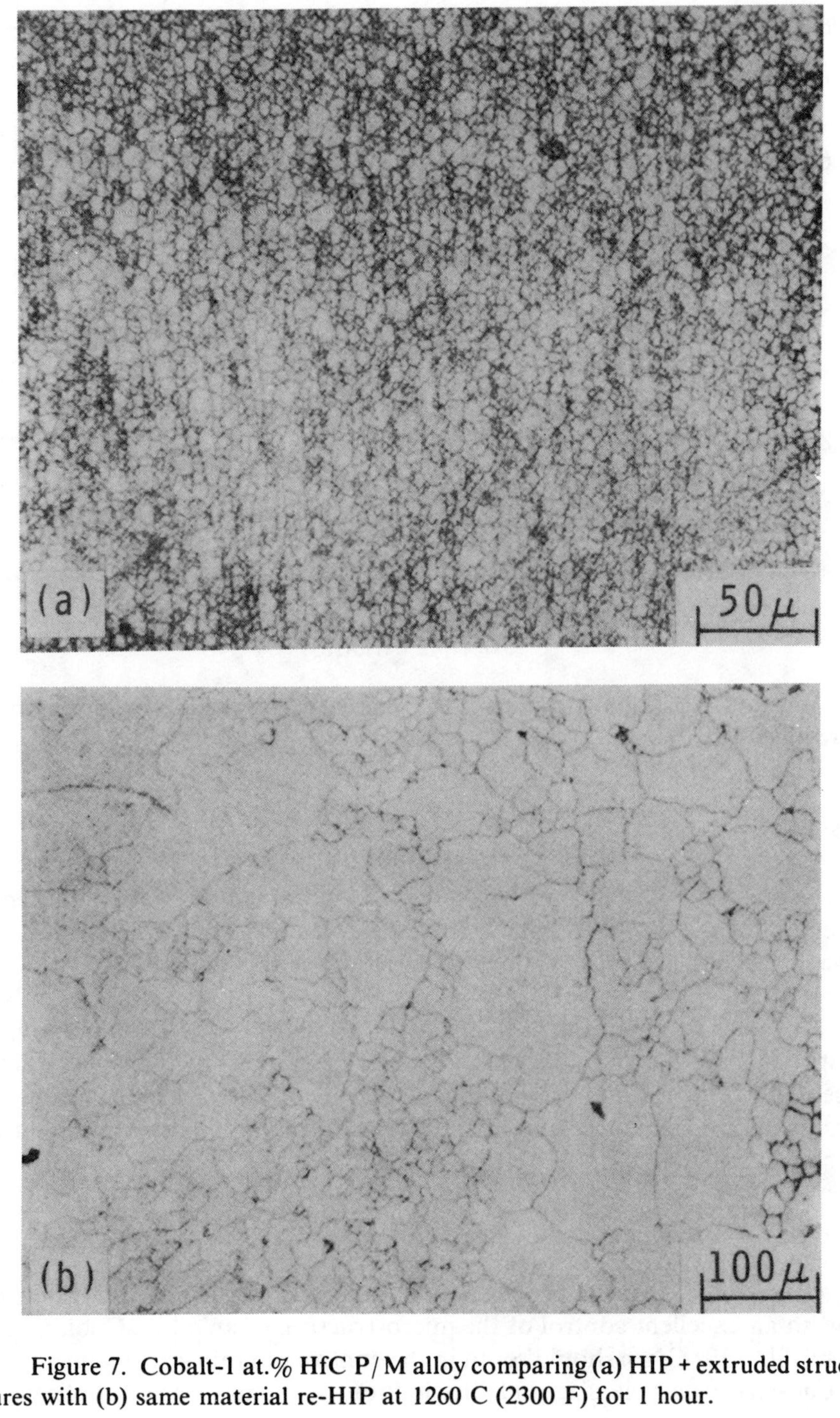

Figure 7. Cobalt-1 at.% HfC P/M alloy comparing (a) HIP + extruded structures with (b) same material re-HIP at 1260 C (2300 F) for 1 hour.

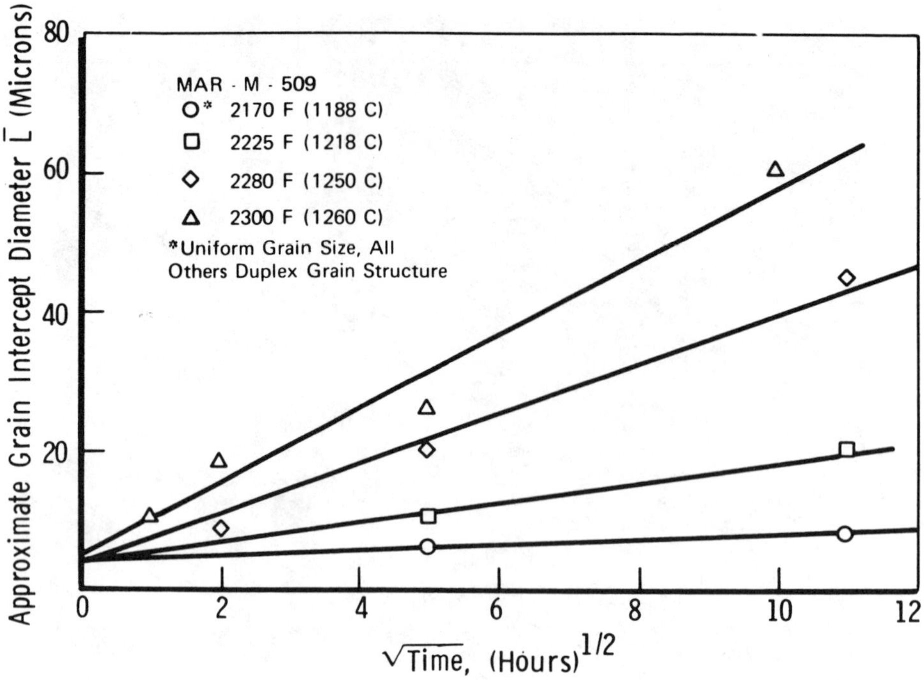

Figure 8. Grain growth of fine grains adjacent to coarse grains in duplex structures MAR-M509.

Again, significant improvements are observed in yield and tensile strength for the extruded condition, along with large increases in ductility. The HIP product, as noted above, shows low ductility and relatively low tensile strength.

In stress-rupture tests at 982 C (1800 F), as with the IN-100 alloy, the best results are shown by the cast alloy and the poorest by the highly ductile, essentially superplastic, fine-grained extruded alloy. Grain coarsening attempts have been successful in improving the high-temperature properties, as shown in Table 8.[31]

DISCUSSION

It is evident that the proper use of atomized and quenched powders can lead to an excellent control of the microstructure of an alloy. Cooling rates from 10^1 to 10^6 C/sec should result in the control of grain size, inclusion, and second-phase particle size and distribution. In some special applications,

Table 7. Room-Temperature and 982 C Properties of MAR-M509

Condition	0.2% YS Ksi	0.2% YS Kg/mm^2	UTS Ksi	UTS Kg/mm^2	Elong., %	RA, %
As cast (4000 μ g.s.)	80	56.7	115	81.8	2.5	—
As extrd. (3-4 μ g.s.)	130	92.6	192	136.0	14	12
Extrd., recryst. (10-20 μ g.s.)	94	66.6	166	117.5	11	8
HIP (5-6 μ g.s.)	81	57.3	108	76.5	1.1	0.4

Table 8. Stress-Rupture Properties at 982 C for MAR-M509

Condition	Stress for 200-Hour Life Ksi	Stress for 200-Hour Life Kg/mm^2	Elong., %
As cast (4,000 μ g.s.)	17	12	8-10
As extrd. (5 μ g.s.)	2.5	1.77	50-120
(4 and 500 μ mixed g.s.)	7.5*	5.32	10

*At 1010 C (1850 F)

higher cooling rates may be desired to achieve increases in solubility of solutes, to produce metastable intermetallic compounds, and even to produce amorphous structures. Such structures would then be converted to stable phases but with particle sizes, shapes, and distributions which are not otherwise attainable.

The improvement in ductility and hot plasticity of complex and highly alloyed material should encourage use of the powder approach to higher alloy contents. For instance, increases in Al + Ti in the IN-100 alloy and of carbon and carbide former elements in the MAR-M509 alloy are feasible.

A number of other fruitful research areas are worth mentioning.

1. Improvements in atomization are still necessary, both to control the powder size and distribution and also to limit the retained or trapped argon in inert gas atomization or hydrogen in the soluble gas-vacuum method. Cost improvements are definitively required.

2. Control of the powder surface films is needed when the hot-pressed condition is desired as the final processing step.

3. A better understanding of the influence of the HIP parameters on the properties of the HIP billet is required.

4. Optimum hot-working conditions to shape a billet and to control the grain size must be determined.

5. Thermomechanical treatments should be devised to take advantage of the high ductility and the superplastic potential of the totally new structures. Solution and precipitation processes have to be adapted because of the metastability of compositions and structures of the rapidly quenched materials.

6. Alloys should be developed to take advantage of the increased ductility and hot plasticity. In this respect, alloy chemistries may be modified for the HIP state and for the hot-worked product.

In summary, there are a number of interesting technological developments that make the processing of high-performance alloys from quenched powders look feasible and attractive. The long-range opportunities are great, and these new processes should not be looked at as mere laboratory curiosities.

ACKNOWLEDGMENT

The authors are indebted not only to their students and associates for the data in this presentation, but to NASA, ARPA, and INCRA for support of various phases of the research.

REFERENCES

1. Matyja, H., Geissen, B. C., and Grant, N. J., "The Effect of Cooling Rate on the Dendrite Spacing in Splat-Cooled Aluminum Alloys", *J. Inst. Metals,* **96,** 30 (1968).
2. Sarin, V.K., and Grant, N.J., "Cu-Zr and Cu-Zr-Cr Alloys Produced from Rapidly Quenched Powders", *Met. Trans.,* **3**, 875 (1972).
3. Duwez, P., Willens, R.H., and Klement, W., Jr., "Continuous Series of Metastable Solid Solutions in Silver-Copper Alloys", *J. Appl. Phys.,* **31**, 1136 (1960).
4. Duwez, P., and Willens, R.H., "Rapid Quenching of Liquid Alloys", *Trans. Met. Soc. AIME,* **227**, 362 (1963).
5. Klement, W., Jr., Willens, R. H., and Duwez, P., "Noncrystalline Structure in Solidified Gold-Silicon Alloys", *Nature (London),* **187,** 2267 (1960).
6. Duwez, P., Willens, R.H., and Crewdson, R.C., "Amorphous Phase in Palladium-Silicon Alloys", *J. Appl. Phys.,* **36**, 2267 (1965).
7. Predecki, P., Mullendore, A.W., and Grant, N.J., "A Study of the Splat Cooling Technique", *Trans. Met. Soc. AIME,* **233**, 1581 (1965).

8. Dean, W. A., and Spear, R. E., "Controlled Solidification—Discussion", *Strengthening Mechanisms—Metals and Ceramics,* J. J. Burke, N. L. Reed, and V. Weiss, Eds., Syracuse, N.Y.: Syracuse University Press (1966), 268.

9. Ruhl, R.C., "Cooling Rates in Splat-Cooling", *Mat. Sci. Eng.,* **1**(6), 313 (1967).

10. Ruhl, R.C., and Cohen, M., "Splat Quenching of Iron-Carbon Alloys", *Trans. Met. Soc. AIME,* **245**, 241 (1969).

11. Anatharaman, T.R., and Suryanarayana, C., "Review: A Decade of Quenching from the Melt", *J. Mat. Sci.,* **6**, 1111 (1970); and "Reply to a Comment on a Decade of Quenching from the Melt", *J. Mat. Sci.,* **7**, 351 (1972).

12. Burden, M.H., and Jones, H., "Determination of Cooling Rate in Splat-Cooling from Scale of Microstructure", *J. Inst. Metals,* **98**, 249 (1970).

13. Jansen, C.H., "Splat-Cooled Aluminum-Rich Transition Metal Alloys: Supersaturation, Annealing Behavior and Atmospheric Corrosion", Ph.D. dissertation, Massachusetts Institute of Technology, 1971.

14. Ruhl, R.C., Giessen, B.C., Cohen, M., and Grant, N.J., "Metastable Body-Centered Cubic Phases in the Vanadium-Nickel and Nickel-Indium Systems", *Mat. Sci. Eng.,* **2**(6), 314 (1968).

15. Giessen, B.C., "Constitution of Non-Equilibrium Alloys After Rapid Quenching from the Melt", *Developments in the Structural Chemistry of Alloy Phases,* B.C. Giessen, Ed., New York: Plenum Press (1969), 227.

16. Lebo, M.R., "Structure Refinement and Property Control in Aluminum Alloys through Rapid Quenching of the Melt", Ph.D. dissertation, Massachusetts Institute of Technology, 1971.

17. Rohl, R.C., Giessen, B.C., Cohen, M., and Grant, N.J., "New Microcrystalline Phases in the Nb-Ni and Ta-Ni Systems", *Acta Met.,* **15**, 1693 (1967).

18. Strivastave, P.K., Giessen, B.C., and Grant, N.J., "A Noncrystalline Pt-Sb Phase and Its Equilibration Kinetics", *Met. Trans.,* **3**, 977 (1972).

19. Duwez, P., and Lin, S.C.H., "Amorphous Ferromagnetic Phase in Iron-Carbon-Phosphorus Alloys", *J. Appl. Phys.,* **38**, 4096 (1967).

20. Rastogi, P. K., and Duwez, P., "Rate of Crystallization of an Amorphous Iron-Phosphorus-Carbon Alloy", *J. Non-Cryst. Solids,* **5**, 1 (1970).

21. Chen, H.S., and Turnbull, D., "Thermal Properties of Gold-Silicon Binary Alloy Near the Eutectic Composition", *J. Appl. Phys.,* **38**, 3646 (1967).

22. Ray, R., Giessen, B.C., and Grant, N.J., "New Non-Crystalline Phases in Splat-Cooled Transition Metal Alloys", *Scripta Met.,* **2**, 357 (1968).

23. Revcolevschi, A., and Grant, N.J., "Study of a Splat-Cooled Cu-Zr-Monocrystalline Phase", *Met Trans.,* **3**, 1545 (1972).

24. "Proceedings of the International Conference on Metastable Metallic Alloys", M. Paic, Ed., *Fizika,* **2**, Suppl. 2 (1970).

25. Grant, N. J., "Specialty Methods of Powder Atomization", *Powder Metallurgy for High-Performance Applications,* J. J. Burke and V. Weiss, Eds., Syracuse, N.Y.: Syracuse University Press (1972), 85.

26. Itagaki, M., Giessen, B. C., and Grant, N. J., "Supersaturation in Rapidly Quenched Al-Rich Al-Si Alloys", *Trans. ASM,* **61,** 330 (1968).

27. Erhardt, K., and Grant, N.J., "Behavior of 2024 Aluminum in Low-Cycle Fatigue at Low Strain Rates as a Function of Temperature", *Fracture 1969: Proceedings of the Second International Conference on Fracture,* Brighton (England), April 1969, P.L. Pratt, Ed., London: Chapman and Hall (1969), 702.

28. Pelloux, R.M., Erhardt, K., and Grant, N.J., "Applications of the Scanning Electron Microscope to Electron Fractography", *Proceedings of the Third Annual Scanning Electron Microscope Symposium,* O. Johari, Ed., Chicago: Illinois Institute of Technology Research Institute (1970), 281.

29. Moskowitz, L.M., "Properties of a Nickel-Base Superalloy Processed by Powder Metallurgy", M.S. thesis, Massachusetts Institute of Technology, 1972.

30. Widmer, R.W., "Coarse Powder Techniques", *Powder Metallurgy for High-Performance Applications,* J.J. Burke and V. Weiss, Eds., Syracuse, N.Y.: Syracuse University Press (1972), 69.

31. Robinson, R.K., personal communication.

Microstructural effects of splat cooling a high-speed steel

I. R. Sare and R. W. K. Honeycombe

A metallographic investigation of splat-cooled M1 high-speed steel is described. Apart from general microstructural refinement, the splat-cooling process retains in solid solution much greater amounts of the alloying elements than conventional quenching processes. This stabilizes the austenite to such an extent that the M_s temperature is depressed close to or below ambient temperature and no martensite is formed. The morphology of the matrix phases, δ-ferrite and austenite, is shown to depend critically upon the specimen thickness and hence local cooling conditions. Owing to the high solute supersaturation, subsequent tempering of the splat-cooled steel leads to precipitation of the vanadium-rich carbide MC on {100} planes in the δ-ferrite, in sharp contrast to the molybdenum-rich M_2C carbide which is normally responsible for secondary hardening in conventionally processed high-speed steels. In the splat-cooled steel, M_2C appears only at the solute-rich interdendritic boundaries. Sintering studies were carried out to consolidate splat-cooled material into a more useful form. The presence of high concentrations of carbon in solution in the as-splatted material depresses the solidus temperature so much that partial melting occurs upon sintering at 1200°C, while the excess dissolved carbon permits the precipitation of M_2C at 1100°C. The practical application of the knowledge gained from this study to possible future developments in high-speed steel technology is discussed.

Manuscript received 1 September 1978; in final form 19 November 1979. I. R. Sare, PhD, was and R. W. K. Honeycombe, PhD, DSc, FIM, is in the Department of Metallurgy and Materials Science, University of Cambridge. Dr Sare is now with CSIRO Division of Materials Science, Production Technology Laboratory, Adelaide, South Australia.

The most important single factor responsible for producing the desired mechanical properties of a high-speed steel is the optimization of the carbide distribution.[1] This is not merely a function of hardening techniques and temperatures, but starts with the solidification process itself. The complex solidification sequence[1-4] produces a microstructure which possesses serious alloy segregation, rendering the steel unsuitable for immediate use in cutting tool applications. The long freezing range makes dendritic coring an inevitable consequence of conventional casting practice, and the formation of primary carbide networks during solidification and cooling to ambient temperatures results in a coarse particle dispersion in the as-cast microstructure.

Since the carbide distribution is largely determined by the extent of alloy segregation during solidification, attempts have been made to develop alternative processes which utilize high rates of cooling to minimize this segregation. The powder metallurgy route has been extensively investigated[5-11] and has been shown to produce considerable microstructural refinement in high-speed steels. In one such technique, the ASEA/Stora process,[9] the liquid steel is atomized by high-pressure gas jets and the droplets cooled at a controlled rate to produce a spherical powder which is subsequently consolidated by hot isostatic pressing. The resulting steel has a finer, more uniform carbide dispersion than a conventionally produced high-speed steel, and tool-life tests are claimed to show a considerable improvement in mechanical properties. More recently, however, it has been found[11] that although powder metallurgy processes can produce extremely fine and uniform microstructures in high-speed steels, they do not necessarily improve the cutting properties over conventionally produced steels.

The cooling rates which can be achieved in the production of alloy powders by atomization are limited to about 10^4–10^5 K s^{-1}, the limiting factor being the rate at which the layer of hot gas or vaporized liquid from the cooling medium in contact with the specimen can be removed. To achieve higher rates of cooling, heat removal by conduction into a solid heat sink must be used. This is the basis of the splat cooling process, in which a thin liquid film is rapidly quenched against a solid substitute of high thermal conductivity. The very high cooling rates attainable with this technique (typically 10^6–10^8 K s^{-1}) should produce finer primary carbide dispersions or even complete solid solubility in high-speed tool steels.

Two investigations of tool steels splat cooled from the liquid state have been reported, but in neither case was complete solubility retained to room temperature. Arai and Komatsu observed[12] optically resolvable carbides in high-speed and die steels rapidly quenched from the liquid state, albeit far more finely dispersed than in commercial forged material. However, the cooling rates achieved by their rapid cooling methods would not have been as high as those customarily expected for splat-cooling processes owing to the large splat thicknesses obtained of 0·5 to 1·5 mm. The crystallization process in the high-speed steels was deduced to occur by the formation of primary austenite followed by an austenite plus carbide eutectic, most of the austenite later transforming to martensite as cooling continued to the ambient temperature. In regions where the cooling rate was assumed to be lower, primary δ-ferrite crystallized and remained to room temperature. The carbides formed during solidification of M2 type high-speed steel (6%W, 5%Mo,

4%Cr, 2%V, 0·9%C) were found to be M_2C and MC, in contrast to the M_6C and MC found in conventionally hardened high-speed steels.[2] It was proposed that the change in carbide type occurs at a critical cooling rate above which M_2C forms in preference to M_6C.

In an investigation of splat-cooled T1(18%W, 4%Cr, 1%V, 0·8C) and M2 high-speed steels, Jama and Thursfield[13] detected two distinct structural zones in transverse sections of the splats: δ-ferrite adjacent to each splat surface and austenite with a much coarser microstructure in the central region. No carbide particles were observed by optical microscopy, in sharp contrast to the observations of Arai and Komatsu,[12] although small precipitates were detected in this material by transmission electron microscopy.[14] The absence of optically resolvable carbide particles and the fact that splats were generally an order of magnitude thinner than those obtained by Arai and Komatsu[12] suggest that cooling rates were considerably higher in this work, and therefore that the retention of δ-ferrite is favoured by high cooling rates and not low cooling rates as claimed by Arai and Komatsu.[12] Support for this contention may be drawn from Jones's observation[15] of similar duplex structures in splat-cooled aluminium–iron alloys, in which the outer zone adjacent to each splat surface was shown to have cooled faster than the central zone. The aim of the present work was to study the structures produced in a splat-cooled commercial high-speed steel, AISI M1, as a corollary to a related investigation of pure Fe–Mo–C alloys.[16] Since the material produced by splat-cooling techniques is little more than a laboratory curiosity in the as-quenched condition, the study was extended to an examination of cold-compacted and sintered material.

EXPERIMENTAL

The steel for this investigation was supplied as annealed 6·35 mm rod, which was subsequently cold-swaged to 3 mm diameter. The composition of the alloy is shown in Table I. The surface was cleaned by filing to remove any oxide, and small pieces weighing about 0·25 g were cut from the rod for use in the controlled-atmosphere 'gun' splat cooling apparatus which has been described in detail elsewhere.[17] The methods used for X-ray analysis, scanning and transmission electron microscopy, heat treatment, and differential thermal analysis of splat-cooled material have been described in a previous paper.[16]

Table I Chemical analysis of M1 high-speed steel, wt-%

C	Mo	W	Cr	V
0·77	8·35	1·50	4·15	1·15

In order to consolidate splat-cooled material prior to high-temperature sintering it was first necessary to reduce the flakes to a fine powdery form. Very little reduction could be achieved by grinding in a pestle and mortar, and the following process was adopted. About 0·2 g of splat-cooled material was mounted in transoptic plastic in a mounting press. The plastic mount was then machined in a lathe with the removal of 25 μm per pass and the swarf was collected in a box underneath the lathe tool. The swarf was treated with acetone which dissolved the plastic and left the metallic material behind. The liquid was subsequently filtered and the remaining machined splat-cooled 'powder' dried and degreased in trichloroethylene. The 'powder' was compacted in a 5·5 mm diameter die at 770 MN m^{-2} for 30 s, with two drops of collodion as a binder. The compact was then sealed in a silica capsule under a partial pressure of argon. During evacuation of the capsule before the admission of argon, a gas flame was played over the compact to drive off the solvent in the collodion binder. Sintering was carried out in a horizontal tube furnace controlled to within ±3 K of the desired temperature, and after sintering the compacts were quenched in their capsules in water and mounted for metallographic examination.

RESULTS

X-ray diffraction analysis

X-ray powder diffraction analysis of splat-cooled material from several runs revealed a mixture of austenite and a bcc product in varying proportions, with austenite contents ranging from 10 to 70%. The average lattice parameters of these two phases are presented in Table II. The greater error in the parameter of the austenite is a result of there generally being less austenite than ferrite in most runs, and therefore more uncertainty in the measurement of the weak, diffuse high-angle austenite diffraction lines. The variation in the amount of austenite retained to room temperature was found to be caused by variable cooling rates. Some thicker, less adherent flakes from an experimental run which gave about 40% retained austenite in X-ray powder photographs were examined by X-ray diffractometry. The amount of retained austenite in the larger, more slowly cooled flakes was at least 60%, indicating that the amount of austenite retained to room temperature was inversely related to the mean cooling rate.

Dyson and Holmes[18] have derived an equation to express the effect of alloying additions on the lattice parameter of iron–nickel–chromium austenites. By extrapolation of their results to zero alloy content, Dyson and Holmes suggested that the room-temperature lattice parameter of austenite in pure iron is 3·5780 Å. Ridley[19] pointed out that the parameter of austenite at room temperature should be 3·573 Å on the basis of many previous results on binary alloys extrapolated to zero alloy content. If this latter value

Table II Lattice parameters of the phases in splat-cooled material and water-quenched filings of M1 high-speed steel

Treatment	Lattice parameter, Å		
	Ferrite or martensite	Austenite	M_6C
Splat-cooled material	2·889±0·003	3·637±0·011	—
Water-quenched filings	2·889±0·002	3·606±0·009	11·085±0·027
Calculated from Dyson and Holmes[18] equation	—	3·632	—

and the appropriate regression constants derived by Dyson and Holmes[18] are combined, the following equation is obtained, which should express the effect of the alloying elements in high-speed steel on the lattice parameter of austenite:

$$a_{\text{austenite}}\ (\text{Å}) = 3{\cdot}573 + 0{\cdot}0330\ (\text{wt-\%C}) + 0{\cdot}0031\ (\text{wt-\%Mo}) + 0{\cdot}0018\ (\text{wt-\%W}) + 0{\cdot}0006\ (\text{wt-\%Cr}) + 0{\cdot}0018\ (\text{wt-\%V})$$

Substitution of the amounts of the alloying elements in the high-speed steel (Table I) into this equation gives the value 3·632 Å, in good agreement with the measured lattice parameter.

Filings of the high-speed steel, quenched in a silica capsule into water from 1200°C, had a mixture of three phases: approximately 80% martensite (although no tetragonality was evident), 15% austenite, and 5% M_6C. The lattice parameters of these phases are also listed in Table II. It is clear that the martensite in the filings has about the same degree of supersaturation as the ferrite in the splat-cooled material, while the austenite in the filings is considerably less solute-rich than its splat-cooled counterpart. The value of the lattice parameter obtained for M_6C agrees very well with that quoted by Andrews *et al.*[20] of 11·082 Å.

Metallography of splat-cooled material

An overall view of the microstructure was obtained by examining flakes by scanning electron microscopy. Figure 1 illustrates the detail of the solidification structure on an electropolished and etched flake; it is apparent that there has been considerable solute segregation during solidification, although the micrograph does not indicate whether carbide formation has taken place at the solute-rich cell or dendrite boundaries.

A number of distinct morphologies of the matrix phases was detected by transmission electron microscopy. In unthinned flakes the ferrite exhibited two forms. In the first case the microstructure consisted of elongated grains containing very fine precipitate particles lying in cube directions (Fig. 2). The corresponding electron diffraction pattern showed pronounced streaking in ⟨100⟩ ferrite directions, characteristic of the early stages of both M_2C and MC precipitation.[21] Such areas were observed only infrequently in the thinnest parts of the electron-transparent areas. Invariably the ferrite grains were elongated in the plane of the foil, indicating that heat extraction, and hence solidification, had taken place in that plane.[22]

The second ferrite morphology in unthinned foils is shown in Fig. 3, and is a similar dendritic structure to that in Fig. 1, with heavy segregation of solute to the dendrite boundaries. Within this and other similar areas, no precipitation was observed at the interdendritic boundaries, and in no cases were precipitate reflections evident in the diffraction patterns. The feature marked *A* in the centre of the micrograph is a 'predendritic' area[23] at which the solid phase first nucleated and grew by a 'diffusionless' process. There is then a transition to the dendritic structure which develops from it. Since the dendrites have developed in the

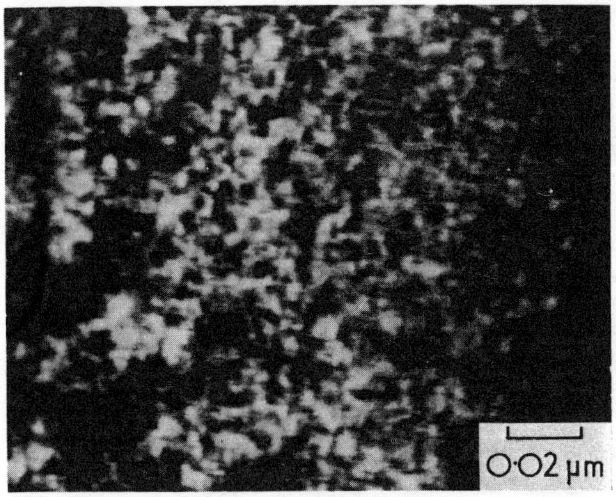

2 Part of elongated ferrite grain in unthinned foil of splat-cooled M1 high-speed steel, showing very fine precipitation of carbide particles. Transmission electron micrograph.

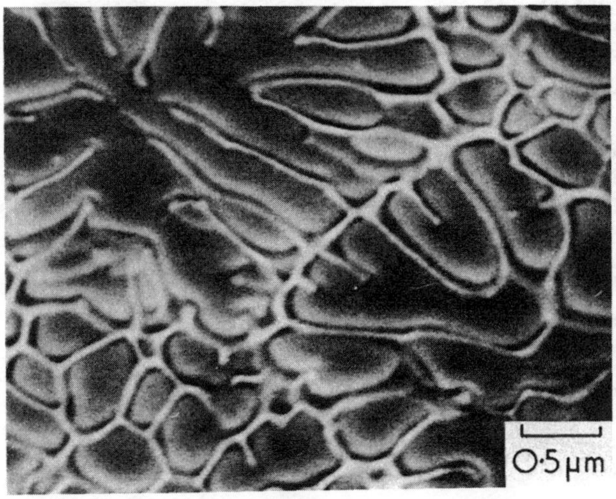

1 Scanning electron micrograph of electropolished and etched surface of splat-cooled M1 high-speed steel.

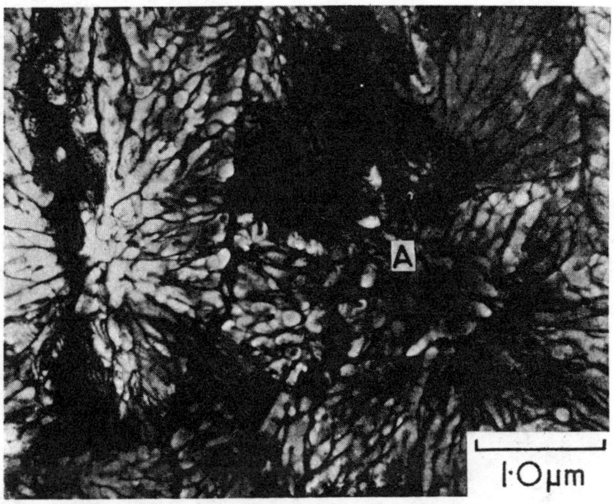

3 Radial growth of ferrite dendrites from pre-dendritic area A in unthinned foil of splat-cooled M1 high-speed steel. Transmission electron micrograph.

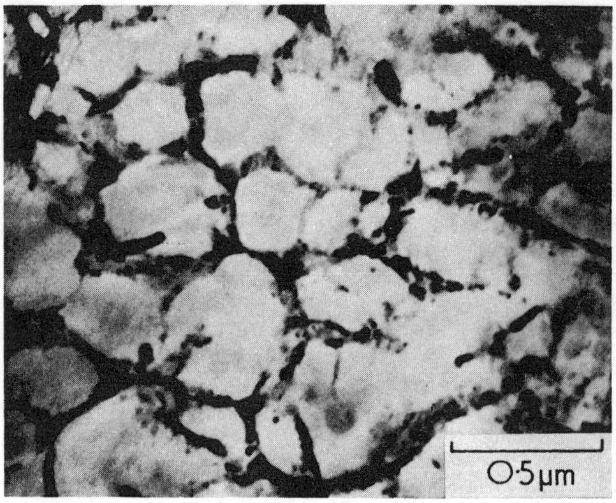

4 Precipitate particles at boundaries of ferrite dendrites in ion-beam thinned foil of splat-cooled M1 high-speed steel. Transmission electron micrograph.

5 Small equiaxed grains of austenite in ion-beam thinned foil of splat-cooled M1 high-speed steel. Transmission electron micrograph.

plane of the foil, it is clear that again growth has occurred within the foil plane and not perpendicular to it.

On thinning the thicker splat specimens a dendritic structure was observed in the ferritic matrix (Fig. 4), but with copious precipitation along the interdendritic boundaries. The precipitate particles produced spots in the diffraction patterns but the interplanar spacings corresponded to both M_6C and $M_{23}C_6$ and so no positive identification could be made. The fact that the precipitate particles were observed only in thinned foils is consistent with the fact that such foils would have been more slowly cooled than unthinned ones, allowing time for the nucleation and growth of the carbides.

Austenite was detected only in thinned foils and, as with the ferrite, two different growth morphologies were observed. Firstly, small equiaxed grains were evident (Fig. 5) with a grain size less than 0·5 μm. The relatively wide boundaries between the grains suggest that they may in fact be solidification cells, but dark-field microscopy showed that the grains are separated by high-angle boundaries. Since, in the usual definition of a cellular solidification structure, the cells within a single grain all have the same crystallographic orientation[24] and therefore cannot be separated by high-angle boundaries, the structural units in Fig. 5 must be separate grains.

Secondly, long parallel dendrites of austenite were observed (Fig. 6(*a*)) with precipitation of unidentified particles at the interdendritic boundaries (Fig. 6(*b*)). The interplanar spacings of these particles calculated from their reflections in diffraction patterns were found to be the same as those in thinned ferritic regions (Fig. 4); therefore the particles are probably either M_6C or $M_{23}C_6$.

Ageing of splat-cooled material

Differential thermal analysis (*DTA*)

Samples from several splat-cooling experiments were heated at $0{\cdot}33\ K\,s^{-1}$ ($20\ K\,min^{-1}$) in a differential thermal

(a) 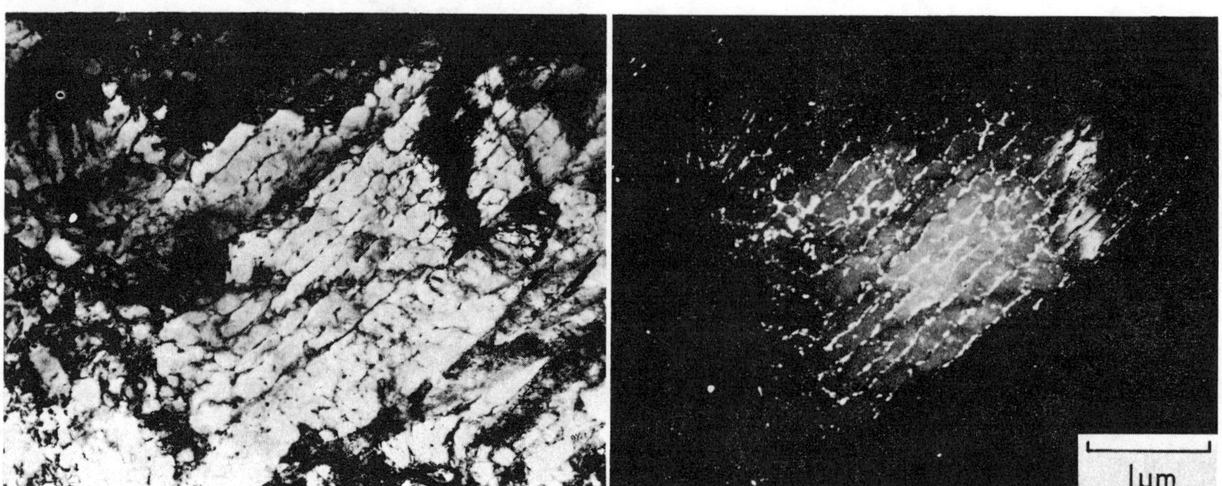 (b)

6 Austenite dendrites in an ion-beam thinned foil of splat-cooled M1 high-speed steel. Transmission electron micrographs: (a) bright-field, (b) dark-field imaged with a precipitate reflection.

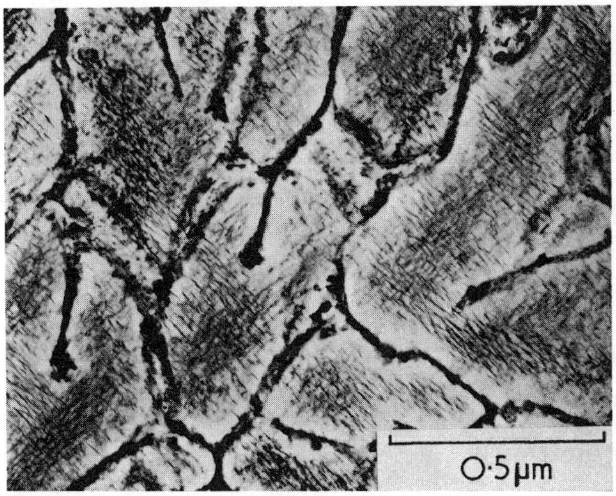

7 Splat-cooled M1 high-speed steel aged 20 h at 600°C, showing fine MC plates on {100} ferrite planes with M_2C at the dendrite boundaries. Electropolished foil. Transmission electron micrograph.

analyser to determine whether any phase changes occurred. None of the samples yielded peaks in the ΔT vs. temperature graph up to 600°C, the upper temperature limit of the furnace. X-ray diffraction patterns taken of these samples after they had cooled to room temperature were identical to those obtained from the as-splatted material. In addition, isothermal ageing experiments were carried out in the range 550 to 700°C in the DTA apparatus. Subsequent X-ray analysis of these samples revealed no austenite disappearance after 20 h at 550°C, but complete disappearance after 1 h at 700°C.

Metallography of aged material

Heat treatments were carried out at 600 and 700°C, and all specimens were subsequently quenched in their silica capsules into water. After 5 h at 600°C, heavy streaking in diffraction patterns with an [001] ferrite zone axis was observed, and the microstructure contained a very fine matrix precipitate superimposed on the dendritic solidification structure. A detailed electron diffraction–dark-field microscopy study[25] revealed that the matrix precipitate particles were plates of the vanadium-rich carbide MC obeying the Baker–Nutting[26] orientation relationship with the ferrite, namely:

$$\{100\}\ \text{MC} \parallel \{100\}\ \text{ferrite}$$
$$\langle 100\rangle\ \text{MC} \parallel \langle 110\rangle\ \text{ferrite}$$

The interdendritic precipitate was found to be the molybdenum-rich carbide M_2C obeying the Pitsch and Schrader[27] orientation relationship, namely:

$$(2\bar{1}.0)\ M_2C \parallel (100)\ \text{ferrite}$$
$$(01.0)\ M_2C \parallel (01\bar{1})\ \text{ferrite}$$
$$(00.1)\ M_2C \parallel (011)\ \text{ferrite}$$

Figure 7 shows the microstructure in the overaged condition after 20 h at 600°C; the edges of the MC plates are visible on {100} ferrite planes, surrounded by the interdendritic M_2C network.

After 1 h at 700°C considerable coarsening of the MC had occurred (Fig. 8), and precipitation of large M_6C particles was evident. No retained austenite decomposition product analogous to the alloy pearlite found in an aged splat-cooled Fe–10%Mo–0·5%C alloy[16] was detected metallographically, although the X-ray/DTA work showed that all the austenite had disappeared after 1 h at 700°C. It is possible, however, that the products of retained austenite decomposition in this alloy closely resemble the aged ferrite, in which case they would not have been distinguishable from it.

Sintering of splat-cooled material

The aim of producing sintered compacts of the splat-cooled high-speed steel was twofold:

(i) to produce suitable samples for microhardness tests;
(ii) to examine the microstructure of the splat-cooled material after high temperature sintering, and to compare it with that of conventionally hardened specimens.

Owing to the high hardness of the splat-cooled flakes, crushing in a pestle and mortar had no size-reduction effect, and the tedious method described above had to be used to produce a powder suitable for sintering. Although considerable particle reduction was achieved by this latter method, the powder obtained was never really fine enough to produce dense sintered compacts. As a result the microhardness testing had to be abandoned, for whenever a test was carried out the indenter simply pushed the material on the surface into the void space underneath, and a true measure of the hardness was not obtained.

Metallography of sintered material

Since the austenitizing temperature of M1 high-speed steel in commercial practice is 1200°C,[28] this temperature was chosen initially for the sintering treatment. Figure 9(*a*) is an optical micrograph of the splat-cooled material sintered for 5 h at 1200°C, and for comparison Fig. 9(*b*) is an optical micrograph of conventional material given the same heat treatment. It is obvious that partial melting has occurred in the sintered material, resulting in the solidification of a eutectic structure in the regions which became liquid at the sintering temperature. The remainder of the microstructure in Fig. 9(*a*) shows the formation of martensite plates in the austenite grains which did not melt at 1200°C. The black

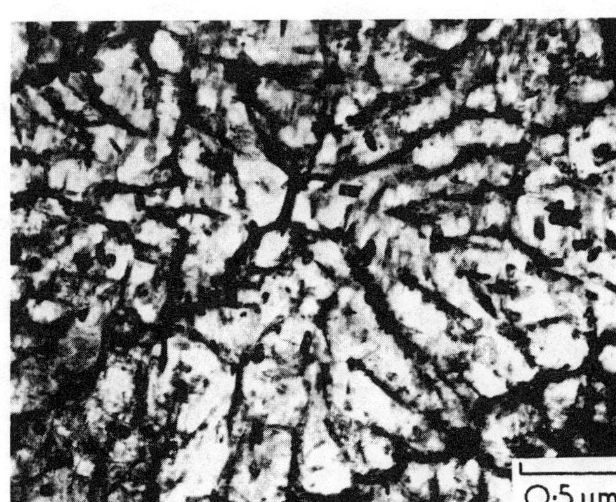

8 Splat-cooled M1 high-speed steel aged 1 h at 700°C, showing coarsening of the precipitate particles. Electropolished foil. Transmission electron micrograph.

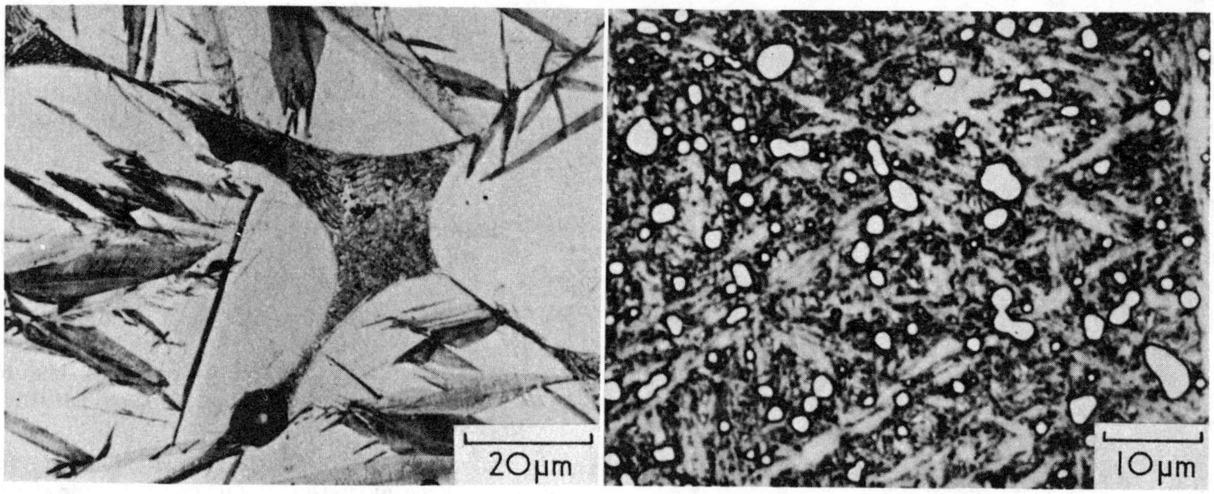

(a) (b)

9 Optical micrographs of M1 high-speed steel sintered 5 h at 1200°C: (a) splat-cooled material (etched in 15% sodium metabisulphite) and (b) conventional material (etched in ferric chloride and hydrochloric acid).

spheres within the eutectic network are voids formed by the thermal contraction of the eutectic liquid as it solidified. The microstructure of the conventional material (Fig. 9(*b*)) shows large undissolved carbide particles (principally M_6C, with a small amount of MC also) in a matrix of martensite and retained austenite.

The fact that partial melting occurred during the 1200°C sintering treatment reflects the increasing carbon supersaturation obtainable by splat cooling. In other words, although the overall carbon content is the same in both the splat-cooled and conventional material, the presence of undissolved carbides in the latter means that its matrix carbon content will be considerably less than in the former. Thus, since the effect of higher dissolved carbon contents is a lowering of the solidus temperature, the splat-cooled alloy will begin to melt at a lower temperature than its conventionally treated counterpart.

In order to overcome this problem a lower sintering temperature was chosen, and Fig. 10(*a*) shows the microstructure of a compacted splat-cooled sample sintered for 20 h at 1100°C. The microstructure from a piece of the conventional material given the same heat treatment is illustrated in Fig. 10(*b*). At this sintering temperature the microstructures of both specimens appear very similar, except for the morphological difference between the undissolved carbides. In the sintered specimen the carbides are mostly rod shaped. Selective etching techniques were used to identify the carbides, and Fig. 11 shows the results of using an electrolytic chromic acid etch on the splat-cooled and conventional specimens. This etchant attacks M_2C rapidly, MC less rapidly, and M_6C not at all,[29,30] and therefore it is apparent that the sintered splat-cooled specimen contains a mixture of M_2C, M_6C, and MC undissolved carbides while the conventional material contains only M_6C and MC.

An X-ray diffractometer trace was made of the splat-cooled and sintered compact in order to determine whether martensite and retained austenite were present in the same proportions as in the conventional material. Visual inspection of the trace of the sintered compact gave the

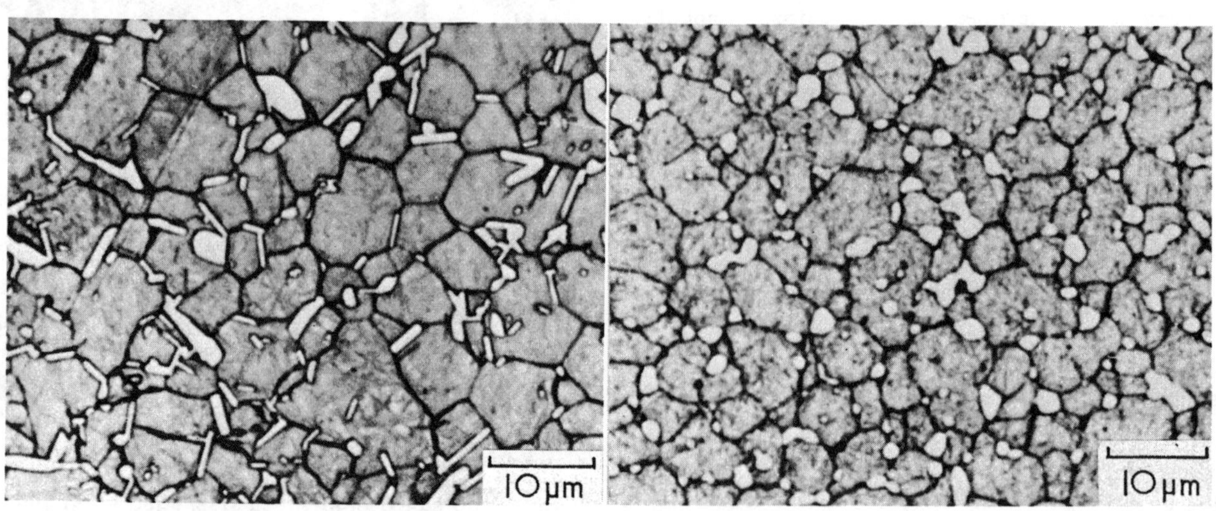

(a) (b)

10 Optical micrographs of M1 high-speed steel sintered for 20 h at 1100°C and etched in 15% sodium metabisulphite: (a) splat-cooled material, (b) conventional material.

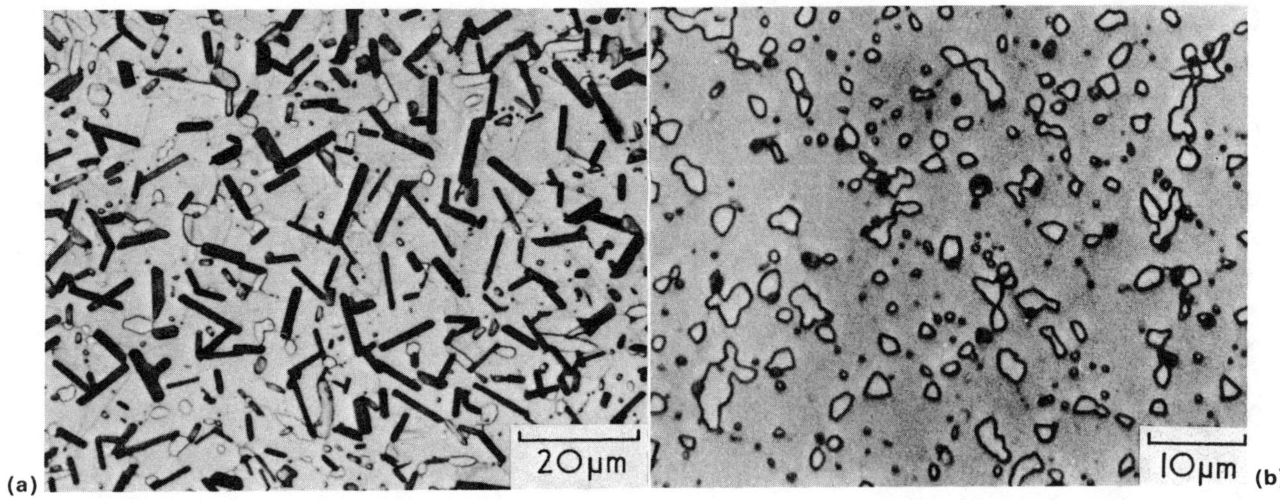

11 Optical micrographs of M1 high-speed steel sintered for 20 h at 1100°C and etched electrolytically in 10% chromic acid: (a) splat-cooled material, (b) conventional material.

following relative intensities of the phases detected: austenite – very strong, martensite or ferrite – strong, M_2C – strong. Hence, despite the long holding time at 1100°C, the austenite in the sintered specimen must still be richer in austenite-stabilizing carbon than that in the conventional material, since the austenite content of the latter is only 15 to 30%.[28]

As far as the mechanical properties of high-speed steels are concerned, undissolved M_2C is known to be deleterious. Bahmiller, in a private communication to Mukherjee and

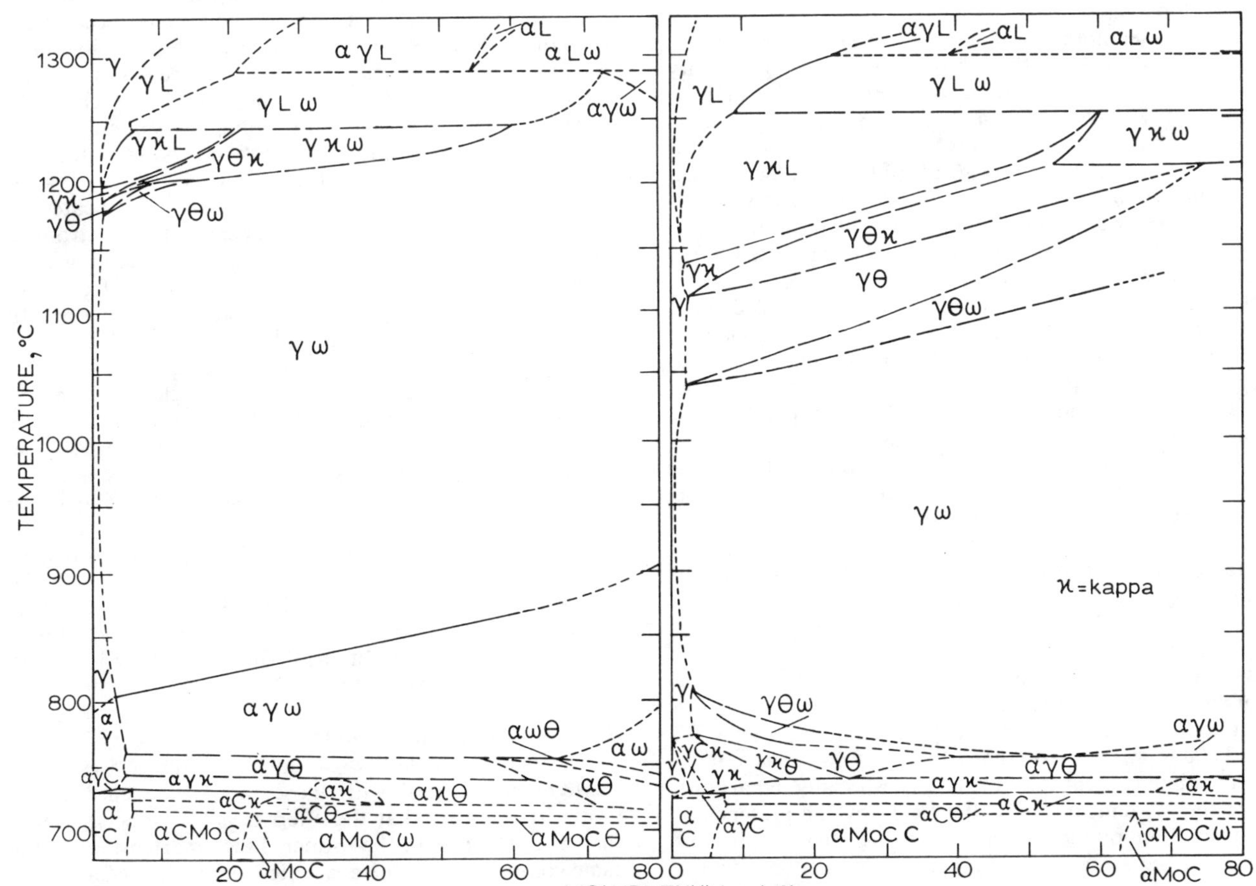

12 Vertical sections at constant carbon content of the ternary Fe–Mo–C system.[30] (a) section at 0·40%C. (b) section at 0·87%C. (γ = austenite, θ = Mo_2C, ω = M_6C).

(a) 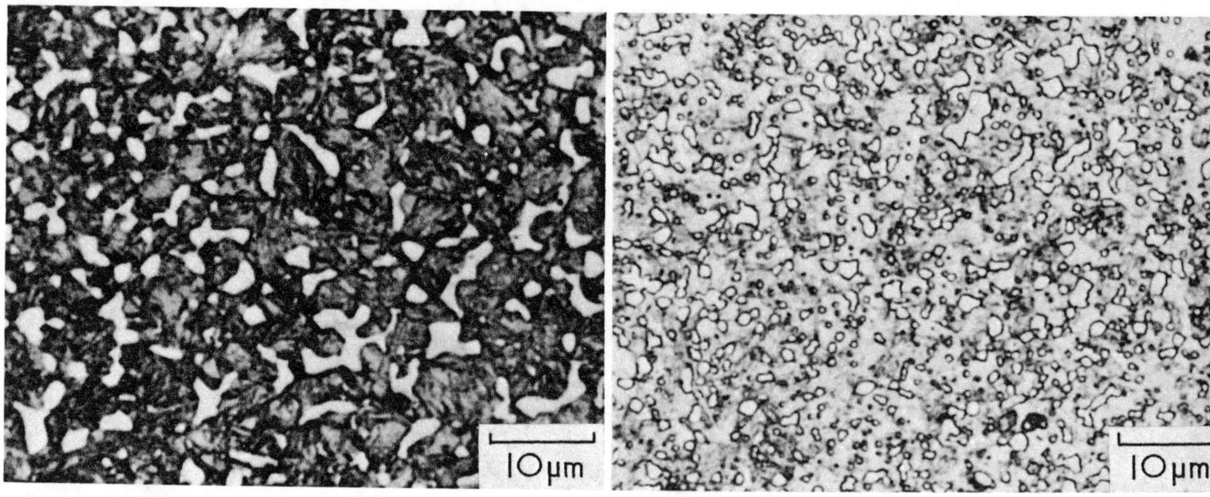 (b)

13 Optical micrographs of M1 high-speed steel sintered for 20 h at 1100°C and etched in 15% sodium metabisulphite: (a) splat-cooled material, (b) conventional material.

Dyson,[31] has reported that M_2C contributes to the lower strength, lower hardening response, and grain coarsening of high-speed steels, and that it should be avoided as one of the primary carbides present before tempering. Thus, in order to eliminate it from the microstructures of the sintered compacts in this investigation, yet another sintering temperature was sought.

Although equilibrium diagrams do not exist for this particular high-speed steel, those determined for ternary Fe–Mo–C alloys are of some help. Figure 12(*a*) and (*b*) shows vertical sections at 0·4%C and 0·87%C, respectively, for steels containing up to 8·0%Mo.[30] It is evident that at the higher carbon content (Fig. 12(*b*)) the three-phase field designated $\gamma\theta\omega$ (γ = austenite, $\theta = Mo_2C$, $\omega = M_6C$) is moved to lower temperatures relative to its position in Fig. 12(*a*). Therefore, in order to avoid M_2C formation, 1000°C was chosen as the next sintering temperature.

Figure 13(*a*) is a micrograph of the splat-cooled alloy sintered for 20 h at 1000°C, and Fig. 13(*b*) is the corresponding micrograph of the conventional material. The rod-shaped M_2C carbides are no longer present in the sintered compact and the microstructures are quite similar, although the undissolved carbides are apparently larger in the sintered material. An X-ray diffractometer trace was made of the sintered compact and, as before, austenite was the most predominant matrix phase. However, after the compact was cooled to −196°C and re-examined by X-ray diffractometry, the proportion of austenite had fallen to about 50%.

DISCUSSION

Nature of bcc product

Comparison of the lattice parameters of the austenite in the splat-cooled material and water-quenched filings (i.e. 3·637 and 3·606 Å, respectively) shows that the former has a much higher solute content than the latter (especially in carbon). Since the M_s temperature is depressed by the presence in solid solution of the principal alloying elements in this steel,[32,33] the austenite in the splat-cooled material would have a much lower M_s temperature, possibly below room temperature. This, coupled with the fact that no martensite was observed in any of the thin foils and by analogy with the results obtained with splat-cooled Fe–Mo–C alloys,[16] leads to the conclusion that the bcc product in the splat-cooled high-speed steel is δ-ferrite only.

In their splat-cooled high-speed steels, Arai and Komatsu[12] observed martensite, δ-ferrite, austenite, and carbide, but their results are not really comparable with those obtained in the present investigation. They assumed that the cooling rates in their work would have been of the order of 10^6 to 10^7 K s^{-1}, which is clearly not possible since the thickness of their splats was in the range 0·5 to 1·5 mm. According to Ruhl's theoretical predictions[34] such thicknesses would be consistent with cooling rates of only 10^3 to 10^4 K s^{-1}. In the present work, however, cooling rates were of the order of 10^6 to 10^8 K s^{-1} (Ref. 17), and splat thicknesses were as small as about 0·1 μm, as indicated by the existence of numerous areas transparent to electrons at 200 kV. Direct evidence of this difference in cooling rates comes from Arai and Komatsu's observation of carbides sufficiently large to see in the optical microscope. In contrast the carbides present in the splat-cooled high-speed steel in this study were much smaller in size, and could only be observed by electron microscopy. Therefore, martensite was observed in Arai and Komatsu's work because the cooling rate was sufficiently slow to permit considerable solute depletion of the austenite by carbide precipitation, with a concomitant rise in the M_s temperature and hence transformation to martensite.

The results of Jama and Thursfield[13] are more consistent with the present results because they observed only δ-ferrite and austenite in their tool-steel splats even though their experiments were carried out in air, which produces lower cooling rates than are obtained when quenching is carried out in an inert atmosphere.[35]

Solute supersaturations

Comparison of the lattice parameters of the phases in the splat-cooled material with those in the quenched filings (Table II) indicate that the δ-ferrite is less supersaturated than the austenite in the former. Since martensite has the same composition as the austenite from which it forms, the lattice parameters measured for the martensite and austenite in the quenched filings reflect the same degree of supersaturation, assuming that no autotempering occurred in the martensite phase. However, when the lattice parameters of the phases in the splat-cooled alloy and quenched filings are compared, the δ-ferrite and the martensite have the same lattice parameter in both materials, whereas the austenite in the splat-cooled material has a much larger lattice parameter than the austenite in the quenched filings. This can only mean that the austenite in the splat-cooled alloy is more supersaturated than the δ-ferrite. This difference in the degree of supersaturation is a result of the solidification sequence of the alloy. The first phase to form is the δ-ferrite which crystallizes as solute-lean solid. Later, austenite forms from the remaining solute-enriched liquid, and therefore it is more highly supersaturated.

Microstructures obtained on splat-cooling

It has been shown[16] that the solidification structure of the δ-ferrite in splat-cooled Fe–Mo–C alloys depends upon the degree of undercooling achieved relative to the equilibrium liquidus–solidus temperature range of the alloy. If this range is small, the undercooling may be sufficient to permit an alloy to solidify without segregation below its solidus temperature, thus maintaining the mean composition of the liquid. Alternatively, if the equilibrium liquidus–solidus temperature range is large, solidification will generally commence above the solidus and considerable solute segregation will occur. Because of the very large freezing range (solidus–liquidus gap) in high-speed steels, the degree of undercooling attainable with splat cooling will not, in general, be enough to permit 'diffusionless' solidification below the solidus temperature. Therefore, solidification will take place in the liquid plus solid phase fields, with the formation of alloy-lean dendrites and the rejection of solute (principally carbon, because of its much higher diffusivity) into the liquid.

The extent to which solute segregation will occur in any given area of a splat-cooled flake depends on the local cooling rate, for this will determine the amount of undercooling possible before solidification commences. If the local cooling rate is relatively low, the liquid will only be undercooled and solidification will commence not far below the liquidus. In this situation a large amount of solute segregation will occur and, as a further result of the slow cooling rate, precipitation at the interdendritic boundaries will have time to take place (Figs. 4 and 6). On the other hand if the cooling rate is relatively high, the liquid will be more undercooled and solidification will begin at a lower temperature, near the solidus. Hence, less segregation of solute will be possible and precipitation at the solute-enriched regions will not occur very readily, if at all (Fig. 3).

If the local cooling rate is very high, the undercooling may be sufficient to suppress the commencement of solidification to a temperature below the solidus, and 'diffusionless' solidification will take place without change of composition.[23] Such a situation has occurred in the area of Fig. 2; there has been no solute segregation during solidification, and the solid has formed with the mean composition of the liquid. That this area had suffered the highest cooling rate was evident from the fact that it was thinner than any other areas observed.

The presence of very fine carbides in Fig. 2 implies that the area in the micrograph cooled relatively slowly after it had solidified. By examining the underside of splat-cooled flakes in the scanning electron microscope, Wood and Honeycombe[36] have observed that very thin electron-transparent areas in splat-cooled foils are often associated with regions of 'lift-off' from the substrate. If the very rapidly-cooled area in Fig. 2 were such a region of lift-off, the observed carbide precipitation can be readily accounted for in the following way. Although the initial cooling rate in this very thin area was high enough to permit 'diffusionless' solidification of the liquid, subsequent solid-state cooling was sufficiently slow to allow the atomic diffusion necessary for the formation of alloy carbide particles to take place, because the area was not actually in contact with the conducting copper substrate. Ruhl[34] has shown that cooling rates do decrease during solid-state cooling, even when the heat transfer coefficient is infinite and ideal cooling obtains. This decrease will be far greater when direct thermal contact with the substrate is so poor that it is non-existent, as in the case of lift-off areas.

The small equiaxed grains of austenite (Fig. 5) must have grown in a more quickly cooled area than the dendrites of austenite (Fig. 6) because, unlike the dendritic regions, they did not contain precipitate particles at their grain boundaries. However, the fact that the grains are not elongated in the plane of the foil like the δ-ferrite grains in Fig. 2 confirms that they come from a thicker, more slowly cooled region than those ferrite grains. Nevertheless, the cooling rate has been high enough to suppress solute segregation effects, and so this structure must be seen as an intermediate stage between the most quickly cooled regions with grains growing in the foil plane, and the more slowly cooled areas exhibiting dendritic growth.

Microstructures obtained on ageing

The most important feature of the ageing studies on the splat-cooled high-speed steel was the precipitation in the matrix of the fcc vanadium-rich carbide MC instead of the usual hcp molybdenum-rich carbide M_2C. The reason for the difference in behaviour between the splat-cooled and conventional high-speed steels is that MC is one of the insoluble carbides which cannot be dissolved at the austenitization temperature in the latter,[2] and therefore it is not available for re-precipitation during tempering. But in the splat-cooled alloy all of the carbides are taken into solution in the liquid state, so that the carbides which can precipitate upon ageing will depend on thermodynamic and kinetic factors alone.

Because there is much more molybdenum than vanadium in this particular high-speed steel, it would be reasonable to assume that the precipitating alloy carbide would be M_2C, particularly in view of the fact that M_2C can dissolve appreciable amounts of vanadium.[37] However, the standard negative free energies of formation of Mo_2C and VC (the pure carbides which correspond to M_2C and MC, respectively) at 600°C are about 32 and 80 kJ mol^{-1}, respectively,[38] and so the formation of MC is more energetically favourable. Several previous studies of lower alloy steels containing both molybdenum and vanadium in solution have shown that MC is the phase which precipitates during tempering, even though the alloy compositions were richer in molybdenum than vanadium (see, for example, Refs. 26 and 29).

M_2C precipitation was observed at the interdendritic boundaries (Fig. 8), which implies that there had been more segregation of molybdenum than vanadium during solidification. However, this may simply reflect the greater molybdenum concentration in the overall alloy composition. In any case, some M_2C precipitation must occur, for the precipitation of all the vanadium as MC would not use up all the available carbon.

Some molybdenum will also take part in the matrix precipitation of MC, for it is known that replacement of vanadium atoms by molybdenum atoms in the MC lattice can occur to an appreciable extent. For example, from the measurement of the MC lattice parameters for various tempering treatments of a Mo–V steel, Baker and Nutting[26] found that the Mo : V ratio in MC can be as high as 1 : 1, while Clayton[40] measured the composition of MC in M2 high-speed steel by electron probe microanalysis and found that the carbide contained 11 wt-% molybdenum.

Sintering

The most significant features of the sintering work were the microstructural changes brought about by there being more carbon in solution in the as-splatted material than in its conventional counterpart. This led to sufficient depression of the solidus temperature of the splat-cooled alloy to permit partial melting upon sintering at 1200°C, and to the formation of M_2C instead of M_6C at 1100°C.

Arai and Komatsu[12] observed that the carbides which formed during splat cooling of their high-speed steels were generally M_2C and MC, and not the expected M_6C and MC. The replacement of M_6C by M_2C was attributed to the increased cooling rate, although the reason for the change was surmised as resulting from the formation of a new metastable phase (i.e. M_2C), analogous to the formation of the ε phase reported by Ruhl and Cohen[41] in splat-cooled Fe–C alloys. It is now apparent that this is not the case, and that the presence of M_2C is attributable to the higher concentration of carbon in solution which, as Fig. 12 indicates, broadens considerably the composition and temperature ranges over which M_2C can form.

Applications to high-speed steel technology

The results obtained with the splat-cooled M1 high-speed steel confirm that otherwise insoluble carbides can be dissolved in the liquid state and partially retained in solid solution upon rapid quenching to room temperature. The size of the carbide particles which do precipitate during solidification and cooling is sufficiently small to indicate the considerable advantage splat cooling has over the powder metallurgy processes developed recently for high-speed steel production. In order that this advantage is not destroyed, however, it is apparent from the sintering work that better consolidation techniques need to be used, which do not permit the growth of large carbide particles at high temperatures. Consolidation by hot forging or hot extrusion would allow lower temperatures to be used, with less accompanying solute supersaturation loss from the matrix than was found during the simple high-temperature sintering process. Much more striking advances are likely to be made if the consolidation stage can be combined with the rapid quenching process, as demonstrated by the plasma-jet spraying technique,[42,43] although even here further consolidation is necessary since, in aluminium–vanadium alloys, the deposit had a density of only 70% that of aluminium.

The most important feature of this study, however, is not simply that highly supersaturated solid solutions can be produced in conventional high-speed steel compositions, but that splat cooling in general is able to dissolve phases which are insoluble in the solid state, and to produce relatively homogeneous microstructures. Therefore, the major technological advance likely to be made by a splat cooling process is not that better tool steels will be produced from existing compositions, but that tool steels will be produced from totally new compositions which do not require the expensive alloying additions currently used. It has been stated[44] that the 'Achilles heel' of high-speed steels is the cost of their alloying elements, which generally amount to about 70 to 80% of the total cost of a tool-steel product. Hence the principal object of new development work must be to reduce the amount of these expensive alloying elements used and to maintain, at the same time, the desired mechanical properties.

Obviously many alloying elements, both singly and in various combinations, would have to be investigated before the most economically viable high-speed steel, with the desired mechanical properties, would be found. Consideration, however, would have to be given to the following factors when designing a new high-speed steel composition for production by a splat-cooling process.

(i) The matrix phase after quenching from the liquid state to room temperature should be as close as possible to 100% martensite. If the matrix contains δ-ferrite the degree of hardness ultimately obtained during tempering will be less than that obtainable by tempering of martensite, as shown by Arai and Komatsu[12] in their work on splat-cooled high-speed steels. The only way of subsequently reducing the amount of δ-ferrite would be a long high-temperature treatment during which time much of the initial solute supersaturation will be lost. If the matrix contains retained austenite, multiple tempering treatments will be required to deplete the austenite of solute and permit transformation to martensite on re-cooling. Furthermore, the austenite may be so supersaturated that abnormally long tempering times will be needed to induce this conditioning process. In order to inhibit the formation of δ-ferrite and the retention of untransformed austenite by splat cooling, careful balancing of the ferrite-stabilizing elements with austenite-stabilizing carbon would be required.

(ii) The liquidus–solidus temperature range should be small so that the undercooling will be sufficient to permit 'diffusionless' solidification below the solidus, thus avoiding solute segregation effects. Unfortunately the compositional requirements which favour a small liquidus–solidus gap may be incompatible with the requirements of having sufficient alloying elements present to produce the necessary volume fraction of precipitate, and hence desired mechanical properties.

(iii) The composition should be balanced stoichiometrically for the carbides which will form during tempering in order to ensure the complete utilization of the substitutional alloying elements and carbon, leaving sufficient excess of the former for solid-solution strengthening. Balancing of the composition in conventional high-speed steels has been shown to provide optimum secondary hardening characteristics and resistance to tempering.[45]

(iv) Consideration should be given to the addition of very strong carbide-forming elements, such as tantalum and niobium, which raise the hardness peak during secondary hardening and reduce the rate of overageing and subsequent softening[46] in steels containing only one species of precipitating alloy carbide. In a recent study of this phenomenon, Dunlop[47] found that although small quaternary additions of titanium to Fe–V–C alloys did produce this effect, additions of niobium gave anomalous results. This was attributed to the inability of solution treatment to dissolve all the niobium-rich carbide particles. The splat-cooling process provides the opportunity to obtain complete dissolution of even the most insoluble (in the solid state) carbides, and therefore the very desirable effect of small additions of the extremely strong carbide-forming elements can be utilized to the full.

CONCLUSIONS

1. Splat cooling of M1 high-speed steel produces supersaturated solid solutions of δ-ferrite and austenite,

with the precipitation of particles at the interdendritic boundaries in thicker sections. The austenite is more supersaturated than the δ-ferrite as a result of solute partitioning during solidification, and its M_s temperature is depressed to a level near or below room temperature.

2. The morphologies adopted by the two-matrix phases in this alloy reflect the local cooling rate, and range from segregation-free grains at the highest cooling rates, through dendritic structures exhibiting solute segregation but no carbide formation, to dendritic structures containing considerable amounts of carbide at the solute-rich interdendritic boundaries.

3. Retained austenite in the splat-cooled high-speed steel transforms only after prolonged isothermal ageing at temperatures in excess of 550°C. The solute supersaturation in the δ-ferrite is relieved by the precipitation of MC on {100} planes, satisfying the Baker–Nutting[26] orientation relationship between V_4C_3 and ferrite. At the same time M_2C precipitates at the interdendritic boundaries.

4. During sintering of the splat-cooled high-speed steel at 1200°C, partial melting occurs as a result of the solidus temperature depression produced by the high dissolved-carbon content. The high carbon concentration retained in solid solution by splat cooling favours the precipitation of M_2C during sintering at 1100°C. Sintering at 1000°C produces large M_6C particles dispersed in a matrix of martensite and retained austenite.

5. The splat cooling process can retain higher concentrations of alloying elements in solid solution than conventional solution treatment plus quenching processes. It therefore offers the possibility of producing entirely new tool-steel compositions, perhaps with less expensive alloying additions than those currently used.

ACKNOWLEDGEMENTS

The authors wish to thank Osborn Steels Ltd, Sheffield, for providing the M1 high-speed steel used in this investigation. One of us (IRS) is grateful for the award of a CSIRO Postgraduate Studentship.

REFERENCES

1. J. T. BERRY: 'Recent developments in the processing of high-speed steels'. **1970:** Greenwich, Conn. (Climax Molybdenum).
2. G. HOBSON and D. S. TYAS: *Met. Mater.*, 1968, **2**, 144.
3. F. A. KIRK, H. C. CHILD, E. M. LOWE, and T. J. WILKINS: *J. Iron Steel Inst.*, 1971, **209**, 606.
4. A. FISCHER and E. KOHLHAAS: *Prakt. Metallogr.*, 1975, **12**, 393.
5. S. B. BRANDSTEDT: *Precis. Met.*, 1969, April, 52.
6. A. R. JOHNSON and T. C. HAMAKER: *Met. Eng. Q.*, 1969, **9** (3), 6.
7. R. JOHANSSON and S. E. ISAKSSON: *Powder Metall. Int.*, 1970, **2**, 49.
8. E. J. DULIS and T. A. NEUMEYER: 'Materials for metal cutting', 112. **1970:** London (Iron Steel Inst.).
9. P. HELLMAN, H. LARKER, J. B. PFEFFER, and I. S. STROMBLAD: 'Modern developments in powder metallurgy' (ed. H. H. Hausner), vol. 4, 573. **1971:** New York (Plenum).
10. E. HORN: *Prakt. Metallogr.*, 1976, **13**, 221.
11. L. RADEMACHER and H. W. MÜLLER-STOCK: *ibid.*, 1977, **14**, 181.
12. T. ARAI and N. KOMATSU: *Tetsu-to-Hagané*, 1972, **58**, 899.
13. S. A. B. JAMA and G. THURSFIELD: Tube Investments Research Laboratories, Report No. 322, 1972.
14. J. V. WOOD: PhD dissertation, University of Cambridge, 1974.
15. H. JONES: *Mater. Sci. Eng.*, 1969, **5**, 1.
16. I. R. SARE and R. W. K. HONEYCOMBE: *J. Mater. Sci.*, 1978, **13**, 1991.
17. J. V. WOOD and I. R. SARE: 'Proc. 2nd Int. Conf. on Rapidly Quenched Metals' (ed. N. J. Grant and B. C. Giessen), Section 1, 87. **1976:** Cambridge, Mass. (MIT Press).
18. D. J. DYSON and B. HOLMES: *J. Iron Steel Inst.*, 1970, **208**, 469.
19. N. RIDLEY: *ibid.*, 1971, **209**, 396.
20. K. W. ANDREWS, D. J. DYSON, and S. R. KEOWN: 'Interpretation of electron diffraction patterns', 2nd edn. **1971:** London (Hilger).
21. D. RAYNOR, J. A. WHITEMAN, and R. W. K. HONEYCOMBE: *J. Iron Steel Inst.*, 1966, **204**, 349.
22. J. WOOD and I. SARE: *Metall. Trans. A*, 1975, **6**, 2153.
23. H. BILONI and B. CHALMERS: *Trans. Metall. Soc. AIME*, 1965, **233**, 373.
24. G. A. CHADWICK: 'Metallography of phase transformations'. **1972:** London (Butterworths).
25. I. R. SARE: PhD dissertation, University of Cambridge, 1975.
26. R. G. BAKER and J. NUTTING: 'Precipitation processes in steels', Publ. No. 64, 1. **1959:** London (Iron Steel Inst.).
27. W. PITSCH and A. SCHRADER: *Arch. Eisenhüttenwes.*, 1958, **29**, 715.
28. J. T. BERRY: 'High-performance high-hardness high-speed steels'. **1970:** Greenwich, Conn. (Climax Molybdenum).
29. D. J. BLICKWEDE, M. COHEN, and G. A. ROBERTS: *Trans. ASM*, 1950, **42**, 1161.
30. R. F. CAMPBELL, S. H. REYNOLDS, L. W. BALLARD, and K. G. CARROLL: *Trans. Metall. Soc. AIME*, 1960, **218**, 723.
31. T. MUKHERJEE and D. J. DYSON: *J. Iron Steel Inst.*, 1972, **210**, 203.
32. W. STEVEN and A. G. HAYNES: *ibid.*, 1956, **183**, 349.
33. K. W. ANDREWS: *ibid.*, 1965, **203**, 721.
34. R. C. RUHL: *Mater. Sci. Eng.*, 1967, **1**, 213.
35. C. H. JANSEN: PhD dissertation, Massachusetts Institute of Technology, 1971.
36. J. V. WOOD and R. W. K. HONEYCOMBE: *J. Mater. Sci.*, 1974, **9**, 1183.
37. G. STEVEN, J. J. HAUSER, T. A. NEUMEYER, and J. M. CAPENOS: *Trans. ASM*, 1969, **62**, 180.
38. J. F. ELLIOTT and M. GLEISER: 'Thermochemistry for steelmaking', vol. 1. **1960:** Reading, Mass. (Addison-Wesley).
39. D. RAYNOR: PhD dissertation, University of Sheffield, 1967.
40. D. B. CLAYTON: *Metalurgia*, 1965, **71**, 247.
41. R. C. RUHL and M. COHEN: *Acta Metall.*, 1967, **15**, 159.
42. M. MOSS, D. L. SMITH, and R. A. LEFEYER: *Appl. Phys. Lett.*, 1964, **5**, 120.
43. M. MOSS: *Acta Metall.*, 1968, **16**, 321.
44. B. FREDRIKSSON: 'Molybdenum 1973', 69. **1974:** Toronto (Noranda).
45. G. STEVEN, A. E. NEURENBERG, and T. V. PHILIP: *Trans. ASM*, 1964, **57**, 925.
46. A. K. SEAL and R. W. K. HONEYCOMBE: *J. Iron Steel Inst.*, 1958, **188**, 343.
47. G. L. DUNLOP: PhD dissertation, University of Cambridge, 1974.

Ingots about to be preheated prior to reduction at Alcan's Oswego facility. Many of the newer aluminum alloys require extremely careful process control to preserve critical microstructures. Overheating and overaging lead to recrystallization that can ruin uniformity. In some cases, process restrictions can add 25% to 100% premium to the price of these alloys.

Many of the new grades of aluminum emerging today are high-purity offshoots of conventional alloys, such as this Alcan 7010-T73651 plate. In this potential structural aircraft alloy, iron and silicon content are held to extremely low tolerances compared to parent Alloy 7050. Benefits of the new alloy include improved toughness up to 350°F.

New aluminum alloys sighted on the horizon

Rising stars promise high strength-to-weight ratios and excellent corrosion resistance

By John C. Bittence, Senior Editor

Aluminum alloys can be made 20% stronger and 10% to 15% lighter than the best aluminum grades commercially available today. Based on these figures, aluminum could probably beat high-strength steels in the race for automotive applications and even outperform composites in some aircraft applications, according to aluminum producers.

And that's exactly what many researchers are hoping for.

New, complex alloy systems, some of which are made from rapidly solidified powder, promise to boost the room-temperature yield strength of aluminum to at least 60,000 psi (414 MPa). High-temperature performance goals are even more ambitious. Look for new grades that can endure 450°F (232°C) in the near future. A bit further down the road will be alloys that maintain 35,000-psi yield strength (241 MPa) and 18,000-psi 0.2% creep strength (124 MPa) for 100 hr at 650°F (343°C). In addition, expect better fatigue life, corrosion resistance, and notch fracture toughness in tomorrow's aluminum alloys.

Primarily, researchers are examining two ways of achieving these goals. The first, based on conventional "ingot" metallurgy, involves developing new binary and tertiary alloys. A great deal of this work has centered around the aluninum-lithium alloy system. The second thrust in aluminum-alloy research is devoted to the newly emerging rapid-solidification technology. Techniques are now being developed to roll, forge, or extrude parts and raw material from alloy powders produced by rapid solidification.

Lithium and others

Aluminum alloys containing substantial amounts of lithium are not easy to process. For this reason, Al-Li alloys were

Machined aluminum panels at Reynolds Metals Co. will be used for the Space Shuttle fuel tank. Such applications typify today's requirements for high-performance aluminum alloys. To meet aerospace demands, researchers are seeking new alloys with better fatigue life, higher strength-to-weight ratio, better corrosion resistance, high strength up to 650°F, and better notched fracture toughness.

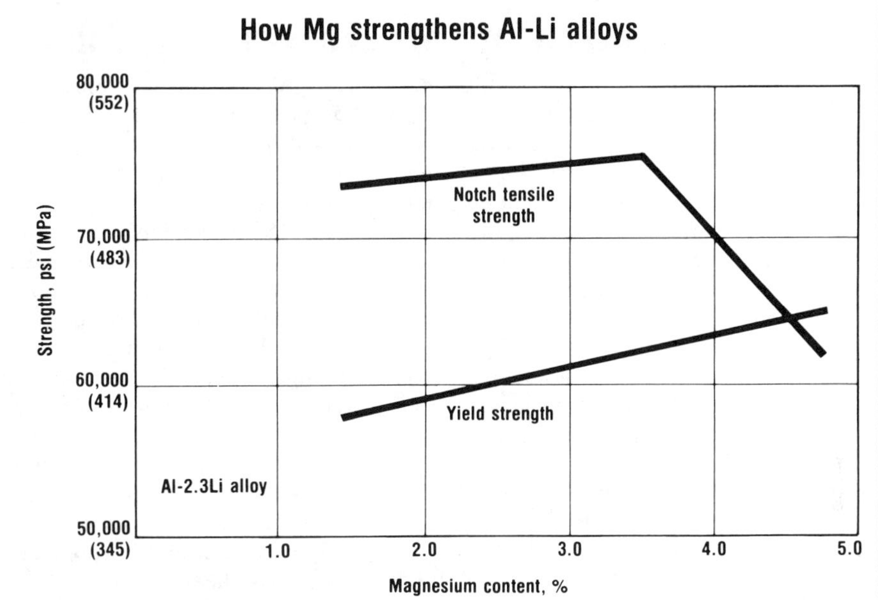

considered impractical a decade ago. For instance, lithium increases the oxidation rate of molten aluminum and absorbs hydrogen into the metal during processing. Both oxygen and hydrogen are detrimental contaminants in aluminum alloys. In addition, an Al-Li cast ingot is prone to cracking as it is reduced down to mill forms. Such cracks can be extremely serious in aircraft applications, making these alloys inherently unreliable.

On the other hand, lithium, the lightest naturally occurring metal, reduces the density of aluminum while increasing its stiffness and elastic modulus. The weight-saving potential of Al-Li alloys could cut in half the fuel used in an average aircraft according to a Battelle-Columbus study. And even though the cost premium for these alloys is currently between 25% and 100% — primarily due to the restrictive chemistry and processing requirements — the Al-Li metals are still cheaper and more efficient than composites.

To overcome the difficulties of producing Al-Li alloys, producers are developing vacuum melting and processing techniques. In one Reynolds Metals process, the alloy is melted in a vacuum-degassed argon chamber and poured into an argon-atmosphere mold.

Additional metals are added to the alloy to help reduce cracking and enhance certain mechanical properties. While metals such as Ga, Sc, and Ag are used in some experimental alloys, more common materials include Cu, Mg, Mn, Co, Cd, Zn, and Ce. A recent study at Reynolds, in fact, concluded that the Al-Li system will probably need at least one additional element, possibly copper, in order to be successful.

Rapidly solidified PM

While conventionally cast ingots cool slowly (at rates measured in hours), powders, under proper conditions, can be cooled at rates approaching millions of degrees per second. For aluminum, this rapid solidification can "trap" a greater amount of alloying metal in solution. For example, 500 times more cobalt and about 200 times more iron can be alloyed with aluminum by rapid solidification. In addition, elements that would not normally alloy with aluminum under conventional conditions can be coaxed into alloy systems by rapid solidification. By normal ingot metallurgy, approximately eight elements are used in alloy systems with aluminum. By rapid solidification, at least 20 elements can be used. Many of these elements are present as dispersions of high-modulus intermetallics and solutes that help strengthen the metal.

While techniques are currently under development for continuously producing extrusions and other forms directly from rapidly solidified powder, most alloy powders today are processed by isostatic pressing. In a typical sequence, powders are cold-isostatic pressed to about 80% theoretical density and subsequently hot-reduced to essentially 100% density. In some processes, cold-pressed billets are hot-extruded; in others, preforms are hot-die forged into net-shape parts.

Even conventional alloys have better properties when produced by powder techniques. For example, high-ratio ex-

Source: Materials Engineering, June 1980, 32-35

Aluminum alloys worth waiting for

While research into unusual aluminum alloy systems has been under way in Europe for a decade now, development projects in this country are relatively new. Concerns over the energy shortage, uncertain titanium supplies, plus increasingly favorable economics of powder metallurgy have finally encouraged research into new aluminum alloys in the U.S. Although dozens of different grades are under development, a handful appear to be most promising. Some of them are described here:

Superpure 7XXX series — Designated 7150, this alloy is an offshoot of Alloy 7075. Chemical purity is held to much closer tolerances at the high end of the limits for 7075, and tramp elements are held to a minimum. The result is 10% increase in strength, better corrosion resistance, and lower fatigue-crack growth rate. Fatigue strength is equivalent to the parent alloy. This new grade is currently being produced in limited quantities of sheet, plate and extrusion by Alcoa to Boeing specifications. Produced in either of two thermal treatments, T651 and T6511, the new grade carries a cost penalty of at least 25% over standard 7075-T6.

And another 7XXX — Designated 7010, another high-purity alloy is being developed by Alcan as an improvement over 7050. Content of the new grade is 6.3% Zn, 2.47% Mn, 1.8% Cu, 0.13% Zr, with minimum amounts of Fe, Si, Ti, Cr, Mn and other impurities. Si and Fe are especially minimized to boost toughness. This alloy, which maintains good properties at temperatures up to 350°F (177°C), is being evaluated for possible aircraft applications. Yield strength at room temperature, 250° and 350°F (121°, 177°C) for 2-in.-thick (50mm) plate is 63,200, 55,900, and 46,000 psi (434, 384, 371 MPa) respectively. Elongation drops from 15% to 13% in the longitudinal direction. These values are measured after T73651 treatment.

Two 2XXX hopefuls — Designated 2224 and 2324 by prime developer, Boeing Commercial Airplane Co. and supplier Alcoa, these alloys are designed to offer several improvements over conventional 2024-T3. Generally, strength, fatigue resistance, and fracture toughness are better, while fabricability and corrosion resistance are equivalent to 2024. The 2224 grade, after T3511 treatment, is processed into extrusions, while 2324, after T39 treatment, is produced as sheet and plate.

Both grades require special process control to maintain an unrecrystallized microstructure, which is necessary for the improved properties. Consequently, the alloys are from 25% to 100% more costly than 2024. Thermal treatment T39 involves a number of steps: First, the metal is solution treated and quenched, then pre-aged, cold rolled, stretched, and naturally aged.

A "pinch" of magnesium — Magnesium is an effective strengthener of aluminum. A number of Mg-bearing aluminum alloys are under development, although those discussed here have not as yet received an official designation. Magnesium, added to Al-Li alloys, provides solid-solution strengthening. In additions up to about 3.5% Mg, notch tensile strength and yield strength increase with increasing amounts of Mg. Above that point, notch strength begins to drop. Toughness of Al-Li-Mg alloys produced to date is lower than conventional high-strength alloys, but is expected to improve as high-purity processing techniques are developed.

Without Li, the Al-Mg alloy system can be effectively strengthened by thermomechanical processes. Alloys containing 10% Mg, or 10% Mg plus 0.5% Cu, subjected to solution treatment and warm isothermal rolling at 570°F (300°C) with reductions of 8:1 or more, show substantial property improvement. An additional cold rolling step further increases tensile strength. Unfortunately, careful process control is necessary. Recrystallization during treatment could destroy homogeneity and uniformity of the microstructure and diminish most of the alloy's properties.

Squeezing in strength — In an experimental process called strengthening by fractional melting, or SFM, strength of conventional 7XXX alloys can be boosted by squeezing. In this process, cast ingots containing a larger amount of solute than required in the final microstructure are subjected to a moderate pressure over a ceramic filter in an oven. As the temperature increases to just over the solidus, partial melting occurs. The excess solute is forced out of the interdendritic spaces of the ingot and drips through the filter. The remaining metal is the billet, which is subsequently worked and heat treated.

The results of this strange technique are impressive: Ultimate strengths were increased from 36,000 psi (250 MPa) to as high as 99,000 psi (692 MPa) and yield strength from 11,000 psi to 92,000 psi (100 to 643 MPa). Elongation is 10% after SFM. Improved strength and ductility are caused by extremely homogeneous microstructure with high solute content and low residual second phase.

Rapid-quenched PM — A high-temperature alloy consisting of Al, Mg, Zn, Cu and Co is produced by hot pressing powder that has been quenched at 10^5 to 10^6 degrees/sec. The 3300-lb (1500 kg) billets are then extruded at a 12:1 reduction or hot-die forged into net-shape parts. Designated CT90 and CT91 by Alcoa and U.S. Army Armament R&D Command, these alloys are scheduled for pilot manufacture into plant-scale extrusions in the near future. Other tests are being carried out to evaluate these alloys as replacements for titanium turbine fan blades.

PM with iron — A series of Al-Fe alloys containing Cr, Ni, Co, or Ce, made from rapidly quenched powder, shows excellent high-temperature strength. In cooperation with Air Force Materials Labs, Alcoa is evaluating 14 such alloys. Most promising were Al-8.2 Fe-1.8 Co and Al-8.0 Fe-3.4 Ce. These alloys retain 26,600 psi and 25,500 psi (182 and 175 MPa) ultimate tensile strength at 650°F (343°C). As yet unidentified intermetallic phases in the microstructure are believed to contribute to the high-temperature strength and creep resistance of these alloys. Parts are made by cold pressing pre-forms to 80% density, then hot pressing to 100% density at either 600°F or 700°F (316° or 371°C). For testing, discs are hot upset to 40% reduction in height.

PM with copper — Alloys in the Al-Cu-Mn and Al-Cu-Mg systems have good corrosion resistance, tensile strength, and notched fatigue resistance. When produced from rapidly quenched powders, parts from these alloys are suitable for intermediate-temperature structural service. Now in early stages of development by NASA and Lockheed-California, these alloys are designed as replacements for presently available aluminum alloys for supersonic cruise vehicles.

Preliminary data from the labs

	Yield strength, 1000 psi (MPa)	Ultimate tensile strength, 1000 psi (MPa)	Elongation, %
Ingot products			
Al-2.8 Li	65.7 (453)	n.a.*	n.a.
Al-3.5 Mg-2.28 Li	62.6 (521)	n.a.	n.a.
Al-4.9 Mg-1.8 Li	53.3 (367)	n.a.	n.a.
7075 SFM (Strengthened by fractional melting)	90 (620)	96.4 (664)	10
7010-T73651 (2-in. plate, longitudinal)	63.2 (435)	72.3 (498)	15
Al-8.3 Mg-0.4 Mn-0.1 Cr -0.12 Fe (Plate, hot-rolled, stretched, welded)	30 (207)	n.a.	n.a.
PM products			
Al-5.4 Zn-2.5 Mn-1.0 Cu-2.3 Co (Hot extruded at 700°F, T6 treatment at 870°F)	90 (620)	96.4 (662)	5.3
Al-8.8 Zn-2.5 Mn-1.0 Cu 0.8 Co (Hot extruded at 650°F, T6 treatment at 870°F)	103 (710)	106.5 (734)	6.7
Al-8 Fe-3.4 Ce (Forged at 600°F)	82 (566)	87.2 (600)	3
Al-8.2 Fe-1.8 Co (Forged at 600°F)	81.3 (559)	83.7 (579)	1
Al-10 Mg-0.5 Cu (Warm isostatic rolled)	n.a.	75 (520)	12
Al-10 Mg-0.5 Cu (Same as above with final cold reduction)	n.a.	90 (620)	6

**n.a. = not available*

trusions (around 45:1) from prealloyed powders of 7XXX and 2XXX alloys have 25 to 30% higher minimum tensile strengths than conventional ingot-metallurgy materials. An experimental 7XXX alloy with 0.8% Co has 40% higher notched fatigue strength than conventional 7050 alloy ingot at 10^7 cycles. Fatigue-crack growth is slightly lower for the PM alloy as well, primarily because high residual stresses in the nonstress-relieved PM structure help prevent cracks from growing.

The accompanying table and data, compiled by *Materials Engineering* from numerous sources, reviews some of the most promising aluminum alloys under development. ■

Source: Materials Engineering, June 1980, 32-35

Aluminum-Lithium Alloys: New Materials for Tomorrow's Technology

by T. H. Sanders, Jr. and E. A. Starke, Jr.

Recently developed alloys of aluminum and lithium, the lightest metal of all, offer a combination of desirable properties of special interest for aerospace applications.

Editor's Note: The following article describes an outstanding contribution to the technology of aluminum-lithium alloys. To give the readers of FOOTE PRINTS some idea of what a full-dress technical paper looks and reads like, and to provide the metallurgists among our readership with data on some recent leading-edge developmental work in their field, we are presenting the Sanders/Starke report exactly as received from the authors. For those unfamiliar with the discipline of metallurgy and its often necessarily arcane language, we have prefaced the article with a summary elucidating the gist of the paper and appended a short glossary explaining some of its difficult terms.

Reprinted with permission from Foote Prints, Vol. 44, 1981, 11-17, © 1981 Foote Mineral Company

(Title picture) The Navy's most sophisticated reconnaissance aircraft, the RA-5C Vigilante incorporates an aluminum-lithium alloy in its structure. Developed and produced by the Columbus Aircraft Division of North American Rockwell Corporation (NR), the Mach 2 twin-engine multi-sensor aircraft is capable of all-weather reconnaissance missions involving high-altitude supersonic or very low altitude high-speed penetrations. Primary mission of the Vigilante is tactical reconnaissance.

Photo courtesy of Rockwell International

SUMMARY

It doesn't take much lithium—only 2 to 3 weight percent—to significantly improve aluminum's specific modulus and tensile strength. However, lithium additions decrease the ductility of aluminum alloys, and this property must be improved before they can be commercially utilized. The mechanisms associated with the low ductility are being identified, and improvements are being made by selected alloying and processing. These recent advances make Al-Li-X alloys very attractive for aircraft structures in a world plagued with increasing fuel consumption and decreasing reserves.

Introduction

Recent aerospace system trade-off studies have shown that low density and high modulus are the material properties most desirable for weight reduction of future aircraft.[1] The desired combination of properties cannot be obtained with presently used aluminum alloys and processing methods, and extensive research is currently being focused on methods of improving the specific modulus of these materials.[2,3] However, one stringent requirement is that improvement in specific modulus not be accompanied by a significant degradation of other properties.

Lithium is the lightest metal found in nature and its addition to aluminum significantly reduces density. Aluminum-lithium alloys offer interesting combinations of low density, high elastic modulus, and high strength; properties which are especially attractive for aerospace applications. However, these alloys have also shown commercially unacceptable ductility and fracture toughness values.[4,5]

Lithium Additions—Beneficial and Adverse Effects

The high elastic modulus and yield strength, as well as the low ductility and fracture toughness of aluminum-lithium alloys, are a consequence of the metastable, ordered, and coherent δ' phase (Al_3Li). When aluminum-lithium alloys are quenched from the single-phase region and subsequently aged in the two-phase field, decomposition of the supersaturated solid solution occurs by the homogeneous precipitation of δ' throughout the matrix (Figure 1). During plastic deformation, the coherent δ' particles are sheared by moving dislocations, indicated by the arrows in Figure 1. The strength increase associated with δ' is due to the magnitude of their resistance to dislocation shearing. This depends on the degree of order of the particles, coherency strains, modulus difference between the matrix and the particle, and surface, chemical and/or stacking fault energy effects.[6] In any event, the

Source: Foote Prints, Vol. 44, 1981, 11-17

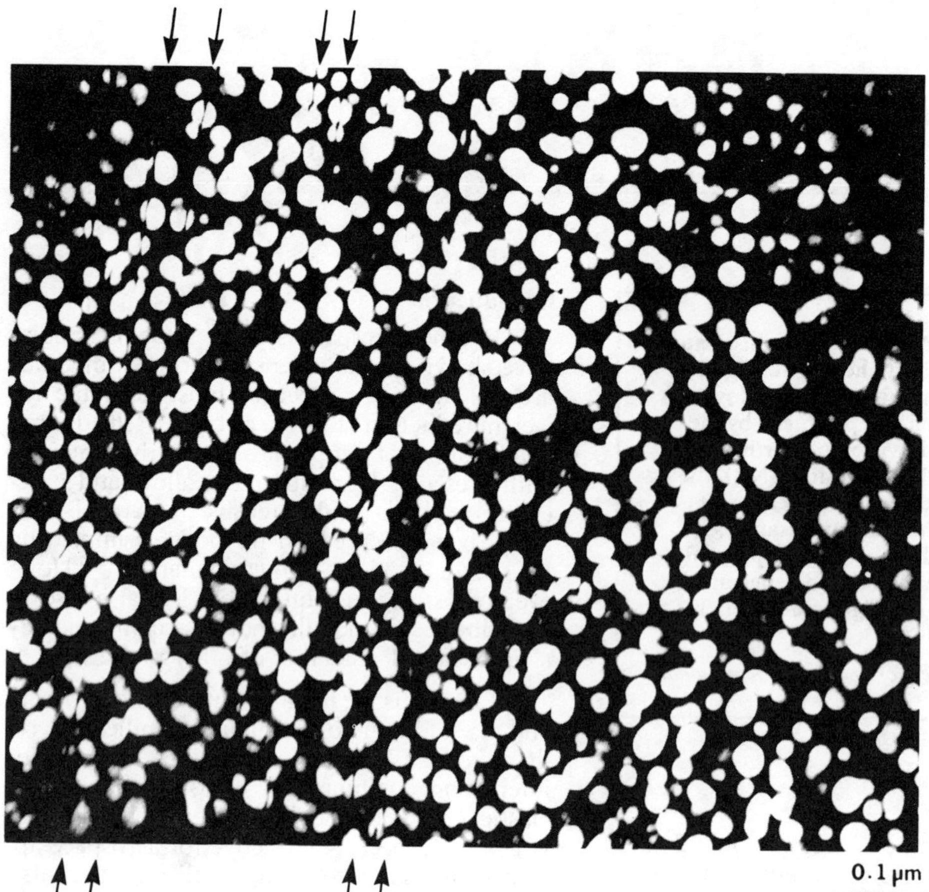

Figure 1. Dark field transmission electron micrograph (TEM) showing shape of Al_3Li precipitates. Arrow indicates where coherent precipitates have been sheared by moving dislocations.

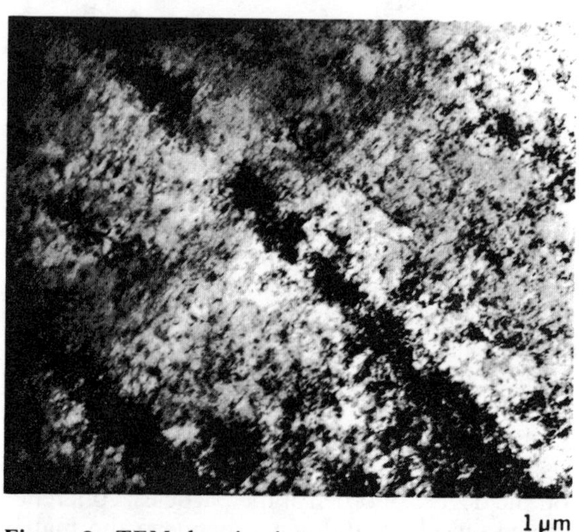

Figure 2. TEM showing intense deformation bands in a deformed aluminum-lithium binary alloy.

associated strength is reduced when the coherent, ordered Al_3Li particles are sheared by dislocations.[7] This results, successively, in a local decrease in resistance to further dislocation motion and a concentration of slip (Figure 2). Such concentrated slip results in premature fracture in the intense slip bands, or produces stress concentrations across grain boundaries which enhance intergranular fracture (Figure 3).

Although Al-Li alloys obtain their beneficial mechanical properties from homogeneous precipitation of δ', heterogeneous precipitation of the equilibrium δ (AlLi) phase is always possible below the solvus temperature, and occurs concurrently with the homogeneous precipitation. Since all boundaries may lower the surface and strain energies associated with the precipitate, nucleation of the stable incoherent phase occurs preferentially at grain boundaries. In aluminum-

Figure 3. Scanning electron micrograph (SEM) of aluminum-lithium binary showing intergranular fracture resulting from strain localization in intense slip bands.

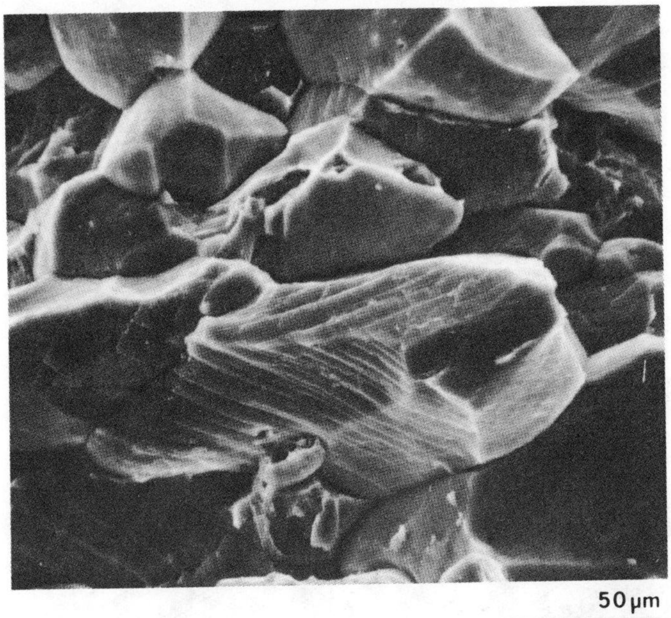

lithium alloys, this results in a solute-depleted precipitate-free zone (PFZ) adjacent to the grain boundary. This zone is weaker than the matrix and can be the site for preferential deformation which leads to high stress concentrations at grain boundary triple points. Cracks can then nucleate either at these triple points or at grain boundary precipitates, and propagate intergranularly within the PFZ (Figure 4).

Improving Ductility by Minimizing Strain Localization

The degree of softening that occurs when dislocations shear the coherent precipitate, and the magnitude of the stress concentrations produced at grain boundaries due to the intense slip bands and deformation within the precipitate free zone, all depend on the number of dislocations that can be accommodated on a given slip

Figure 4.
(a) TEM showing presence of PFZ and GBP in aluminum-lithium binary alloy,

(b) SEM of associated intergranular fracture.

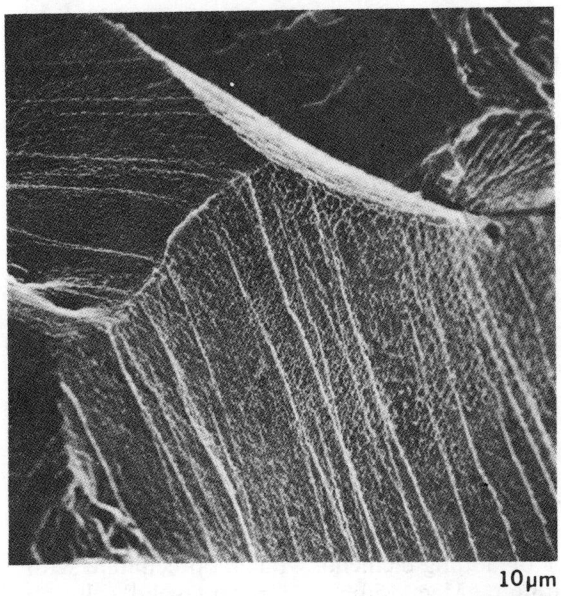

plane. Consequently, reducing the grain size is a very effective method of preventing early crack-nucleation due to strain localization.[7] This reduces the slip distance, and lowers the stress concentrations across grain boundaries and at grain boundary triple points. The fracture mode can likewise change from a low-energy intergranular to a higher-energy transgranular mode. A small grain size can be achieved, for example, by thermal mechanical treatments[8], or by applying a powder metallurgy process.[1] A current research and development program,[2] supported by DARPA, is investigating the use of rapid solidification techniques for improving the mechanical properties of aluminum-lithium-X alloys. The rapid solidification approach is designed to reduce the grain size; reduce or eliminate segregation effects; extend the solubility of lithium; and refine the dispersoid particle size.

Some commercial alloys have alloying additions (for example manganese and chromium in 2XXX and 7XXX aluminum alloys, respectively) that form small (0.1 to 0.2 μm) incoherent dispersoids during solidification and high-temperature homogenization treatments.[9] These small intermetallic compounds are primarily used for controlling the grain size and shape. However, they also disperse slip and inhibit the formation of intense slip bands. Therefore, plastic deformation is more homogeneous, and early crack nucleation due to intense slip bands is then avoided. The size of the dispersoids is important since larger ones may themselves act as stress concentrators thus reducing ductility and fracture toughness. As mentioned in the previous paragraph, the rapid solidification technique reduces the dispersoid size as well as the grain size.

Figure 5. Metallurgist at the Fracture and Fatigue Research Laboratory, Georgia Institute of Technology, tests one of the new aluminum-lithium alloys developed by the authors.

Since strain localization depends on the presence of soft regions resulting from dislocations shearing coherent precipitates, or precipitate-free zones, it may be minimized by strengthening the matrix (and PFZ) by solid solution strengthening or by adding elements which coprecipitate with lithium. Magnesium is a potent solid solution strengthener in aluminum alloys and aluminum-lithium-magnesium alloys show commercial promise.[10] In addition to this beneficial effect, magnesium suppresses the $\delta' \rightarrow \delta$ transformation and improves the corrosion resistance in overaged aluminum-lithium alloys.[11] Elements which have limited solid solubility, e.g., copper,

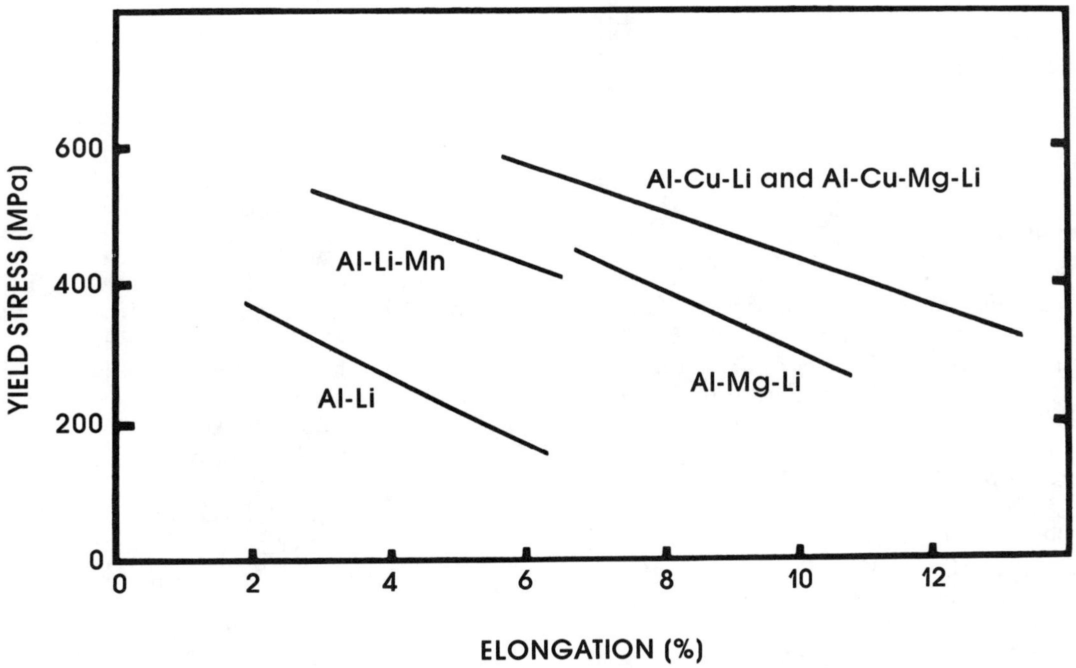

Figure 6. Schematic of strength-ductility relationships for various aluminum-lithium-X alloys.

can have a great strengthening effect and can improve the homogenization of deformation when coprecipitated with lithium. For example, when added to aluminum-lithium-magnesium alloys, copper precipitates as the S′ phase, increases the strength, homogenizes slip, suppresses the $\delta' \rightarrow \delta$ transformation at grain boundaries, decreases the PFZ width and increases the strength near the grain boundaries.

A schematic showing the improvements in strength-ductility relationships that have been recently obtained in aluminum-lithium alloys using the approaches described in this paper is given in Figure 6. The new alloys show high strength, high elastic modulus, good ductility, low density, and superior fatigue crack growth and corrosion resistance. The recent improvements make these alloys very attractive for aircraft structures, and it appears that the promise offered by these materials is close to being realized.

Glossary

DARPA—Defense Advanced Research Projects Agency.
Dispersoid—finely distributed particles of relatively insoluble constituents.
Heterogeneous precipitation—nonuniform precipitation.
Higher-energy transgranular mode—fracture through the grains.
Homogeneous precipitation—uniform precipitation.
Low-energy intergranular mode—fracture along grain boundaries.
Matrix—the continuous phase.
Solute—depleted precipitate free zone region of the matrix which contains no precipitates.
Solvus—a solid solubility limit.
Specific modulus—ratio of elastic modulus to density.

References

1. R. E. Lewis, D. Webster and I. G. Palmer, *A Feasibility Study for Development of Structural Aluminum Alloys From Rapidly Solidified Powders for Aerospace Structural Applications,* Lockheed Palo Alto Research Laboratory Final Report, Contract F33615-77-C-5186, Technical Report No. AFML-TR-78-102, July 1978.
2. Air Force Contract F33615-78-C-5203 (DARPA Order 3417).
3. Air Force Grant AFOSR-78-3471.
4. T. H. Sanders, Jr. and E. S. Balmuth, *Met. Prog., 113,* (1979) 37.
5. T. H. Sanders, Jr., *Naval Air Dev. Cent. Final Report,* June 1979 (Naval Air Systems Command Contract N62269-76-C-0271.
6. P. M. Kelly, *Int. Met. Rev., 18,* No. 172-C (1933) 31.
7. E. A. Starke, Jr. and G. Lütjering in M. Meshii, ed, *Fatigue and Microstructure,* ASM, Metals Park, Ohio (1979) 205.
8. E. A. Starke, Jr. in *1980 Year Book of Science and Technology,* McGraw-Hill Company, New York (1980) 218.
9. E. A. Starke, Jr., *Mat. Sci. Eng., 29* (1977) 99.
10. Br. Patent 1,172,736 (December 3, 1969) to I. N. Fridlyander.
11. P. Niskanen, T. H. Sanders, Jr., M. Marek and J. G. Rinker, "The Influence of Microstructures on the Corrosion of Al-Li, Al-Li-Mn, Al-Li-Mg, and Al-Li-Cu," Proceedings of the First International Al-Li Conference, Stone Mountain, GA, AIME (in press).

T. H. Sanders, Jr. is Research Scientist, Fracture and Fatigue Research Laboratory (FFRL), Georgia Institute of Technology, where he is responsible for research programs in aluminum alloy development, microstructure/properties relationships, and solidification. He holds B.S., M.S., and Ph.D. degrees from Georgia Tech. Prior to joining the staff of FFRL, he was a scientist in the Physical Metallurgy division of the Alcoa Technical Center.

E. A. Starke, Jr. is Director, Fracture and Fatigue Research Laboratory and Professor of Metallurgy, Georgia Institute of Technology. He received his B.S. from Virginia Tech, his M.S. from the University of Illinois and his Ph.D. from the University of Florida; all in metallurgical engineering. Professor Starke has been with Georgia Tech for seventeen years.

Microstructures of Rapidly Solidified Aluminum Alloy Submicron Powders

C. G. LEVI and R. MEHRABIAN

The microstructures of electron transparent submicron aluminum alloy powders produced by an electrohydrodynamic process are described. The observations are coupled with thermodynamic, kinetic, and heat flow concepts to deduce the thermal history and solidification mode of the powders. The range of microstructures observed includes: homogeneous plane-front solidified single crystals; cellular crystals; and powders containing blocky segregates, multiple grains, and twins. In general, a decrease in the size of the submicron particles decreases the amount of segregation. This observation appears to be caused by an increase in undercooling prior to nucleation which leads to higher liquid/solid interface velocities. On the other hand, the same trend promotes multiple nucleation and increases the incidence of twin crystals. Liquid/solid interface breakdown from planar to cellular in the middle of the powders is correlated with solute "trapping" during recalescence and rejection of solute ahead of the solidification front after recalescence. The latter manifests itself as a solute depleted zone in the plane-front to cellular transition region. Finally, some of the microstructural observations lend credence to the recent extension of the morphological stability theory to rapid solidification.

I. INTRODUCTION

RAPID solidification processing (RSP) commonly results in progressive departures from the microstructures produced by conventional casting, generally in the direction of increased refinement and chemical homogeneity. These modifications are determined by the transport processes occurring at the liquid/solid interface, and hence, are critically dependent on the growth rate (*i.e.*, local undercooling). It is now acknowledged that RSP effects may be achieved even under moderate or slow cooling conditions if the liquid is substantially undercooled prior to nucleation.[1,2] However, the benefits of this approach may be optimized only through a fundamental understanding of the solidification mechanisms and much needed characterization of the undercooled microstructures.

Atomization is unique among the rapid solidification processes because the interactions with molds, substrates, or other particles are practically absent. Conceivably, then, most of the liquid droplets could achieve high undercoolings prior to nucleation, provided the purity level is adequate, because the active nucleation catalysts will be isolated into a small fraction of the total volume. This concept was proven experimentally by Turnbull *et al*[3,4,5] by slowly cooling emulsions of metal droplets. More recently, Perepezko *et al*[2,6] developed a refined version of Turnbull's technique and measured maximum undercoolings ranging between 0.26 T_M for lead to 0.50 T_M for gallium. In general, it was found that the achievable undercooling was a function of the surface coating (emulsifer) and the droplet size distribution. Furthermore, experiments in the Sn-Bi[6] and Cu-Ni[5] systems revealed that undercoolings comparable to those of the pure metals can be obtained in these alloys across the entire phase diagrams.

C. G.LEVI, formerly a Research Associate at the University of Illinois, is now with the Metallurgical Process Engineering Department at ATISA-ATKINS, S. A. de C. V. in Mexico City, Mexico. R. MEHRABIAN is Chief of the Metallurgy Division, National Bureau of Standards, Washington, DC 20234.

Manuscript submitted April 1, 1981.

Recent developments in vacuum atomization *via* electrohydrodynamics (EHD) have provided a tool for the controlled generation of powders below 1 μm in diameter.[7] These powders, while not of commercial importance, are especially suited for fundamental studies for two reasons. First, they are likely to undercool substantially due to their size and the minimization of surface films. Second, because they are electron-transparent – especially those of light metals like aluminum – it is possible to analyze the evolution of a *complete microstructure* from nucleation until the end of solidification without recourse to thinning procedures.

This paper deals with the characterization of some novel microstructures typical of aluminum alloy submicron powders produced by the EHD process. The observations are coupled with thermodynamic, kinetic, and heat flow concepts to elucidate the thermal history and solidification mode of the powders and their effects on the microstructure.

II. HEAT FLOW DURING SOLIDIFICATION OF UNDERCOOLED METALS

The thermal history of powder solidification can be readily described on the enthalpy-temperature diagram for a pure metal as shown in Figure 1. Assuming that temperature differences inside the droplet are negligible (Newtonian conditions), the enthalpy (H) of a system at temperature (T) and fraction solid (g) is given by:

$$(H - H_{SM}) = [\Delta H_M + C_L(T - T_M)](1 - g) + C_S(T - T_M)g \quad [1]$$

where H_{SM} and ΔH_M are the enthalpy of the solid (reference point) and the heat of fusion at the melting temperature T_M, C_L and C_S are the specific heats of the liquid and solid, respectively. Eq. [1] can be written in terms of the dimensionless enthalpy, $\psi = (H - H_{SM})/\Delta H_M$, and temperature, $\theta = C_L(T - T_M)/\Delta H_M$, as

$$\psi = (1 + \theta)(1 - g) + (C_S/C_L)g\theta \quad [2]$$

The two terms on the right-hand side of Eq. [2] represent the

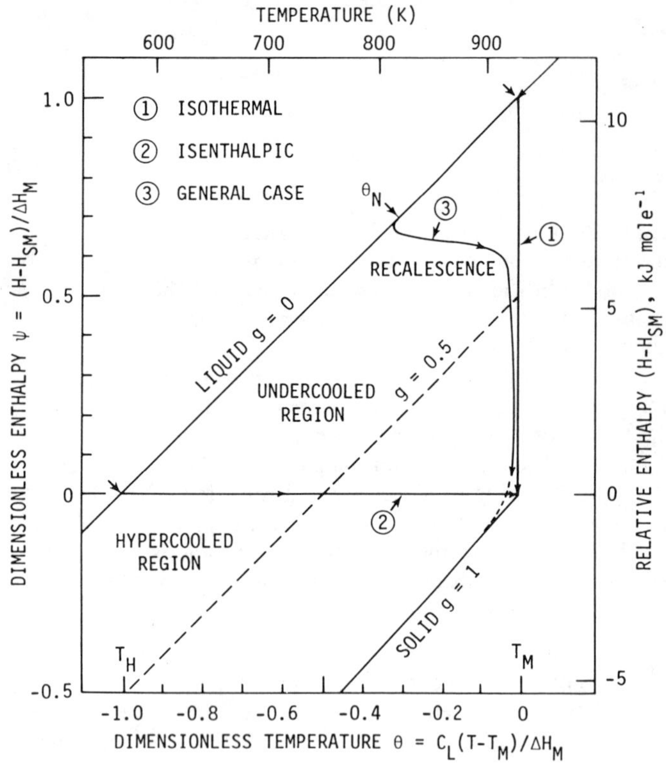

Fig. 1—Enthalpy-temperature diagram showing possible solidification "paths". θ_N is the dimensionless temperature at the moment of nucleation.

enthalpies of the liquid and solid fractions, respectively. For a solid volume ($g = 1$) the dimensionless nodal enthalpy ψ is always negative or zero since $\theta \leq 0$, whereas for volume of superheated liquid ($g = 0, \theta > 0$) ψ is larger than unity. Since constant fraction solid lines in Figure 1 intersect the melting temperature axis ($\theta = 0$) at $\psi = 1-g$, a volume of partially solid material at T_M will have an enthalpy equal to its fraction of liquid.

The system can conceivably follow any "path" of decreasing enthalpy in the undercooled region ($\psi < 1$) starting at the nucleation temperature ($\theta_N < 0$) on the liquid line, and ending at some point on the solid line. A vertical "path" represents isothermal solidification, of which a limiting case is the no-undercooling situation ("path" one in Figure 1). A horizontal "path" represents isenthalpic (or adiabatic) solidification. If nucleation can be suppressed until an amount of enthalpy at least equivalent to the heat of fusion has been removed (*i.e.*, $\psi_N \leq 0$), the system can undergo complete solidification without further heat extraction ("path" two in Figure 1). The region of the diagram for which $\psi \leq 0$ is known as "hypercooled".

A more general case in which the droplet is undercooled ($-1 < \theta_N < 0$) and concurrent heat extraction occurs during solidification has been analyzed by the authors in a recent publication.[8] The thermal history is represented by "path" three in Figure 1, where two distinct solidification regimes can be identified. The first one is the recalescence or "rapid solidification" stage, where the undercooled droplet is absorbing most of the liberated heat of fusion. In this regime, solidification is always "rapid" regardless of the heat loss to the surroundings. The second regime is one of a slower growth, and develops after the droplet undercooling has been largely relieved, thereby limiting the progress of solidification to the conditions imposed by the external heat flow. This regime usually involves temperatures close to the melting point of the metal and is roughly isothermal.

The thermal history reflects the competition between the external rate of heat extraction and the recalescence rate (release of latent heat) which is proportional to the solidification rate (change in fraction solid with time, dg/dt). The latter is in turn a product of the liquid/solid interface velocity (kinetic dependent) and its area. Predominance of the recalescence rate translates into a shallow negative slope for the ψ-θ curve in the "rapid solidification" stage. As the heat extraction process is able to relieve some of the heat of fusion being transformed into sensible heat, the slope of the curve becomes steeper (more negative) and eventually positive if the external heat flow is dominant.

Results of our heat flow studies on powder solidification[8] revealed that for typical atomization conditions the external cooling plays only a minor role during the recalescence stage of the thermal history. Therefore, the "rapid solidification" effects are usually limited by the undercooling achieved before nucleation occurs. Newtonian conditions, likely to occur in submicron particles, yield the maximum volume fraction of "rapidly solidified" material for a given initial undercooling. However, as the particle size increases, the liquid/solid interface tends to recalesce earlier than the rest of the droplet – this effect is more pronounced with faster attachment kinetics. The development of non-Newtonian conditions then requires higher undercoolings than the Newtonian case in order to produce similar volumes of "rapidly solidified" material.

III. THERMODYNAMIC AND KINETIC CONSIDERATIONS

Various types of microstructural modifications have been produced in aluminum alloys *via* RSP.[9] Among them, extension of the equilibrium terminal solid solubility – even beyond the eutectic composition – is of particular interest to this paper and is briefly discussed.

Increased solid solubility *via* rapid solidification permits higher alloying levels resulting in superior strength-ductility combinations in a number of alloy systems. For example, enhanced solid solution hardening and dispersion strengthening after aging have been reported for Al-Cu compositions beyond the equilibrium solubility limit.[10] The possibility of precipitation hardening in Al-Fe base alloys has been demonstrated,[11] and considerable interest exists in the substantial ductilities attainable in high-silicon aluminum alloys.[12,13]

Baker and Cahn[14] have discussed the thermodynamics of nonequilibrium solidification in a binary alloy. With the aid of free energy (F) *vs* composition (X) curves, they defined the domain of possible interface compositions for isothermal and steady state solidification. They concluded that the maximum solid composition X_S that can be formed from a liquid of composition X_L at the temperature of interest is determined by the point at which the liquid and solid free energy curves intersect. This is known as the $X(T_O)$ composition and the locus of all these points forms a curve between the liquidus and the solidus of the equilibrium phase diagram. Partitionless transformations can occur only below the T_O curve ($T < T_O$ for a given X_L). This involves an increase in the chemical potential of the solute on crossing the interface and therefore implies that the two species have to solidify in a "cooperative" manner. The solute is then "trapped" by a reaction which produces an overall decrease in the free energy of the system. Kinetic considerations will further decrease the critical temperature for partitionless solidification below the T_O value.[15]

If nucleation is suppressed until the temperature is below the solidus, the first solid to form is of composition very similar to the liquid.[16] However, heat flow calculations for atomized droplets[8] show that the subsequent solidification is nearly adiabatic, limiting the extent of homogeneous or partitionless solidification by the reheating of the interface to above a temperature for the onset of segregation. The latter is characterized by a critical temperature T_C, whose value falls presumably between the solidus and the T_O curve. The production of a homogeneous solid under adiabatic conditions requires at least an initial undercooling dictated by the following relationship

$$\Delta H_M \leq \int_{T_N}^{T_C} C_L \, dT \qquad [3]$$

where T_N is the nucleation temperature. The maximum value of T_N that satisfies the above relationship is the "hypercooling" temperature T_H.

In general, for nucleation to be suppressed until a desired undercooling is reached in an assembly of volume V, the total number of nuclei N, given by

$$N = V \int_{T_M}^{T_N} [J(T)/\varepsilon] \, dT \qquad [4]$$

should be less than unity. In this expression, J is the temperature-dependent nucleation frequency and ε is the cooling rate.

The heat transfer to the surroundings is then more important before rather than during solidification. Hirth[17] has estimated the cooling rates necessary for a particle to satisfy Eq. [3] before the number of nuclei given by Eq. [4] becomes unity. His calculations show that homogeneous solidification appears feasible for alloys of iron and cobalt but not for those of nickel, copper, or aluminum, under the cooling rates achievable in current atomization processes.

Another subject of importance in dealing with undercooled microstructures concerns the stability of the moving interface. For most cases of low interface velocities, the principle of constitutional supercooling has been widely used as an adequate guideline to determine the growth conditions that result in unstable interfaces during alloy solidification.[18] This principle is basically an application of thermodynamics to determine if the liquid ahead of a moving interface is undercooled with respect to its composition. The stability parameter emerging from this theory is the ratio of the temperature gradient in the liquid ahead of the moving interface G_L to its velocity R. Greater stability is promoted by increasing G_L/R.

The theory of morphological stability[19,20] approaches the problem from a kinetic point of view, solving the thermal and diffusion field equations about the interface under steady state conditions. It also assumes that local equilibrium prevails at the solidification front. The resulting criterion identifies three factors contributing to the overall stability of the interface. The solute field created by segregation is destabilizing, but the capillarity forces oppose the growth of a perturbation. The thermal field, as determined by the average gradient

$$G_{SL} = 2(k_S G_S + k_L G_L)/(k_S + k_L) \qquad [5]$$

will be stabilizing if G_{SL} is positive, and destabilizing in the opposite case, where k_S and k_L are the thermal conductivities of the solid and liquid, respectively, and G_s is thermal gradient in the solid at the interface. The constitutional supercooling criterion predicts instability for all cases where $G_L < 0$; however, the morphological theory may predict stability if G_S is sufficiently positive to make the average gradient term also positive.

Recent studies have compared the predictions of both theories for the case of rapid solidification.[21,22,23] The first two studies[21,22] extended the morphological stability theory and indicated that two effects are important in rapid solidification phenomena. The first one is known as absolute stability and arises because only short wavelength perturbations are important at high velocities, and these are stabilized by surface energy. The net result is that given a sufficiently high velocity, the interface will be stable for any value of G_L (assuming that G_{SL} is positive). The second effect is caused by departure from local equilibrium at the interface. Coriell and Sekerka[21] concluded that most departures from local equilibrium enhance the interface stability, especially in the case where the partition coefficient tends to one. Baker[24] analyzed the behavior of the partition coefficient as a function of interface velocity for various models and indicated that given a sufficiently high velocity, the partition coefficient in most alloys tends to unity. Cohen *et al*[23] mapped the regimes of stability in a G_L *vs* R plot and indicated that increasing the cooling rate and decreasing the solute content will promote planar interfaces.

IV. SPECIMEN PREPARATION

The aluminum alloy powders used in this investigation were prepared by the electrohydrodynamic (EHD) atomization process.[7] Most of the studies were focused on the Al-Si system, using alloys of three and six wt pct Si, as well as a high purity (99.999 pct) aluminum standard. Some supplementary observations were made on Al-4.5 wt pct Cu. The alloys were prepared from aluminum ingot (99.97 wt pct), Cu plate (99.95 wt pct) and Si lumps (99.98 wt pct) by melting in a clay-graphite crucible, degassing with chlorine, and casting into graphite molds.

Atomization Technique

The EHD method for droplet generation[7] involves the application of very intense electric fields on a conductive liquid at the tip of a capillary nozzle (emitter). The molten metal, contained in a pressurized ceramic reservoir, is given a positive potential (usually +10 kV) while delivered to the nozzle, which is typically of 50 μm in diameter. The emitter is exposed to a negative (−1 kV) annular electrode (extractor), concentrating the electric field on the very small liquid meniscus. The interaction of the electrostatic stresses with the surface tension forces produces a liquid spike that breaks into charged droplets. These are then accelerated by the voltage difference and fly through the orifice in the extractor to be collected on a target after solidification. Velocities above 200 ms^{-1} have been reported for particles below

1 μm in diameter, with calculated times of flight lower than 2.5 ms.[7]

The atomizer is enclosed in a vacuum chamber (10^{-5} torr), thereby restricting the heat loss from the droplets to radiation and the heat transfer coefficient ($h < 100\ W\ m^{-2}\ K^{-1}$). Maximum cooling rates prior to solidification are estimated to range from $\sim 100\ K\ s^{-1}$ to $\sim 10^4\ K\ s^{-1}$ at the onset of hypercooling, for a 1 μm droplet of aluminum with an emissivity of one. Actual values may be lower by more than one order of magnitude if we consider that emissivity values reported in the literature for solid aluminum surfaces range from 0.05 to 0.3.[25]

EHD powders produced ranged in size from less than 100 nm to about 100 μm. They were collected by deposition on a 125 μm thick plastic tape (of the type used to prepare replicas for electron microscopy) placed on the target. By appropriate manipulation of the process variables, the substrate was covered with a monolayer consisting mostly of submicron powders. Particles larger than ~10 μm were usually limited in number, presumably because they bounced off the tape. The latter loose material was gathered from the chamber at the end of the experiment.

Figure 2 shows a typical sample of powders as collected on the plastic substrate. They are nearly spherical and have smooth surfaces, although some appear to have solidified on the tape. Mainly intended as a laboratory tool, this process is able to produce only a few grams of material per hour. Nevertheless, it has yielded a substantial number of samples under reproducible conditions, making it especially suitable for fundamental solidification studies.

TEM Specimen Preparation

Aluminum powders below 1 μm in diameter are transparent to 200 kV electrons, and therefore lend themselves to conventional TEM analysis without thinning. EHD specimens were prepared by cutting the tape into 3 mm discs and coating the discs on the deposition side with a 20 to 50 nm film of evaporated carbon. The substrate was then dissolved in acetone and the powders remained attached to the carbon film. This was carefully rinsed in isopropyl alcohol and placed on a 3 mm, 400 mesh copper grid. The powder density was commonly about 100 particles per grid opening ($\sim 10^3\ \mu m^2$).

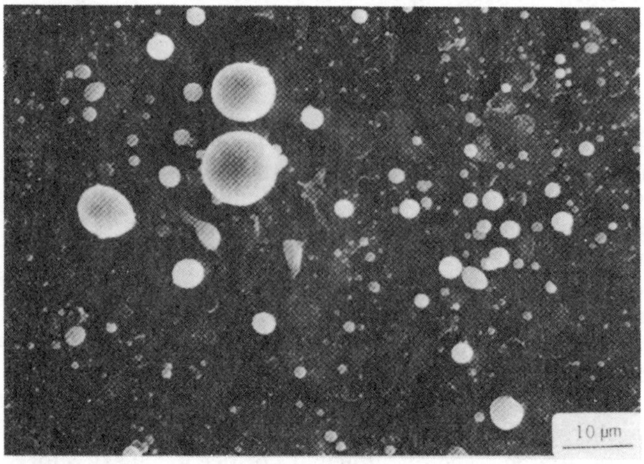

Fig. 2—SEM view of Al-4.5 wt pct Cu powders produced by electrohydrodynamic (EHD) atomization process.

V. RESULTS AND DISCUSSION

The observations resulting from the analysis of EHD powders are presented in two categories depending on whether they pertain to their segregation or their nucleation behavior. For the purposes of the ensuing discussion, we will call *segregates* those second phases originating from the melt, whereas *precipitates* will refer to phases formed after solidification.

Most of the Al-Si EHD produced powders conform to a variant of the basic segregation pattern shown in Figure 3. The microstructure exhibits a region (A) containing a uniform distribution of very fine precipitates or clusters of Si which, according to the electron-diffraction evidence, have an orientation relationship (111)Si//(111)Al to the primary α-aluminum phase. Dark-field imaging of precipitates slightly coarser than those in the present figure suggest that they are either needles or platelets on the {111} α-aluminum planes. This finding is consistent with the decomposition kinetics of Al-Si solid solutions; the plates evolve from

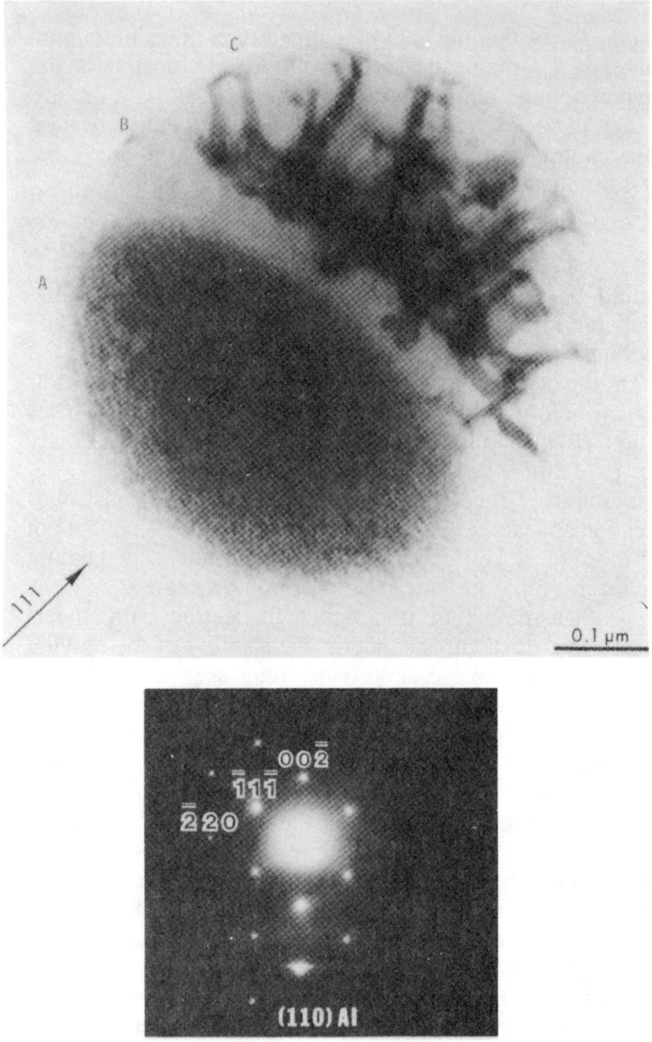

Fig. 3—Typical segregation pattern for an Al-6 wt pct Si EHD submicron powder (1 μm > $2r_0$ > 0.5 μm) showing three distinct regions. The initial "rapid solidification" produces a supersaturated zone (A), which later breaks into cells (C) after the droplet recalesces. (B) is a transition zone where solute builds up in the liquid in front of the interface. The growth axis is close to a <111> direction.

spherical clusters and needles that form in the early stages of precipitation.[26-30] It was concluded that this zone was originally a supersaturated solid solution.

Region (A) is followed by a precipitate-free zone (B), and the change in microstructure clearly delineates the solidification front. The latter is symmetric about the growth axis, which was consistently found to be close to a <111> direction.

Finally, the particle in Figure 3 also has a well developed cellular region (C), with segregate spacings on the order of 0.1 μm. A similar structure in an Al-4.5 wt pct Cu powder, shown at a different angle in Figure 4, reveals the cellular pattern more clearly.

Figure 5 illustrates the general effect of particle size on the extent of the different zones in the powders. As the powder diameter decreases, the cellular structure becomes less defined 5b, and eventually disappears, leaving only a few segregates 5c. At the lower end of the size distribution, the particles appear almost featureless 5d, showing only the supersaturated region (A). The intensity of the silicon spots in electron diffraction was also reduced with diminishing diameter, although the crystallographic relationship noted above prevailed.

The structures in Figure 5 are by no means restricted to a particular size range. For example, segregate-free powders as large as 1 μm were found in some instances, and segregated silicon was detected in some of the smallest particles. The trend illustrated in this figure was established from the TEM analysis of over 400 powder particles ranging in size from 0.1 to 1.0 μm for each Al-Si composition. The results, presented in Figure 6, confirm that increasing the particle size or the solute content enhances the incidence of segregation. Since not all of the segregated powders exhibit a cellular pattern (see, for example, Figures 5c and 7), the occurrence of the latter is also plotted in Figure 6.

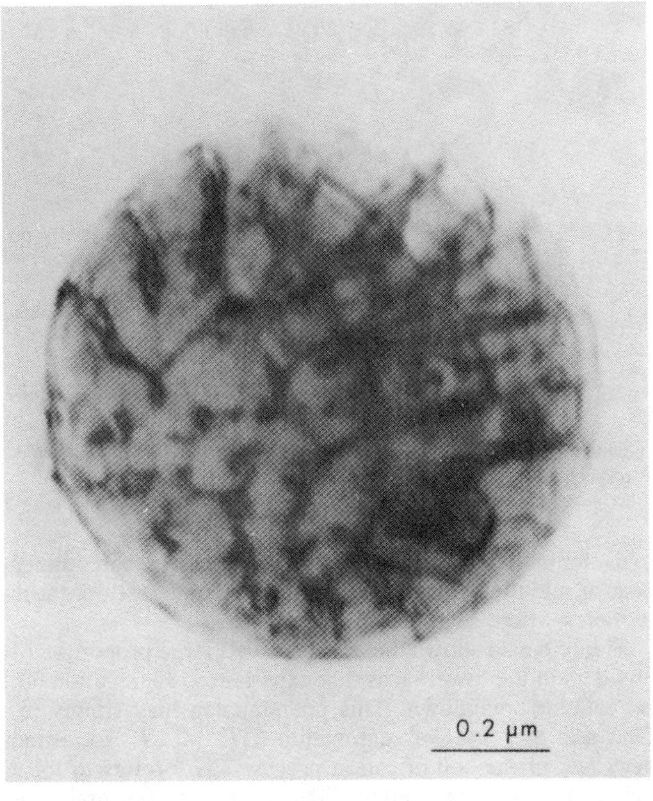

Fig. 4—Cellular structures observed in Al-4.5 wt pct Cu EHD powders.

The previous microstructures reveal that substantial undercoolings must have been reached in the liquid droplets prior to solidification, which is believed to proceed as follows. In general, nucleation occurs at the surface, as evidenced by the growth direction. The subsequent "rapid solidification" into a highly undercooled liquid produces a homogeneous zone (A) of extended solubility. As the droplet recalesces, the interface velocity and undercooling are reduced, thereby decreasing the driving force for solute "trapping". When the interface temperature surpasses some critical temperature for the supersaturation process; T_C, solute is rejected ahead of the solidification front. The interface is eventually rendered unstable and breaks into cells.

The extent of homogeneous solidification can be rationalized in terms of the heat flow concepts previously discussed. Assuming that thermophysical properties are reasonably constant, the behavior of an alloy growing in a plane front mode without solute partitioning is expected to resemble that of the pure metal.

Eq. [3] for constant specific heat indicates that the required nucleation temperature for adiabatic homogeneous solidification is

$$T_N \leq T_H = T_C - \Delta H_M / C_L \qquad [6]$$

where T_H is now the alloy hypercooling temperature.

As noted earlier, the onset of segregation probably occurs at a temperature between the solidus T_S and the T_o curve. For purposes of illustration, we have arbitrarily assigned to T_o the average value of T_L and T_S from the Al-Si phase diagram, and estimated on this basis an upper bound to the hypercooling temperature T_H; see Table 1. The lower bound to T_H corresponds to $T_C = T_S$. The values of θ were calculated by substituting T_L for T_M in its definition and hence are proportional to the undercoolings. The values in this table indicate that dimensionless temperatures on the order of -1.18 ($\Delta T_N/T_L = 0.47$) and -1.36 ($\Delta T_N/T_L = 0.55$) may be necessary to eliminate segregation in aluminum alloys with three and six wt pct Si, respectively.

A Newtonian heat flow model described by the authors[8] may be used to estimate the values of fractions solidified below different critical temperatures (*i.e.*, the extent of supersaturation). The results are shown in Table 2 for different dimensionless nucleation temperatures θ_N and two heat transfer coefficients. For an $h = 20\ W\ m^{-2}\ K^{-1}$, typical of radiation cooling (EHD atomization process), the recalescence is essentially adiabatic. It is concluded from this table that the "rapid solidification" effects are diminished when the critical temperature decreases as a result of increasing the solute content. In addition, the achievable undercooling prior to nucleation is likely to increase as the droplet size decreases, and the initial zone of supersaturation is consequently extended until it covers the entire powder. This is illustrated in Figures 5 and 6.

Table 2 also indicates that even for a high heat transfer coefficient of $h = 2 \times 10^5\ W\ m^{-2}\ K^{-1}$ such as that expected from very efficient convection (*e.g.*, Rapid Solidification Rate (RSR) process), the external cooling is expected to extend the "rapid solidification" regime only slightly over that of the EHD powders. It should be noted

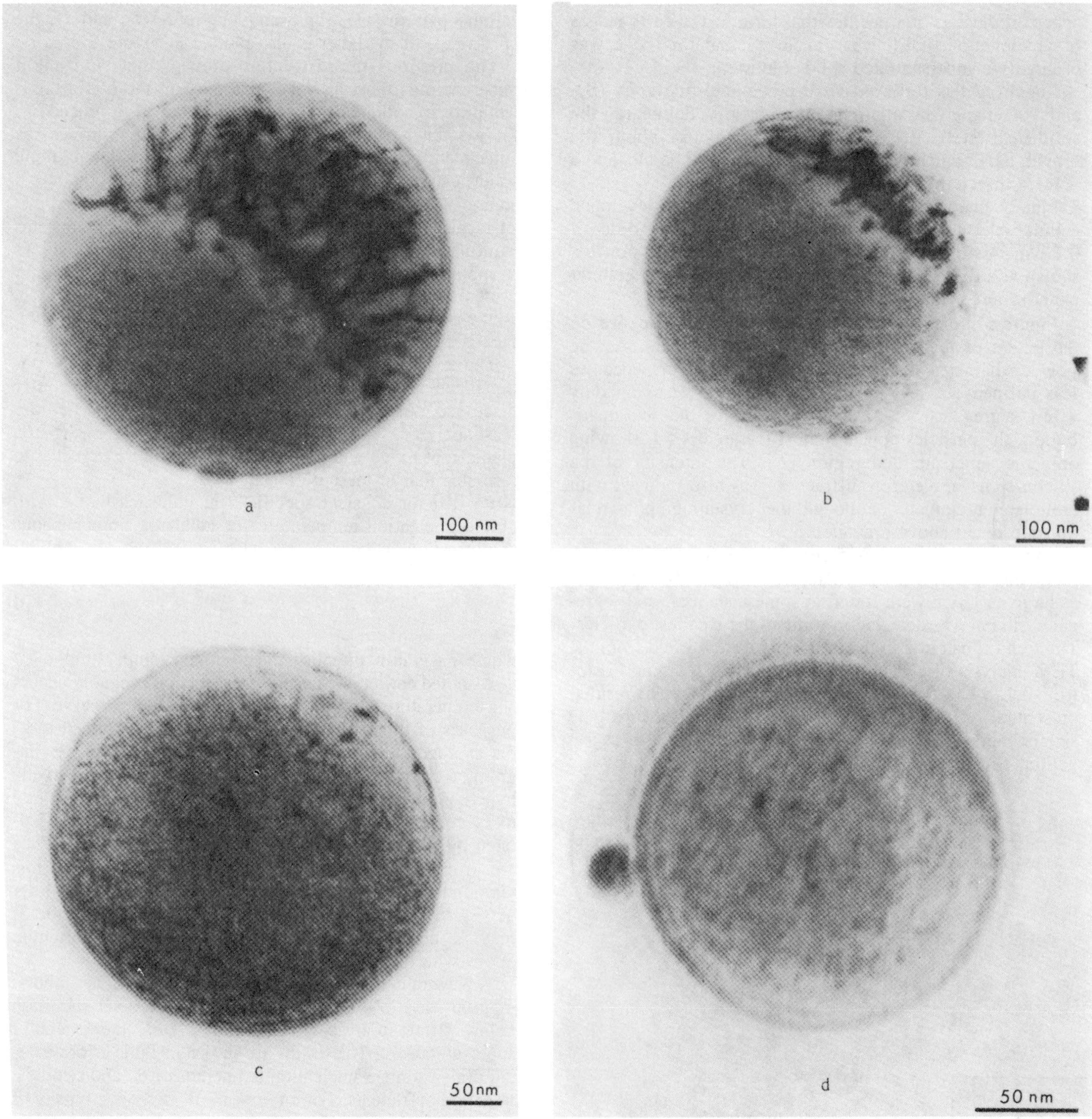

Fig. 5—Transition from a partially segregated structure to a homogeneous one as powder particle size decreases in the Al-6 wt pct Si alloy. The diameters of the powders shown are (*a*) 570 nm, (*b*) 470 nm, (*c*) 220 nm, and (*d*) 170 nm, respectively.

that the data presented are calculated on the basis of the Newtonian cooling model, are independent of powder particle size, and give an upper limit for $g(\theta_c)$.* Under the general non-Newtonian cooling conditions, the liquid/solid interface recalesces much faster than the overall droplet temperature (used in the Newtonian model) and higher initial undercoolings are required for equivalent values of $g(\theta_c)$.[8] Finally, one may also conclude that undercoolings near or into the hypercooled regime were obtained for those particles where segregation was absent.

Figure 6 also shows the existence of a large proportion of powders in the lower sizes that experienced segregation but no cellular breakdown. This phenomenon may simply reflect that the onset of segregation ($\Delta T^* < \Delta T_C$) occurred very late in the solidification process. ΔT^* refers to local undercooling at the liquid/solid interface in a general non-Newtonian heat flow model. On the other hand, interface

*$g(\theta_c)$ denotes the fraction of the particle solidified during recalescence from the initiation of nucleation at θ_N to a critical temperature θ_c.

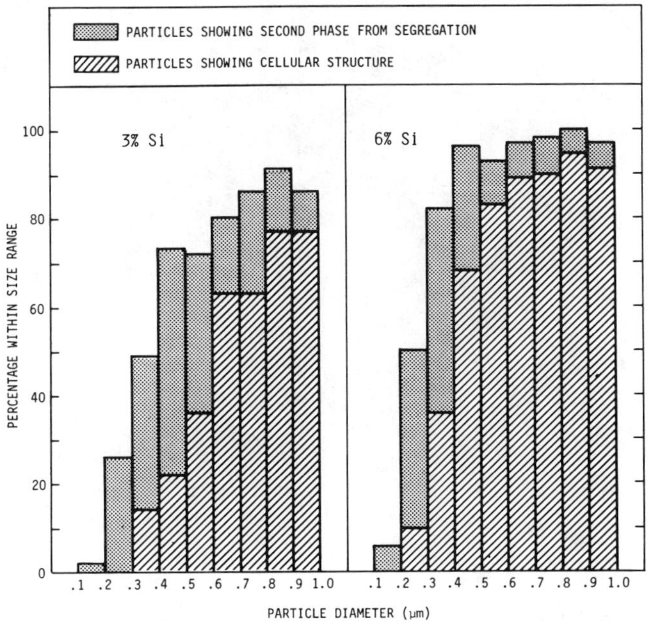

Fig. 6—Segregation behavior of aluminum-silicon submicron powders produced by the EHD atomization process. Results are from examination of over 400 particles in each alloy composition.

Table 1. Effect of the Alloy Composition on the Critical and Hypercooling Temperatures†

		3 pct Si		6 pct Si	
		T	θ	T	θ
Liquidus	T_L	913	0	893	0
Critical	$T_C(T_0)$	848	−0.18	762	−0.36
	$T_C(T_S)$	782	−0.36	631	−0.72
Hypercooling	$T_H(T_0)$	484	−1.18	398	−1.36
	$T_H(T_S)$	418	−1.36	267	−1.72

†Estimated from the equilibrium phase diagram, T_S values are from the metastable solidus. $T_0 = (T_L + T_S)/2$.
Dimensionless temperature $\theta = C_L(T - T_L)/\Delta H_M$.

Table 2. Extent of "Rapid Solidification" in Al-Si Alloys as Fraction Solidified below a Critical Temperature $g(\theta_C)$†

		Critical Temperature θ_C		
$h(Wm^{-2}K^{-1})$	θ_N	−0.18	−0.36	−0.72
20	−1.0	0.80	0.62	0.26
(EHD)	−0.75	0.56	0.38	0.03
	−0.50	0.31	0.13	—
	−0.25	0.07	—	—
2×10^5	−1.0	0.92	0.67	0.28
(RSR)	−0.75	0.67	0.43	0.05
	−0.50	0.43	0.20	—
	−0.25	0.21	—	—

†Newtonian cooling is assumed.

breakdown may also be prevented by the stabilization of short wavelength perturbations due to surface tension effects[21]; the smallest perturbation that can grow is comparable to the particle size.

A small proportion of the powders (<five pct) exhibit quite a different type of segregation microstructure, where large faceted silicon second phases appear distributed along the particle surface, as in Figure 7. Diffraction analysis did not reveal any particular orientation relationship with the matrix, but rather strong reflections of individual silicon crystals. The results indicate that these phases were probably segregation products, and that the supersaturated zone (A) typical of most submicron particles was apparently absent in these powders. Furthermore, this anomalous segregate morphology was not observed in the Al-4.5 wt pct Cu alloy powders, and its occurrence was more common in the larger particles and the higher silicon content alloy.

It was concluded that powders showing the faceted silicon were unlikely to have reached substantial undercooling prior to nucleation, and hence were extensively segregated. Although the absence of cells could be ascribed to interface stabilization by surface tension effects, it is not clear why most of the particles of equivalent size were cellular. In addition, the distribution of segregates was not consistent with an axisymmetric growth pattern, in which silicon should be accumulated on one side of the powder.

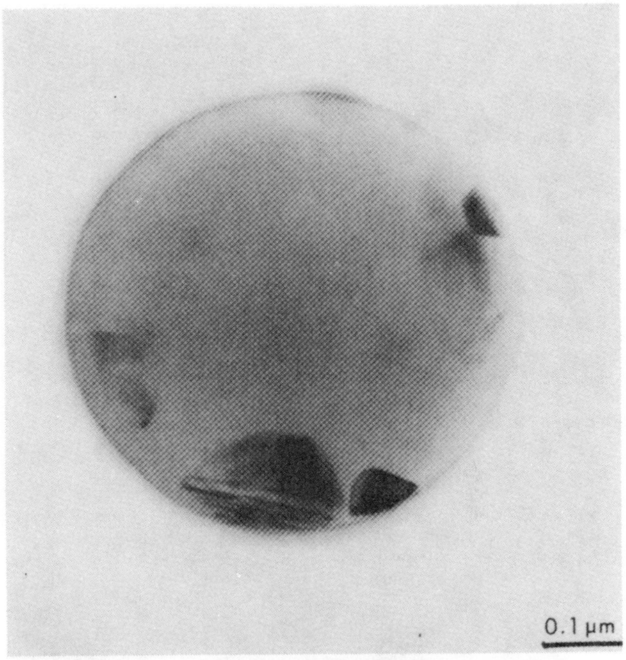

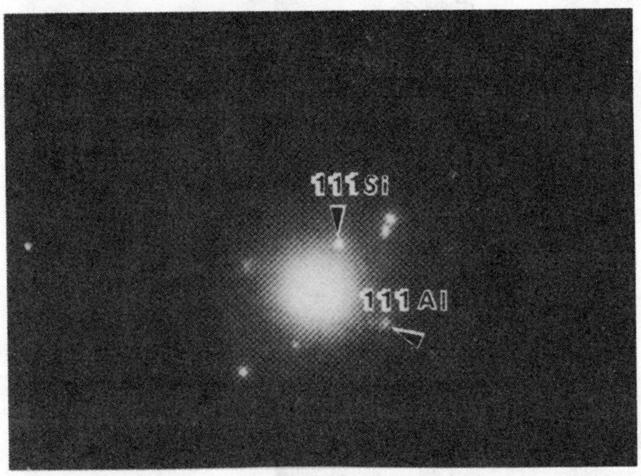

Fig. 7—Representative TEM view of anomalous morphology of silicon segregates observed in some of the Al-Si powders. Reflections of Si and Al with no apparent relationship are mixed in the diffraction pattern.

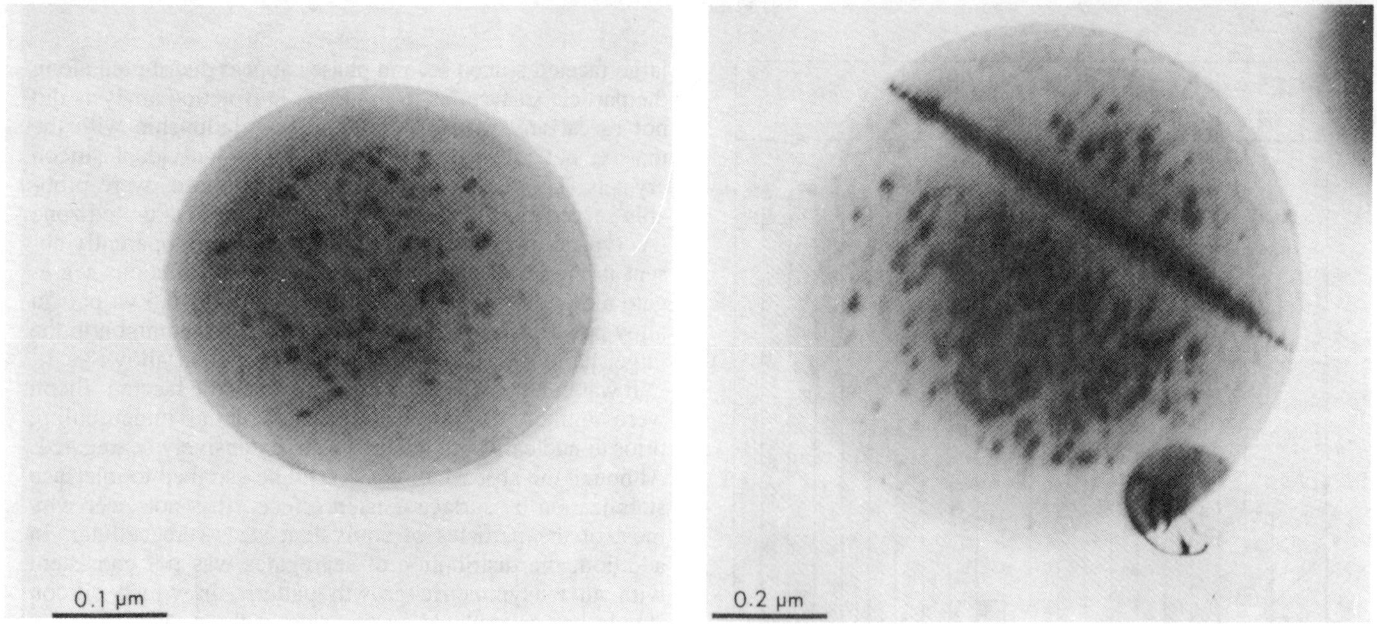

Fig. 8 — Typical structures in high purity aluminum (99.999 pct) submicron EHD powders. (*a*) shows a monocrystalline particle, and (*b*), a set of twin crystals.

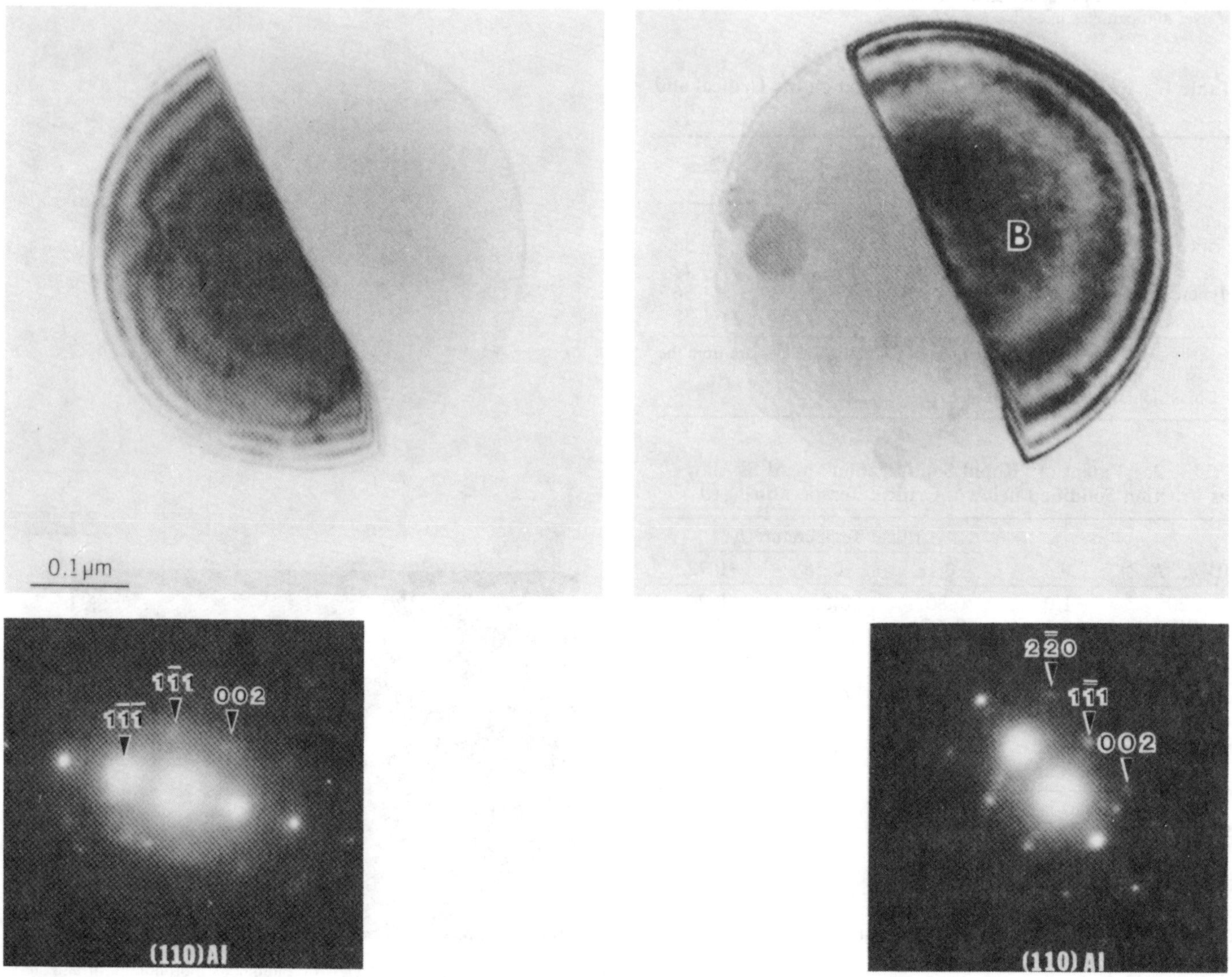

Fig. 9 — Typical polycrystalline powder of Al-3 wt pct Si alloy where two nucleation events occurred simultaneously in the droplet and formed an impingement boundary.

High purity aluminum submicron powders, such as those in Figure 8, do not exhibit any evidence of cellular structures that could have originated from thermal instability due to the negative temperature gradients in front of the interface. This observation lends support to the fact that the breakdown observed in the alloys was mainly of constitutional nature. It could also be argued that thermal cells in a pure metal would be indistinguishable after solidification, unless they are decorated by impurities or strained areas caused by crystalline misorientation. Nevertheless, pure aluminum particles show some platelike features that can be interpreted either as clusters of impurities or vacancy condensation loops. The existence of denuded zones near the surface and crystal boundaries suggests that vacancies are involved in the formation of these features.

The microstructural analysis also revealed that most EHD powders are monocrystalline, indicating that nucleation was usually limited to a single event per droplet. A substantial proportion of them, however, consist of two or more crystals which could be either randomly oriented with incoherent boundaries, as in Figure 9, or twins, as in Figure 10. It was concluded that each random grain originates from an independent nucleation event, and its boundaries result from impingement with other growing crystals. On the other hand, twinned crystals normally exhibit {111} coherent boundaries, and it is unlikely that they were independently nucleated.

The nucleation behavior of Al-Si EHD submicron powders is quantitatively summarized in Figure 11. In general, the proportion of polycrystalline material, the number of nucleation events per particle, and the incidence of twinning all increase as the powder diameter decreases below 1 μm. It was therefore concluded that larger undercoolings promote the occurrence of twins and multiple nucleation. This is consistent with the qualitative observation that the polycrystalline powders are commonly less segregated than their monocrystal counterparts of equivalent size.

The incidence of multiple nucleation can be accounted for in terms of Eq. [4]. Reducing the particle size decreases the volume ($V \propto r_0^3$) and increases the cooling rate ($\varepsilon \propto 1/r_0$), thereby reducing N roughly with a fourth power of the radius. On the other hand, $J(T)$ increases with higher undercoolings ΔT_N, and therefore opposes the volume effect on N. Since $J(T)$ behaves approximately like a step function of ΔT_N, small undercoolings (such as those expected in large droplets) might not affect its magnitude significantly, and the number of grains would decrease with particle size. However, when substantial undercoolings are achieved (as in the submicron powders) $J(T)$ may increase rapidly with ΔT_N, which varies inversely with r_0. There is conceivably a size range in which the increase in nucleation frequency overcomes the volume effect, enhancing the probability for multiple nucleation.

The occurrence of twins is a phenomenon more difficult to rationalize than multiple nucleation. Twins are considered rather rare in aluminum,[31] although they have been reported to occur frequently in evaporated films,[32] presumably due to stacking errors in the faceted growth from the vapor. Furthermore, most aluminum twins reported in the literature show boundaries with coherent and incoherent sections.

Twins observed in the EHD-produced powders are definitely formed during solidification, as evidenced by the segregation pattern in Figure 10. Note that an incipient cell

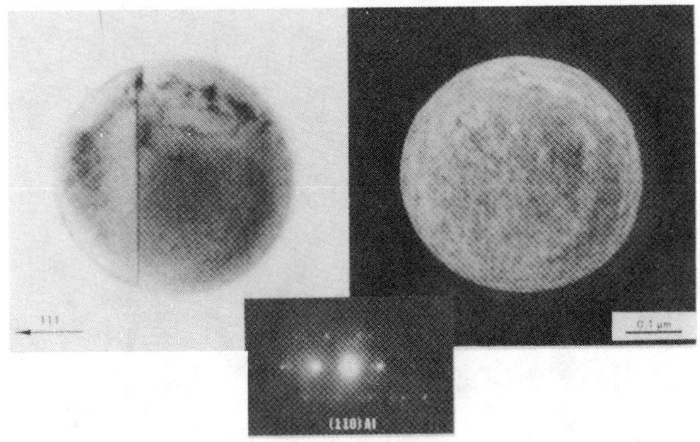

Fig. 10—TEM of twin crystals in an Al-6 wt pct Si alloy powder. The segregation pattern suggests that nucleation occurred at a point on the line intersection of the powder surface and the twin boundary.

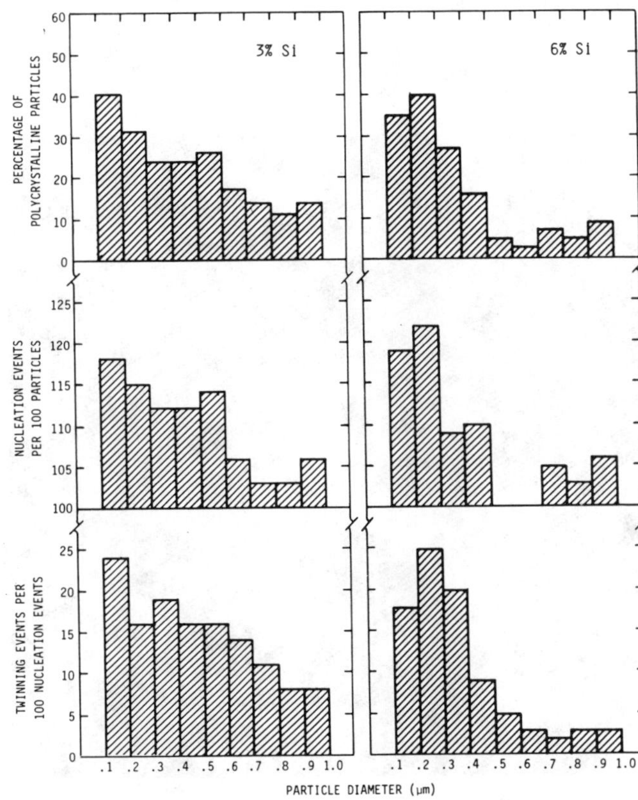

Fig. 11—Nucleation behavior of aluminum-silicon powders produced by the EHD atomization process. Results are from examination of over 400 particles in each alloy composition.

within the larger grain has penetrated the smaller one, disrupting the otherwise planar boundary. In those twins where a growth direction could be identified, the nucleation site apparently lies on the junction of the twin boundary and the droplet surface. It was then concluded that twins originate from a stacking "accident" in the nucleation process, and the boundary so created propagates later as the solid front advances.

Although it is anticipated that the preservation of the twin boundary will be energetically more favorable than its degeneration into a noncoherent type, its apparent perfection is remarkable in view of the high solidification rates involved in these powders. Furthermore, twin boundaries are

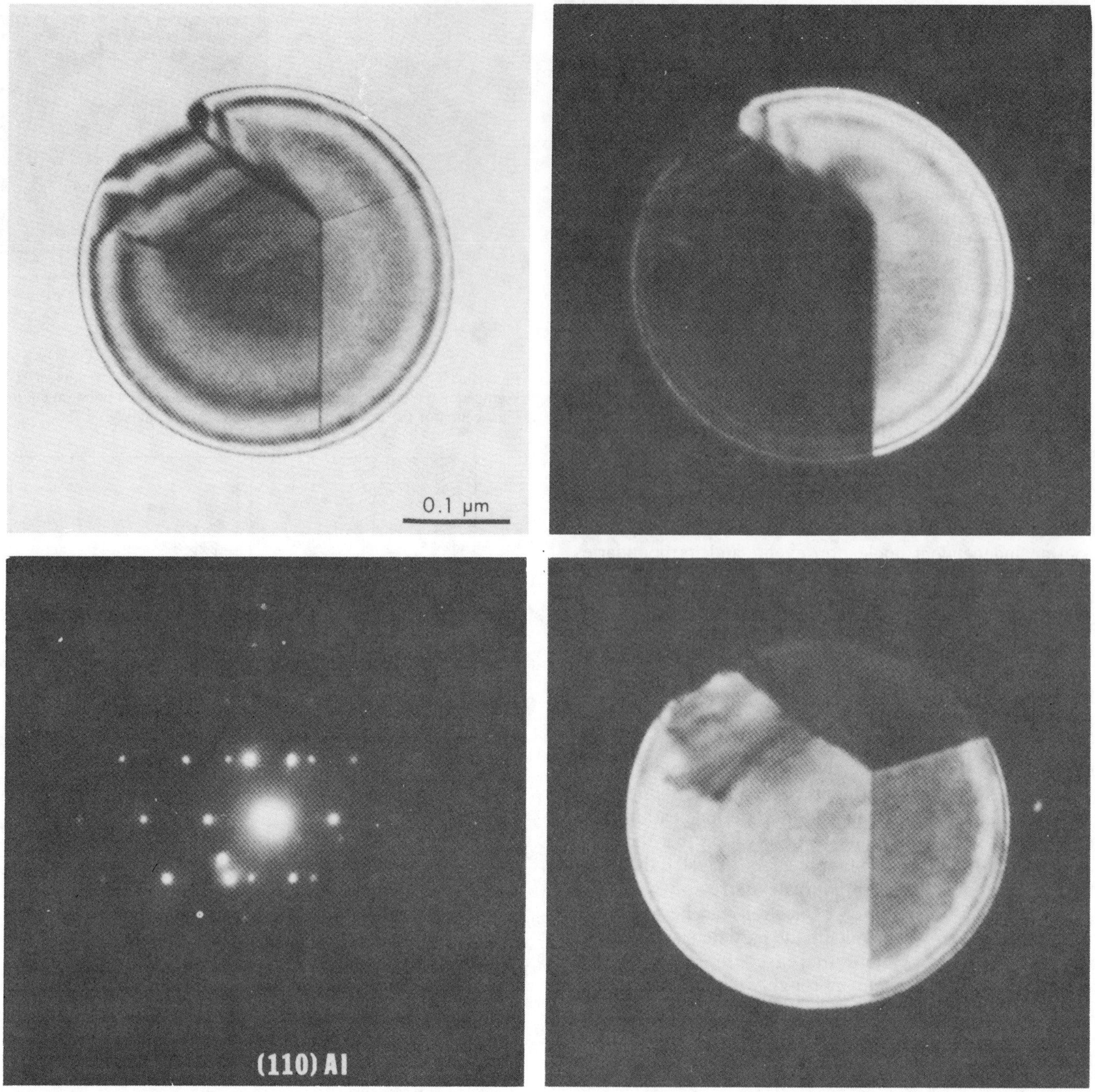

Fig. 12 — TEM images of multiple twins in an Al-3 wt pct Si alloy powder. Two of the boundaries are {111} planes and seem perfectly coherent. A smaller grain with a partially faceted boundary is also evident but out of contrast.

sometimes preferentially decorated with precipitates or may show denuded zones in their vicinity, (*e.g.*, Figure 8). If the boundary were perfectly coherent at the atomic level, these effects could be explained by assuming that the segregation conditions where the boundary emerges into the liquid are different from those at the rest of the interface. It could also be argued that small irregularities occur at the boundary during growth, and their further rearrangement in the solid state (to reduce surface energy) can accommodate vacancies or solute atoms.

A small fraction of the powders exhibiting this phenomenon (<10 pct) show more than one twin boundary (see Figure 12) with as many as four crystallographically related grains being detected in some cases. In addition, twins with boundaries showing incoherent segments were also found. These variations of the structure seem unaccountable at the moment if we ascribe the origin of the twins only to the nucleation process.

VI. SUMMARY

The availability of submicron powders generated by the electrohydrodynamic atomization process has provided a suitable tool for fundamental solidification studies. In addition to having the potential to achieve substantial undercoolings prior to nucleation, these powders are electron-transparent and permit the microstructural examination of the complete solidification structure of a small casting.

The present investigation revealed that decreasing the particle size – hence increasing the achievable undercoolings and liquid/solid interface velocities during growth – reduced the extent of solidification segregation. The consequent supersaturation of the primary phase eventually results in homogeneously solidified powders with silicon contents as high as six wt pct.

The observation of interface breakdown from planar to cellular in the middle of powder particles coupled with a solute depleted zone lends credence to the hypothesis that plane-front homogeneous solidification was achieved due to the substantial undercoolings prior to nucleation and the high interface velocities during recalescence.

Some powder particles showed segregation but no cellular breakdown. This may be evidence of interface morphological stability due to surface tension effects, since the temperature gradient ahead of the interface is expected to be negative throughout solidification and therefore favor the growth of perturbations.

It was also found that increasing undercooling by reducing particle size promotes multiple nucleation and the incidence of twin crystals in submicron powders. The latter are believed to originate mainly from stacking "accidents" in the nucleation process. The twin boundaries appear to be coherent, although a few instances of incoherent segments were observed. In general, twins and polycrystalline powder particles show less segregation than their monocrystal counterparts.

ACKNOWLEDGMENTS

This investigation was sponsored by the Defense Advanced Research Projects Agency and monitored by the Office of Naval Research under contract N00014-78-C-0275. The program manager was Dr. E. C. van Reuth and the technical monitor was Dr. B. A. MacDonald. The authors are indebted to Dr. J. Perel and Mr. J. F. Mahoney of Phrasor Scientific Inc. for the production of the EHD atomized powders. Finally, enlightening discussions with Drs. J. W. Cahn, W. J. Boettinger, S. R. Coriell, and R. J. Schaefer of the National Bureau of Standards; Prof. R. F. Sekerka of Carnegie-Mellon University; and Prof. H. L. Fraser of the University of Illinois are gratefully acknowledged.

REFERENCES

1. R. Mehrabian: *Rapid Solidification Processing: Principles and Technologies*, R. Mehrabian, B. H. Kear, and M. Cohen, eds., Claitor Publishing Division, Baton Rouge, LA, 1978, p. 9.
2. J. H. Perepezko: *Rapid Solidification Processing: Principles and Technologies II*, R. Mehrabian, B. H. Kear, and M. Cohen, eds., Claitor Publishing Division, Baton Rouge, LA, 1980, p. 56.
3. D. Turnbull: *J. Chem. Phys.*, 1952, vol. 20, p. 411.
4. J. H. Hollomon and D. Turnbull: *Prog. Metal Phys.*, 1953, vol. 4, p. 333.
5. D. Turnbull and R. E. Cech: *J. Appl. Phys.*, 1950, vol. 21, p. 804.
6. J. H. Perepezko, D. H. Rasmussen, I. E. Anderson, and C. R. Loper, Jr.: *Solidification and Casting of Metals*, Proc. of Conference held at the University of Sheffield, July 18 to 21, 1977, The Metals Society, London, 1979, p. 169.
7. J. Perel, J. F. Mahoney, P. Duwez, and B. E. Kalensher: *Rapid Solidification Processing: Principles and Technologies II*, R. Mehrabian, B. H. Kear, and M. Cohen, eds., Claitor Publishing Division, Baton Rouge, LA, 1980, p. 287.
8. C. G. Levi and R. Mehrabian: Submitted for publication, *Metall. Trans. A*, in press.
9. H. Jones: *Aluminum*, 1978, vol. 54, No. 4, p. 274.
10. K. O. Krishnanand and R. W. Cahn: *Rapidly Quenched Metals*, N. J. Grant and B. C. Giessen, eds., M.I.T. Press, Cambridge, MA, 1976, p. 67.
11. G. Thursfield and M. J. Stowell: *J. Mater. Sci.*, 1974, vol. 9, p. 1644.
12. C. Panseri and M. Paganelli: *Alluminio*, 1968, vol. 37, p. 387.
13. N. J. Grant: *Fizika*, 1970, vol. 2, No. 2, p. 16.1.
14. J. C. Baker and J. W. Cahn: *Solidification*, American Society for Metals, Metals Park, OH, 1971, p. 23.
15. M. Hillert and B. Sundman: *Acta Met.*, 1977, vol. 25, p. 11.
16. J. W. Christian: *The Theory of Transformations in Metals and Alloys*, Oxford University Press, Oxford, England, 1965, ch. 14.
17. J. P. Hirth: *Metall. Trans. A*, 1978, vol. 9A, p. 401.
18. W. A. Tiller, J. W. Rutter, K. A. Jackson, and B. Chalmers: *Acta Met.*, 1953, vol. 1, p. 428.
19. W. W. Mullins and R. F. Sekerka: *J. Appl. Phys.*, 1964, vol. 35, p. 444.
20. R. F. Sekerka: *J. Appl. Phys.*, 1965, vol. 36, p. 264.
21. S. R. Coriell and R. F. Sekerka: *Rapid Solidification Processing: Principles and Technologies II*, R. Mehrabian, B. H. Kear, and M. Cohen, eds., Claitor Publishing Division, Baton Rouge, LA, 1980, p. 35.
22. J. W. Cahn, S. R. Coriell, and W. J. Boettinger: *Laser and Electron Beam Processing of Materials*, C. W. White and P. S. Peercy, eds., Academic Press, New York, NY, 1980, p. 89.
23. M. Cohen, B. H. Kear, and R. Mehrabian: *Rapid Solidification Processing: Principles and Technologies II*, R. Mehrabian, B. H. Kear, and M. Cohen, eds., Claitor Publishing Division, Baton Rouge, LA, 1980, p. 1.
24. J. C. Baker: "Interfacial Partitioning During Solidification", Ph.D. Thesis, M.I.T., 1970, ch. 5.
25. *Thermophysical Properties of Matter*, TPRC Data Series, Thermal Radiation Properties: Metallic Elements and Alloys, Y. S. Touloukin and D. P. DeWitt, eds., I.F.I. Plenum Data Corporation, New York, NY, 1970, vol. 7, p. 2.
26. K. M. Entwistle, P. Fuller, and I. Brough: *Acta Met.*, 1978, vol. 26, p. 1055.
27. L. F. Mondolfo: *Aluminum Alloys: Structure and Properties*, Butterworths, London, England, 1976, p. 373.
28. M. Itagaki, B. C. Giessen, and N. J. Grant: *Trans. Quart. ASM*, 1968, vol. 61, p. 330.
29. S. K. Bose and R. Kumar: *J. Mater. Sci.*, 1973, vol. 8, p. 1795.
30. H. Matyja, K. C. Russel, B. C. Giessen, and N.J. Grant: *Metall. Trans. A*, 1975, vol. 6A, p. 2249.
31. C. M. Sargent: *Trans. TMS-AIME*, 1968, vol. 242, p. 1188.
32. M. J. Olney and G. C. Smith: *Acta Met.*, 1955, vol. 3, p. 205.

Rapid Solidification Rate Processing and Application to Turbine Engine Materials

R.J. Patterson II,* A. R. Cox,* and E. C. van Reuth**
*Pratt & Whitney Aircraft Group
Florida
**Defense Advanced Research Projects Agency
Virginia

INTRODUCTION

The ability to minimize chemical segregation, eliminate the formation of massive phases, and increase the solubility of alloying elements has been a goal in materials research for some years. In the area of casting, improvements were found, to some degree, through reduction of the size of the casting and through refinements to the control of the solidification process; dendrite arm spacing, for example, was found to decrease[1,2] as solidification time decreased. Widespread interest in rapid rates of solidification, however, dates to 1960, when Klement et al.[3] reported that one could bypass crystallization on solidification of at least some alloy systems if the rate were high enough.

In this paper, we will review the techniques for forming rapidly solidified materials. We will then restrict the balance of the paper to the Pratt & Whitney Aircraft "RSR" powder process, with which we are most familiar. Microstructural features of the powder and of the consolidated materials will be highlighted, properties of some alloys will be discussed, and applications seen for these materials will be reviewed.

RAPID SOLIDIFICATION PROCESSING

The rate of solidification depends partly on the rate of heat extraction, which depends not only on the nature of the process but also on the physical properties and dimensions of the materials. While no specific quench rate conveniently separates "rapid" techniques from "conventional" techniques, the typical realm of interest reported in the literature[4] ranges upward from about 10^4K/s.

Polk et al.[4] conveniently divide the processes into two broad categories: those which involve atomic deposition, and those which involve rapid solidification from the melt. Atomic deposition techniques include various sputtering and vapor deposition processes which form an alloy by consecutive deposition of atomic layers onto a substrate. Rapid solidification techniques, many and varied, produce materials ranging from thin ribbon to powder to deformed particulate called "splat." The ribbon is typically formed by melt extraction and melt spinning, in which a thin foil of material is solidified on the surface of a rapidly rotating drum; variations include roller quenching, in which molten material is injected between counterrotating drums, and the pendant-drop technique, in which a molten droplet is brought into contact with one (or more) rotating drums. While asymmetric flake, or splat, may be formed by the previous processes (for example, by serrating the surface of the drum), earlier studies of rapidly quenched materials were based on the "gun" and the "piston and anvil" techniques. In the former, a small droplet is projected at high velocity against a cold substrate, while in the latter, a droplet is trapped between approaching substrates. Other techniques such as plasma deposition (in which a plasma gun is used to form and accelerate many droplets) and self-quenching (in which a thin layer on the surface of a material is melted by a sharply focused high-energy beam) have also been developed. Overviews of these techniques are available from a number of references.[4-7]

The production of rapidly solidified materials in the form of powder differs from the techniques discussed previously in that quenching is accomplished not by conduction to a substrate, but rather by conduction and convection to a fluid (and in some cases, by radiation). A variety of techniques have been developed, including the more conventional gas and water atomization methods in which a stream of the molten alloy is disintegrated into droplets subsequently quenched by the fluid.[8] In vacuum atomization, a melt supersatured with a soluble gas is introduced to a large vacuum chamber, whereupon the gas comes out of solution and disintegrates the melt into droplets.[9] In the former processes, particles can be large and, depending on materials properties, cooling rates may be substantially lower than 10^4 K/s. To improve the situation, more recent techniques have been developed which rely largely upon substantial reduction in particle size to achieve reduced particle volume and increased surface-to-volume ratios and, hence, increased quench rates. Such methods include ultrasonic atomization and electrostatic atomization.[5,10] While substantial improvements to quench rate have been gained by reducing particle size, these techniques have thus far suffered from low production rates.

An alternative procedure would be to produce the droplets mechanically. Such techniques include the rotating electrode process (REP) and the rapid solidification rate process (RSR). In REP, a precisely machined bar of the desired material is rotated in a chamber; the end of the bar is melted with the fluid being flung off by centrifugal forces to form the ultimate powder.[10,11] Typically, particle size is moderate due to speed limitations on the rotating bar, and chemistry may vary if the bar is not homogeneous. The technique has the advantage of ceramic-free melting, so refractory and reactive alloys may be readily produced in the form of powder.

The RSR process also uses rotary atomization, but in this case a molten alloy is poured at controlled rates onto a high-speed (typically 400 revolutions per second) disc; after atomization, the particles are quenched by high-velocity helium.[5,12] Quench rates vary with particle size and alloy but are typically greater than 10^5 K/s for powder finer than

The views and conclusions in this report are those of the authors and should not be interpreted as necessarily representing the official policies, either expressed or implied, of the Defense Advanced Research Projects Agency, U.S. Air Force, or the U.S. Government.

Reprinted with permission from Journal of Metals, September 1980, 34-39, © 1980 The Metallurgical Society of AIME about 100 μm diameter.[13] The process has the advantage of achieving reasonable quench rates while producing powder at commercially viable rates, but with the present crucible melting system it is now limited to nonrefractory and nonreactive melts. Recent improvements to the system include a closed-loop gas recirculation system, increased melt capacity, and higher atomization speed (which results in finer powder).

HEAT TRANSFER

Experimental determination of the quench rate occuring during rapid solidification is difficult, at best, and results are confined to various splat techniques.[6] Analytical considerations of the problem are found in the literature for both splat quenching and spherical powder.[5,14-17] All are approximations to a greater or lesser extent since a variety of factors such as the degree of undercooling achieved and the progression of nucleation and growth are not precisely known and, in the case of nucleation and growth, are difficult to include in mathematical models. The models are thus somewhat limited but are useful in estimating the significance of factors such as particle size or splat thickness, the thermal properties of the melt and quench medium, the degree of undercooling, and so on. In general, these efforts indicate that quench rates greater than 10^4 K/s can be achieved for most materials if powder diameter (or splat thickness) is kept below about 100 μm and, in the case of powder, if the particles have moderate velocity in a high conductivity fluid (such as helium).

MICROSTRUCTURAL FEATURES OF RSR POWDER

The response of an alloy to rapid solidification depends not only on the rate of quenching but also on the degree of undercooling achieved before nucleation, which is known to depend upon alloy chemistry. At Pratt & Whitney Aircraft, three alloy powders produced by the RSR process were examined to determine the influence of particle size and chemistry on internal microstructure. The alloys selected were MAR-M200, B1950, and Monoloy 444; nominal compositions are given in Table I. These nickel-base superalloys are similar except for the levels of carbon and boron. Precision sieves (having a tolerance of ±2 μm) were used in closely spaced pairs to separate particles at well-defined sizes; the fractions were then mounted and examined with the electron microscope to determine general structure and secondary dendrite arm spacing. The variation of the latter is shown in Figure 1; note that the spacing seems to be inversely correlated with carbon content.

Two different structures have also been found in the superalloy powders produced at Pratt & Whitney. One is the classical dendritic structure; the other seems to consist of discrete crystals and has come to be called "microcrystalline." Representative structures, obtained by scanning transmission electron microscopy (STEM) on thinned particles, are shown in Figure 2. The occurrence of these structures was also found to depend on chemistry and quench rate (particle size), and the results obtained are summarized in Table II. The formation of the microcrystalline form is believed due to the degree of undercooling achieved before nucleation begins; and since nucleation is sensitive to temperature, time, and chemistry it is not surprising that smaller particles, which experience higher quench rates, show larger percentages of this form.[18,19] The forms are also interesting because of the differences in segregation which have been found; results for a modified nickel-molybdenum-aluminum alloy are shown in Figure 3.

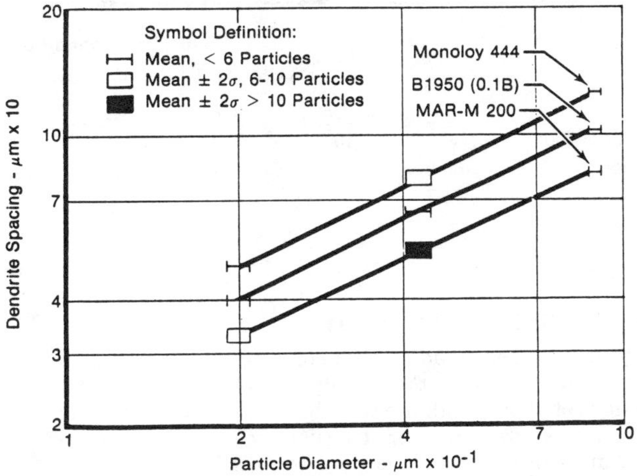

Figure 1. Variation of secondary dendrite arm spacing with particle size and alloy chemistry.

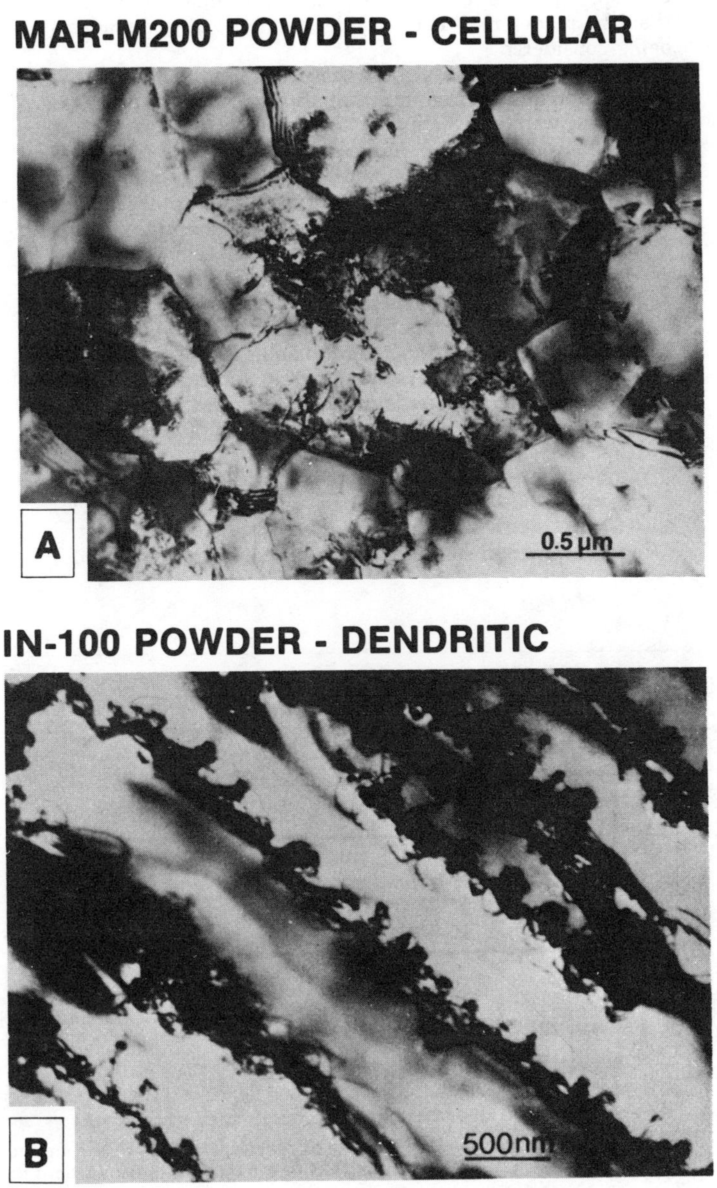

Figure 2. Thin foil STEM of (A) a microcrystalline superalloy particle; (B) a dendritic superalloy particle. Primary arm is seen running from center top, with secondary arms running toward the lower right. Note the carbides and associated dislocation loops in the interdendritic regions. ▶

Table I: Composition of Alloys, wt. %

	Ni	Co	Cr	Al	Ti	C	B	Zr	W	Cb
MAR-M200	60.3	10.0	9.0	5.0	2.0	0.15	0.06	12.5	1.0	
B1950	60.2	10.0	9.0	5.0	2.0	0.01	0.10	0.06	12.5	1.1
Monoloy 444	70.7	--	9.0	5.2	2.1	--	--	--	12.0	1.0

RSR SUPERALLOY STUDIES

Alloy development studies were conducted with the rapid solidification process for the purpose of achieving high concentrations of alloying additions without deleterious effects of macro/micro-segregation or transient compound formation during solidification. The work was confined to the use of the precipitation hardening class of superalloys, with particular emphasis placed on alloys of high (>60%) γ' content and on alloys based on the Ni-Al-Mo ternary. The objective of the study was to define the feasibility to alloy through rapid solidification for improving the creep resistance of materials used for jet engine blades and vanes operating in the temperature range of 760−1100°C. Table III lists typical compositions of the two types used in this evaluation in order that the reader may better understand the alloys concepts being considered.

With respect to alloying and the subsequent effect of rapid solidification on suppression of phase reactions, the most obvious ones were complete suppression of the γ' precipitation in all alloys, and α Mo suppression in the Ni-Al-Mo series of compositions. This is in contrast to the same alloys processed in a more conventional manner where, typically, not only do the γ' and α Mo phases precipitate, they do so with significant differences relative to alloy segregation and are accompanied by massive eutectic γ-γ'. In the high γ' alloys, where up to ~78% of the alloy was subsequently heat treated to γ', conventional solidification results in eutectic phases about 7−10 vol.% of the total alloy. Similarly for α Mo occurrence, conventional solidification processing of the Ni-Al-Mo alloys used in the evaluation causes α Mo precipitation which accounts for more than 50% of the total Mo starting concentration.

We had hoped that the rate of solidification used in this study would be sufficiently fast to suppress reactions associated with the interstitials carbon and boron. Such was the case with particles solidifying in the microcrystalline mode; however, the dendritic powders continued to exhibit the typical monocarbide and boride reactions normally associated with compositions of these types.[20] Mean monocarbide/boride precipitate size was measured to be about 0.1 μm in the dendritic powders. Because we were unable to isolate the microcrystalline form from the dendritic, further work was based on interstitial reactions observed in the dendritic materials.

Powder consolidation and metalworking operations were carried out with the rapidly solidified powders, which included hot isostatic pressing (HIP), extrusion, isothermal forging, and combinations thereof. Consolidation was commenced in all trials by pouring −140 mesh (105 μm) powders into stainless steel containers, which were initially evacuated to approximately 10^{-6} torr, and sealed after filling.

Table II: Particles Found to be Microcrystalline, %

Alloy	Particle Size 90μm	62μm	43μm	20μm
MAR-M200	1.0	--	0.09	37.7
B1950	2.7	--	34.7	50.4
Monoloy 444	3.7	15.8	27.1	66.4

Table III: Experimental Alloy Compositions, At.%

	Ni	Cr	Al	Mo	W	C
High λ'	72	10	17	--	--	1
	68	11	17	--	3	1
Ni-Al-Mo	72	--	17	11	--	--
	72	--	17	9	2	--

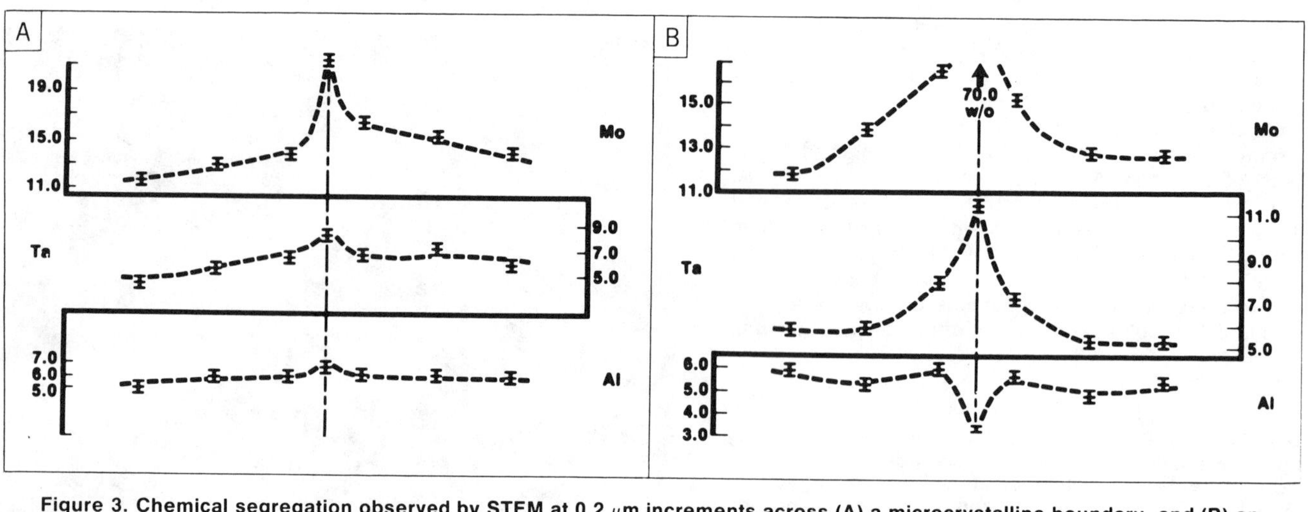

Figure 3. Chemical segregation observed by STEM at 0.2 μm increments across (A) a microcrystalline boundary, and (B) an interdendritic region, in modified Ni-Mo-Al particles. Vertical scales in weight percent.

HIP and HIP + isothermal forging were satisfactory for consolidation and forming, and were best accomplished in the temperature range of 0.8−1.0 of the principle secondary phase solvus. Recrystallization was not controlled by so doing, however, in that apparently those particles referred to as microcrystalline would undergo superplastic deformation during consolidation and deform to fill the interstices, resulting in little, if any, deformation of the coarse dendritic particles. During isothermal forging, the same result occurred. Subsequent annealing cycles at high temperatures (~0.9 T) were effective in eliminating the cast structure but at the expense of grain coarsening, a condition we were trying to avoid. HIP consolidation followed by extrusion, or direct extrusion of the loose powders, led to a consolidated product with a uniform recrystallization pattern. The extrusion conditions we judged best included temperatures again in the range of 0.8−1.0 of the principle secondary phase solvus and reduction ratios in excess of 8 to 1. Grain sizes about ASTM 8−14 (depending on actual parameters) were achieved when processed this way.

Phase changes resulting from the metalworking cycles, as well as response to continuing heat treatment, were evaluated for each class of alloys subsequent to consolidation. Precipitation of the γ' phase occurred in all alloys as a result of the thermal exposure associated with working. Coalescence and dissolution rates, as the temperature was raised to the phase solvus, appeared slower than in typical cast structures. For the Ni-Al-Mo series of compositions, this was attributed to an experimentally determined increase in refractory element concentration (up to 9 wt.% Mo in the Ni-Al-Mo series) in the precipitated phase above that measured in casting equivalents (up to 5 wt.%). For the high γ' alloys, it was attributed to absence of γ-γ' eutectic and total utilization of the γ' forming elements. In all alloys, final dissolution temperatures were accurately repeatable to within limits of measurements (better than 10°C).

Carbide and boride changes were closely associated to the specific composition involved. Typically, it was determined that growth kinetics during various short time (~ 1 h) thermal excursions to the melt point resulted in volume increases up to and beyond 30 times the starting particle sizes, with little change occurring because of additional time at temperature. For some alloys, such as those involving WC, it was possible to effect dissolution of the phase entirely prior to remelt; in others, such as those which reacted to TaC, coalescence continued beyond initial alloy melting.

The Ni-Al-Mo alloys, where Mo concentrations were less than 20 wt.%, reacted to a γ-γ'-Ni$_x$Mo upon consolidation and subsequently underwent secondary phase dissolution as the temperature was brought near T_m. Alloys in excess of about 20 wt.% Mo reacted to a γ-γ'-Ni$_x$Mo-αMo, with α Mo persisting beyond initial melting. The series of reactions which occur during heating and reprecipitation were vastly different from those observed in casting, and the reader is referred to References 21 and 22 for a comprehensive treatment.

The incipient melt temperatures of all alloys initially processed as rapidly solidified powders were higher in every case than identical compositions processed as castings. The typical increase was about 75−100°C. For those alloys in which secondary phase dissolution could be achieved prior to melting, it was possible to effect abnormal grain growth, or secondary recrystallization, by annealing slightly above the phase solvus. Further, it was possible to control the grain growth by zone annealing so that aligned grain structures of the type shown in Figure 4 could be produced. The crystallization orientation of the high γ' alloys was <011> subsequent to zone annealing, <111> in the Ni-Al-Mo alloys.

Creep-rupture mechanical property testing was the principal screening method used to evaluate the effects of modified alloying relative to the program objective of better creep resistance. Aligned grain samples, with the grains

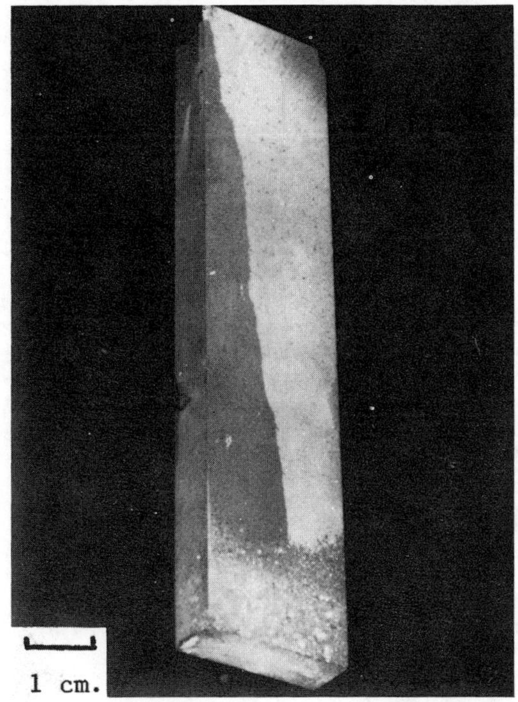

Figure 4. Aligned grain structure in Ni-Al-Mo alloy.

aligned parallel to the direction of applied stress, were used on all tests. Test temperatures covered the range noted previously, and stresses were varied from 690 MPa at the low end to 93 MPa at the high end of the temperature range.

The high γ' series of alloys was heat treated, after establishing abnormal grain growth, to re-solution the γ' phase, followed by somewhat arbitrarily selected 1080°C/4 h/AC + 870°C/12 h/AC age cycles. Test results showed that these alloys were essentially equivalent in creep strength to alloys now used for jet engine turbine airfoils. Alloy thermal stability, in post-test evaluation, showed no evidence of phase degeneration or reaction other than an expected secondary phase coalescence at the high temperature end of the test range. The most interesting feature of this series was that the alloys exhibited outstanding oxidation resistance. In both dynamic and static tests conducted up to 1150°C, the oxidation resistance of these alloys was judged to be 10 times better than that obtained from alloys currently used in the engine. These data reflecting creep and oxidation resistance are shown in Figures 5 and 6.

The Ni-Al-Mo series were heat treated in the manner described above. In testing, this series of compositions attained creep strength levels about 40−80°C higher than possible with current alloys. One alloy, a Ni-7%Al-14%Mo-6%W composition achieved the best overall strength properties. The results are shown in Figures 7 and 8. The strength was attributed to the γ-γ'-Ni$_x$Mo partitioned structure and phase dispersion which remained stable under all conditions of testing.

APPLICATIONS

The alloys referred to above are presently being studied for use in hollow turbine airfoils in which flat wafers with complex passages for cooling are joined together and subsequently machined to final configuration. The configuration is being used because of inherently high cooling effi-

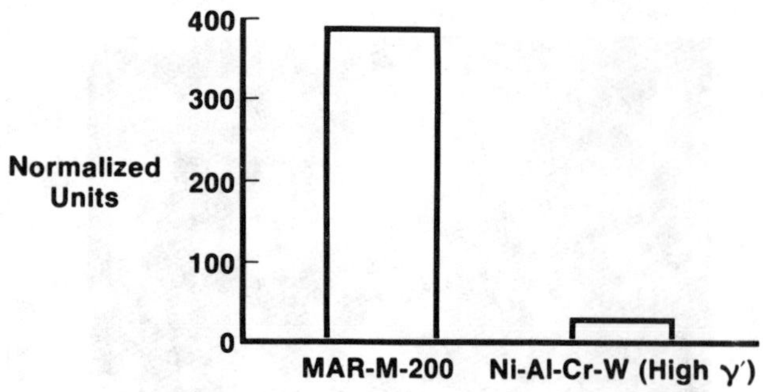

Figure 5. Oxidation degradation of current alloy (MAR-M200) and high λ′ alloy after 370 h at 1080°C.

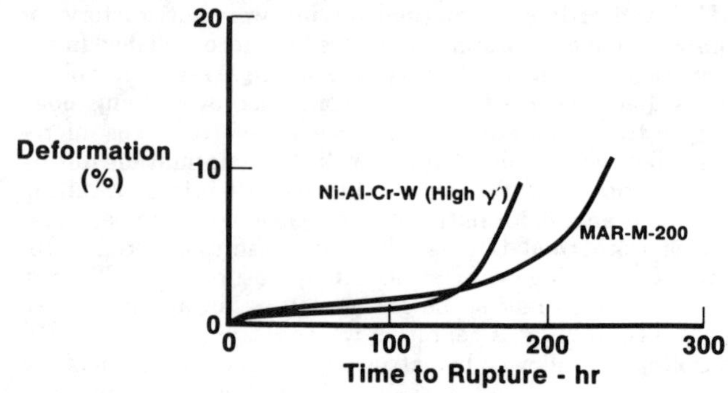

Figure 6. Creep deformation of current alloy (MAR-M200) and high λ′ alloy at 99 MPa and 1037°C.

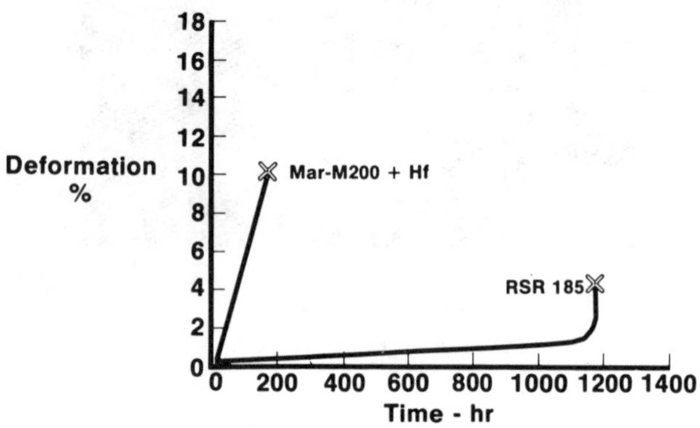

Figure 7. Creep deformation of current alloy (MAR-M200) and Ni-Al-Mo alloy (RSR 185) at 209 MPa and 982C.

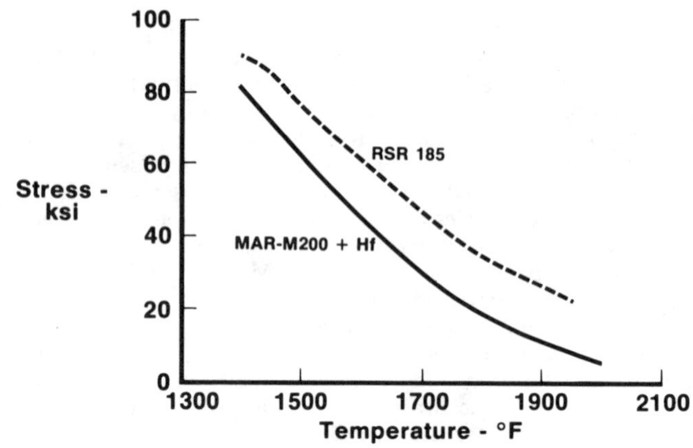

Figure 8. Comparison of 1% creep resistance of current alloy (MAR-M200) and Ni-Al-Mo alloy (RSR 185).

Figure 9. Wafer turbine blade (A) schematic, and (B) part produced from Ni-Al-Mo alloy.

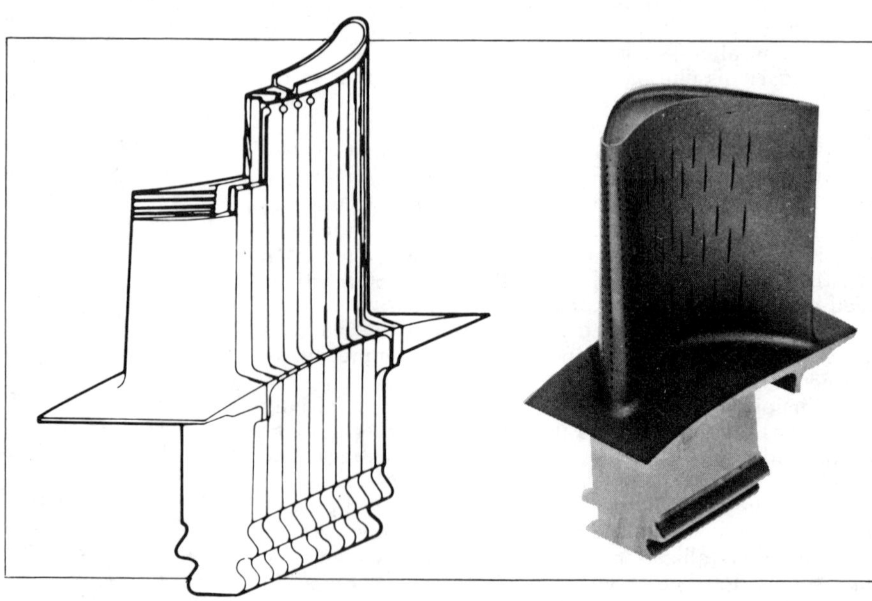

ciency, a requisite for advanced jet engines, and because the concept lent itself effectively to fabrication from wrought, powder alloy. The wafers are prepared as flat stock, cooling details machined, assemblies bonded together, heat treated to produce an aligned grain structure, and machined by electrochemical means to provide the internal plenum chamber and the aerodynamic shape. The appearance of the final turbine blade configuration is shown in Figure 9. Analytical studies indicate that the combination of advanced materials in wafer airfoil configuration can withstand major increases in turbine inlet temperature which, in turn, can be applied for greater thrust-to-weight capability for jet engines.

CONCLUSIONS

The extension of rapid solidification processing to superalloy compositions allowed alloy modifications which were characteristically outside the range possible by current superalloy process methods. Eutectic regions were eliminated in high γ' alloys, so that higher additions of γ' partitioning elements could be added. BCC Mo occurrence in the Ni-Al-Mo-type alloys was controlled to produce effectively dispersed phases. Incipient melt temperatures were high to enable effective heat treat, and adaptability of the powders to subsequent fabrication was achieved. Higher strength levels than current materials, as well as better second order properties such as oxidation resistance, resulted because of the effective alloying obtained with rapid solidification. Combined with new techniques for fabricating and designing turbine airfoils, these materials promise improved performance features for advanced engines.

ACKNOWLEDGMENTS

This work is being sponsored by the U.S. Defense Advanced Research Projects Agency, Arlington, Virginia, and the (USA) WAL/Materials Laboratories, Wright-Patterson Air Force Base, Ohio. The authors thank Dr. Hamish Fraser for his work on the characterization of microcrystalline and dendritic powders.

References

1. H. Matyja, B.C. Giessen, and N.J. Grant, "The Effect of Cooling Rate on the Dendrite Spacing in Splat-Cooled Aluminum Alloys," **J. Inst. Metals, 96** (1968) p. 30-32.
2. P.A. Joly and R. Mehrabian, "Complex Alloy Powders Produced by Different Atomization Techniques: Relationship Between Heat Flow and Structure," **J. Mat. Sci., 9** (1974) p. 1446-1455.
3. W. Klement, R.H. Willens, and P. Duwez, **Nature (London),** 187 (1960) p. 869-870.
4. D.E. Polk, B.C. Giessen, and F.S. Gardner, "State-of-the-Art and Prospects for Magnetic, Electronic and Mechanical Applications of Amorphous Metals," **Mat. Sci. Eng.,** 23 (1976) p. 309-316.
5. National Materials Advisory Board, *Amorphous and Metastable Micocrystalline Rapidly Solidified Alloys: Status and Potential* (NMAB 358), National Technical Information Service, Springfield, Virginia, May 1980.
6. H. Jones, "Splat Cooling and Metastable Phases," **Rep. Prog. Phys.,** 36 (1973) p. 1425-1497.
7. N.J. Grant and B.C. Giessen, eds., *Rapidly Quenched Metals,* MIT Press, Cambridge, Massachusetts, 1976.
8. W.L. Batten, "Apparatus for Atomizing Molten Metal," U.S. Patent 2,956,304 (1960).
9. J.M. Wentzell, "Methods for Powdering Metals," U.S. Patent 3,510,546 (1970).
10. A. Lawley, "An Overview of Powder Atomization Processes and Fundamentals," **Int. J. Powder Met. and Powder Technology,** 13 (3) (1977) p. 169-188.
11. A. R. Kaufmann, "Method and Apparatus for Making Powder," U.S. Patent 3,099,041 (1963).
12. P.R. Holiday and R.J. Patterson, "Apparatus for Producing Metal Powder," U.S. Patent 4,078,873 (1978).
13. A. Lawley, "Preparation of Metal Powders," *Annual Review of Materials Science,* Vol. 8, Annual Reviews Inc., Palo Alto, California, 1978.
14. R.C. Ruhl, "Cooling Rates in Splat Cooling," **Mat. Sci. Eng.,** 1 (1967) p. 313-320.
14. C.G. Levi and R. Mehrabian, "Heat Flow in Atomized Metal Drops," **Met. Trans., 11B** (3) (1980) p. 21-27.
16. P.R. Holiday, A.R. Cox, and R.J. Patterson II, "Rapid Solidification Effects on Alloy Structures," p. 246-255 in *Rapid Solidification Processing Principles and Technologies,* ed. R. Mehrabian, B.H. Kear, and M. Cohen; Claitor's Publishing Division, Baton Rouge, Louisiana, 1978.
17. M.R. Glickstein, R.J. Patterson II, and N.E. Shockey, "Rapid Solidification Effects of Micron-Size Droplets," p. 46-63, ibid.
18. J.P. Hirth, "Nucleation, Undercooling and Homogeneous Structures in Rapidly Solidified Powders," **Met. Trans., 9A** (3) (1978) p. 401-404.
19. J.H. Perepezko and D.H. Rasmussen, "Discussion of 'Nucleation, Undercooling and Homogeneous Structures in Rapidly Solidified Powders'," **Met. Trans., 9A** (10) (1978) p. 1490-1492.
20. R.D. Field and H.L. Fraser, "Microstructural Observations of Metal Powders Using Analytical Electron Microscopy," p. 270-277 in *Rapid Solidification Processing Principles and Technologies,* ed. R. Mehrabian, B.H. Kear, and M. Cohen; Claitor's Publishing Division, Baton Rouge, Louisiana, 1978.
21. R.G. Bourdeau and J.B. Moore, ibid., p. 334-346.
22. E.H. Aigeltinger, S.R. Bates, R.W. Gould, J.J. Hren, and F.N. Rhines, ibid., p. 291-305.

ABOUT THE AUTHORS

R. J. Patterson, II is an Engineering Specialist with the Pratt & Whitney Aircraft Group, Government Products Division. He received his BS in Physics with honors from the University of Florida. Since joining Pratt & Whitney in 1973, he has been active in research associated with powder metallurgy, quench from melt processes, and crystallographic and phase changes resulting from rapid solidification. He is co-inventor of the Pratt & Whitney rapid solidification process.

A. R. Cox is Supervisor, Advanced Materials Technology, Pratt & Whitney Aircraft Group, Government Products Division. He received his BS in Chemical Engineering and MS in Physical Metallurgy from Georgia Institute of Technology. Since joining Pratt & Whitney in 1965, he has been active in the engineering development of new materials and material processes for jet-engine components.

E. C. van Reuth is Director for Materials Sciences at DARPA. Prior to his present position, he was affiliated with the Naval Ship R&D Laboratory at Annapolis, Md., from 1964 to 1972 as Head of the Metals Physics and Composite Metals Branches and later as Assistant to the Technical Director. During this period he was a Fellow at the Johns Hopkins University. In 1966, he spent a year's sabbatical at the University of Leiden in the Netherlands. Prior to his positions at Annapolis, he was Assistant Professor of Metallurgical Engineering at Virginia Polytechnic Institute and Staff Metallurgist with the Martin-Marietta Corporation. He holds BS and MS degrees from Virginia Polytechnic Institute and a PhD from the University of Illinois. He has authored over 40 publications, is the co-editor of three books, and has lectured throughout Europe and in Russia.

MICROSTRUCTURE OF RAPIDLY SOLIDIFIED POWDERS

R. D. Field, A. R. Cox,* and H. L. Fraser**
Department of Metallurgical and Mining Engineering
University of Illinois
Urbana, Illinois 61801

Individual rapidly solidified Ni-based powder particles have been studied in the transmission electron microscope (TEM) in the conventional (CTEM) and scanning (STEM) modes of operation. Information on morphology and microchemistry has been obtained and is related to solidifying conditions.

INTRODUCTION

One of the new techniques for the processing of superalloys which has been explored with increasing enthusiasm in the past decade is the use of rapidly solidified powders in the fabrication of superalloy components. The goal of the present study is to characterize rapidly solidified Ni-based powders in terms of microstructure and associated microchemistry and to relate these observations back to the solidifying conditions.

In order to study these powders on a scale appropriate to their fine microstructures it is necessary to resort to the transmission electron microscope (TEM). Therefore, one must have a technique for the preparation of thinned samples of small particles. Such a technique was developed and, using TEM in the conventional (CTEM) and scanning (STEM) modes, morphological, structural, and chemical observations were made on a very fine scale. In this paper, experimental results will be presented for Ni-based alloy powders produced by rapid solidification.

---

*Government Products Division, United Technologies, West Palm Beach, FL 33402
**Now at UTRC, Silver Lane, East Hartford, CT 06108

EXPERIMENTAL PROCEDURE

The Ni-based superalloy powders for this study were supplied by the Pratt and Whitney Aircraft Government Products Division in West Palm Beach, Florida. They were produced by rotating disk atomization and helium quenching. The powders discussed in this paper are of a Ni-based $14^{\omega}/o$ Mo, $6.0^{\omega}/o$ Ta, $5.8^{\omega}/o$ Al alloy.

Thin foils suitable for transmission electron microscopy were produced using a specially devised technique (1). By electroplating a nickel binder on either side of a layer of powders and subsequently thinning the resultant composite foil using conventional techniques it was possible to produce suitably thinned particles.

A secondary electron image of a typical thinned powder particle is shown in Fig. 1. The particle is held by the matrix of electroplated nickel binder and several other powder particles can also be seen in the perimeter of the micrograph. Perforation has occurred in the particle and areas adjacent to the hole are sufficiently thinned for CTEM and STEM work (∿50-500nm).

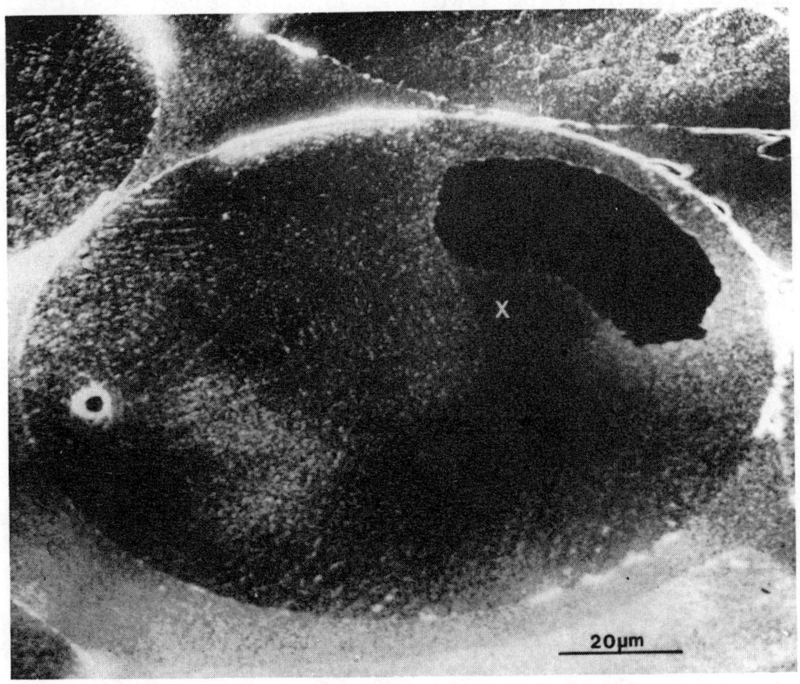

Fig. 1. Secondary electron image of a typical thinned powder particle. Areas adjacent to hole (designated with an "x") are sufficiently thinned for STEM analysis.

RESULTS

Morphological Observations

Three different morphologies were observed in the rapidly solidified powders of several alloys. Results for IN100 and Mar M200 are reported elsewhere (1,2).

The most common morphology was the branched dendritic solidification structure, where both primary and secondary dendrites were observed with growth occurring in the cube directions of the crystal. A second phase was present in the interdendritic regions. The second morphology found had a cellular growth pattern. This differed from the branched dendritic powders in that no secondary dendritic arms were present. Precipitation of an intercellular second phase was also observed in these powders. The third morphology observed had a microcrystalline solidification structure consisting of many individual equiaxed grains with no apparent precipitation in the grain boundaries. The absence of the second phase and the evidence of multiple nucleation sites makes this structure fundamentally different from those in the dendritic and cellular particles.

In Fig. 2 a rare observation of all three of these morphologies in a single powder particle of a Ni-Mo-Ta-Al alloy is shown. Examination of this particle indicates the following

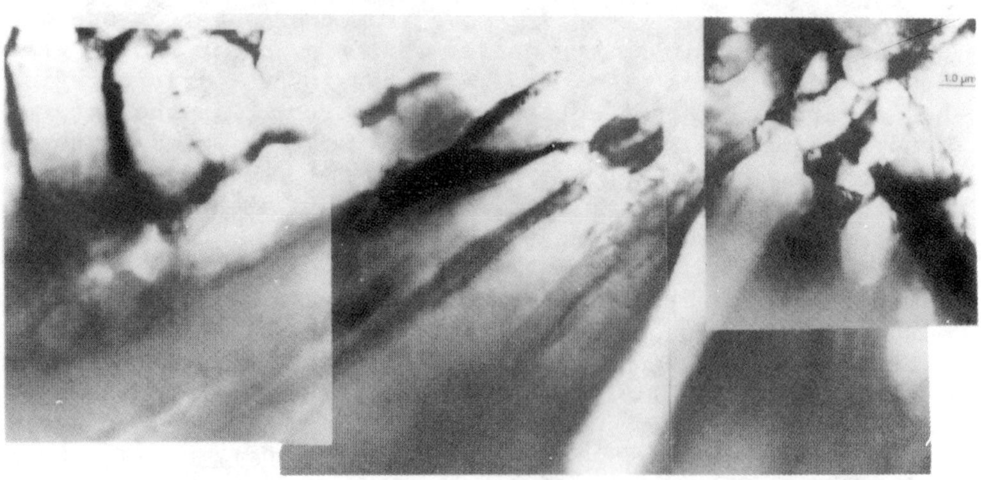

Fig. 2. Transmission electron micrograph of a Ni-Mo-Ta-Al powder particle showing three different morphologies in a single particle.

solidification sequence. Initially several simultaneous nucleation events occurred in the liquid droplet giving rise to the microcrystalline region. This was followed by the development of cellular growth from this area. Further from the initial nucleation region secondary arms evolved growing from the advancing cells. The remainder of the powder solidified in a branched dendritic manner. The mechanisms of these transitions in growth morphology will be examined further in the discussion section.

Microsegregation

In order to further understand the origin of rapidly solidified structures in the powders, microsegregation profiles were measured for the dendritic and microcrystalline powders. The fineness of the microstructure (secondary dendrite arm spacings of approximately 1μm) called for the analytical capabilities of the electron microscope where, in the STEM mode, spatial resolution for elemental analysis by characteristic x-ray emission lines is on the order of 30nm.

In Fig. 3 micrographs from two different Ni-Mo-Ta-Al alloy powder particles are shown. In Fig. 3a, a dendritic morphology is observed with interdendritic precipitation visible. Figure 3b is of a single grain in a microcrystalline specimen. Note that there is no evidence of precipitation in the grain boundaries. In Fig. 4 are shown segregation profiles for the two specimens. These data were obtained by collecting a series of x-ray spectra in the STEM mode from the center of one dendrite arm or grain to the center of the adjacent arm or grain. Severe segregation of Mo and Ta to the interdendritic region has occurred. The Al concentration decreases in the interdendritic region. It is likely that this is caused by the rejection of Al from the second phase as it solidifies. This hypothesis is supported by the slight enrichment of Al adjacent to the interdendritic precipitate.

In the microcrystalline sample segregation of Mo and Ta also occurs during solidification, though not to as great an extent as in the dendritic case. Despite the fact that a concentration gradient from the center of the grain to the grain boundary is observed, second phase precipitation is not evident in the grain boundaries.

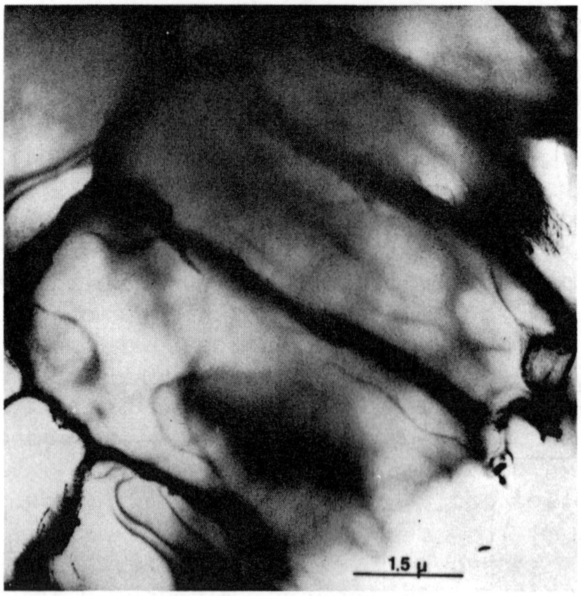

Fig. 3. Transmission electron micrographs of two different powder particles of a Ni-Mo-Ta-Al alloy.
a) dendritic morphology

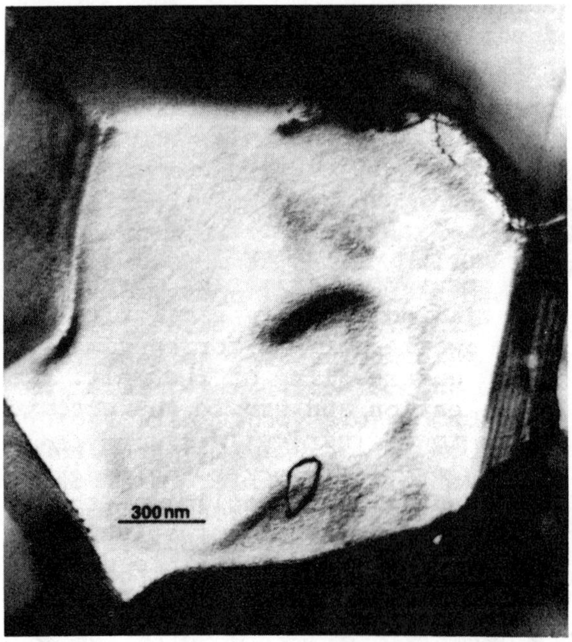

Fig. 3b. Single grain in a microscrystalline morphology particle

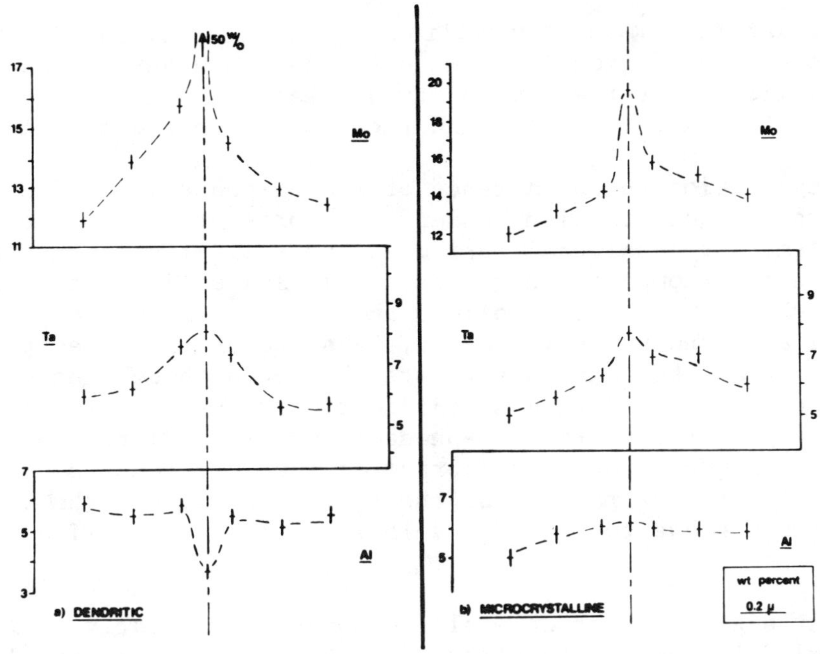

Fig. 4. STEM x-ray data for specimens shown in Fig. 3.
a) dendritic: points taken from center of one secondary dendrite arm to the center of the adjacent arm. Center line indicates interdendritic precipitate.
b) microcrystalline: points taken from center of one grain to center of adjacent grain. Center line indicates grain boundary.

DISCUSSION

The three different solidification morphologies observed in rapidly solidified powders are similar to those reported by Kattamis and his coworkers (3-5) for specimens solidified after various degrees of undercooling of the melt. Kattamis observed a change in nucleation and growth characteristics from branched dendritic to cylindrical(cellular) to spherical (microcrystalline) morphology with increasing undercooling. This suggests that substantial undercooling of the molten powders occurs prior to solidification and that this is responsible for the microstructures observed. Since undercooling is expected to increase with decreasing powder size (6,7) a study of morphologies for various powder size ranges is being preformed.

In the powder shown in Fig. 2, initial undercooling

resulting in a microcrystalline region being formed is decreased as heat of fusion is given off. As the undercooling and solid-liquid interface velocity decrease, the growth morphology changes to cellular and then to branched dendritic.

Segregation in the absence of second phase precipitation in the microcrystalline powders also presents an interesting problem. If local equilibrium at the solid-liquid interface is assumed, composition of the liquid and solid at the interface should follow the solidus and liquidus lines respectively until a second phase is formed. Absence of this precipitation in the grain boundaries suggests that even though heat of fusion given off during solidification raises the temperature of the powder into the two-phase (liquid-solid) region, giving rise to segregation, equilibrium at the interface is not maintained. Solute trapping of the type reported by Cahn (8,9) takes place due to the high velocity of the solid-liquid interface.

Although the microcrystalline morphology clearly is formed by multiple nucleation sites of the solid phase in the liquid powder, it is uncertain whether this is the result of homogeneous nucleation or the activation of a wide spread heterogeneous nucleation agent. According to calculations by Hirth (10) sufficient undercooling for homogeneous nucleation in nickel should be practically unobtainable. However, the author also indicates that these calculations are not reliable since values of parameters used in the equations were not well known.

Kear et al. (11) have also observed what they term microcrystalline structures in powders produced by rotating disk atomization and helium quenching. They explained these structures as resulting from partial solidification on the disk, noting that the occurrence of powders with this morphology decreased with increasing superheat in the melt. In this explanation, the liquid powders have preexisting nuclei picked up from the mushy zone on the disk resulting in a rheocast type structure. Close examination of the structures observed in their study reveal that they are fundamentally different from those termed microcrystalline in the present work. Second phase precipitation is clearly visible in the grain boundaries of the specimens used in their analysis whereas this is not observed in the microcrystalline structures of this study.

CONCLUSION

Some disagreement exists as to whether rapidly solidified

powders are substantially undercooled prior to solidification. In the case of the powders studied in the present work strong evidence for undercooling has been found. Firstly, solidification morphologies identical to those observed in bulk undercooling experiments have been observed. In the case of the microcrystalline morphology, the multiple nucleation events responsible for the structure as well as the single phase nature of the powders strongly indicates extreme undercooling prior to solidification. The substantial difference between the segregation profiles of the microcrystalline and dendritic microstructures coupled with the absence of grain boundary precipitation in the former also suggest fundamentally different solidification mechanisms caused by different undercoolings.

Further studies of a Ni-27^{w}/o Mo binary alloy are in progress in order to further clarify the effects of undercooling and particle size on the solidification morphology and resultant microsegregation characteristics in rapidly solidified powders.

REFERENCES

(1) R. D. Field and H. L. Fraser, Met. Trans, 9A, Jan. 1978, p. 131.

(2) R. D. Field and H. L. Fraser, in Rapid Solidification Processing, Principles and Technologies (Proceedings: International Conference on Rapid Solidification Processing, Preston, VA), Nov. 1977, p. 270.

(3) T. Z. Kattamis and M. C. Flemings, Trans. Met. Soc., 236, Nov. 1966, p. 1523.

(4) T. Z. Kattamis and R. Mehrabian, J. Vac. Sci. Tech., 11, No.6,Nov/Dec. 1974, p. 1118.

(5) T. Z. Kattamis, J. Cryst. Growth, 34, 1976, p. 215.

(6) T. A. Joly and R. Mehrabian, J. Mat. Sci., 9, 1974, p. 1446.

(7) C. Acrivos, J. Mat. Sci., 11, 1976, p. 1752.

(8) J. C. Baker and J. W. Cahn, Acta Met., 17, May 1969, p. 575.

(9) J. C. Baker and J. W. Cahn, in Solidification (ASM), 1971, p. 155.

(10) J. P. Hirth, Met. Trans., 9A, March 1978, p. 401.

(11) B. H. Kear, P. R. Holiday, and A. R. Cox, Met. Trans., 10A, Feb. 1979, p. 191.

INDEX